About the CD

Take a tour through the inside of a cell. Travel with Darwin aboard the *Beagle*. Perform Mendel's pea experiments. Transcribe DNA, translate RNA, and assemble proteins. Embark on these and many other extraordinary biological adventures when you launch the *BioInquiry* CD, and move beyond the textbook reading with sound, video, interactive exercises, a self-contained chemistry review, and links to the World Wide Web. The fully integrated *BioInquiry* CD opens up to a new generation of students the wonders of biology—the concepts, the experiments, the excitement—by appealing to multiple learning styles and by taking full advantage of this rich electronic medium. Planned and designed at every step in conjunction with the text, the *BioInquiry* CD is much more than a supplemental study aid; it is an integral part of the *BioInquiry* learning system, designed to help teachers teach and students learn. Enjoy your adventure!

About the Web Site

BioInquiry web pages link to web sites around the world and direct your exploration of the latest biological research.

Three Easy Steps to Access the Web Site

1. Go to: http://www.wiley.com/college/bioinquiry
2. Enter your name
3. Enter the password: bioinquiry

NSF

DIGITAL FROG
INTERNATIONAL

BioInquiry

Learning System 1.0

Making Connections in Biology

Nancy L. Pruitt
Colgate University

Larry S. Underwood
Northern Virginia Community College

William Surver
Clemson University

JOHN WILEY & SONS, INC.

New York • Chichester • Weinheim • Brisbane • Singapore • Toronto

EXECUTIVE EDITOR: David Harris

EDITOR: Joseph Hefta

DEVELOPMENTAL EDITOR: Suzanne Thibodeau

MARKETING MANAGER: Clay Stone

PRODUCTION EDITOR: Sandra Russell

COVER AND TEXT DESIGNER: Dawn L. Stanley

ILLUSTRATION EDITORS: Edward T. Starr & Sigmund Malinowski

ILLUSTRATIONS: Michelle R. North-Klug

PHOTO EDITOR: Hilary Newman

COVER PHOTO: ©Frans Lanting/Minden Pictures, Inc.

To
Jon, Nathan, and Danny
Sally
Greg, Jeff, and Andy

This book was typeset in 10/12 ITC Garamond Light by Color Associates, Inc. and printed and bound by Von Hoffmann Press, Inc. The cover was printed by Lehigh Press Lithographers, Inc.

The paper in this book was manufactured by a mill whose forest management programs include sustained yield harvesting of its timberlands. Sustained yield harvesting principles ensure that the number of trees cut each year does not exceed the amount of new growth.

This book is printed on acid-free paper. ∞

Library of Congress Cataloging in Publication Data:
Pruitt, Nancy L.
     Bioinquiry learning system 1.0: making connections in biology /
Nancy L. Pruitt, Larry S. Underwood, William Surver
       p.   cm.
     Set includes CD-ROM and references to a World Wide Web site—cf.
CIP pref.
     ISBN 0-471-19280-5 (cloth : alk. paper)
     1. Biology.   I. Underwood, Larry S.   II. Surver, William.
III. Title.
QH308.2.P78   1999
570—dc21   99-38867
                                                          CIP

Printed in the United States of America.

10 9 8 7 6 5 4 3 2 1

Several years ago, a student enrolled in one of our introductory biology courses wrote the following comment on a course evaluation form: "Why do I need to memorize the fact that there are 22,000 species of bryophytes? What does that really teach me about biology?" Ordinarily instructors try not to dwell on wisecrack comments that students write on course evaluations, but this comment had a certain resonance. Biology—its excitement, its insights, its connections—had, after all, attracted the instructor to a lifetime of learning. Why was it being taught as if it were nothing more than a compendium of facts? What, indeed, does knowing the number of bryophyte species teach a student about biology? The time had come for a good hard look at how we teach introductory courses.

A Time for Change

At about that time, several prestigious agencies and organizations, charged with the task of evaluating science education in America, were coming to similar conclusions. In 1990, delegates to the joint Sigma Xi- and NSF-sponsored "Wingspread" conference held in Racine, Wisconsin urged instructors to re-examine their entry-level courses, and to conceive new approaches that "explore the development and use of concepts, minimize memorization, and actively engage the student in learning."[1] Introductory courses were beginning to be taken seriously.

A few years later, the Wingspread theme was again at the fore at the 1996 Shaping the Future conference, sponsored by the National Science Foundation and the National Research Council. The conclusions of that conference were published in an often cited report titled *Shaping the Future: New Expectations for Undergraduate Education in Science, Mathematics, Engineering, and Technology.* The report suggested that every student, not just those destined to become scientists, can learn and understand science. To that end, the

conference report stated, entry-level science courses should "offer a curriculum engaging the broadest spectrum of students" and "use technology effectively to enhance learning."[2] This goal, they emphasized, would entail rethinking and revising both the content of entry-level courses and the methodologies used to teach them.

Many of us teaching introductory biology embraced the challenge of Wingspread and *Shaping the Future*, and began the exciting process of evaluating and redesigning our introductory courses. But it seemed that the teaching materials available were moving in the opposite direction. Textbooks were getting longer and more fact-laden. The accumulated knowledge of biology was not explained as an integrated whole, but instead was presented in short modules, an approach that misrepresents biology as a series of "soundbites." The computer materials—compact disks and web pages—that accompanied many textbooks were more of the same: regurgitations of facts presented in the books. There had to be a better way.

Drawn together by our mutual concern, the authors of *BioInquiry* met for the first time in April 1996. We mapped a strategy for developing teaching materials that both told the engaging story of discovery in biology and incorporated fully integrated computer technology to enhance learning. We wanted to avoid intimidating students or bogging them down with unnecessary detail. We wanted our materials to be accurate; we wanted them to be timely; but most of all, we wanted *BioInquiry* to convey the fun and excitement of one of the most rapidly-moving, fascinating fields in science. When students finish with *BioInquiry*, we told ourselves, they would be informed citizens capable of understanding the scientific method and making rational judgments on the many biology-related issues that we will face in the coming millennium. We submitted a proposal to the National Science Foundation—the very organization that had led the cry for change—and were granted funds to develop our ideas (Division of Undergraduate Education, grant no. 97-52454).

---

[1]Sigma Xi, 1990. Entry-level Undergraduate Courses in Science, Mathematics and Engineering. Research Triangle Park, N.C.

[2]NSF (National Science Foundation). 1996. Shaping the Future: New Expectations for Undergraduate Education in Science, Mathematics, Engineering, and Technology. Document number NSF 96-139.

The Core Concepts

Among the recommendations of the Wingspread report was the proposal that entry-level courses should "encompass carefully selected topics from accumulated knowledge and experience with processes of investigation."[3] But the delegates stopped short of suggesting what that list of topics should be. If *BioInquiry* was to be a new approach altogether, it would be up to the authors to decide where such a list should begin and what topics it should include.

Bold new approaches require difficult decisions, and in this regard the authors made several. For example, we agreed that the real story of modern biology began with Darwin's theory of evolution by natural selection, and introductory courses should not teach otherwise. Hence our story begins with Darwin and the enduring question of why there are so many different living things. Because Darwin's theory provided an intellectual environment ripe for the rise of classical genetics, our chapter on evolution (Chapter 2) guides students through a process of discovery to Mendelian genetics (Chapter 3). From Mendel's garden came insights that led biologists to discover cells (Chapter 4), chromosomes (Chapter 5), and ultimately DNA and modern molecular biology (Chapter 6).

Continuing on with this intellectual theme, the connection between Darwinian evolution and Mendelian genetics leads to population genetics, explored in Chapter 7. A description of the great diversity of life (Chapter 8) that results from population phenomena, and energy constraints (Chapter 9) that have given rise to the many ways in which organisms have evolved to adapt to the challenge of life, follow next. Then we turn to animal physiology (Chapter 10), human physiology (Chapter 11), and plant form and function (Chapter 12). We end with chapters on animal behavior (Chapter 13), population ecology (Chapter 14), and community ecology (Chapter 15) to illustrate how organisms interact with each other and their environment.

The decision to organize in this way means that chemistry and the properties of the atom are neither our starting point nor chapters by themselves. Instead, chemistry is introduced in the text where it is most relevant, and it is also compiled in a series of visual exercises on the *BioInquiry* CD. Students can access the Chemistry Review section on the CD whenever they need to enlighten an important biological concept. Students and instructors reviewing our materials applauded this treatment of the subject.

Another difficult decision was which concepts to include and which to omit or abbreviate. We broke from tradition, for example, by spending several pages of the textbook on the conceptual topic of the cell cycle, but providing only the level of detail on the topics of DNA replication, transcription, and translation that we believe is required for non-majors. Those wishing to learn about the detailed mechanisms of these processes are directed to the *BioInquiry* web-based exercises. Likewise, the textbook chapter on cellular energetics develops many important ideas about energy, but directs the student to the CD and the web site to learn the detailed steps of photosynthesis, glycolysis, and the Krebs cycle.

These tough and sometimes controversial decisions were made in the spirit of educational reform, such as that championed by the National Science Foundation and the National Research Council.

Making Connections in Biology

The organization of chapters in *BioInquiry* emphasizes methods and theoretical foundations of ideas and minimizes isolated facts—an approach we believe makes biology more meaningful and accessible to introductory students. Concepts are given a place in the history of ideas and are connected historically and intellectually to illuminate their importance.

First and foremost *BioInquiry* stresses the integration of ideas by making connections among the important ideas that form our understanding of the living world. Throughout *BioInquiry* we have tried to give students an appreciation of the debates and difficulties that characterize the advance of scientific understanding and the satisfaction of seeing how new theories bring together—in other words, *connect*—our ideas.

Second, we have worked hard to ensure that the three components of *BioInquiry* work together to complement the central ideas of the field. This three-pronged pedagogy—textbook, CD, and web site—is a significant departure from more traditional text-centered materials that include supplementary CDs and web sites. For example, in *BioInquiry* students find that topics covered on both the CD and the web site are not identical, but are closely related, to the material in the textbook. The content on the three different media was developed simultaneously, is organized in the same way, and shares the same look. But we have used each medium to connect and enrich the central ideas in new ways. As such, it is not appropriate to call *BioInquiry* a textbook; it is a learning system.

The Textbook

The *BioInquiry* textbook is intended to be readable, introductory, and motivational. Where appropriate it is historical. Here students get a sense of the origin of major ideas in biology. The reading style is engaging to encourage students to read. Icons in the margins direct students to related topics on the CD. Each major section of the text ends with a "Piecing It Together" section where key concepts are summarized. Throughout the text, key terms are in boldface type the first time they are used in each chapter. These terms are

---

[3]Sigma Xi, op. cit.

defined in the glossaries found at the end of the text and on the CD. Each chapter ends with a set of end-of-chapter review questions to assist students in self-assessment.

The Compact Disk

For many introductory students, learning is enhanced by visualizing biological concepts. Compact disk technology, with its capacity for animation, interactivity, and step-by-step progression, is uniquely suited to this purpose. The CD incorporates the latest technologies in virtual reality that allow students to go inside a cell, inside a membrane, watch mitosis, or analyze data from key historical experiments. Interactivity is encouraged by allowing students to design virtual experiments and analyze their data. Students work through difficult concepts at their own pace, one step at a time. They can back up and repeat steps if necessary, and speed through steps where possible. The CD also walks students through key experiments, such as Mendel's work with sweet peas and Pasteur's experiments disproving spontaneous generation. Students are guided along paths of discovery, asked to make decisions, and to analyze data—to encourage interactivity.

The Web Site

Included in the text and identified with a WWW icon are WebModules, sets of questions that lead students beyond the text and CD to *BioInquiry* web pages (**http://www.wiley.com/college/bioinquiry**). Each web page begins with a background paragraph followed by a series of open-ended but directed questions that can be answered by exploring the World Wide Web. Web sites from around the world are linked to our web pages, affording the opportunity to explore up-to-the-minute advances in biological research. From the *BioInquiry* web pages, students can download worksheets to record their findings and conclusions. When a worksheet has been successfully completed, the student will have learned a new facet of the topic introduced in the text.

Each component complements the other two, adding texture and detail. Instructors wishing to explore a concept in depth can use all three components; others may wish to use only the CD or the textbook for some topics. The organization and the look of all three components are similar, such that students will not get lost. Students' energies are focused on ideas and not materials. How they are presented remains the prerogative of the instructor.

Self-Assessments

There are several pathways to self-assessment built into *BioInquiry*. Questions at the end of each chapter ask students to recall what they have learned and to synthesize the ideas covered in the text. After each topic on the CD, students are presented with a summary paragraph drawing together the lessons of the topic, The summary also includes review

questions or directs students to our web site. Each Web-Module includes a Self-Test section allowing students to evaluate their progress as they work through the materials.

The Context-Based Approach

There are many ways to teach introductory biology successfully, and many innovative approaches are being used today. We applaud those pioneering instructors who have begun using context-based courses at the introductory level, in which students learn the ideas and approaches of biology by focusing on a single topic for the entire course—biotechnology, for example, or endangered species. While this approach is not for everyone, we recognize that, as the sheer volume of information in biology continues to expand, context-based courses for non-majors may soon become an important pedagogical tool. On our web site we offer gateways to two different units for designing a context-based course: BioTechnology and The Endangered Cheetah. Each of these web units comprises a broad network of over one hundred linked pages relating to their single topic. As students are directed from page to page, exploring the complex questions that arise with each new page, they discover the underlying concepts in the life sciences as well as why these topics are so timely and important.

A Community Effort

The movement for reform in introductory biology has been widespread, and we understood from the start that *BioInquiry* would have to be a community effort. To that end, we tested individual chapters with hundreds of students at colleges and universities nationwide. We took their suggestions to heart, revamping and rewriting every chapter. We recruited faculty from all kinds of teaching institutions to review our work. We presented our ideas at national conferences of biology teachers, and followed up on discussions emerging from those presentations.

We are keenly aware of the deep debt of gratitude owed to all of those who have supported us during the development of *BioInquiry*. First among them is the crack team of editors at John Wiley and Sons. David Harris, the biology editor at Wiley throughout most of this project, was perhaps the most influential force in the development of *BioInquiry*. David's sound judgment and good will kept us on task. He seemed to know exactly when to let us be creative—and when to rein us in. David can also take much of the credit for fostering the warm friendship and mutual respect that has developed among the three of us, and for the atmosphere of trust that has existed between the authors and the publisher from the very start. We are deeply grateful to have had him as our editor.

Not surprisingly, Wiley also realized David's talents, and toward the end of the project he was promoted within the organization. We were fortunate to be put in the capable hands of Joe Hefta. Although a latecomer to the project, Joe has been unwaveringly enthusiastic. We wish to express our gratitude to Joe for seeing us through to the completion of *BioInquiry*.

We are also indebted to Suzanne Thibodeau, our developmental editor. Suzanne's uncanny attention to detail and her wonderful way of gently pushing us to meet our deadlines have contributed to the timely production of what we believe is a high-quality learning system. We are grateful, too, to Suzanne's team of editors, including Joan Kalkut, Deena Cloud, Elmarie Hutchinson, and Jennifer Mullin, who have worked to ensure that the text and art are coordinated and accurate. It goes without saying that any errors that remain are entirely the responsibility of the authors.

Also at Wiley, Sandra Russell did an excellent job coordinating the production of the book. Dawn Stanley created an innovative, attractive design. Hilary Newman, our photo researcher, found just the right photographs to complement the text. Edward Starr, Sigmund Malinowski, and Michelle North-Klug were our talented team of creative artists, interpreting our sometimes indecipherable sketches and turning them into a first-rate art program. Marketing researcher Carl Kulo traveled to many campuses to learn what students need, and marketing managers Catherine Beckham and Clay Stone worked tirelessly to get our message out. Editorial assistants Catherine Donovan and Thomas Hempstead cheerfully responded to our many needs and requests.

The creative team at Digital Frog International (DFI), who produced our CD, worked with such good cheer that it was sometimes easy to forget the long, difficult hours they invested in *BioInquiry*. One look at the CD, however, serves as a reminder of DFI's commitment to excellence in educational media. We are grateful for their many contributions, especially those of Jane Recoski, Jeff Warner, Simon Clark, and Sarah Clark.

Many biologists read various drafts of *BioInquiry* and gave us the kind of constructive criticism that every project of this scope needs. We are especially indebted to Professor Sandra Winicur, University of Indiana at Southbend, who read every draft of every chapter and gave sound advice on all of them—always with remarkable insight and good humor. We also owe a special debt to the following reviewers:

David J. Asai Purdue University

James Averett Nassau Community College

Sarah F. Barlow Middle Tennessee State University

Erika Barthelmess Vanderbilt University

Paul J. Bottino University of Maryland

Sandra Berry-Lowe University of Colorado, Colorado Springs

Annalisa Berta San Diego State University

David Betsch Bryant College

Kathleen Burt-Utley University of New Orleans

Ronald Edwards Valdosta State University

Orin G. Gelderloos University of Michigan, Dearborn

Judith Goodenough University of Massachusetts, Amherst

Robert Husky University of Virginia

Florence Juillerat Indiana University-Purdue University at Indianapolis

Gerald Karp University of Florida, Gainesville

Anne S. Lumsden Florida State University

Margaret A. Lynch Tufts University

Charles H. Mallery University of Miami

Deanna McCullough University of Houston, Downtown

Randall McKee University of Wisconsin, Parkside

Betty Molloy Georgia Perimeter College, Lawrenceville

Mark Neilsen University of Utah, Salt Lake City

Judy M. Nesmith University of Michigan, Dearborn

Gary Olivetti University of Vermont, Burlington

Charles Rodell Saint John's University

Thomas D. Seeley Cornell University

Millard Susman University of Wisconsin, Madison

John Thompson Texas A & M University, Kingsville

John Welborn Mississippi State University

The professors who conducted the beta testing of the CD gave us invaluable feedback; their enthusiasm about the CD and the learning system approach greatly sustained us:

James Averett Nassau Community College

David Betsch Bryant College

Art Buikema Virginia Polytechnic Institute and State University

Jean DeSaix University of North Carolina, Chapel Hill

Anne Donnelly State University of New York, Cobbleskill

Richard Falk University of California, Davis

Rita Farrar Louisiana State University

Leland Grim Rainy River Community College

Thomas Lancraft St. Petersberg Junior College

Kelly Luguire Northern Virginia Community College

Anne S. Lumsden Florida State University

Charles Mallery University of Miami

Deanna McCullough University of Houston, Downtown

Betty Molloy Dekalb College, Central Campus

Finnie Murray Ohio University

Joseph Pavelites University of Houston, Downtown

William Simcik Tomball College

William Wischeson Louisiana State University

And finally, we gratefully acknowledge the students who participated in focus groups in the beta testing stage of *BioInquiry*; their responses were insightful and enlightening:

Louisiana State University: Lizette Biel; Donifo K. Denka; Karen Kalia; Steve Malik; Jon Rabuk; Rinat Salikhov.

University of Houston, Downtown: Mhmuda Akhter; Jon Ebert; Cyndee Mitchell; Lisa Nguyen; Jorge Ocampo, Jr.; Maria Ortega; Jutamas Saeuwapamong; Alfredo Sanchez; Travis Smith; John Tamuno.

DeKalb College: Gwinnett Campus: Marshall Arnold; Daniel Foley; Lorelei Mitchell; Bradley Wache.

Indiana University—Purdue University, Indianapolis: David Boyce; Sharon Delay; Jennifer Elliot; Marcia Pilon; Tina Stephanoff.

University of Miami: Maria Castenada; Darrin Davis; Erin Davis; Maggie Eidson; Eboni Faux; Sunnia Gharib; Mae Hyre; Kenneth Karner; David Lalli; Michael Larkin; Matthew Lawrence; Noah Lee; Armando Morales; Daniel Murphy; Simone Myers; Melissa Ordenes; Max Quinones; Sonique Sailsman; Deborah Sampson; Cassidy Smith; Eunice Smith; Andrea Tejera; Michael Vasher; Guillermo Vildosola; Marcello Zinn.

University of North Carolina—Chapel Hill: Matt Bardill; Jamie Braxton; Melissa Caldwell; Samlanchith Chanthvong; Jill Crouchley; Huong Dang; Callie Fischer; Wesley Gaines; Melissa Hart; Vivian Huang; Crystal Ivey; Gary Judson; Teal Lewis; Angie Mabe; Lisa McGinley; April Melton; Tracy Monroe; Laura Nasrallah; Joe Parker; Sarah Peters; Lea Ray; BreAnn Reynolds; Lindsey Schaefer; Brownley Elizabeth Senn; Melissa Singer; Bill Speas; C. L. Toney; Lauren Venable; Whitney Wells; Camille Williams; John Wright.

BioInquiry is really the beginning of a process of creating a new generation of materials that will evolve with technology, discovery, and the students themselves. In that spirit of educational reform we invite you to share your thoughts and criticisms with us. Please fell free to contact us with your comments at <npruitt@wiley.com>.

NANCY L. PRUITT
LARRY S. UNDERWOOD
WILLIAM SURVER

About the Authors

Photo: John Hubbard

Nancy L. Pruitt received a bachelor's degree in biology from Gettysburg College, a master's degree in biology from Wake Forest University, and a Ph.D. from the zoology department at Arizona State University. For the past 17 years, she has taught in the biology department at Colgate University, where her courses include introductory biology, cell biology, and human physiology. Nancy has authored numerous research articles, several with undergraduate student coauthors, about the cellular adaptations of cold-blooded animals to changes in temperature. She has also written a student workbook to accompany a textbook on cell and molecular biology. In addition, she is an active member of the Council on Undergraduate Research and the National Association of Biology Teachers, and is director of the Howard Hughes Undergraduate Science Education program at Colgate University. She also organizes an annual program of off-campus study for undergraduate biology students at the National Institutes of Health in Bethesda, Maryland. Above all, however, Nancy is devoted to bringing the excitement of biology to her students, who have twice nominated her for Professor of the Year at Colgate.

Nancy lives in Hamilton, New York, with her husband, Dr. Jonathan Jacobs, who is a professor of philosophy, and their two young sons, Nathan and Danny.

Larry S. Underwood received a bachelor of arts degree in biology from the University of Kansas, a master's degree in science education from Syracuse University, and a Ph.D. in zoology from The Pennsylvania State University. He has taught biology at all levels, including middle-school, high school, and college. For the past 15 years he has taught undergraduate courses in introductory biology, environmental science, methods of field biology, teacher recertification courses in biology, and anatomy and physiology at Northern Virginia Community College, Annandale. For 22 years, Larry lived in Alaska where he held several administrative, research, and teaching positions, primarily with the University of Alaska. While there, his research interests centered on cold adaptation of Arctic homeotherms and Arctic human ecology. He has published numerous research papers and five books. Larry is a passionate birder and is active in community affairs, especially those involving environmental concerns. In 1974 Larry was elected Fellow of the Arctic Institute of North America, and in 1975 he was elected Member of the New York Explorer's Club. In 1999, he received a NISOD Excellence in Teaching Award.

Larry lives in Manassas, Virginia, with his wife Sally, who is a family therapist and gardener. Between them they have four children and six grandchildren.

William Surver received a bachelor's degree from Saint Francis College and a Ph.D. in biology and genetics from the University of Notre Dame. For the past 20 years, he has taught introductory biology and human genetics at Clemson University, where he also serves as chair of the biology program, which is nationally recognized for its innovations in the teaching of introductory biology. Bill has been a pioneer in the development and implementation of technology in the classroom. He is the author of several videotape programs and two biology videodisks and has presented numerous workshops at national meetings. Bill has been the recipient of several grants and contracts to fund such activities as implementing technology in the classroom, increasing minority participation in science, and improving the science literacy of public school teachers. He is an active member of the National Association of Biology Teachers and the Association of Land Grant Universities. Bill is a dedicated teacher and is devoted to quality teaching. He has been acknowledged for his teaching distinction and is the recipient of a Clemson University Excellence in Teaching Award.

Bill lives in Clemson, South Carolina, where he is an avid gardener and landscaper. He is the father of two sons who currently reside in California.

Brief Contents

Contents

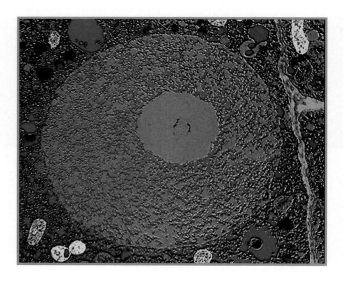

Contents

Biology:
What Is Biology?

Next to music and art, science is the greatest, most beautiful, and most enlightening achievement of the human spirit. —Karl Popper, 1968

—Vignette 1—April 26, 2011

Schafer wasn't surprised by his doctor's diagnosis. Diabetes was part of his family. His grandfather's heroic fight with the disease—amputation of both legs, eventual blindness—was the stuff of family legend. At every family gathering, parts of the legend were sure to be told, with more or less embellishment. There were other diabetics in the family, too. Now, it was Schafer's turn.

"Any questions?" the doctor asked.

He hadn't any, really. He had studied diabetes all his life. He had done his first report on the subject in seventh grade. He pretty well knew what the disease was, what it affected, and how it was treated.

"No, I guess not," he answered. "What's next?"

"Well, your treatment will involve three trips to the clinic. At your first, which we'll try to schedule early next week, we'll remove about a needleful of marrow cells from your thigh bone. These will be sent to our laboratory, where they will be reengineered, genetically. The cells will be given the genes they need not only to synthesize insulin, but to monitor blood sugar levels and release insulin as needed. The laboratory will culture those cells. They will be grown until there is about a pint of them.

"On your second trip, we will replace most of the marrow from one thigh bone with your genetically modified cells. That treatment could take two or three hours; it'll be your longest.

"Your last trip will be only for a blood sample to be sure you are in fact getting your insulin. That should be all there is to it."

Pretty much what he expected.

"And this will be a cure, right?"

Chapter opening photo—Someday we may live on the Moon.

"There's about a 3% chance we'll have to repeat the procedure in 30 years. Otherwise, we're 99.9% sure you'll be cured of diabetes for life."

"Well, let's do it."

"Good. Stop by the nurse's station and make your appointments. See you next week."

As he walked the long hall to the nurse's station, Schafer thought, "My bout with diabetes won't make much of a family legend." He smiled.

—Vignette 2—November 7, 2021

"Wow," Dani thought, "I want to do this!" She downloaded the application form from her TV and began to fill it out.

She was about to become part of the biggest and most ambitious experiment in the history of science. Global warming was affecting world climates and human affairs. Increased instances of violent storms, unpredictable drought–flood cycles, and noticeable rises in sea level were occurring everywhere. Warm climates were rapidly drifting poleward, north and south. The effects were spreading to wild plants and animals. Plant communities, especially forests and mature grasslands, could not keep pace with the changing climate. The richest of Earth's plant communities were dying worldwide and taking animals with them. Something had to be done.

The year before, representatives from 75 countries had signed the Toronto Accords, agreeing to a massive effort. If plant communities could not keep pace with global warming by themselves, human intervention was called for. Governments agreed to produce and distribute seedlings and seed stocks. Local governments—in Dani's case, her state and county—would coordinate efforts, identify critical sites, and distribute plants. Actual fieldwork would be done by volunteers like Dani working together, led by newly hired coordinators. Volunteers would not only put out plants, they would care for them for at least two years.

As a volunteer, Dani would not only become a master gardener, something she had wanted to do all her life, but most importantly, she would be part of a worldwide effort to change Earth for the better. Eagerly, she sent off her application.

—Vignette 3—July 17, 2026

The bad news arrived a little after noon in the form of an e-mail memo to all staff. After last year's losses, a disappointing first quarter had been followed by an even more disappointing second quarter. A complete company reorganization was necessary. Staff should expect significant downsizing. After a few minutes, Sydney left the knot of shocked staff that gathered in the coffee room to return to her own office. There, door closed, she stared glumly out the window.

Could she survive a downsizing? Probably. Her annual evaluations were consistently "good" to "excellent." Her work on the Rico Project had gotten her an "Outstanding!" She could probably survive.

But what about next time? The whole industry was suffering. Was this only the first of a series of significant downsizings? Slowly, reluctantly, Sydney came to a decision. It was time for a career change. This was not an unexpected development. She had been told as early as high school to expect several in the course of her professional life. But this one had come on so fast, with so little warning. It was a little bit frightening.

Career change? To what? Nothing leapt instantly to mind. She needed information. On her computer, she punched up her Personal Mole—in effect, her own individualized Internet search engine—to get information. "Give instructions," the brochure had said, "in everyday English, and your Mole, individualized to your personal needs and personality, will surf all four existing Internets, separate the useful from the useless, and deliver back the information you need."

She spoke distinctly into her computer's microphone, "Get reports and information on promising careers for the next 10, 20, and 30 years."

An anxious hour later, she checked for results. Her faithful Mole had brought back 256 hits. Eagerly, she looked over the first 10:

"Average Starting Salaries Down in the Technical Fields, Except for the Life Sciences."

"Opportunities in the Health Care Industry Still on the Rise."

"Entrepreneurs Needed for Biotech Firms."

"Restoration Ecology—A Wide-Open Field for the '20s and Beyond."

"Worldwide Shortage of Computer Modelers with Life Sciences Background."

"If You've the Will; if You've the Skill. Opportunities for the Daring—Bioprospecting in the Tropics."

"Seven of the Top 10 Most Promising Career Possibilities Are in the Life Sciences."

"Biotech Firm Offers Free Graduate Degrees and 10-Year Guaranteed Contracts to Promising Applicants."

"Science Writers Needed, Especially with Backgrounds in the Life Sciences."

"Developing Countries Need Agrobiologists and Applied Ecologists—Badly."

One thing was clear to Sydney as she asked for the next 10 hits. She'd need a refresher course in biology. Top priority.

1.1 WHY STUDY BIOLOGY?

Today the above vignettes may seem like science fiction—more promise and prospect than reality. The use of bioengineering to treat human disease is still in its infancy. The reality of global warming is hotly debated and professional opinions differ as to its effects. Except in the health fields, biology-oriented careers are at best only starting to rise. One thing seems clear, however: the 20th century will be remembered, in part, for advances in the life sciences and life-science-related technologies. The 21st century will build upon those advances in ways that will touch us all (Figure 1-1). Why study biology? Here are two important reasons.

In Big Advance in Cloning, Biologists Create 50 Mice

Indicates More Rapid Progress Than Imagined

By GINA KOLATA

(b)   *(c)*

Figure 1-1. Biology in the 21st century. A basic understanding of biology is important to everyone for at least three reasons: *(a)* For a growing number of people, biology will provide numerous opportunities for rewarding and challenging careers. *(b)* Bird watchers, gardeners, and pet owners are only a few of the people who use biology in their leisure time activities. *(c)* Understanding biology can help us understand and solve some of the most contentious problems facing society.

(a)

1.1.1 Biology Is Relevant

Biology and biotechnology will be put to use in solving some global problems and improving our lives in important ways. Indeed, they already have been. Let's look at some concrete examples.

1. Great advances have been made in the treatment of once-dreaded diseases. In the 19th and 20th centuries, medical science focused on controlling diseases caused by pathogens—viruses, bacteria, and parasites. As a result, many of these are routinely cured or avoided by antibiotics, immunizations, or other treatments. For most of us today, the flu is mainly a nuisance; it is not life threatening. But the war with pathogens is far from over. Recently, we have seen many pathogens reemerging after developing resistance to our most powerful drugs. It may well be that in coming decades we will see a resurgence of diseases now under control. Citizens need to understand that such an eventuality is natural and predictable. In Chapter 2 we will encounter some of the reasons.

   Emphasis in the medical sciences is shifting to include those diseases related to physiological malfunction or failure. Yesterday, diabetes caused by failure of the pancreas was deadly. Today, it can be controlled with carefully regulated, daily injections of insulin. Tomorrow, diabetes may be curable. Yesterday, kidney failure was fatal. Today, dialysis, involving specialized treatments and equipment, replace failed kidneys indefinitely (Figure 1-2). Tomorrow, it may be possible to stimulate the body to regrow healthy kidneys. Yesterday, cancer was a death sentence. Today, many cancers can be controlled with surgery and chemotherapy. Tomorrow, there may be a cure for cancer.

   It is safe to say that nearly everyone reading these words will, if they have not already, contract, be treated for, and recover from diseases that our older citizens remember as fatal. The number of diseases that can be routinely treated is on the rise.

2. Bioengineering also promises to touch the lives of every person. For example, agricultural biologists are initiating a new Green Revolution—a massive, multinational effort to develop new varieties of crop plants. Some varieties will have genetically enhanced capabilities to resist pests. Others are being developed with enhanced

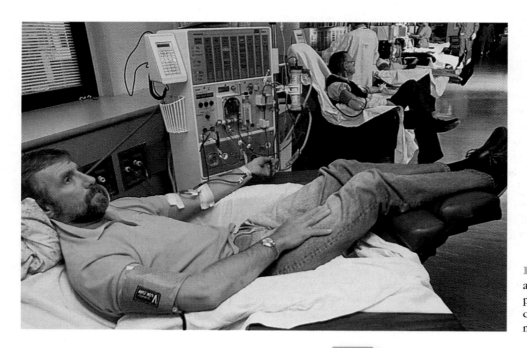

Figure 1-2. Technology, such as the kidney dialysis machine pictured here, can bring under control diseases that were formerly fatal.

cold tolerance and drought resistance so that they can be grown in colder, drier regions. If successful, this Green Revolution could help alleviate world hunger. Citizens need a background in biology to understand the promises and limitations, costs and benefits of these efforts.

3. After a millennium of unparalleled growth, the human population is expected to level off sometime in the next century. Awareness of the need to do so is growing worldwide. This leveling off will not be easy or quick. We probably have to expect that our human population will at least double one more time. But if population growth is to stop there, tough decisions will have to be made at every level of society, from individuals and couples, to nations, and beyond.

    In the first half of the next century, we will see an increasing and pressing need for difficult decisions on where to locate new cities, new farm land, and natural parks and sanctuaries in which to preserve wildlife. Wild areas that could in the past be left more or less alone will have to be carefully monitored and managed. Additional natural areas will have to be restored and maintained. We will witness the dawn of the Age of Applied Ecology. It will impact everyone, and tough decisions will need to be made by an informed, thoughtful public.

4. Emerging technologies, especially computers, will bridge the gap between biology and technology. Added emphasis on biology-related technology will create new career opportunities. Some opportunities in the biotech field will be mostly oriented toward laboratories. Others, especially in applied ecology, will be more field oriented. Many careers will straddle both the lab and the field.

The future of biology is bright. Its impact on everyone is certain. As its promise becomes reality, it will be vitally necessary for every thoughtful human to have a good background in biology.

1.1.2 Biology Can Be Controversial

If I choose to limit the size of my family, what methods of birth control shall I use? How much money are we willing to spend to save endangered species? Is it ethical to use human fetal tissue in biomedical research? Is it right to save the spotted owl at the cost of jobs for loggers, builders, and factory workers? At what age does a human fetus become a human being? Are irradiated foods safe to eat? Are there dangers in cloning animals? Should humans be cloned? By introducing genes from one organism into another, are we creating new species? Are we playing God?

Questions like these illustrate another reason for studying biology: biology and the technologies it gives rise to are often embroiled in controversy. Sometimes controversy arises strictly among biologists, as we shall see in Chapter 4 when we discuss the cell theory. These kinds of controversies are a vital part of science, where new ideas are generally met with skepticism until enough supporting evidence is amassed on one side of the controversy to be convincing. We will discuss this more fully later. But quite often biologists find themselves drawn into controversies that are not strictly biological. Notice that only one of the above questions—on the safety of irradiated food—can be resolved by doing experiments and gathering information. All the others have elements that lie outside of biology, involving economic, moral, ethical, and religious considerations, and biology—indeed, science—cannot by itself answer these questions. But biology can contribute to understanding the problems. Biologists clarify the questions by identifying options or describing impacts. What would be the ecological impact if bald eagles became extinct? Biologists can answer that question. How much should we spend to save the bald eagle? This is an economic and perhaps a moral question. Biology, by itself, cannot provide an answer for that one.

Look over the questions in the above paragraph again. They are important questions that need to be answered. Who should provide answers? Those who could be called "authorities?" An informed general populace? A background in biology can help whoever makes decisions—another important reason to study biology.

1.2 WHAT IS BIOLOGY?

Biology is, of course, the study of life. It is a branch of science and a way of understanding nature. Biology is also a human endeavor. Biologists strive to understand, explain, integrate, and describe the natural world of living things. Let's start with a basic understanding of exactly what biology entails.

1.2.1 What Is Life?

Before you read any more, take a moment to write down your definition of life.

You probably found that defining life is not all that straightforward. Let's remember that: life is difficult to define. There are certain aspects of life that lie beyond the science of biology, and biologists are content to leave those questions in the hands of others. Questions like "What is the meaning of life?" or "Why should there be life?" are fascinating but best left to philosophers and theologians.

Biology focuses on a different set of questions, those that deal with *how life works*. For biologists, life is that set of characteristics that distinguishes living organisms from inanimate objects (including dead organisms). Living organisms

- ◆ Are highly organized, complex entities
- ◆ Are composed of one or more cells
- ◆ Contain a blueprint of their characteristics, that is, a genetic program
- ◆ Acquire and use energy
- ◆ Carry out and control numerous chemical reactions
- ◆ Grow in size and change in appearance and abilities
- ◆ Maintain a fairly constant internal environment
- ◆ Produce offspring similar to themselves
- ◆ Respond to changes in their environment
- ◆ May evolve into new types of organisms

Many inanimate systems are also highly complex, well organized, and may possess some of these characteristics (Figure 1-3). Uniquely, living organisms possess *all* of these characteristics simultaneously. Find an entity with all of these characteristics, and you will have found a living thing—an object for biological study.

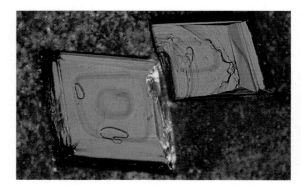

Figure 1-3. Crystals, such as these of sodium nitrate, share some of the characteristics of life. They are complex and highly organized. However, they do not share all of the characteristics of life; therefore, they are not considered living. (magnified 80 times)

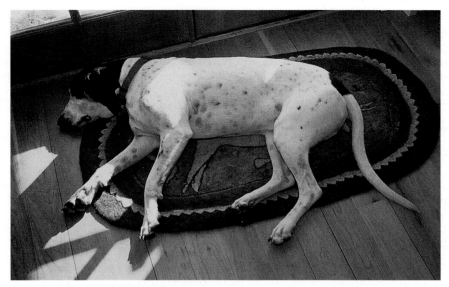

Figure 1-4. If you push on a sleeping dog, what will it do? A common goal in science is to be able to predict future events. In biology, precise predictions are sometimes next to impossible. Biologists often concentrate on describing optional responses to a particular situation and the likelihood that particular options will occur.

1.2.2 Biology Is a Branch of Science

Science is a way of knowing the natural world. We use the term "science" in two contexts. First, science is an activity—what scientists do. But science is also the body of knowledge that derives from that activity. The two are inexorably linked: scientific knowledge accumulates only as a result of scientific activity. In this way, biology is like the other kinds of science, such as physics, chemistry, and geology, to name just a few.

In other ways, biology differs from other kinds of science. For one thing, it is younger. The physical sciences—physics, mathematics, astronomy—got their start in the 1500s. The *science* of biology got its start much later. Many believe that the modern science of biology actually began in the mid-19th century, with the discovery of evolution by natural selection.

For another thing, the subject matter of biology—life—is fundamentally different from that of the other sciences. For example, push on a ball and what happens? It starts to roll. Furthermore, if you know how much it weighs, how hard you push on it, and similar factors, you can predict how fast it will move and how far it will go. Now, predict what will happen if you push on a sleeping dog. Any number of things could happen. The best you could hope for would be to describe a set of possible responses. This illustrates a fundamental difference between biology and the physical sciences. Biology often deals with options (Figure 1-4). It is much less absolute. Biological phenomena are more probabilistic than are physical ones.

1.2.3 Biology Is Integrated with Other Sciences

Generally, we think of the branches of science, such as physics, chemistry, geology, mathematics, and biology, as distinct entities. Certainly, what they cover is different. Physics deals with matter, energy, motion, and force. Chemistry deals with molecules and their properties and transformations. Geology deals with the Earth. Biology deals with life. But beneath these differences there are connections. The knowledge of one group of scientists impinges on and is important to other groups of scientists. The organisms that biologists study are subject to the same laws of physics as the rest of matter.

All are pulled by gravity, have mass, harness and utilize energy, and perform work. Many of the properties of an organism can be explained in terms of chemical reactions. The energy you are using to hold and read this book comes largely from the chemical breakdown of glucose in your cells. To understand the evolutionary history of organisms, we turn to geology and look for signs of previous life forms and clues to their activities recorded in rocks as fossils.

Biologists are constantly confronted with such questions as the following: How big is it? How many of them are there? How much is being produced? How probable is it that my results are reproducible? Mathematics is the tool biologists use to record their observations, analyze their data, and express their conclusions.

As we explore biology, we will need to explore some aspects of physics, chemistry, and geology, and we will make frequent use of numbers. There is only one natural world, and it is studied by all scientists.

1.2.4 What Are the Major Theories of Biology?

The science of biology is as diverse as the living forms with which it is concerned. But there are unifying themes, or "big ideas," that emerge from this diversity to make sense of it all and provide a framework for understanding biology. *BioInquiry* is organized around the most important of these big ideas. In the following paragraphs, we will introduce the unifying themes of biology and the people whose names are connected with them. As we shall see in *BioInquiry*, however, all of these ideas had long histories and cannot truly be credited to a single person. The names associated with the ideas are of the people who first tested them scientifically. Today, after decades of confirmation, they are the major theories of biology.

Evolution by Natural Selection

Charles Darwin was among the first to propose a scientifically testable explanation for the diversity of life, or, in other words, why there are so many different kinds of living things on Earth. Darwin's theory of **evolution**[1] by **natural selection** has two parts. First, **species**, or specific kinds of organisms, change or evolve over the generations. Second, the mechanism for this species change is called natural selection. Both aspects of Darwin's theory are discussed in Chapter 2.

Today, evolution by natural selection is accepted by nearly all scientists as biology's most important theory. Indeed, most agree that contemporary biology began when Darwin published his book *The Origin of Species* in 1859. More than any other single idea, evolution ties together and interrelates all of the other ideas and theories of biology.

Inheritance

At the same time that Darwin was formulating his theory of biological evolution, Gregor Mendel, a monk living in a monastery in Austria, was grappling with another concept: how are traits inherited by offspring from parents? The principles of inheritance proposed from his experiments using common pea plants have been successfully applied to all organisms. Mendel proved that the traits or characteristics of organisms pass from one generation to the next (Figure 1-5) by means of hereditary "factors," now called **genes**. Darwin and Mendel published their first significant findings within a few years of each other, but unlike Darwin, Mendel's results were largely ignored for the first 35 years. They were rediscovered in the early part of the 20th century and have since become one of the foundation stones of modern biology. Chapter 3 explains Mendel's findings; Chapter 7 puts them in the context of Darwin-

---

[1]Words in boldface are defined in the glossary at the end of the text.

Figure 1-5. Offspring inherit characteristics from their parents. Offspring often resemble parents, but they are seldom identical to them. Gregor Mendel was the first to describe principles that explain these seemingly paradoxical observations.

ian evolution and shows how intimately connected these two ideas—evolution and inheritance—really are.

One of the triumphs of 20th century biology is how much we have been able to extend our understanding of genes and inheritance. Indeed, it was Gregor Mendel's results that gave rise to the modern science of genetics and molecular biology developed in Chapters 5 and 6.

Cells

Matthias Schleiden and Theodore Schwann, contemporaries of both Darwin and Mendel, proposed the cell theory, which states that all organisms are composed of cells and that all cells come from preexisting cells. They realized that the cell is the smallest unit capable of exhibiting all of the characteristics of life. Unlike Darwin's and Mendel's theories, this one owes its discovery to technology. Cells could not be observed until after the microscope was developed and refined around the beginning of the 17th century. Thereafter, people could see things not visible to the unaided eye. The history and implications of the cell theory, as it came to be known, are described in Chapter 4.

Biological Classification

Not all of biology's main ideas were first articulated in the 19th century. Biological classification started much earlier. Biologists deal with millions of species. Organizing and keeping track of this many "items" is a daunting task. In the late 18th century, Carolus Linnaeus distinguished himself by classifying living organisms according to their similarities and differences. With Darwin's theory of evolution by natural selection nearly one hundred years later, classification schemes came to be based less on similarities and differences in form and more on evolutionary relatedness among species. Species that diverged from the same ancestors were grouped into the same categories. This evolutionary approach is how we organize and classify organisms today. Chapter 8 describes biological classification.

Bioenergetics

The energy that powers life, **bioenergetics**, operates according to the same rules that govern energy in the inanimate universe. Antoine Laurent Lavoisier's experiments of the late 18th century helped to place the chemistry of life into the context of a larger understanding of chemistry and bioenergetics. Some of the important principles that have emerged from those experiments, including the unique network of chemical reactions in cells called metabolism, are described in Chapter 9.

Homeostasis

In the mid-19th century, Claude Bernard realized that organisms function best when their internal conditions are maintained within rather narrow limits. Organisms tolerate widely varying external conditions by maintaining stable conditions internally, a condition known as **homeostasis**. The manner in which they do so constitutes the study of physiology. We address some of the many strategies different organisms have evolved to maintain a constant internal environment in Chapters 10, 11, and 12.

Ecosystems

Organisms interact with each other and with their environments. Changes in any part of the biological community or the physical environment invariably cause changes in other parts. This concept of **ecosystems** recognizes that organisms do not exist alone, but are part of populations of similar beings, communities comprising many different living things, and environments that include important nonliving features as well. This is the youngest of biology's major ideas, a product largely of the 20th century. Unlike the other major theories of biology, this one has no readily identifiable parent. Rather, it was forged more slowly by a widely separated and diverse group of specialists. It is perhaps the most complex of biological concepts, fusing together not only major ideas of biology, but other sciences as well. The concept is the backbone of ecology, a topic addressed in Chapters 13, 14, and 15.

1.3 HOW IS BIOLOGY STUDIED?

When you stop to consider the scope of biology—from molecules to ecosystems—it might seem unlikely that one approach could guide the study of it all. But in fact, at all levels of study, the activities of biologists can be described in five key words: **observation**, **questioning**, **hypothesis**, **testing**, and **explanation** (Figure 1-6). These activities, taken in this order, summarize not only how biologists learn about life, but how all scientists study nature. They define the scientific method: the set of procedures that form the rational approach to studying the natural world. You could easily define each of these activities based on your own experience, but perhaps putting them in the context of an ongoing investigation will help you to see how they apply to working scientists.

Figure 1-6. The development of a scientific theory rests on a five-tiered foundation. The basic tier comprises observations made of natural phenomenon. Observations lead to questioning. Does what we observe fit with our expectations of what should occur? Especially if the answer is no, observations and questions lead to tentative, testable, alternative explanations, that is, to hypotheses. Now follows intense testing, conducted ideally by both the framer of the hypothesis and others. Testing may lead to refinement of the hypothesis and an eventual explanation of the hypothesis. Only after extensive testing and widespread acceptance within the community of scientists can an idea in science reach the level of theory.

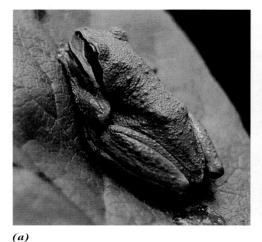 

Figure 1-7. *(a)* A healthy adult Pacific tree frog. *(b)* Typical deformities of Pacific tree frogs include extra, missing, and misshaped limbs.

(a)   *(b)*

In 1995, schoolchildren on a field trip in Minnesota made an alarming discovery. Many of the young frogs they found were suffering from deformities: misshaped limbs, extra limbs, missing limbs. They reported what they had found to local authorities, who not only verified their discovery, but extended it. Many ponds in Minnesota were hosting frogs with similar deformities. The discovery was widely reported in both scientific journals and the popular media. Alarmingly, others elsewhere found the same thing. Within a few years, deformed frogs, toads, and salamanders had been reported in 42 of the United States and in numerous sites in Canada. A million-dollar-plus series of investigations was initiated to find the cause.

One of these investigations was initiated by Pieter Johnson, then an undergraduate student at Stanford University. He undertook the project, working with Paul Ehrlich, a world-famous ecologist who had been conducting environmental investigations for many years. Between 1996 and 1998, Johnson surveyed 35 ponds in central California and found severely deformed Pacific tree frogs (Figure 1-7) in 4 of them. It was on these frogs that Johnson focused his attention.

1.3.1 The Scientific Method In Biology

Observation

All science begins with observation. An observation can be something entirely new, never before reported, that the scientist wishes to investigate and ultimately explain. Often, observations involve a new way of looking at things, or the astute realization that the natural world is at odds with ideas that are currently accepted.

Efforts to find the cause of frog deformities started with observations made by schoolchildren and quickly spread to the scientific community. Johnson's observations started with a relatively large number of ponds, but his concerns eventually narrowed to four ponds of critical interest.

In science, observations lead to questions.

Questioning

Asking questions is at the heart of science. Scientific questions are those that can be answered by experimentation or by direct observations of the material universe. Biologists use their observations and experience to ask "how" something happens, or "why" something appears or acts as it does.

Biologists search for a cause for frog deformities centered on three possibilities: (1) Was the cause of frog deformities a buildup of chemical pollutants in the watery environment in which they reproduce? Chemicals are well-known causes of deformities in other animals. Most frog deformities were observed in a most critical stage in life—the

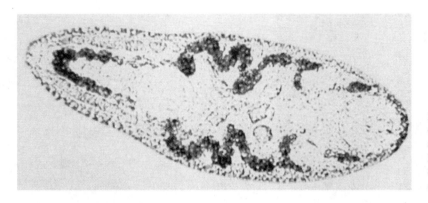

Figure 1-8. According to research conducted by Pieter Johnson, a parasitic flatworm called a trematode is a likely cause of hind limb deformities observed in young Pacific tree frogs in central California.

transition from tadpole to adult. (2) Could the deformities be related to degradation of the ozone layer? Near the outer reaches of the atmosphere, the ozone layer protects life from the sun's incoming ultraviolet (UV) radiation. In recent years, a serious degradation of the ozone layer has been observed, triggered by human activities (this topic will be discussed more fully in Chapter 15). In other organisms, including humans, UV radiation can cause skin cancers and birth defects. Frog eggs, floating near the surface of ponds, might be particularly susceptible to elevated UV radiation. (3) Could frog deformities be caused by parasites? Parasites such as worms, bacteria, and fungi cause deformities in other organisms, although how they do it is largely unknown. Frogs are known to host large numbers of such parasites.

Initially, Johnson asked, what is the cause of deformities observed in the Pacific tree frogs found in four specific ponds? He quickly ruled out the possibility of chemical pollutants. Waters in the ponds were free of pesticides, heavy metals, and other chemicals. He narrowed the question further after making an additional critical observation: in addition to the Pacific tree frog, those ponds were home to a particular kind of snail. The snails are not remarkable, except that they are the intermediate hosts of a parasitic flatworm called a trematode (Figure 1-8). These flatworms have extremely complex life cycles; they alternate living between snails and frogs or other vertebrates. Could trematodes be the culprit?

Hypothesis

To answer a scientific question, scientists construct a tentative explanation, or hypothesis. A hypothesis is conjecture, a possible answer to the how or why questions that have been posed. Scientists adhere to a set of assumptions when they formulate new hypotheses. First, they accept that the world is a real place that can be studied objectively.

Second, scientists' hypotheses are based on the belief that the world is neither chaotic nor dependent on a metaphysical or supernatural realm. Hypotheses that depend on factors outside the material world cannot be tested and as such are not considered scientific.

Third, scientists believe that the events and phenomena of the material world have causes. Understanding causes means understanding why the world is the way it is and not some other way. That is what science is about.

Fourth, it is a principle of scientific ideas that the simplest explanation that adequately accounts for all the observations is to be preferred over other, more complex explanations. This guiding principle is called Occam's razor, named for William of Occam, a medieval philosopher and scholar. Occam stated that where several hypotheses are possible, one should always choose the simplest one.

Guided by these assumptions, scientists generate hypotheses that can be tested. Hypotheses are always tentative. They remain so until rigorously tested and found to be consistent with both the original observations and usually many new ones. A scientist should be ready to abandon any hypothesis or any explanation when a better one—

one more consistent with what has been observed—is proposed. This is one important distinction between science and religion.

As Pieter Johnson proceeded in his investigation, he formed a hypothesis that fit the criteria: the trematodes were causing the deformities in Pacific tree frogs in four ponds in central California.

Testing

When a nonscientist imagines the work of science, it is usually this phase—the testing phase—that is pictured. Indeed, testing is the activity that occupies most of a working scientist's time. Systematic observations, controlled experiments, and detailed studies are all ways of testing the validity of scientific hypotheses.

Testing begins when a scientist makes certain predictions related to a hypothesis, then performs controlled tests to determine the accuracy of the predictions. Scientific predictions are not the same thing as foretelling the future. Fortune telling, a decidedly unscientific activity, is when one makes predictions about events that will happen in the future outside the context of controlled experimentation. Scientific predictions are logical predictions. In other words, within the context of a given hypothesis, given such-and-such a constellation of factors (the controlled experiment), then such-and-such an outcome can be expected. If the outcome under these controlled conditions is different from that predicted, then the hypothesis is not supported; it may not be accurate. If the outcome matches the prediction, then the hypothesis is supported and its validity comes one step closer to being accepted. Hypotheses usually require many successful tests before scientists are willing to accept them as valid.

How did Pieter Johnson set about testing the validity of his hypothesis? First, he dissected hundreds of affected frogs and found trematodes clustered around their malformed limbs. This was good indirect evidence of cause and effect, but he wanted something more convincing. He and a group of friends conducted further tests. First, they collected 200 Pacific tree frog eggs from a site north of their study region from which no deformed frogs had been reported. These were returned to the laboratory, hatched, and raised in individual containers. Next, they collected parasites from their study ponds. This involved numerous intense, late-night efforts. The parasites emerge from their snail hosts mainly between 10:00 PM and dawn. Next, they exposed the tadpoles to varying concentrations of parasites and waited to see what happened (Figure 1-9).

Tadpoles exposed to no parasites experienced no deformed limbs; those exposed to parasites often did. Furthermore, the higher the concentration of parasites, the higher the incidence of defects. The evidence seemed overwhelming: trematodes cause deformities in Pacific tree frogs. Johnson and his friends published the results of their study in the April 30, 1999 issue of *Science*, one of the most prestigious scientific journals in the world.

Figure 1-9. Testing the importance of trematodes as the causative agent of hind limb deformities in developing frogs called for raising individual tadpoles in 1-liter-sized containers.

Fate of Frogs What is the current status of frogs and other amphibians around the world? What factors other than trematodes might be affecting their numbers?

Explanation

A scientific explanation is nothing more than the best hypothesis—the one that has passed the widest and most comprehensive series of tests—about a natural phenomenon. In other words, a scientific explanation is a mature hypothesis. In science, all explanations are subject to review and reconsideration when new evidence comes to light or better explanations are proposed. Scientists pay close attention to data when deciding on the most valid explanation.

Typically, scientists receive new information skeptically. Before results from the Johnson study can be fully accepted, other scientists in other laboratories conducting similar experiments will have to get similar results. Assuming that will happen, the next question will be, How universal are the results? Can trematode or other parasitic infections explain deformities seen in other frogs, toads, and salamanders from other regions? Some people doubt it. Carol Meteyer, an investigator with the U. S. Geological Survey in Madison, Wisconsin, finds that trematodes are not observed in her affected frogs until after limb buds start to develop. Perhaps trematodes cause deformities in certain frogs in California, but not in other species nor in other regions.

Other scientists are not ready to abandon other possible causes altogether. Perhaps pollutants or UV exposure increase the susceptibility of developing frogs to parasites. Most authorities doubt that any one cause, acting alone, can explain all frog deformities. But one thing is clear. As a result of the work of Pieter Johnson and his friends, investigators worldwide will be taking a renewed, hard look at the role of parasites as a cause of frog, toad, and salamander deformities.

With consistent confirmation, a lack of exceptions, and time, a scientific explanation may be elevated to the status of **theory**. Nonscientists use this term very differently than scientists. In general, nonscientific usage, the term "theory" describes pure speculation, in the absence of evidence. In science, however, a theory is a demonstrable or well-established principle, a proposition for which there is overwhelming supporting evidence.

Many explanations in biology, for example, Darwin's theory of evolution by natural selection, have passed so many rigorous tests that most biologists consider them to be true. They have earned the status of theory. We will encounter several widely accepted theories in this book, some of which were introduced above. Other biological explanations have not been elevated to the status of theory. Some have been discarded because they have been replaced with better explanations that more fully account for observations. Others are still being tested. We have described a few of these, too, in the chapters that follow so that you can begin to get a feeling for how our understanding of biology progresses.

Where Are We Now?

In *BioInquiry*, we look back on biology's illustrious past to find a history that has given us remarkable insights into the workings of the living world and unprecedented opportunities to understand ourselves and our place in that world (Figure 1-10). We also see that that history is not over. Many explanations have withstood the test of time and are described in these pages as widely accepted. Others are still being debated. And there are still a few phenomena for which we have no satisfactory explanations. BioInquiry will introduce you to many important ideas in the life sciences, but you will get the most from these ideas if you first learn of the many features of the BioInquiry learning system designed to help you learn biology.

Figure 1-10. The future of biology is bright. Those who work in the field, or use biology as a basis for hobbies and leisure time, can expect their use of biology to be rewarding and exciting.

The *BioInquiry* textbook is organized around some of the most important questions in biology. In the text, you will read about how those questions have been approached; the essence of the answers we currently accept; and, in some cases, the exciting stories of discovery that have enlightened our understanding of the living world. Concepts are spelled out in clear language that conveys not only the current state of the field, but the important ways in which biological ideas are connected and related to each other. In the margins of the text are icons directing you to explore the other two components of *BioInquiry*: the compact disk (CD) and the *BioInquiry* web site.

The CD is organized in the same way as the textbook, but you may be surprised to find that the CD is not just a reiteration of what you have read. The CD is designed to animate the concepts you learn about in the textbook. Using video clips and virtual reality, the CD can bring a process to life, and in real time. Explorations on the CD allow you to repeat some of the key experiments that have informed our current understanding of the living world. Should a concept in the text seem difficult, simply pop the CD into your computer and watch the concept unfold before your eyes. If you are a visual learner, you will find the CD to be indispensable. Regardless of what kind of learner you are, you will find the CD is fun to use. Now would be a good time to take the Quick Tour on the CD that explains how the program works.

**Fast Find
Quick
Tour**

The *BioInquiry* web site is for those wishing to pursue a topic in depth. Throughout the text, and identified by a WWW icon, are WebModules, questions that ask you to consider new aspects of the topics at hand. These WebModules direct you to pages on the *BioInquiry* web site and to the resources you need to learn more. Download the exploration worksheet and use it to record your findings as you explore the web to learn more about a topic that interests you.

How will you know if you are learning the key points? *BioInquiry* has several ways to assess your progress as you move through the materials. Each textbook chapter ends with a series of review questions that ask you to synthesize what you have learned. Each unit on the CD ends with a summary paragraph reviewing the major concepts that it was designed to teach. And the *BioInquiry* web site includes a self-assessment tool that will keep you apprised of your progress in learning.

All three components are designed to keep you focused on ideas. While there are many facts that have been learned about life, it is the ideas that make biology exciting and fun. That is the part of biology that *BioInquiry* is designed to portray.

Chapter 2

Evolution: Why Are There So Many Living Things?

Nothing in biology makes sense except in the light of evolution.
—Theodosius Dobzhansky, 1973

—Overview—

Evolution is one of the most important concepts in all of science. In biology, it has become a unifying principle, that is, a concept of such central importance that nearly all other facets of the field are directly related to it and influenced by it. The above quote by the American geneticist Dobzhansky captures the esteem with which most biologists hold evolution.

For over 150 years, the concept of evolution has stimulated vast controversy. Most disagreements have come from nonscientists who have examined evolution in the light of their own fields. Some found it useful to adopt parts of the concept out of context. For example, some politicians and economists have focused on one precept of evolution, "survival of the fittest," to justify big countries or companies gobbling up little ones. As we shall see, survival of the fittest is not synonymous with ruthlessness. Sometimes the least conspicuous individuals are the ones most likely to survive. Others have found the theory of evolution conflicting with their belief in a higher order. Theologians and philosophers seem split in their opinions of the theory; some see no conflict and can accept evolution easily, while others find it more difficult to do so. Interesting as these controversies are, they are distracting to our purpose. Keep in mind that biological evolution is a biological theory. It was not intended to be useful outside of science.

When the theory of evolution was first proposed, it also sparked controversy within science and even within biology. Disagreements were particularly keen. This may be surprising in light of what we learned in the first chapter. There we described a series of orderly, straightforward steps through which science supposedly passes, from

Chapter opening photo—Darwin's theory of evolution is the major cornerstone of modern biology.

observations of nature, to collection and analysis of data, to interpretation and proposal of tentative explanations, to extensive testing of hypotheses, to their refinement into accepted theories. Surely such a rigorous methodology should lead to irrefutable understanding. However, this seldom happens. Science is a human endeavor, and scientists are human. As scientists move through the process of discovery, their perceptions of nature change. Old ideas are replaced with new ones. But scientists are also consistently skeptical of new ideas. Each confronts the challenge of a new idea with "Where's the data?" and "How do your data fit with my data?" Sometimes the same data are interpreted differently by different scientists. Furthermore, the more revolutionary a new idea is, the greater the demand for confirming data. No scientific concept more clearly illustrates the difficulties faced by a new idea than the theory of evolution in its early days.

Even today, evolution is both widely accepted and hotly debated. How is this possible? Nearly all of today's biologists readily accept the **central proposition of evolution**, that all living things are descended from common ancestors, and that all living things can change and give rise to new species. Controversies arise when scientists attempt to explain how this might happen. New data pertaining to evolution are being collected all the time. Slowly, our understanding of the process shifts, and changes but it always pushes forward. New refinements are regularly proposed. Some are rejected; some are accepted; all are questioned.

In this chapter, we will explore evolution in some detail. We will examine what the theory says, the struggle it went through gaining acceptance, and how it might be changing. Even if you find the theory disturbing, you should nevertheless understand its principles, which nearly all scientists, and indeed most theologians, accept as the best-supported, scientific explanations for the diversity of life. Remember that having your ideas challenged is one of the purposes of education. Remember that scientists are both skeptical of and open to new ideas. We invite you to adopt the same attitudes as you approach the theory of evolution.

2.1 WHERE DID THE IDEA OF EVOLUTION COME FROM?

Like so many major ideas in science, the theory of evolution has a rather long history. Although often attributed to Charles Darwin, it is more than the work of one individual. Many before Darwin had asked, Why, indeed, are there so many different kinds of living things?

2.1.1 The Idea of Evolution Preceded Darwin

The question of why there are so many different forms of life dates back to ancient times. Then, unlike today, answers to deep and difficult questions were not based on observation. Philosophers in ancient times explained nature in terms of what they imagined to be true, often using elaborate mental constructs. If there were conflicts between what could be observed in nature and what was imagined to be true, these philosophers sided with the latter. Throughout much of human history, answers to our most perplexing questions were passed orally from generation to generation in the forms of myths, stories, and dances. Many of these were lost.

By the Middle Ages, it was widely believed that organisms could be arranged in a "ladder of life." God occupied the top rung, followed in order by angels, men, women and children, and then other organisms. The simplest and least complex organisms occupied the lowest rungs. The system was thought to be perfect and complete as imagined. There was no need for species to change. Ever. No room for evolution here.

Beginning in the 1500s, feelings of discontent began to grow among philosophers

Figure 2-1. A footprint cast of an *Iguanodon* dinosaur that lived in the British Isles 120 million years ago. By comparing teeth of the now-extinct *Iguanodon* and the much smaller present-day Iguana, lizards of the tropics, Georges Cuvier concluded they were related. Later he concluded that extinct organisms were victims of natural catastrophes.

of Europe. They could not ignore the discrepancies that they saw between the accepted explanations of the natural world and what they observed in nature. Long-accepted concepts were being questioned, reframed, replaced. A few philosophers discovered the importance of observation. For example, in the early 1500s, Copernicus proposed that Earth was not the center of the universe, as was then believed, but that it circled the sun as did other planets. About the same time, chemists discovered that matter was made of more elements than fire, water, earth, and air. In the 1790s, the English scientist Sir Isaac Newton described gravity as a force that connects all matter in the universe. Although ancient people were certainly familiar with the effects of gravity, they had no name for it and couldn't imagine its extent.

Similar waves of discontent began to affect long-accepted perceptions about the diversity of life. In the early 1800s, the French scientist Georges Cuvier brought the study of **fossils** or signs of previous life found in rocks, to the level of science (Figure 2-1). He was struck by the differences he found when he compared fossils from one rock layer compared with those in layers immediately above and below. The line separating one layer from another seemed sudden. He hypothesized that natural catastrophes such as floods periodically wiped out existing species, which were then replaced by new ones.

Other geologists saw the fossil record differently. In 1830, the English geologist Charles Lyell proposed the theory of uniformitarianism, which held that geologic changes occur slowly. Furthermore, what we see happening today can be used to explain what happened in the past. In other words, understanding the present is key to understanding the past. Because of erosion and other forces, transitions between one rock layer and another may be destroyed or may appear deceptively sudden. What looks like an instantaneous change from one rock layer to the next in fact may have occurred over millions of years. Organisms living in the periods of transition—those lost to erosion and so forth—have simply vanished without a trace.

The perceptions of some biologists also began to change. In the 1700s, the French biologist Comte Georges Louis Leclerc de Buffon and Erasmus Darwin, Charles Darwin's grandfather, separately proposed that species could be modified and changed with time. In effect, each proposed the concept of evolution, but their ideas failed to be widely noticed. In 1809, another French biologist, Jean Baptiste de Lamarck, published his theory of the inheritance of acquired characteristics. He suggested that organisms could acquire new characteristics during their lifetime. He suggested that giraffes, for example, could lengthen their necks by constantly stretching for the uppermost leaves of tall trees. Furthermore, once acquired, these characteristics could be passed on to the giraffes' offspring. In this way, species could change; species could become new species; species could evolve. Today we know that Lamarck's theory was incorrect; it has been largely replaced by Darwin's theory of evolution and Mendel's theory of inheritance.

Cast of Characters What were some of the contributions of Cuvier, Buffon, Erasmus Darwin, Lyell, and other early natural scientists in the 17th, 18th, and early 19th centuries?

2.1.2 Even as a Child, Darwin Was Interested in Biology

Darwin's father was a prominent physician; his mother was the daughter of Josiah Wedgwood of porcelain ware fame. On February 12, 1809, Charles Darwin was born into a life of relative ease. His grandfather, Erasmus Darwin, poet, philosopher, and naturalist, died when young Charles was only nine. Like his grandfather, Charles had a love of nature and an interest in things biological. He was particularly interested in beetles. Outside of natural history, he showed little interest in his early education.

When Darwin came of age, it was assumed he would become a physician like his father. These were the days before anesthesia. After witnessing several operations in which conscious patients had to be forcibly held down, he decided medicine was not for him. Well, if he didn't want to fix bodies, perhaps he could save souls. After obtaining a B.A. degree, he was enrolled in Christ Church College of Cambridge University, but he showed little inclination for the ministry. Fortunately, a family friend intervened and secured Darwin a position as ship's naturalist on a British survey ship, the HMS *Beagle*, commanded by Captain Robert FitzRoy. At first reluctant to see him go, Darwin's father finally relented. Perhaps a five-year trip at sea would help Charles "find himself."

2.1.3 Many of Darwin's Ideas about Biology Were Shaped by a Voyage around the World

Fast Find 2.1a Darwin's Voyage

The *Beagle* set sail on December 27, 1831. The purpose of the expedition was to sail around the world surveying little-known coastal areas, especially the east and west coasts of South America (Figure 2-2). Darwin's responsibilities were to collect specimens and information on organisms encountered along the way.

It was not an easy voyage for Darwin. He suffered from seasickness. For most people on long sea voyages, sea sickness passes within a few days as they acquire their "sea legs." But Darwin never found his sea legs; his sickness was nearly chronic. It drove him ashore at every opportunity. There he found not only brief relief but made extensive observations and collected thousands of specimens. As the months passed, a new idea about biology began to take shape in his mind.

Everywhere the *Beagle* sailed and Darwin looked, he saw confirming evidence that species can change. In South America, he found numerous fossils, among them the extinct relatives of peccaries and armadillos (Figure 2-3). Why were contemporary organisms different from their extinct ancestors? He found beetles—thousands of them. Indeed, in one day he collected 69 species in a small area and commented in his journal, "It's enough to disturb the composure of the [biologist's] mind to contemplate the future dimension of a complete catalog." The beetles he found were not the same as the ones he had grown up with in England. Why were these beetles different from those of his homeland?

Could this diversity be because of evolution? How might the process work? Slowly,

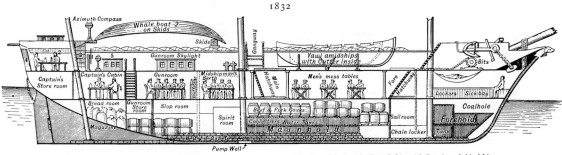

H.M.S. BEAGLE

MIDDLE SECTION FORE AND AFT

1832

1. *Mr. Darwin's Seat in Captain's Cabin* 2. *Mr. Darwin's Seat in Poop Cabin with Cot slung behind him*

3. *Mr. Darwin's Chest of Drawers* 4. *Bookcase* 5. *Captain's Skylight*

(a)

(b)

Figure 2-2. Many of Charles Darwin's ideas began to develop on his five-year voyage around the world sailing on the HMS *Beagle*. At every opportunity, Darwin went ashore to take notes and gather specimens. *(a)* A cutaway drawing of the ship, showing various cabins, including Darwin's (upper left, above the captain's storeroom). *(b)* Map showing the voyage of the *Beagle*.

tentative answers came to Darwin. Some he shared with Captain FitzRoy, himself a well-read English gentleman. FitzRoy was a devout traditionalist. To him, species just were, having been created by God and placed on Earth. He challenged Darwin on every count. On his trip around the world, Darwin was not only getting new ideas and specimens. From Captain FitzRoy he was being introduced to the challenges his new theory would elicit. The education that had eluded him formally at home came to him forcefully at sea.

In 1836, the *Beagle* returned to England, and Darwin began in earnest the work of his life. He had much to think about and many details to work out. He knew his ideas would be challenged. In 1839, he married, and three years later he withdrew to the village of Downe, 16 miles (26 kilometers) from London. There, he continued to work. His

(a)   *(b)*

Figure 2-3. In South America, Darwin compared fossil ancestors and living examples of *(a)* peccaries and *(b)* armadillos.

task was to demonstrate two major ideas. First, he would show that closely related species evolve from and share common ancestors, and second, he would propose a mechanism that would explain how this process of evolution works. By 1842, he had worked out most of the details. Still, he would not be rushed. He sought more data, more confirming evidence.

2.1.4 Some of Darwin's Contemporaries Came to the Same Conclusions

In 1858, a 20-page letter arrived from a friend with whom Darwin had been communicating for years. Alfred Russel Wallace had spent a year in Brazil and several in the East Indies, collecting specimens, sending them to England, and thinking long and hard about evolution. While he was in the throes of a malarial attack, Wallace saw his ideas of a mechanism for evolution crystallize. In only two feverish days, he wrote his ideas and sent them off to Darwin, requesting that the letter be circulated among their peers in England.

To Darwin, the letter was a near disaster. Wallace had come to the same conclusions that Darwin had concerning an evolutionary mechanism. Darwin was getting scooped! Desperate, he asked his friends Charles Lyell, the geologist, and Joseph Hooker, a botanist, for advice. In 1842, Darwin had written, but never published, a summary of his work. This clearly predated Wallace's letter. Hooker and Lyell arranged that Darwin's summary and Wallace's letter would both be presented to the Linnaean Society of London, a prestigious organization of scientists, on July 1, 1858. Now spurred into action, Darwin published his complete theory a year later in his book *The Origin of Species*. After more than 20 years, the theory of evolution was finally available to the general public. The entire edition sold out the first day. As we shall see, reviews were mixed.

2.1.5 Darwin and Wallace Were the Fathers of Evolution and Pioneers in Methodology

Although others had conceived various elements of the theory of evolution long before, the contributions of Darwin and Wallace were nonetheless significant. They were the first to coherently formulate a theory for which evidence had been growing for a long time. That they both could independently and almost simultaneously come to the same point intellectually, even though separated physically by half a world, suggests that the time was ripe for such a synthesis.

They were also pioneers in a relatively new method of describing nature. Starting in the

1700s, the overall goals of scientists were to explain, understand, and describe nature *based on observation*. First, a lot of observations about nature were gathered, often from all over the world. Next, conclusions and explanations were sought that were consistent with all of the observations, not just some of them. Henceforth, explanations of nature were based solely on what was observed. It was this process of drawing conclusions based on observations that led Darwin and Wallace to their theory of evolution.

Piecing It Together

1. The idea that species could change through time did not originate with Darwin and Wallace but their contributions to science were nonetheless significant.
2. Darwin and Wallace based their conclusions on carefully gathered and analyzed observations of nature.

2.2 HOW DID DARWIN ACCOUNT FOR SPECIES?

In his book, *The Origin of Species*, Darwin deals with two separate, but closely related concepts. First, he presents evidence that evolution has, indeed, occurred. Species can and do change. Past species have given rise to today's. Closely related species share common ancestors. As we have seen, the idea of evolution did not originate with Darwin. Others had stated its essence before. What distinguished Darwin's writings was his voluminous and overwhelming evidence. He found evidence for evolution in the fossil record, in his extensive travels, and in comparisons of structure and form in numerous species of plants and animals.

Fast Find 2.2a Darwin's Theory

The second major concept of Darwin's *The Origin of Species* was his explanation of how the process of change occurs. Here, Darwin went well beyond the description of nature to the proposal of a mechanism for evolution, a process that results in established species giving rise to new species. This mechanism he called **natural selection** or the differential survival and reproduction of individuals with certain, advantageous characteristics. It was an idea totally original with Darwin (and, of course, Wallace). In Darwin's day, the idea of natural selection was clearly a hypothesis. Today, nearly 150 years after its introduction into science and with voluminous supporting evidence, natural selection is a well-established biological theory, even though some of its details are still being worked out.

How does the process of natural selection work? Let's examine what Darwin actually said (Figure 2-4), using his words quoted from later editions of *The Origin of Species*.[1] Visit the CD for further details.

2.2.1 First Observation

Populations Have the Potential to Increase Exponentially

"Every being, which during its natural lifetime produces several eggs or seeds, must suffer destruction during some period of its life..., otherwise, on the principle of geometrical increase, its numbers would quickly become so inordinately great that no country could support the product." (page 90)

---

[1]Quotes taken from the 1993 Modern Library Edition published by Random House, Inc.

(a)   *(b)*

DARWIN'S OBSERVATIONS AND DEDUCTIONS

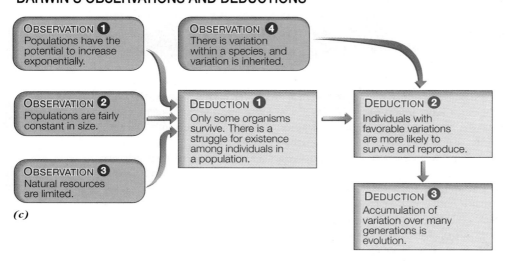

OBSERVATION 1
Populations have the potential to increase exponentially.

OBSERVATION 4
There is variation within a species, and variation is inherited.

OBSERVATION 2
Populations are fairly constant in size.

OBSERVATION 3
Natural resources are limited.

DEDUCTION 1
Only some organisms survive. There is a struggle for existence among individuals in a population.

DEDUCTION 2
Individuals with favorable variations are more likely to survive and reproduce.

DEDUCTION 3
Accumulation of variation over many generations is evolution.

(c)

Figure 2-4. *(a)* Portrait of Charles Darwin at age 40. *(b)* It was at this desk that Darwin wrote *The Origin of Species*. *(c)* The major points of Darwin's theories of evolution and natural selection can be summarized into four observations and three deductions.

For a species to persist through time, it is necessary that its population remain relatively constant. This means that over time births should be matched with deaths. For sexually reproducing species, each reproducing pair should, on average, produce two surviving young. In other words, during their reproductive lifetimes, each pair should, on average, replace themselves. Populations in which pairs consistently fail, on average, to replace themselves dwindle. Eventually they become extinct. Populations in which pairs consistently produce, on average, more than two young will eventually overpopulate. Only populations in which the members of each reproducing pair simply replace themselves, on average, during their reproductive lives will remain stable through time. (In Chapter 14 we will learn much more about population growth and dynamics.)

Notice the emphasis on "on average." Every couple need not produce exactly two young. Stability requires that every pair that produces three offspring should be matched by a pair that produces only one; every pair that produces four offspring must be balanced by two pairs producing one or one pair producing none; and so forth. In any case, stability requires that each couple replace itself, on average. Darwin saw that, for all species, the capacity to increase is always greater than two.

In making this point about population potential, Darwin was extending the thoughts of Thomas Robert Malthus, an early 18th century economist and clergyman. Malthus was concerned with the rate at which human populations increase and the effects of such increases. Already, to his eye, the world was becoming overcrowded with humans. He saw that human population growth was fast outstripping the ability to produce food. Obviously, this discrepancy between increasing numbers of people and constant amounts of available food was creating serious problems. According to Malthus, there was a constant struggle for survival, and, war, famine, and disease were the inevitable outcomes that would control human populations.

Malthus was concerned with humans; human populations always tend to increase. Darwin observed that this is potentially true for all populations of organisms. Here are some examples:

◆ During her lifetime, a female ocean sunfish (Figure 2-5) produces and releases into the ocean an estimated 28 million eggs that mature males then fertilize. Eventually, the adults die. Some offspring survive. How many should survive, on average, for the population to neither increase nor decrease?

◆ Under optimal conditions, a pair of fruit flies, tiny gnats that sometimes invade homes and buzz around fruit (much more about them in Chapter 3), lays and fertilizes 200 eggs. In just 21 days, each pair of surviving flies from that initial batch can produce additional lots of 200 eggs of their own. If none die, in just 17 generations the mass of flies would exceed the mass of Earth. Obviously, this doesn't happen, but the fruit fly's capacity to increase is astronomical.

◆ Baleen whales may have the slowest reproductive rates of any animal. Typically, females start reproducing in their third year, produce only one calf about every two years thereafter, and live for 20 years. Thus, a pair of baleen whales can produce eight calves during their lifetimes—six more than is required for replacement.

Darwin observed that for all species the capacity to increase is always greater than what is required for replacement.

Figure 2-5. Potentially, ocean sunfish can produce enormous numbers of offspring. A large adult may weigh a ton and measure 11 feet across.

2.2.2 Second Observation

Populations Are Fairly Constant in Size

"This geometrical tendency to increase must be checked by destruction at some period of life." (page 93)

Darwin was aware that although species have great capacities to increase, generally they do not do so. We shall see in Chapter 14 that factors regulating population numbers are complex and most were not well studied until well after Darwin's time. One factor that Darwin was aware of led him to his third observation.

2.2.3 Third Observation

Natural Resources Are Limited

"The inhabitants will have been numerous... and will have been subjected to severe competition." (page 39)

Here, Darwin recognized the importance of competition in limiting survival. Malthus had earlier proposed that food was a limited resource for humans. Darwin extended that idea for other species and other resources. For any species, there is only so much space. Likewise, there is only so much food or nutrients. There is only so much shelter. As populations increase, more and more individuals depend on and need to exploit the fewer and fewer resources available. This is the essence of competition.

2.2.4 Deduction One

Only Some Organisms Survive. There Is a Struggle for Existence among Individuals in a Population

"Can we doubt that individuals having any advantage, however slight, over others, would have the best chance of surviving and of procreating their kind?" (page 108)

Darwin's first two observations present a paradox: populations have a consistent ability to increase, but they generally fail to do so. His third observation suggested an explanation that lead to his first deduction to explain who will survive and reproduce.

It occurred to Darwin that many individual differences might contribute to an organism's ability to survive. Among rabbits, if their survival strategy is to outrun foxes, individuals with longer legs and bigger muscles are more likely to survive. Young marine animals that can find their mothers when they've been separated are more likely to survive than those who cannot. Among many species of deer, more aggressive bucks with larger antlers are more likely to collect and defend harems of does than meeker, smaller antlered males.

All these animals, then, possess characteristics that enhance survival. Characteristics that enhance survival are called **adaptations**. The success of any species can be described in terms of its adaptations.

2.2.5 Fourth Observation

There Is Variation within a Species and Variation Is Inherited

"No one supposes that all the individuals of the same species are cast in the same actual mold. I am convinced that the most experienced naturalist would be surprised at the number of the cases of variability." (page 67)

Darwin observed that differences exist within all populations and all species. At first glance, it may appear that "if you've seen one red-winged blackbird, you've seen them all," but this is not true. There are subtle differences in size, coloration, and even shape.

Figure 2-6. Individual differences are also important to scientists. Patterns of color and irregularities along the trailing edge of tail flukes are used to distinguish individual humpback whales. Such data allow scientists to trace individual migratory travels and, for females, reproductive histories.

By taking careful measurements, biologists have found rather large differences in wing length, for example. Some blackbirds have relatively long wings, whereas others have relatively short ones.

In some species, individual differences allow for individual identity and recognition (Figure 2-6). For example, many marine animals, including mammals and sea birds, come ashore to bear young. Periodically, females leave their young, go to sea, and feed. After some time they return, only to find that their offspring have moved. Their problem is to somehow search for and find their offspring in a crowd of several thousand similar individuals. Almost invariably they find them. Subtle individual differences in coloration, vocalization, smell, and behavior allow females and offspring to reunite.

In general, individual differences affect survival. The individual differences that Darwin noticed in all species led him to two important conclusions.

2.2.6 Deduction Two

Individuals with Favorable Variations Are More Likely to Survive and Reproduce

"Natural Selection; Or Survival of the Fittest." (title of Chapter 4, page 107)

Darwin envisioned nature as a "struggle for existence." Individuals who possess the best adaptations will probably survive. Those who do not will probably die. Those that survive will most likely reproduce and pass on their successful adaptations to their offspring. The fastest rabbits get away; the slowest rabbits are fox food. On average, faster foxes are more likely to be successful hunters than slower ones. Fast foxes and fast rabbits possess adaptations that help them survive. The same is true of plants, of course. Plants whose leaves are foul tasting or toxic are most likely to be avoided by plant-eating animals, are most likely to survive, and will most likely pass on their ability to produce foul-tasting leaves to the next generation (Figure 2-7).

Now we come to the crux of Darwin's theory. He proposed a mechanism to explain how evolution might occur and gave that mechanism a name. Here are his own words from *The Origin of Species*: "This preservation of favorable variations and the rejection of injurious variations, I call NATURAL SELECTION."

Darwin, and indeed all scientists at the time, knew little of how probability and chance related to biology. With the advantage of hindsight, we can see the value of these

Figure 2-7. Warning! The leaves of *Dieffenbachia* are laced with poisons. Such toxic substances are a powerful deterrent to would-be predators.

factors. Darwin says that the fit will survive. But that is not always the case. Consider the following scenario. A pack of hungry wolves spots a bull caribou on a mountainside in central Alaska. They give chase. With his superior strength and stamina, the alpha male wolf, leader of the pack, streaks ahead of the others toward the bull. With a toss of antlers and a snort, the bull heads off, up slope. He is in the prime of life and easily outdistances the wolves. But then, disaster. Inadvertently, he steps in a ground squirrel burrow and breaks a leg. He can no longer run but hobbles on. Easily, the lead male overtakes the bull, gets above him on the mountainside, and turns him down slope toward the lagging pack. The bull may well have been the most highly adapted caribou in all of Alaska, but he was unlucky. No theory of natural selection can predict who will step in a hole.

And isn't it similarly possible for the slowest, weakest individual to somehow avoid all dangers and survive? Natural selection may not explain every situation. There will always be exceptions.

But natural selection looks beyond exceptions. It says that the most highly adapted are *most likely* to survive and have offspring. The less well adapted are *most likely* to be selected against.

2.2.7 Deduction Three
Accumulation of Variation over Many Generations Is Evolution

"Natural selection leads to divergence of character...[F]rom so simple a beginning endless forms most beautiful and most wonderful have been and are being evolved." (page 649)

Darwin saw that natural selection can, in time, change species. Consider a population of rabbits whose survival strategy is to outrun foxes. Initially, the population is a mixture of short-legged, slow rabbits; long-legged, fast rabbits; and gradations in between. Foxes that discover this population will have a relatively easy time at first, feeding mainly on short-legged, slow rabbits. Most long-legged rabbits survive and have offspring that tend to have long legs and be fast. After a few generations, most short-legged rabbits disappear. The population is no longer what it was. Now, foxes are forced to feed on other rabbits. They still tend to cull out those individuals with the shortest legs, but the hunting is more difficult. These are not the same rabbits the foxes started with. Eventually, most of the rabbits will be long-legged and fast. They are still rabbits, but the characteristics of the population have changed.

Up to this point, Darwin's observations and deductions were rather straightforward and easily observed or imagined. But in stating that natural selection could lead to new species, he faced a problem. For a theory as revolutionary as evolution, Darwin would have liked more than indirect evidence. The idea would be more convincing if he could point to an irrefutable example, a case in which characteristics diverged so much that a new species was born. Darwin could find no such example. Now, with nearly 150 years of additional observations and evidence, we can cite concrete examples of evolution, which we will return to in later sections of this chapter and in later chapters.

But before we can look for evidence that species evolve, let us first clarify what a species is. In Chapter 1, we defined "species" as "distinct forms of life." That definition seems simple enough, but problems arise. Blue geese and snow geese look distinctly different, yet are considered to be **color phases** of the same species (Figure 2-8). Mule deer and white-tailed deer closely resemble each other, yet they are considered two separate species. To understand why, we need a more precise definition of species.

Species are considered distinct if they do not interbreed in nature. Even though their ranges overlap, mule deer and white-tailed deer do not interbreed. Snow geese and blue geese do. But there are more problems:

Horses and donkeys interbreed, yielding mules. Are they one species? No, two, because mules are sterile. "Interbreed in nature" implies that the resulting offspring will themselves be able to reproduce.

What about dogs and wolves? They can be distinguished by the shape of their muzzles and where they are found. They interbreed if given the opportunity, and their offspring are fertile. A strict application of our definition would probably lump them into one species. We continue to think of wolves and dogs as distinct species more as a matter of convenience than strict biology.

Why is it so difficult to define species? Why are there so many examples that challenge our definition? Surely, something as basic and fundamental as "distinct forms of life" ought to be straightforward and easy to define. That's what the ancient philosophers thought. They saw species as fixed and unchanging. Darwin gave us a framework to explain why Earth's "distinct forms of life" are not always distinct. He observed individual differences within all species. He saw that characteristics within populations change and that, eventually, new species arise, but not suddenly and not at the same rate in all cases. Blue geese and snow geese are examples of individual differences that exist within a single species; they may have only started to differentiate. Wolves and dogs may be a bit further along in the process. Horses and asses have proceeded to the point

(a)   *(b)*

Figure 2-8. Although they look dissimilar, snow geese *(a)* and blue geese *(b)* are considered the same species. They readily interbreed, and their offspring, which can be of either variety, are capable of reproducing when mature.

where their offsprings' reproduction is affected. Mule deer and white-tailed deer have fully arrived; they have fully differentiated into distinct species.

Evolution is a process—according to Darwin, a gradual separation of populations, driven by natural selection, resulting in new species. Exceptions and borderline cases are consistent with the process. But Darwinian evolution takes time, thousands of generations, perhaps. Species are distinct forms of life only after the process has proceeded beyond a certain point. While it's happening, Darwinian evolution is slow, tedious, and barely detectable.

Nearly Species: What are some other examples of color phases in plants or animals, either in species that readily hybridize or in closely related species that cannot hybridize?

Piecing It Together

To summarize, these are the major tenets of Darwin's theory of evolution:

1. The capacity for populations to increase is always greater than what is required for replacement.
2. Individual differences exist within all populations.
3. Some individuals are better able to survive in a given environment than others; they pass on the characteristics that help them survive to their descendants.
4. Characteristics within populations change.
5. Natural selection favors the better adapted and selects against the less well adapted.
6. If the changes within a species become great enough, new species result.

2.3 HOW DOES THE PROCESS OF EVOLUTION WORK?

Darwin and his proponents could directly observe no species transforming into new species, although they did have considerable indirect evidence to support their theory. In the years that followed Darwin's discovery, biologists built on this foundation and amassed overwhelming indirect evidence for both evolution and natural selection. What is the nature of that evidence?

2.3.1 A Great Deal of Indirect Evidence Supports the Theory of Evolution

To support his theory, Darwin leaned heavily on evidence from animal and plant husbandry, practices that had flourished throughout the world for some time. He noted how farmers have been able to "improve" domesticated animals and plants by selecting for breeding those individuals that had the characteristics the farmer wanted. Thus, starting with rather small-bodied cattle that could barely produce enough milk to feed calves, farmers worked in two directions. One group selected cattle for their size and ended up with Herefords and Anguses—large-bodied breeds noted for their steaks and roasts. Other farmers selected for breeding those cows

Figure 2-9. In *The Origin of Species*, Darwin stressed the importance of selective breeding in developing domesticated animals and plants. The *Auroch*, now-extinct ancestor of modern cattle, once roamed Europe and may have been one of the first cattle breeds to be domesticated. Selectively breeding cattle who produced the most meat resulted in modern beef cattle. Similarly, selectively breeding cattle who produced the most milk resulted in modern dairy cattle.

who produced the most milk. Within a few generations they developed the dairy breeds (Figure 2-9).

Next, Darwin wondered if the same process of selection could happen just as well in natural species. He concluded most definitely that it could. Darwin's idea of natural selection was soon encapsulated as "survival of the fittest." Nature came to be seen— not by Darwin, but by his contemporaries—as a place of constant struggle where the strong conquer the weak, the fastest outstrip the slowest, and brilliance outsmarts foolishness. To quote the poet Tennyson, a contemporary of Darwin, "Nature red in tooth and claw."

We now know, with over a hundred years of hindsight, that natural selection does not always favor the fastest, strongest, biggest, or smartest. Among rabbits, for example, those that always run when confronted with foxes expend lots of energy. Survival might be better achieved by rabbits that can stay stock still and be overlooked by foxes. Here, natural selection would favor a new set of characteristics, favoring those individuals that stay calm in the face of crisis, emit the least amount of odor, and blend in most perfectly with their backgrounds.

In which direction does natural selection push rabbits? In later chapters, we will discuss some of the factors that shape natural selection. For now it is enough to understand that the many factors are apparently random, difficult to predict, and not under the control of the species.

Fossils have also yielded a wealth of indirect evidence to early proponents of evolution. They continue to do so. One of the most convincing examples involves the evolution of the modern horse. The earliest fossil horse looked nothing like its modern descendants. Roughly 60 million years ago, *Hyracotherium* was a fox-sized, forest-dwelling creature with teeth adapted for browsing. It had four toes on its forefeet. Each toe was capped with a tiny hoof. It would not have been a particularly fast run-

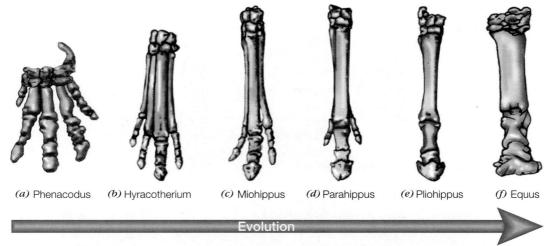

(a) Phenacodus   (b) Hyracotherium   (c) Miohippus   (d) Parahippus   (e) Pliohippus   (f) Equus

Evolution

Figure 2-10. Evolution of the modern horse involved a 60-million-year odyssey. Limbs experienced a reduction in the number of toes on each foot and a modification of the remaining toe into a hoof. Other significant skeletal changes included an increase in overall size, changes in the shape and size of the head, and changes to the teeth. Basically, evolution of the modern horse involved the transition of a rather small, not particularly fast forest browser into a relatively large, fast-running, grassland-dwelling grazer. (Adapted from W.K. Gregory, *Emerging Evolution*. New York: Macmillan, 1951. Courtesy of the American Museum of Natural History.)

ner. How could this animal evolve into today's horse? Darwin's allies could point to an extensive set of fossil horses demonstrating a gradual increase in overall size, a transition to teeth adapted to grazing, a reduction in the number of toes, and an enlargement and lengthening of the one remaining toe into the modern-day hoof (Figure 2-10). Today, even more fossil horse skeletons have been discovered, and it appears that the evolution of the horse was in fact more complex than the gradual, linear, straightforward progression imagined in Darwin's time. That is, the same lineage also gave rise to donkeys and zebras. But nothing seen today refutes the basic premise that modern horses evolved from ancient horses by the accumulation of numerous small changes. Furthermore, the fossil record provides numerous additional examples of evolution.

Telltale Fossils: What are some examples of fossils that provide evidence of evolution?

Fast Find 2.3c Homologous Structures

Another line of evidence supporting evolution involves **homologous structures**. These are *structures, often dissimilar in form and function, that have underlying structural similarities.* Among vertebrates, forelimbs show great variations in form and function. Consider those of lizards, chickens, dolphins, cats, horses, and humans. Each type of forelimb has its own specific use and obvious and significant differences. But dig deeper and you'll find great similarities. The bone structures of all these forelimbs are similar—one end of a single, long bone is connected to the shoulder, and the other end is connected to two additional long bones lying side by side. Smaller bones follow, forming in turn wrists, palm bones, and paws. How can we explain these similarities? To biologists, the most logical explanation is that all these vertebrates evolved from a common ancestor—a now extinct amphibian—with a generalized forelimb. (See Figure 2-11 and also go to the CD for examples and pictures.) Depending on the specific environment, over time natural selection exerted pressure on the generic forelimb, giving rise to the variations seen today. In marine environments, flippers evolved. Wings evolved in birds. All bird wings, including those of chickens, evolved from a common ancestor. For moles the demands of living underground favored short, squat, strong forelimbs. The forelimbs of most terrestrial vertebrates and some fish have similar bone structure; all are homologues. Numerous other examples of homologous structures can be found among a variety of organisms.

Another body of supporting evidence comes from **analogous structures**. *These are structures that have similar form and function, but are structurally quite different.* The wings of insects bear a superficial resemblance to the wings of vertebrates (birds and bats). Among the differences, insects lack bones. There are also similarities. Flight requires a smooth surface in the shape of an airfoil that produces lift and some means of locomotion. In both vertebrates and insects, these requirements are met in a single structure, the wing, which lifts and propels. No terrestrial animal evolved propellers (until humans invented them). What part did natural selection play in the evolution of wings? The wings of insects and those of vertebrates evolved independently from different structures, but they achieve the same function. As a result of natural selection, extensions of insects' externalized skeletons (their exoskeletons) evolved to form a variety of different kinds of wings. In vertebrates, wings evolved through modifications of the

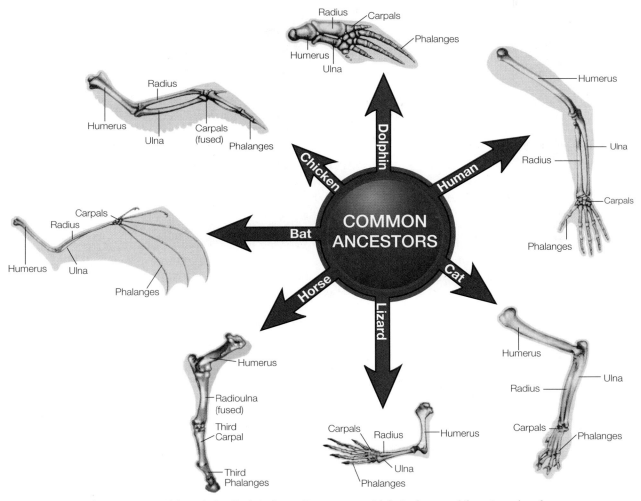

Figure 2-11. Although forelimbs of vertebrates vary widely in form and function, they have similar skeletal elements. The upper portion of the forelimb consists of a single large, long bone, which meets two smaller long bones in the lower portion of the forelimb. These in turn connect to varying numbers of smaller, irregularly shaped bones. Because of their similar structure, vertebrate forelimbs are homologous. All vertebrates derived from an ancestral fish. All mammals evolved from a common reptilian ancestor.

Dolphin

Shark

Figure 2-12. Sharks and dolphins are examples of convergent evolution. They belong to widely different vertebrate groups—sharks are fish, whereas dolphins are mammals—yet they live in similar environments and have evolved similar characteristics. Note their similar overall shape and coloration, location and shape of dorsal (back) fins, and shape (but not orientation) of tail flukes.

forelimbs. Natural selection shaped features the organisms already had (insect exoskeleton, vertebrate forelimb) into superficially similar structures (wings) and overcame an environmental challenge faced by both. Many additional examples of analogues can be found among vertebrates. Unrelated organisms evolving similar adaptive characteristics is called **convergent evolution**. For another example, compare the similarities between dolphins and sharks (Figure 2-12). Less familiar are numerous convergences found between ancient reptiles and mammals, such as the analogous body shapes of ichthyosaurs and dolphins.

Additional evidence for evolution comes from the study of life's chemistry. We will look at that chemistry in more depth in later chapters. The point here is that the chemicals of life reflect life's complexity. Let's look at one example. **Proteins** are seen in every organism and are among the most complex compounds known. Proteins are made of smaller units, called **amino acids**, coupled together to form long, complex chains rather like letters strung together to form words. Whereas words typically consist of strings of less than 10 letters, proteins often comprise strings of hundreds of amino acids. Now, the proteins of each organism are unique. My proteins are slightly different from yours. Our proteins are different from those of dogs. The proteins of mammals are different from those of other vertebrates, and so on. But the proteins in humans are nearly identical, barely different from those of dogs, and quite different from those of fish. The proteins of two species of bony fish are nearly identical. In other words, the more closely related two organisms are, the more closely their proteins resemble each other.

Such observations are completely consistent with the theory of evolution. Indeed, the individual differences noted by Darwin start as minute differences in proteins. Change a protein slightly; the result may be to change an organism slightly. If these differences in protein structure become sufficiently great, new species will result. In later chapters, we will discover numerous examples of biochemical similarities between organisms. We will find that the study of proteins and other chemical components of life give us biochemical evidence for evolution. For now, let's move away from chemistry to examine the process of speciation in greater detail.

2.3.2 Speciation May Occur When Populations Become Isolated

How does the process of **speciation**—species giving rise to new species—work? Describing a typical scenario is difficult. Every species' evolutionary story is unique. Our goal in this section is to examine some possible scenarios and to see the common themes around which species spin their own histories. Expect exceptions. Each species is unique.

First, an example. The Hawaiian Islands are surrounded by the Pacific Ocean, approximately 2,100 miles (3,400 kilometers or km) west of California, 2,500 miles (4,000 km) south of Alaska, and 3,400 miles (5,400 km) east of Asia. These islands originated as underwater volcanoes. Over millions of years, lava mounded up on the ocean floor. Eventually, lava broke the surface, became islands, and continued to build. At first, the islands would have been devoid of life, but not for long. Persistent winds move past the islands, east to west. Ocean currents move in the same general direction. These air and ocean currents along with birds, served as vehicles, bringing the first seeds of flowering plants to the islands. Most seeds would have fallen short or missed the islands and perished. But some were successful. Indeed, a success rate as low as one new introduction every 70,000 years would have been sufficient to bring an estimated 275 species of flowering plants to the islands. From these pioneers, over 1,000 new species evolved, most of which are found nowhere else on Earth. Spectacular examples are several species of silversword (Figure 2-13), closely related to the tarweeds of southwestern North America.

How did the pioneers from America evolve into new species? All species occupy a finite range. For some, the territory they occupy is small. In southwestern United States, for example, a species of pupfish is found only in a single hot spring. Throughout the

Figure 2-13. *(a)* Silverswords are spectacular plants found only in the Hawaiian Islands. *(b)* They are closely related to the tarweeds of western North America.

(a)   *(b)*

world, oceanic islands are often the exclusive homes of certain species. Most species, however, occupy larger ranges, but none lives everywhere. What stops them? What restricts the Shenandoah salamander to a few high mountains in northern Virginia? What keeps the eastern cottontail rabbit out of western Kansas?

Simply put, it's a matter of adaptations and tolerances. Species possess the necessary adaptations to tolerate the conditions found in their home ranges and lack the adaptations to tolerate conditions outside. Shenandoah salamanders can only tolerate the extreme, dry conditions of north-facing rocky slopes. Outside of their optimal territory, conditions may be too wet, too hot, or too windy. New territories may lack key food resources, have too much competition, or have too many predators. Often, what stops species from expanding territories is a combination of limiting factors: their range of tolerances, the strength of their adaptations, and the nature of their environment.

Within its optimal territory, a species environment is easily tolerable. Here, food, shelter, mates, and all other requirements for life are found. Here natural selection continues to favor the most highly adapted, and successful adaptations are further honed. Populations tend to rise. Eventually, a limit is reached beyond which species cannot easily grow. All possible territories are now occupied, competition intensifies, predators and perhaps diseases increase, food sources are stretched to the limits. The most highly adapted continue to be successful and give rise to highly adapted offspring, but populations become as large as they can, and territories become full and crowded.

Within these crowded populations, individual differences continue to exist. Any individuals who can tolerate new conditions, outside the norm for the species, would be at an advantage. They, perhaps, could move into new territories where the stresses of overcrowding are less intense. These movements need not be extensive—only a few miles (kilometers) or even feet (meters) in some cases.

Movement into new territories does not always occur, of course. Perhaps there are no individuals who can tolerate new conditions. Perhaps the barriers keeping new individuals out are simply too intense. In the Hawaiian Islands, for example, there are no native conifers, a group of plants well represented in the Americas.

Seeds that successfully colonized Hawaii had to cross vast expanses of ocean, most starting from Polynesia (Figure 2-14). Only a few survived the trip, and fewer still were

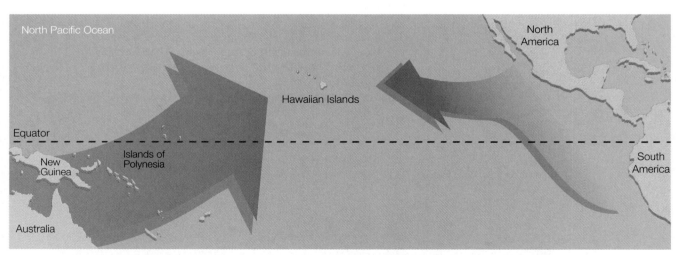

Figure 2-14. The ancestors of all terrestrial plants growing in the Hawaiian Islands had to come from somewhere else. The islands started as volcanoes on the ocean floor. Over millions of years, the volcanoes grew, reached the surface, and became islands. They were never connected to any mainland. Most pioneer seeds came from sources lying south and west of the islands. Only about 20 percent came from the east.

able to tolerate environments found on the islands. But some did survive and evolved into the plants we see there today.

To reach and be successful in a new territory, three opportunities must be realized. First, there needs to be *geographic opportunity* for newcomers. The physical barrier that originally kept species out must be surmountable. Often, this barrier is the easiest problem to solve. Some organisms have a relatively easy time crossing barriers. Birds, for example, have little trouble crossing mountain ranges totally insurmountable to snakes. Luck or chance may also play a part. A once-in-a-millennium drought dries a river, and for a single season a barrier is down. Some organisms are carried by other organisms. Fish eggs hop from watershed to watershed carried on ducks' feet. Seeds are carried into new territories clinging to a mammal's fur or a bird's feathers. Many organisms float vast distances across oceans on logs. Insects and seeds are carried by wind. However they do it, geographic opportunities are realized.

Second, there must be a *physiological opportunity* to enter new territories. Once in a new place, an organism must find the physical environment to at least be tolerable. Fortunately for many species, environments often change. Sometimes when they do, conditions once intolerable to a given species become tolerable. A territory once too hot for a particular species cools. If representatives of the species can ever get to the new territory, they can survive. There are limits, of course. Most species cannot move too far away from their optimal conditions. Saltwater fish would not be expected to tolerate life in a freshwater lake, but they might be able to tolerate life in an **estuary**, a bay or inlet where freshwater and saltwater mix.

Finally, there must be an *ecological opportunity*. Newcomers cannot tolerate intense competition for critical resources with other, better-adapted organisms. A woodpecker newly arrived in a forest cannot be successful if other kinds of woodpeckers are already present.

Only when geographical, physiological, and ecological opportunities are simultaneously realized will newcomers have a chance at surviving new environments. Then, territories can expand.

Conditions in the new environment may not be exactly the same as conditions in the old. The first few seasons are critical for an expanding population. Usually their numbers are few and their adjustments to the new environment imperfect. Natural selection enters once again. Over time, those individuals most tolerant of the new conditions are favored. They survive, have offspring, and pass on their successful adaptations. The least tolerant are selected against. Over time, characteristics within the new population change.

Has speciation occurred? Not as long as the two populations can interbreed. What happens next depends on how completely they are separated. If the geographic barriers are breached fairly regularly, speciation will probably not occur. Travelers, moving in either direction, can interbreed with those already present, mixing and blending characteristics, both those useful in the old environment and those useful in the new (more on this in Chapter 3). However, if the two populations are more or less completely isolated, as in the case of the Hawaiian plants, speciation may occur. If breeding is restricted within isolated populations, the effects of natural selection can intensify. As individuals in the new population become more and more adapted to new conditions, and as the old population remains tolerant of and adapted to the old environment, populations drift apart. Eventually, they may become so different that even chance opportunities at interbreeding are not successful. Then and only then will speciation have occurred. Even though they share a common ancestor, today's silverswords of Hawaii and the tarweed of southwestern United States cannot interbreed.

This is an example of **allopatric speciation**—*speciation that occurs between isolated populations*, that is, populations that originate in different territories, (*allo*=different; *patric*=native land). A key factor is geographic isolation. In the years immediately following Darwin, this was thought to be the most common scenario within which speciation occurs. But it may not be the only possibility.

2.3.3 Can Speciation Occur between Populations Whose Ranges Overlap?

Geographic isolation may not be an absolute prerequisite to speciation. Under certain conditions, speciation may occur among populations occupying the same territory. Let's look at another example.

In the northeastern and north central United States, the goldenrod gall fly depends on goldenrods for its reproduction (Figure 2-15). Adults emerge in spring and live only about 10 days, during which time they mate. Females lay one egg each in the unopened terminal leaf bud of the plant. When they hatch, the larvae burrow several millimeters down into the stem, where they take up residence. Their presence stimulates the plant to grow galls, or swollen areas of plant tissue where the larvae feed, grow, and eventually overwinter. Timothy Craig and his colleagues have found two distinct populations of goldenrod gall flies in the north central United States. There, these flies make use of two species of goldenrod: tall and late.

Although the two species of goldenrod often live in the same field, so that flies feeding on tall goldenrod are sometimes within a few centimeters of those feeding on late goldenrod, the life histories of the flies are not the same. On late goldenrod, fly larvae make larger galls and hatch and breed earlier than those on tall goldenrod. Each population attracts its own predators. "Late" flies (those living on late goldenrod) make relatively large galls that attract the attention of chickadees and downy woodpeckers. "Tall" flies (those living on tall goldenrod) make somewhat smaller galls that are fed upon by small wasps that burrow into the galls to lay their eggs. Birds ignore small galls; wasps are incapable of burrowing into large galls. Thus, "tall" flies avoid birds; "late" flies avoid wasps. The two populations of goldenrod gall fly seem to be reproductively isolated and are experiencing differing pressures from natural selection.

Craig and his colleagues have found that when given a choice, "tall" males prefer to mate with "tall" females, and again when given a choice, "tall" females prefer to lay eggs

(a)   *(b)*

Figure 2-15. Goldenrod gall flies depend on goldenrod for their reproduction. Females *(a)* lay eggs on goldenrod stems, which develop into galls *(b)* in which larvae develop into adults. Two species of fly may be evolving, each depending on separate species of goldenrod.

on tall goldenrod. Likewise, "late" males prefer to mate with "late" females, who prefer to lay eggs on late goldenrod. These mating factors tend to maintain reproductive isolation between the populations.

Hybrids between the two populations have lower survival rates than purebred individuals of either population. On the other hand, those hybrids that do survive can themselves reproduce and show no preferences for mates or egg-laying sites. Furthermore, if given no choice, "tall" males will breed with "late" females, who, when forced to, will lay eggs on tall goldenrod. (The converse is also true.)

Have they become two species? Not yet. (Why not?) But the two populations are at least partially isolated reproductively. We can easily imagine that if the separation continues, and if abilities to discriminate appropriate mates and egg-laying sites improve, they could become separate species in the future.

Speciation among populations whose ranges overlap is termed **sympatric speciation**, or speciation that occurs within one territory, (*sym*=same). Isolation is once again essential, only here the two populations become isolated by occupying different **habitats** (specific area occupied by a species) within a given **range** (the larger geographic area occupied by a species). What appears to be overlapping is not so when seen at the insect's level.

Is sympatric speciation, in fact, possible? How often does it occur? These are, at present, unanswered questions about which biologists disagree. It is important to note that the controversy does not question the basic premise of evolution or natural selection. It is more or less an argument over details. Let's think of it as a possibility whose verification awaits further studies. Here are some possibilities to look for.

Any number of factors could isolate overlapping (sympatric) populations. Some are related to space. Especially in crowded populations, individuals try to live in whatever territories they can occupy, limited by their tolerances, adaptations, and the specific environmental conditions they experience. Occasionally, segments of an original population may become isolated from others of their kind within their existing ranges and may speciate. For example, a population of insects living near the ground may become isolated from individuals of the same species living in the tree tops if no individuals live in between.

Among animals, breeding is often preceded by elaborate courtship displays. Such displays facilitate bonding between individuals who otherwise stay apart. Often, more than bonding is involved. In some cases, females are physiologically aroused by displaying males. Egg production does not occur until they have witnessed courtship displays. Activity that one group finds highly stimulating may fail to impress another. Now, if the courtship displays of one population fail to impress individuals of another, the two populations are isolated. Speciation between them can occur.

Under certain circumstances, sympatric speciation can occur nearly instantaneously; that is, two populations of a given species can become reproductively isolated in a single generation. Understanding the mechanisms behind this form of speciation requires an understanding of genetics, which we will study in detail in Chapter 3, and then we'll return again to the topic of instantaneous speciation in Chapter 6.

Allopatric and sympatric speciation may sometimes occur together. Perhaps the process of separation starts when two populations become geographically isolated (allopatric). Then, for some reason, the barriers dissolve and populations once again overlap (sympatric). If, by this time, the populations are partially isolated by differences between their preferred habitats, courtship displays, breeding season, preferred foods, and so on, speciation may continue sympatrically.

If the populations are not isolated by their differences, they simply merge back into a single species with perhaps a greater degree of individual differences. Color phases within species may arise in this way. For example, Arctic foxes occur in two color phases, white and blue (really gray) (Figure 2-16). Blue foxes are most common on islands in the North Atlantic and North Pacific oceans, but they are some-

(a)   *(b)*

Figure 2-16. Arctic foxes occur in two color phases: *(a)* white and *(b)* "blue." White foxes are more common in mainland areas of Arctic North American and Eurasia. The blue color phase is found mainly in Greenland and on isolated islands in the North Pacific and North Atlantic.

times seen on the mainlands of North America and Eurasia. During relatively warm geologic periods, the islands were probably separated by miles (kilometers) of open ocean. Here blue foxes flourished. When colder geologic periods returned and the oceans once again froze during winter, white foxes and blue foxes intermingled and interbred. Because they can do so, white and blue foxes are two color phases of one species.

2.3.4 Let's Take a Look at Darwin's Finches

Deciphering the evolutionary history of a species is at best a risky and speculative business. Such stories were played out long ago and left few signs of their details. However, working from a detailed knowledge of present conditions and an understanding of how evolution works, it is sometimes possible to make educated guesses. We will do that now for Darwin's finches, a group of species Darwin encountered on his journey around the world.

Thirteen species of Darwin's finches are native to the Galápagos Islands, a group of islands straddling the equator approximately 600 miles (960 km) west of Ecuador (Figure 2-17). An additional species is found on Cocos Island, 450 miles (720 km) north and east of the Galápagos Islands, 300 miles (480 km) offshore from Costa Rica. Like most finches, these are drab-colored, sparrow-sized birds with unremarkable, somewhat grating vocalizations. Only a biologist or bird watcher would find them interesting.

Fast Find 2.3b Darwin's Finches

How did they get to the Galápagos? These islands formed from underground volcanoes millions of years ago. At first they were barren, rocky, and lifeless, but not for long. Plants are the first organisms to pioneer such places, followed closely by insects. These organisms, too, play out fascinating evolutionary stories, which we will skip over to get to the birds.

Worldwide, finches are common, especially in Central and South America. It seems reasonable that the first finches on the Galápagos Islands came from the mainland. We know little about the ancestral mainland finches. They are now extinct. If they behaved like other finches, their numbers fluctuated widely, and because they are good fliers, individuals wandered out of their range with some regularity. But could they cross 600 miles (960 km) of open water? Not a problem for finches. Some species regularly migrate between North and South America by crossing the Gulf of Mexico. A bigger problem than mere distance would be finding a speck of solid ground in a

Warbler Finch
(Insects)

Small Tree Finch
(Insects, fruits, soft seeds)

Medium Tree Finch
(Insects, soft seeds)

Large Cactus Finch
(Insects, seeds)

Vegetarian Finch
(Soft seeds, fruits, leaves)

Large Tree Finch
(Insects)

Woodpecker Finch
(Insects)

Cactus Finch
(Moderate seeds)

Mangrove Finch
(Mostly insects)

Small Ground Finch
(Soft seeds)

Sharp Beaked Ground Finch
(Insects, seeds, nectar)

Medium Ground Finch
(Moderate seeds)

Large Ground Finch
(Hard seeds)

Cocos Island Finch
(Insects, seeds, nectar)

Figure 2-17. Darwin came to realize that the several species of finch on the Galápagos Islands had a common ancestor, even though some had large bills and ate seeds while others had smaller bills and ate insects. Today, scientists recognize 14 species of Darwin's finches (one lives on Cocos Island, north of the Galápagos). All descended from a single ancestral species.

large ocean. We can imagine that some went out to sea, got lost, found no land, and perished. But once—maybe just once—a small flock, a pair, or a female carrying a fertilized egg by chance found an island. Perhaps twice within a few years two individuals with the proper mix of sexes again by chance found an island. Whatever the details, a small number of individuals, wandering out from the mainland, found an island and settled in. They successfully overcame a rather considerable barrier and realized a geographic opportunity.

Next, was there physiological opportunity? For islands in the middle of an ocean, the Galápagos are remarkably desertlike. Compared to other islands, it rarely rains there. Along the western coast of South America, ocean currents flow north from Antarctica. These waters are cold, even at the equator, as they flow past the Galápagos. There is little evaporation and therefore little rain. Predominant winds in the region are from the east—from the mainland—and carry few clouds. Thus, the Galápagos Islands are dry.

And so is the mainland closest to the Galápagos. When we think of South America, we usually think of warm, tropical rain forests. But not along its western coast. In South America, moisture-laden, easterly winds drop rain on the east slopes of tall mountains, leaving a narrow band of dryness along the western coast. If this was the territory of the Galápagos finch's ancestors, they were used to dry climates. To them, the Galápagos may have seemed like home.

Next, they faced the challenge of ecological opportunity. Today, there are other small songbirds on the islands (Figure 2-18). In all cases, they more closely resemble mainland species than do the finches. Perhaps, then, the finches arrived first, at a time when there was no, or at least little, competition. Indeed, once across the ocean, these pioneering individuals may well have found themselves in finch paradise with lots of food, lots of space (relative to their numbers), and little competition. What space and food did they prefer? The best evidence is that the original pioneer finch was a ground-dwelling seed-eater.

For a time, life was good for the pioneering finches. They survived, became successful, and thrived. Their numbers increased. They were isolated from their kind. True, they had successfully crossed open water. But there is no reason to suppose others did so with regularity. In isolation, natural selection took its course. Over time, certain characteristics of the isolated population changed as their adaptations became honed in to the existing environment. We surmise this was the beginning of allopatric speciation.

Islands offer limited space. As time passed and finch numbers increased, it seems likely that competition among them increased. Some finches were forced or just naturally wandered off the island. Most flew out to sea and were lost. But some may have found new islands. In the Galápagos, there are 19 main islands to choose from. Eventually, finches expanded to more than one island. Each island population would be

(a)   *(b)*   *(c)*

Figure 2-18. The Galápagos Islands are home to songbirds other than finches, including *(a)* vermilion flycatchers, *(b)* yellow warblers, and *(c)* mockingbirds. Although they more closely resemble their ancestors than do Darwin's finches, there are subtle differences. Those differences found among the mockingbirds were of particular interest to Charles Darwin.

somewhat isolated from others because each island would be similar to but not identical with neighboring islands. For the finches, there might be differences, say, in available foods. One island might offer an abundance of hard seeds that could be picked up off the ground. Natural selection would favor ground-dwelling individuals with large bills capable of crushing hard seeds. Another island might offer an abundance of insects living off the ground in shrubs. Here, natural selection would favor individuals who were not ground dwellers and who had finer, longer, pointed beaks. In at least partial isolation, sympatric speciation would result in populations gradually drifting apart. On the new island, numbers rose and individuals again wandered. What if some wandered back to the original island? If, by that time, the differences between tree-dwelling, insect-eating finches, and ground-dwelling, seed-eating finches were sufficient to reproductively isolate them from each other, speciation could continue—sympatrically.

So we suppose it was as different islands and different habitats became occupied that different species evolved.

Today, each island of the Galápagos Islands has its own assemblage of species. Each species of finch is specialized, occupying a narrow range of habitats, eating a narrow range of foods, restricted to one or a few islands. Larger islands tend to have more species than smaller ones, but all have at least one species, even Cocos Island.

Piecing It Together

1. The evidence supporting Darwin's theory of evolution is largely indirect. In presenting his evidence, Darwin called on his knowledge of plant and animal husbandry. He reasoned that, in nature, natural selection worked in much the same way.

2. For any species in any environment, certain characteristics are more appropriate than others. Individuals possessing desirable characteristics would be most likely to survive and pass their characteristics on to offspring. Individuals with other, less desirable, characteristics would be most likely to perish. Over time, the characteristics of the species as a whole would change in the direction of the desirable characteristics.

3. For speciation to occur, some segment of the original population must be isolated from the rest; *reproductive isolation is an essential ingredient.* Because environments are seldom the same in any two areas and are constantly changing, natural selection works differently on the two segments of the population. Characteristics most appropriate in one area may not be appropriate in another. Over time, the two populations drift apart and may eventually become different species. Today, biologists recognize two broad scenarios by which separation and evolution occur: allopatric and sympatric speciation.

4. *Allopatric speciation* occurs whenever two populations of a given species are separated geographically. Any physical factor—rivers, oceans, mountains, deserts, and the like—can serve as barriers, as can biological factors, such as absence of food, intense predation, or presence of diseases. Separation by time, such as in the case of *Hyracotherium* and modern day horses, is a special case of allopatric speciation.

5. *Sympatric speciation* occurs whenever two populations of a given species speciate while occupying the same range. They become reproductively isolated by some factor other than space or time.

6. In some complex instances, where several species evolve from a common ancestor, both allopatric and sympatric speciation may occur together. The evolution of Darwin's finches on the Galápagos Islands affords one such example.

Branching Out: How are the honeycreepers of the Hawaiian Islands (Figure 2.19) similar to and different from Darwin's finches on the Galápagos Islands? Are there other groups of organisms on these or other islands that also underwent similar evolutionary patterns?

Figure 2-19. Similar to Darwin's finches on the Galápagos Island, other birds on other islands have also experienced adaptive radiations. The honeycreepers of the Hawaiian Islands provide a fascinating example. Their common ancestor is also thought to have been a finch. How are these birds similar to Darwin's finches? (Hint: Look at the bills.) How do the two groups differ? How do you explain the similarities and differences? (Adapted from G.C. Monroe, *Birds of Hawaii*. Vermont and Tokyo: Charles E. Tuttle, 1960. Used by permission.)

2.4 ARE THERE PATTERNS IN EVOLUTION?

The evolutionary history of every species is unique. For each, a unique set of challenges presents itself, and each species reacts with a unique sets of responses. This presents evolutionary biologists with bewildering complexity. Fortunately, there are patterns in the ways evolution works.

2.4.1 Unrelated Organisms Living in Similar Environments Sometimes Evolve Similar Characteristics

South America is home to the mara. These rodents are not rabbits, yet they share several characteristics with them: long hind legs, long ears, flattened faces. Maras and rabbits live in similar environments and face similar challenges. Solutions found useful to one group are equally useful to the other.

Did natural selection "guide" both groups to similar endpoints? Not exactly. Here is a subtle but crucial point. *Natural selection is a mindless, random, chaotic process. It has no purpose;* there is no design to adapt populations to existing environmental conditions. It guides in no direction other than survival. Yet, within such systems, common patterns—solutions, if you will—often are repeated. A blizzard in Europe looks and feels like a blizzard in South America. Yet, they are not related. They did not result from a common starting point. Similarly, European rabbits and South American maras came to resemble each other through the process of **convergent evolution**. Figure 2-20 is an example of convergent evolution in plants.

Figure 2-20. Convergent evolution is not limited to animals. These plants belong to four different families: cacti, spurge, aster, and milkweed. All live in deserts, where water is scarce and temperatures are often high and fluctuate widely. All have evolved similar adaptations to cope with these conditions. All lack leaves and have thorns and fleshy, green stems.

2.4.2 Closely Related Species May Become More and More Different

Closely related species living in different territories sometimes tend to drift apart. Over time, their characteristics become more and more different until new species result. This is **divergent evolution**. Thus, the small *Hyracotherium* evolved through several intermediate species, not only into today's horses, but also into today's zebras and onagers. Today's "horses" have not only diverged from *Hyracotherium*, but from each other.

2.4.3 Adaptive Radiation Is an Extreme Example of Divergent Evolution

Fast Find 2.4 Evolutionary Patterns

A single species can move into a new territory and evolve into several closely related yet different species. Darwin's finches are a good example of **adaptive radiation**, whereby a number of species evolve from a single, ancestral species. Silverswords and their relatives, discussed previously, are another example. Also in the Hawaiian islands, another group of birds, the honeycreepers, went through a similar process (Figure 2-19).

An extreme example of adaptive radiation occurred among mammals 65 million years ago. Mammals evolved millions of years earlier, but because of intense competition from highly successful dinosaurs who were already specialized to many environments, they remained a relatively minor group. Extinction of the dinosaurs created huge evolutionary opportunities for mammals. They exploded into new forms. Mammals never attained the huge size of dinosaurs, but they came to occupy nearly all the dinosaurs' ecological opportunities.

2.4.4 The Evolution of One Species May Influence the Evolution of Another

Usually, species don't evolve in a vacuum. The evolutionary changes that occur in one creates environmental pressures on adjacent species. *Parallel evolutionary changes that occur simultaneously between interacting species* is what is meant by **co-evolution**.

In Africa, cheetahs and their prey, Thompson's gazelles, are a good example of co-evolution. The cheetah has evolved into the world's fastest mammal, capable of running 80 miles per hour. Their acceleration is exceptional—from a dead stop to 60 mph in three bounds. As we might expect, the Thompson's gazelle is only slightly slower than the cheetah. As the speed of one improved, so did the other. Furthermore, Thompson's gazelles don't run in a straight line. Even at near top speed, they can change direction seemingly instantly. At these speeds, it's relatively easy for the cheetah to lose visual contact with a darting gazelle. Cheetahs evolved two black lines running down the face from just below the eye toward the tip of the nose that serve as a sighting mechanism. Only if the cheetah can keep the darting gazelle "between the lines," by turning its head quickly, will it eventually feed.

As predators evolve, prey evolve. As prey evolve, predators evolve.

Patterns of Evolution: Can you find other examples of convergent evolution? divergent evolution? adaptive radiation? co-evolution? What about examples involving humans?

2.4.5 Darwin Described Evolution as a Slow and Gradual Process

As envisioned by Darwin, natural selection exerts its influence slowly and steadily. Legs gradually lengthen and strengthen. Species gradually gain greater speed. **Gradualism**—evolution of species through a gradual, steady, linear accumulation of small changes—was a pattern greatly favored by early evolutionists, especially the geologist Lyell. It fit neatly into the theory of uniformitarianism, which avowed that geological events happen slowly, too. Sometimes the patterns of change become complex. Sometimes the evolutionary history branches out in several directions. But even within those complex groups, if you trace the history of a single species, you will see a gradual, slow, linear change. This became the accepted, assumed pattern for how evolution occurs.

2.4.6 Evolution May Occur Much More Rapidly Than Previously Thought

Within the last few decades, researchers Niles Eldredge and Stephen Jay Gould have questioned the assumption that evolution has to be a slow, gradual process. They maintain that stasis, or lack of change, is the norm for at least most species most of the time. For millions of years, typically, characteristics within a species stay relatively constant. Then, if evolution does occur, within a few tens of thousands of generations species change, perhaps evolving into new species. This pattern of evolution they termed **punctuated equilibrium** (Figure 2-21). As they see it, evolution is a series of fits and starts rather than a smooth, linear transition. They contend that the linear pattern appears to happen only because the fossil record is incomplete.

Is there evidence for their contention? Eldredge and Gould chose to study the fossil remains of a group of now extinct animals called trilobites (Figure 2-22), which were hard-bodied animals, not unlike modern crabs and insects. Their hard external skeletons are particularly well represented in the fossil record. Looking carefully at many fossilized trilobites from different layers of rocks—different geological time periods—Eldredge and Gould concluded that there was a distinct absence of fossilized trilobites representing intermediate forms.

The theory of punctuated equilibrium has generated a great deal of conversation and new research among evolutionary biologists, and there is certainly no sweeping consensus that the Eldredge and Gould explanation for the evolution of new species is the best one. But this is science at its best. New ideas are proposed based on evidence, and lively debate ensues, which generates new research. A better understanding of biology results.

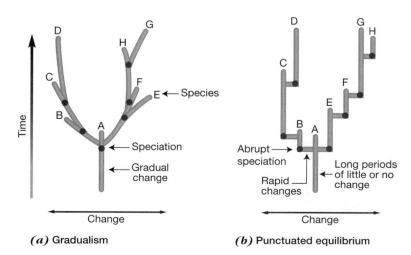

(a) Gradualism

(b) Punctuated equilibrium

Figure 2-21. Gradualism and punctuated equilibrium can be depicted as evolutionary tree diagrams. *(a)* In gradualism, branching is gently angled, indicating slow, steady accumulations of changes over time. Speciation occurs at each branching point. *(b)* In punctuated equilibrium, branching is more abrupt, indicating that speciation has occurred through more sudden accumulation of changes, followed by long periods of time in which few, if any, changes occur.

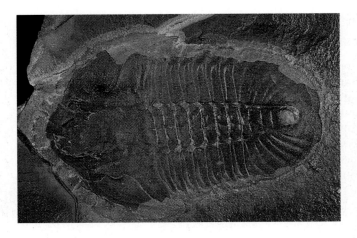

Figure 2-22. The now-extinct trilobites were once a dominant form of life that left numerous fossils. Analysis of these fossils led Niles Eldredge and Stephen Jay Gould to propose punctuated equilibrium as one of evolution's mechanisms.

2.4.7 Evolution Does Not Always Occur

Not all species evolve. For one thing, natural selection can only work on variations that already occur within a population. Variations cannot be created on demand. An example: in many northern regions animals are white. Polar bears, Arctic foxes, snowy owls, gyrfalcons, long-tailed and least weasels, Arctic hares, snowshoe hares, and collared lemmings are white, at least in winter. White fur is camouflage. White hairs are often hollow, filled with dead air, making excellent insulation. As each of these species moved into the Arctic, natural selection favored those individuals that happened to be the lightest in color. These survived, bred, and passed this characteristic on to offspring until the entire population was white and new species evolved.

However, not all Arctic animals are white. Blue-phase Arctic foxes evolved on islands where they live year-round among dark rocks, kept free of snow by wave action and other factors. Arctic ravens, are jet black (Figure 2-23). Why haven't ravens become white? Apparently, there have been no "lightest colored individuals" for natural selection to favor. Thus, they have stayed black.

This is not to say that natural selection has not affected ravens. It has influenced other sources of variation present in ravens. For example, they are unusually large even for birds in the crow family. Large size is advantageous in cold climates because heat is created by body bulk. Indeed, many other Arctic animals are not only white but also the largest of their kind. Ravens are also exceptionally adept at finding food. Some have been known to poke the eyes out of caribou calves, wait until the calves die, and feast. Seen from the perspective of the caribou, such behavior is horrible. But to the raven, such behavior allows exploitation of a food source totally unavailable to other birds.

Figure 2-23. Ravens are year-round residents of Arctic regions throughout North America and Eurasia. Unlike most Arctic animals, ravens are jet black. In spite of their color, they are highly adapted to arctic conditions. Their size (ravens are the largest members of the crow and jay family) discourages heat loss. In addition, they exploit an unusually wide variety of food sources.

Figure 2-24. Coelacanths were once thought to be extinct. Their fossil remains are found in numerous locations, but living examples were unknown. Today, remnant populations are known to occur in the Indian Ocean.

Evolution works best in species with large populations and wide ranges in individual variations. This gives natural selection a number of choices to work on. For example, in the mid-1940s, New York City had a problem with houseflies. They seemed to be everywhere in huge numbers. In 1947, DDT became available, and the city had something with which to fight flies. DDT was widely and indiscriminately sprayed throughout the city. As a result, the fly population plummeted. Then, in the early 1950s, the flies returned. The solution was to spray more and more DDT, but to no avail. The flies came back. These new flies could tolerate DDT and were now free of its effects.

Presumably, within the original, huge population of flies, a few individuals had a natural tolerance to DDT. When other flies were killed, these survived, thrived, and allowed the species to recover. During that time, other insects, with less variation within their numbers, disappeared from New York City. Presumably, their populations lacked the naturally tolerant forms that allowed the flies to survive. Variation within a population is useful. Often, it is key to survival in a changing environment.

Most species are not as fortunate as the houseflies of New York City. Many lack the range of variations seen in the houseflies. Most populations are relatively small. Also, it takes time for natural selection to have its effect. Even punctuated equilibrium is assumed to take tens of thousands of generations before species typically change. There are exceptions, but generally, if the environment changes too rapidly, evolution will not happen. What happens if there are not enough variations within the population to support evolution? What happens if the environment changes too rapidly? Then evolution does not occur.

Some nonevolving species continue to survive unchanged. Hundreds of millions of years ago, the coelacanths (Figure 2-24) roamed the world's oceans. Their fossil remains are found in many places. Today, living coelacanths, apparently identical to fossil forms, live on in restricted, isolated, deep waters offshore from Madagascar. Recently, another population was found near Indonesia. They have not changed for millions of years. As most environments around the world changed, they did not evolve. Their survival depended on finding tolerable, if restricted, habitats.

This is not the usual story. Most nonevolving species become extinct. Sooner or later, the pace of environmental change and the intensity of environmental pressure overwhelm all species' ability to adapt. With the possible exception of some microbes, no species lives forever.

Piecing It Together

1. Evolution is the process by which species change. It is the reason Earth has so many species.

2. Natural selection is the tendency for individuals best adapted to their environments to survive and reproduce and for those less well adapted not to.

3. Environments frequently change, and as they do, species change to survive. To the population and the individuals within it, only survival matters. Speciation comes later, as a byproduct of survival.

4. The course of evolution often follows common patterns:

   ◆ **Convergent evolution:** Unrelated species living in similar environments and facing similar environmental challenges sometimes evolve similar characteristics.

   ◆ **Divergent evolution:** Closely related species living in different environments and facing different environmental challenges sometimes evolve dissimilar characteristics.

   ◆ **Co-evolution:** Often, the evolution of one species affects the evolution of another species.

   ◆ **Adaptive radiation:** Sometimes, several species will evolve from a single, ancestral species.

   ◆ **Gradualism:** Evolution may occur as a slow, gradual process of change.

   ◆ **Punctuated equilibrium:** Evolution may proceed with long periods of relatively little change (stasis) punctuated with short periods of intense change.

2.5 HOW WELL WAS THE THEORY OF EVOLUTION ACCEPTED?

When first proposed, the theory of evolution presented scientists (and others) with a new view of the origins of living things. New ideas are usually accepted by scientists cautiously and skeptically. Evolution was no exception. (Visit the CD for further details.)

2.5.1 After It Was Published, Darwin's Theory Was Hotly Debated

Fast Find 2.5 Acceptance of Evolution

Darwin had little trouble convincing most of his contemporaries of the reality of evolution. The idea that species were not immutable was, after all, not new with him. His own grandfather and others had introduced the idea before Darwin was born. By the time he came along, the idea that species could give rise to new species was relatively easy to accept.

Darwin was much less successful in gaining acceptance for his proposed mechanism. He was convinced that natural selection was the force that drove evolution. For this, he was roundly criticized. "Evolution," his critics seemed to be saying, "may occur, but not the way Darwin proposed."

Criticisms of Darwin centered around three important points. First, Darwin completely lacked direct evidence for natural selection, and his indirect evidence was far from convincing. Today, we frequently decide difficult issues relying solely on indirect, circumstantial evidence. This was not the case in the days of Victorian England. Thinkers in those days were much more likely to demand absolute proof, especially for something as basic as the origin of species.

Not only was absolute proof lacking, in his books Darwin cited only two indirect examples of natural selection. One showed how natural selection might lead to swifter, leaner wolves. The other described how flowering plants gained the ability to reproduce and bees gained the ability to harvest sweet nectar. Even Darwin admitted that these were "imaginary illustrations."

Many of his contemporaries were unconvinced. Adam Sedgwick, a professor who had taught Darwin geology, described the theory of natural selection as "a vast pyramid resting upon its apex, and that apex a mathematical point." To shore up his pyramid, Darwin needed evidence that only time could provide.

Second, Darwin couldn't adequately deal with questions of heredity and variations within populations. None of Darwin's contemporaries fully understood how characteristics were passed on to succeeding generations. It was obvious that offspring resembled parents. But why were they not identical to them? Indeed, by successively weeding out the less well adapted, natural selection should lead to individuals being more and more alike. Eventually, populations should comprise only individuals perfectly adapted to their environments. Every individual should then be identical. In addition, if natural selection had been working on species throughout geologic history, then surely Darwin would be able to find one example of a species in which natural selection had reached its endpoint: a population in which all individual differences had been eliminated, or one species perfectly adapted to its environment. He could not.

Perhaps in Darwin's time it was not obvious that environments frequently change and that therefore the directions in which natural selection operates also change. Also, neither Darwin nor his contemporaries knew the source of individual differences within populations. He needed to supply a mechanism for creating new individual differences to counterbalance natural selection's effect of eliminating them. Examining that evidence will take us deep into Chapter 3.

Third, Darwin got caught up in an argument that has still not been fully resolved. Is evolution a slow, continuous process or a jerky, discontinuous one? At times he seemed to equivocate, describing natural selection as both slow and intermittent. But generally, Darwin opted for slow and steady. In doing so, he sided with Charles Lyell's theory of Uniformitarianism. Lyell argued that geologic changes happen slowly and steadily. Darwin said the same was true of evolution. Opponents to one argument became opponents to the other. Even some of his staunchest defenders balked at accepting that evolution was slow and steady. Thomas Henry Huxley and Francis Galton, influential English scientists of the day, accepted natural selection but thought it had to operate infrequently. Other critics felt if Darwin was wrong about slow and steady, then his whole theory was questionable.

Darwin faced other arguments more successfully. He had little trouble explaining the evolution of gaudy coloration and flamboyant structures among male birds. Male cardinals, for example, are bright red and are frequently the most obvious organism in their environment. Male peacocks have huge, ungainly tails with which they display to females (Figure 2-25). Shouldn't such features also attract the attention of predators and, therefore, be selected against? Darwin saw that more was involved than meets the eye and proposed his theory of **sexual selection**. According to this theory, female birds choose their mates, and they are attracted to males with extreme features. These males are the ones they breed with, and in so doing, the colorful or extreme features are passed on to male offspring.

What about **altruism**, or behavior that has little value to individuals, but benefits the group? Worker bees, for example, are essential to the hive, but they never reproduce and, therefore, never have an opportunity to pass their characteristics on to offspring. Shouldn't natural selection be expected to work against such behavior? How could this behavior evolve? Darwin rightly saw that natural selection could work on entire lineages rather than only on individuals. By caring for those closely related to themselves, family members care for individuals who share their characteristics and, if successful, ensure that those characteristics are passed on to succeeding generations. Not everyone was

Figure 2-25. Peacocks have extremely long tails that they use to attract females. Outside of the breeding season, such tails should be disadvantageous to the male, being cumbersome in dense vegetation and making escape from predators difficult. Darwin recognized the evolution of such characteristics as a special case of natural selection, which he termed sexual selection. Female peafowl choose mates with which to breed. By consistently preferring males with long tails, they ensure that the more extreme example is passed to the next generation.

convinced, and it wasn't until the 1960s that the details of **kin selection**, whereby individuals help relatives raise young (Figure 2-26), was interpreted within the context of natural selection. We'll save the details of these behaviors until Chapter 13.

2.5.2 After Darwin's Death, Debates Continued

While overall acceptance of the basic concept of evolution came relatively quickly, acceptance of natural selection came much more slowly. At the time of Darwin's death in 1882, opponents to natural selection outnumbered proponents two to one. For the next 30 years the arguments raged on. Natural selection lost further ground. Alternative mechanisms and explanations were proposed and then rejected. Convincing evidence to support or refute natural selection simply was not yet available. Such evidence would accumulate in coming decades.

Figure 2-26. Scrub jays illustrate kin selection. Young jays often forego their own reproductive efforts, stay with their parents, and assist in raising subsequent young. By so doing, they enhance survival of closely related individuals with whom they are likely to share many characteristics. Kin selection often evolves in environments where breeding (in this case, nesting) sites are limited. Waiting until a parent dies is the scrub jay's most reliable means of obtaining a high-quality nest site.

Ironically, some key evidence had already been described during Darwin's lifetime.

In an isolated Austrian monastery, Gregor Mendel had initiated what later became a new branch of biology. Some of the basic underpinnings of Mendel's experiments were desperately needed by Darwin and his supporters. But Mendel's work was little understood and was, indeed, lost for over 50 years. His story is told in Chapter 3.

Piecing It Together

1. Immediately after it was proposed, Darwin's theory of evolution by natural selection was not widely accepted. For its time, the idea was too radical. The evidence to support it was largely lacking. In the intervening 150 years, evidence accumulated and today scientists universally accept both the theory and its proposed mechanism (natural selection).

2. In defending his theory, Darwin recognized three special cases of natural selection: sexual selection, in which females choose males with desirable, sometimes extreme, characteristics; altruism, in which individual behavior benefits a social group, often at the expense of the individual; and kin selection, in which individuals help relatives raise young.

Where Are We Now?

Does evolution flow smoothly or does it lurch? Contemporary biologists are far from decided. Existing data are being reexamined and questioned. New data are sought. Debates are ongoing. Biologists continue to disagree. Is evolution being questioned? No. The vast majority of contemporary biologists agree that evolution is real and that it is occurring. What is being argued are some of the details of how it happens. Like so many major ideas in science, the theory of evolution by natural selection continues to be refined. It will never be discarded.

Today's evolutionary biologists have powerful new tools at their disposal. In the past, evolutionary relationships were determined by comparing organisms' external characteristics and internal structures. Generally, organisms that most resembled each other were thought to be closely related. Biologists now decipher relationships using new and much more precise techniques. Biochemical analyses allow comparisons of organisms' basic chemical composition: their proteins and their DNA. Basic assumptions are being reexamined. In most cases, previous conclusions are being verified. In others, new relationships are indicated.

The theory of evolution by natural selection remains one of the most important ideas in science. It is the cornerstone upon which nearly all contemporary biology rests. After roughly 150 years of challenge and evaluation, mainstream biology has accepted both evolution and natural selection. Why?

First, the theory is useful. It explains and ties together whole sets of observations. It explains why there are so many different kinds of organisms on Earth. Each species is now seen as an evolutionary success story, an example of how natural selection and evolution work.

The theory explains why organisms resemble each other and share many characteristics. Even those most distantly related share some characteristics. Every living cell, for example, has the ability to digest glucose. This and other biochemical capabilities may be the only characteristics shared by, say, bacteria and mammals. The fact that they share any characteristics suggests that they might be related, on some level, a point fully compatible with our current understanding of the theory of evolution. The fact that they share relatively few characteristics means that their ancestors diverged long ago.

Evolution and natural selection also explain why species are different. Closely related species live in slightly different environments where the pressures of natural selection are also different. Because these pressures are different, isolated species become more and more different. It's what we see in nature. It's what we've come to expect.

Second, the theory of evolution by natural selection is generally accepted by mainstream biology because it fits within the framework of other theories. Although it shifted scientific thought in a new direction, the theory of evolution did not completely upset science's apple cart. We shall see in succeeding chapters that evolution is not only compatible with biology's other major theories, it is the central idea that holds them together. Indeed, the theory of evolution by natural selection is so pervasive in biology today that the geneticist Theodosius Dobzhansky was led to make the statement with which we started this chapter: "nothing in biology makes sense except in the light of evolution."

It was not always so. In Darwin's time, evolution by natural selection did not seem to fit. It flew in the face of conventional wisdom. It conflicted with too many basic ideas, both within science and outside it. Basically, it went too far, too fast. The story of how the rest of biology caught up is fascinating and will occupy us for much of the rest of this book.

How We Evolved: How did humans evolve?

REVIEW QUESTIONS

1. What is the central proposition of the theory of evolution?

2. How have fossils contributed to our understanding of evolution?

3. How were the conclusions of Darwin and Wallace different from other biologists who preceded them?

4. Distinguish between the theory of evolution and the theory of natural selection.

5. Explain the statement "populations have the ability to increase exponentially."

6. If populations have the ability to increase exponentially, why are they typically constant in size?

7. Explain the statement "natural resources are limited." How is that statement important to the theory of evolution?

8. What did Darwin mean when he said, "there is a struggle for existence among individuals in a population"?

9. What is natural selection?

10. What is a species?

11. How does the existence of homologous structures support the theory of evolution?

12. How does the theory of evolution explain the superficial similarities between insect wings and bird wings? What is this similarity an example of? Why are these wings also different?

13. Which would you expect to be more similar, the proteins of dogs and wolves or the proteins of dogs and trees? Why?

14. For an organism to successfully colonize a new territory, what three opportunities must be realized?

15. How does allopatric speciation differ from sympatric speciation? Give examples of each.

16. How is divergent evolution different from convergent evolution? Give examples of each.

17. Soon after the dinosaurs became extinct, approximately 65 million years ago, mammals proliferated. This is an example of which pattern of evolution?

18. Give some examples of coevolution.

19. How are the concepts of gradualism and punctuated equilibrium different? How are they similar?

20. Do species always evolve into new species? If not, what happens to them? (Hint: There are two possibilities.)

Chapter 3

Mendelian Genetics: How Are Traits Inherited?

It requires indeed some courage to undertake a labor of such far-reaching extent; this appears, however, to be the only right way by which we can finally reach the solution of a question, the importance of which cannot be overestimated, in connection with the history of the evolution of organic forms. —Gregor Mendel, 1866

—Overview—

In this chapter, we will examine how traits are passed from generation to generation—a branch of biology called classical, or Mendelian, genetics. We will see that a modest garden, growing within the walls of a Moravian monastery at about the same time that Darwin was writing *The Origin of Species*, was to change forever our understanding of how traits are inherited. From that same garden would come answers to some of the nagging questions about Darwin's controversial new theory, evolution by the mechanism of natural selection.

Darwin's revolutionary insight was immediately recognized by some as a powerful explanation for the variety of species on Earth. Many people, however, remained unconvinced. For those who held certain religious beliefs, including one stating that all species were put on Earth by a supreme being during one week of creation as set forth in Genesis, no amount of scientific evidence could be convincing. But even those who searched for naturalistic explanations for the diversity of life found reason to criticize Darwin. Some features of natural selection were readily accepted; they were obvious and verifiable by observation and study. For example, one could easily see that most organisms produced more offspring than could possibly survive. And it was evident that traits that enhance the ability of individuals to survive and reproduce are passed on to the next generation. But Darwin's critics were quick to point out that there was no satisfactory explanation for how traits were passed from parents to offspring. The ideas that *had* been proposed about the mechanism of heredity were

Chapter opening photo—Portrait of Gregor Mendel, the Founder of Genetics. Mendel's largely misunderstood paper was published in 1866, less than a decade after Darwin's book, *The Origin of Species*.

highly improbable and not verifiable. Many, including Darwin, mistakenly believed that traits from two parents blended together in their offspring. Thus, the children of a blue-eyed man and a brown-eyed woman would have bluish-brown eyes. Likewise, if the pollen of a red flower landed on a white-flowered plant, the flowers of the offspring should all be pink. But clearly blending does not happen—at least not all the time. Traits of parents may occasionally appear as blended intermediates in their children (more on this in Section 3.3.1), but more often they do not. Some traits are passed on apparently unaltered from parent to offspring; others hide for generations at a time, only to reappear at unpredictable times in family histories. What could account for the seemingly unpredictable behavior of heritable traits as they are passed from generation to generation?

Another problem for Darwin and his supporters was their inability to explain the source of variation among individuals in natural populations. Variation, like heredity, is an essential element of Darwinian selection. In light of the erroneous theory of blending, all organisms of a population should appear quite similar. In fact, if blending were, indeed, occurring, then all variation within a given population would be obliterated within about 10 generations. All individuals would be identical, and evolution would cease. But that is not the case.

Unbeknown to Darwin and his critics, these problems were being addressed in a series of ingenious experiments carried out by an Austrian monk, Gregor Mendel (see chapter opening photograph). Mendel's work was published in 1866. Unfortunately, Darwin died in 1881, unaware of these experiments or their lasting importance to his theory of evolution by natural selection.

3.1 HOW ARE TRAITS PASSED FROM GENERATION TO GENERATION?

Do you look more like your mother or your father? Perhaps you are the spitting image of your Great Uncle Horace. Chances are, while you share some characteristics or traits with each of your parents, you are not a perfect blend of the two. You may even have characteristics that appear in neither of your parents. Clearly, different forms of traits can persist in populations for many generations. They remain whole and unaltered by time and their passage through the individuals that carry them and pass them on. Hereditary traits in all of their varied forms behave as units, moving through time as bits of information.

But living things, be they people, oak trees, grizzly bears, mushrooms, slime molds, or pea plants, are each combinations of perhaps hundreds of thousands of traits. To study them all would have been a monumental task indeed. It would be nearly impossible to find consistencies or rules that govern the inheritance of all of them at once. Gregor Mendel was the first to recognize that traits in individuals are controlled by some type of hereditary units, which he called "factors." He was also the first to describe the passage of these factors through the generations. There were at least three reasons for Mendel's success: (1) he focused on just a few traits—seven to be exact—instead of many traits as others did; (2) he thoroughly documented and quantified all of his experimental results; (3) he chose to study these traits in the garden pea, *Pisum sativum*.

3.1.1 Mendel Discovered That Traits Are Inherited in Discrete Units

Gregor Mendel, born Johann Mendel, was the son of peasant farmers. From a childhood spent on a farm, Mendel understood the value of plant breeding in developing productive varieties of crops. This undoubtedly contributed to his lifelong interest in gardening and horticulture. At the age of 21, Mendel entered the priesthood, taking the clerical name of Gregor by which he is now remembered. In the remote monastery in

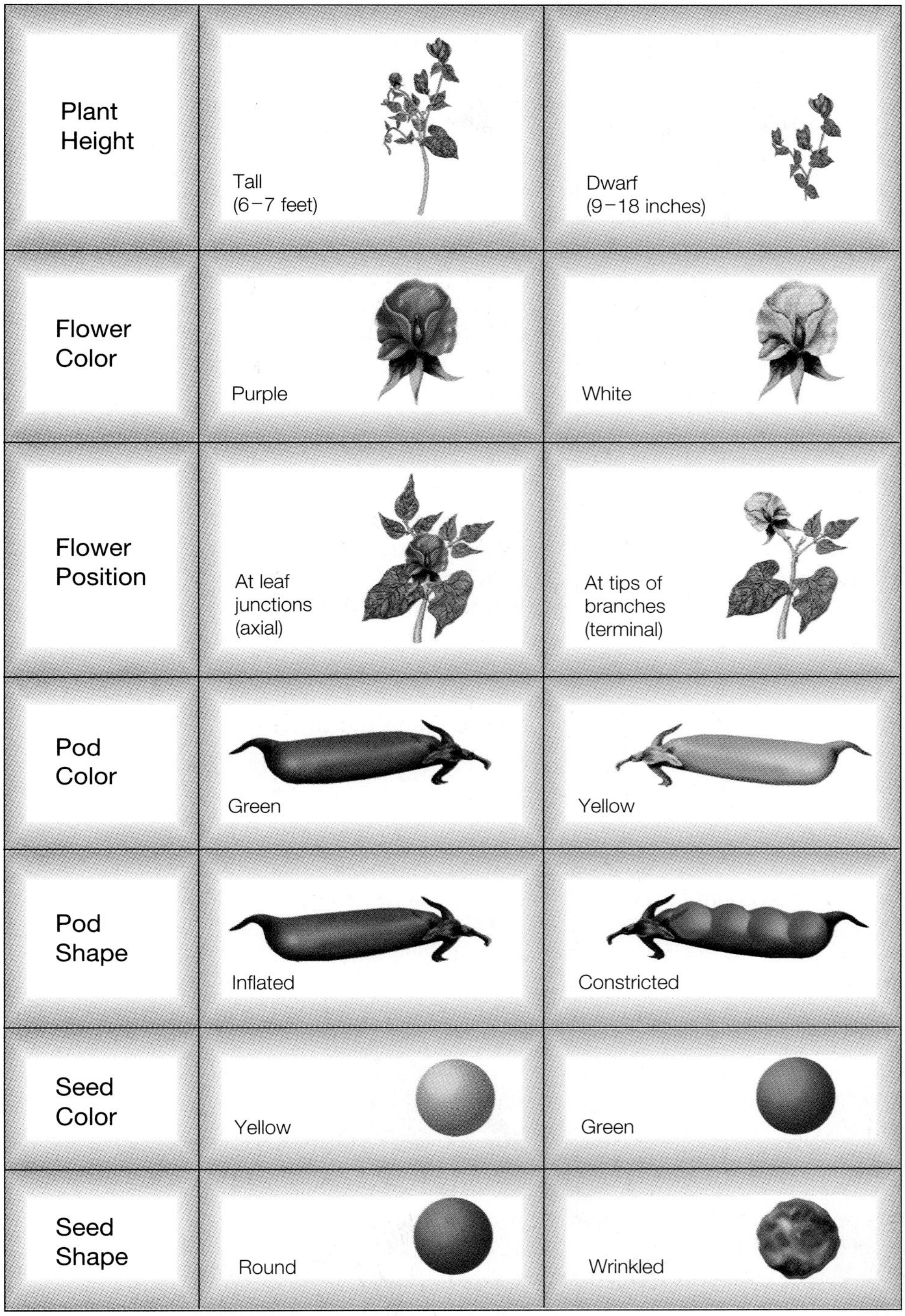

Plant Height	Tall (6–7 feet)	Dwarf (9–18 inches)
Flower Color	Purple	White
Flower Position	At leaf junctions (axial)	At tips of branches (terminal)
Pod Color	Green	Yellow
Pod Shape	Inflated	Constricted
Seed Color	Yellow	Green
Seed Shape	Round	Wrinkled

Figure 3-1. The seven traits that Mendel studied. Note that each trait appears in one of two distinct forms.

what is now the Republic of Czechoslovakia, Mendel had time to indulge his love of plant breeding. He read widely, especially the natural sciences. He was quite aware of the controversial new theory of evolution proposed by his British contemporary, Charles Darwin. He knew of the unanswered questions about heredity arising from Darwin's theory, the very topic Mendel sought to understand. He made the fortuitous choice to study the pea. The pea is an organism that can be easily manipulated in breeding experiments. Even in Mendel's time, pea plants came in many distinct strains or varieties.

Mendel studied traits that each occur in two distinct forms (Figure 3-1). The color of the pea flower, for example, is either purple or white, never purplish-white. The shape

Fast Find 3.1a Pea Plant Traits

of the pea pod is either puffy and inflated or narrow and constricted, never partly puffy, and so on with each of the seven traits he studied. He began by developing true-breeding varieties for each of the seven traits; that is, when bred among themselves, all of the offspring of a given variety were identical to the parent for that trait. For example, one variety produced only purple blooms for many generations, another only white blooms. One variety produced only puffy pods for many generations, another only constricted pods. There were fourteen varieties in all. When he was certain that all of his varieties bred true, he carefully engineered matings between pairs of plants showing different forms of each trait.

In addition to this focused approach, Mendel added yet another new twist: mathematical analysis. He counted the number of young plants that developed the parental forms of each trait and calculated the numerical ratios of offspring showing each form of a trait. In Mendel's day, this application of mathematics to plant breeding experiments was uncommon, to say the least. Let's start with just one trait that Mendel studied and carefully follow his reasoning.

Mendel's First Discovery

Fast Find 3.1b Pea Pollination

When organisms reproduce sexually, both parents produce specialized reproductive cells called **gametes**. Male gametes are **sperm**, and female gametes are **eggs**. When egg and sperm fuse, a process called **fertilization**, a new individual is produced. In flowering plants such as the garden pea, sperm are contained in **pollen**. Eggs are contained in **ovules**, which when fertilized, mature into seeds. Ovules are contained within a structure in the flower called the **carpel** (Figure 3-2). Pea plants are often (but not always) **self-fertilizing**; that is, the sperm-carrying pollen usually lands on the top of the egg-carrying carpel of the same plant. If the flowers are covered to ensure that insects cannot carry pollen from one flower to another, the sperm and eggs from the same plant combine within the bud to give rise to peas—the seeds of the next generation of pea plants.

Mendel manipulated this process by opening the flower buds and cutting off the pollen-bearing structures, called anthers, from some flowers. Using anthers taken from different flowers, Mendel could control exactly which sperm was used to fertilize which

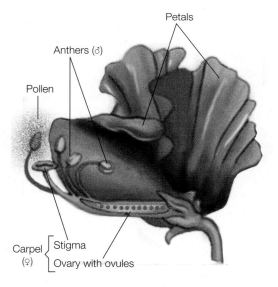

❶ Mendel took pollen from the anthers of a plant exhibiting one form of a trait.

❷ He brushed the pollen onto the stigma of a plant showing a different form of the trait. The anthers of the second plant were first removed to prevent self-fertilization.

Petals

Anthers (♂)

Pollen

Carpel (♀) { Stigma / Ovary with ovules

Amputated anthers

(a) Flower of a pea plant

(b) Cross-fertilization

Figure 3-2. (a) Flower of a pea plant, cut to show male and female flower parts. (b) Using artificial cross-fertilization, Mendel controlled matings between plants.

eggs (Figure 3-2). He started with two varieties of plants, one whose ancestors produced only round, smooth peas and another from a long line of plants that produced only wrinkled peas. He **cross-fertilized** the two varieties. In other words, he used the pollen of one variety to fertilize the egg of the other variety, creating **hybrid** offspring. This was done by brushing the female parts of flowers of the round-seeded line with the male pollen from the wrinkled-seeded variety, and vice versa. If blending were indeed the basis of heredity, as discussed earlier, all of the hybrid seeds from these experimental crosses would have been partially wrinkled in texture, intermediate between the two parental forms. That did not happen. Remarkably, the hybrid peas of this first generation, called the F_1, or **first filial**, generation, were *all* round.

When these hybrid round peas grew into mature plants and were permitted to self-fertilize, some of the many "grandchildren" of the original pure-breeding varieties—the F_2, or **second filial**, generation—were round and some were wrinkled. In fact, 5,474 of the F_2 generation were round, and 1,850 were wrinkled, a ratio of about three round peas for every one wrinkled pea. Within each pea pod of this F_2 generation (a single pod contains about six to nine peas), about 75 percent, or three-quarters, of the peas, were round and about 25 percent, or one-quarter, were wrinkled (Figure 3-3).

What happened when this second generation was allowed to grow and self-fertilize? Mendel planted both the round and wrinkled peas and observed that the wrinkled peas bred true, producing only wrinkled offspring. Likewise, approximately one-third of the round peas bred true, producing only round offspring. Two thirds of the round F_2 peas, however, gave rise to both round and wrinkled peas in the third, or F_3, generation.

Fast Find 3.1c Monohybrid Cross

The results were similar for all seven paired characters. For every cross between parents showing alternate forms of a single trait, the F_1 offspring plants all exhibited just one form of the trait. When the F_1 plants underwent self-fertilization, about 75 percent of the resulting F_2 generation showed one form of the trait, and about 25 percent exhibited the alternate form (Figure 3-3). On your CD, you can try this experiment by crossing white-flowered peas and purple-flowered peas. What color are the blooms of the F_1 generation? Which form of the flower color trait occurs in about 75 percent of the F_2 generation?

Mendel's Interpretations

Perhaps Mendel's real genius was in how he interpreted the results of his experiments. Mendel imagined that each of the contrasting forms of a trait, for example the "roundness" or "wrinkledness" of seeds, was controlled by a hereditary factor. He realized that his results could be best explained by supposing these factors occur in pairs within the individual pea plants. A pea arising from a long line of plants producing only wrinkled seeds has two factors for wrinkled seed texture; likewise, a pea from a family lineage that produces only round seeds has two factors for round seed texture.

When organisms breed, hereditary factors are passed on whole, and usually unaltered, to the offspring. Mendel realized that during reproduction, each parent contributes one hereditary factor for each trait to the offspring. Thus for every trait, each individual has one maternally derived factor and one paternally derived factor—one factor from its mother and one from its father.

When an organism has two identical factors for a trait, as in true-breeding varieties, it is said to be **homozygous** for that trait (the prefix *homo-* means "the same"). If the two factors are different, the organism is said to be heterozygous for that trait (the prefix *hetero-* means "different"). Mendel's true-breeding varieties were homozygous; the hybrid offspring he created were heterozygous.

Mendel's results and interpretations are still relevant today, although our vocabulary is slightly different from his. We use the term **gene** to describe the hereditary information that determines a single trait. Seed texture in peas, for example, is a trait determined by a gene. Flower color is another trait determined by a gene. The different forms that a gene might take—what Mendel referred to as factors—are called **alleles**. Wrinkled and

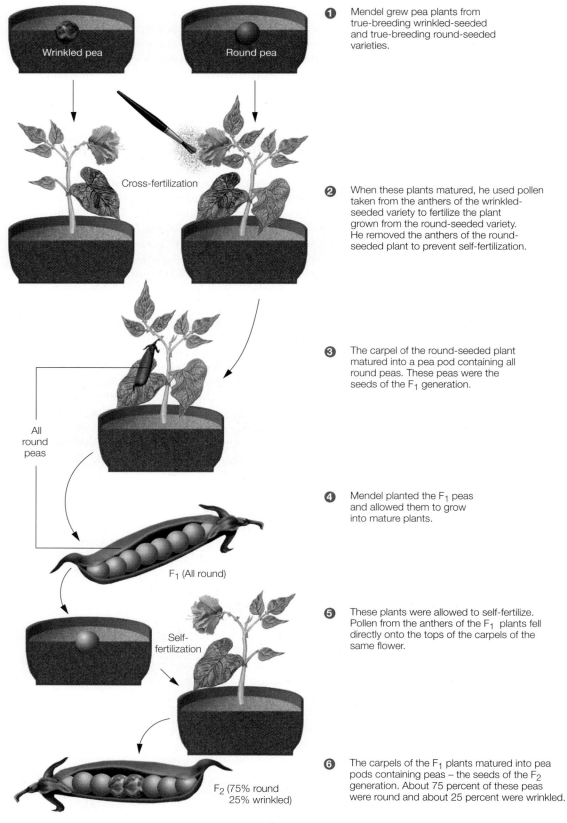

① Mendel grew pea plants from true-breeding wrinkled-seeded and true-breeding round-seeded varieties.

② When these plants matured, he used pollen taken from the anthers of the wrinkled-seeded variety to fertilize the plant grown from the round-seeded variety. He removed the anthers of the round-seeded plant to prevent self-fertilization.

③ The carpel of the round-seeded plant matured into a pea pod containing all round peas. These peas were the seeds of the F_1 generation.

④ Mendel planted the F_1 peas and allowed them to grow into mature plants.

⑤ These plants were allowed to self-fertilize. Pollen from the anthers of the F_1 plants fell directly onto the tops of the carpels of the same flower.

⑥ The carpels of the F_1 plants matured into pea pods containing peas – the seeds of the F_2 generation. About 75 percent of these peas were round and about 25 percent were wrinkled.

Figure 3-3. One of Mendel's crosses between smooth round-seeded and wrinkled-seeded pea plants.

round are alternate alleles for the gene for seed texture; purple and white are different alleles for the gene for flower color. *The relationship between genes and alleles is a fundamental concept of biology.*

In Mendel's peas, the presence of the allele for round seeds was quite visible in the F_1 generation. Indeed all of the F_1 peas were round. But what of the allele for wrinkled seeds that later appeared in the F_2? Where was it hiding in the F_1 peas?

3.1.2 Dominant Traits Mask the Presence of Recessive Alleles

Let's call the allele for round seeds *R*, and the allele for wrinkled seeds *r*. Now, if both parents carry two identical alleles for wrinkled seeds, *rr*, each of the offspring will get an *r* allele for wrinkled seeds from mother and another wrinkled-seed *r* allele from father. All of the offspring will have a double dose of wrinkled-seed alleles, *rr*, and will have wrinkled seeds, just like the parents. These peas, as well as their parents, are homozygous for the gene for seed texture. But in Mendel's first experimental cross, one parent had only wrinkled-seed alleles to give, *rr*, and the other parent had only round-seed alleles, *RR*. One parent could contribute only *r* alleles and the other could contribute only *R* alleles to the offspring. When the peas reproduced, the F_1 generation ended up with two different alleles, *Rr*. They were heterozygous peas, with a wrinkled allele from one parent and a round allele from the other.

Yet in Mendel's experiments, all the F_1 hybrids produced round seeds. There was no sign of wrinkled seeds, even though one of the parents had wrinkled seeds. The seven traits that Mendel studied all showed the same pattern. Mendel called the trait that appeared in the F_1 generation the dominant trait and the trait that was hidden the recessive trait. The allele for the hidden trait was there, whole and unchanged, in the heterozygous individuals, but it was completely masked by the presence of the dominant allele. The recessive form of a trait is only seen when both of the alleles are recessive.

Because some alleles are dominant and some are recessive, you cannot always tell the genetic makeup of an individual by looking at its traits. There may be recessive alleles in the genes masked by their dominant counterparts. For example, a round pea can either be homozygous, having two dominant alleles for round seeds, or heterozygous, having one dominant allele for round seeds and one recessive allele for wrinkled seeds. Either combination would appear the same. We must distinguish between the forms of traits that are expressed—those we can see or otherwise detect—and the combination of the expressed and hidden alleles in the genes. We refer to the traits that are expressed in an individual as its **phenotype**, or those characteristics that are apparent by looking at the individual. The genetic complement, all the alleles within that individual, both dominant and recessive, is called the **genotype**. Some examples of the difference between phenotype and genotype can be found on the CD.

Fast Find 3.1d Test Cross

Try describing your own phenotype with regard to some traits. Do you have a widow's peak or a dimple in your chin? Are your earlobes attached, or are they free? Now try to describe your genotype. You will find that you need more information to accurately assess which alleles you may be carrying. You need to know which alleles are dominant and which are recessive. You need to know which alleles your parents, and perhaps even your grandparents and great grandparents, possessed. If you have children, you can learn more about your genotype by seeing which traits are expressed in them, but even that may not be enough (Figure 3-4). We can learn about the genotypes underlying different phenotypes in plants, flies, even mice by performing controlled crosses between individuals, then calculating the ratio of offspring showing different forms of traits, in much the same way Mendel did. In human populations, where it is impossible to perform controlled crosses, we rely on family trees, called **pedigrees**, to follow the pathways of alleles through generations. Using the CD, you can practice working out the genotypes of individuals by performing controlled virtual crosses.

(a)

(b)

(c)

Figure 3-4. Actors (*a*) Leonardo DiCaprio, (*b*) Matt Damon and (*c*) Parker Posey illustrate some examples of human traits governed by single genes. A widow's peak, seen in all three, is a dominant genetic trait. Compare Leonardo's detached earlobes with Matt's attached earlobes. Detached earlobes are the dominant form of the trait. What other single gene traits do these actors have?

 Family Matters Suppose you are interested in studying human traits, but performing genetic crosses is impossible. How can you learn about the genotypes of humans by studying the phenotypes of extended families? Will such family studies always establish genotypes with certainty?

3.1.3 Mendel's Observations Are Widely Applicable

The principles of classical genetics, established in the mid-19th century by Mendel, are broad in their application and fundamental to our understanding of inheritance. Were they all of what Mendel discovered, his honored position in the history of ideas would be firmly established. But there was more to his work, and that, too, was far-reaching. Let us leave the monastery garden for now, and apply Mendel's principles to more immediate issues of human health and well-being, using as an example the human genetic disease, sickle cell anemia.

Sickle cell anemia is a hereditary disease in which the blood of affected individuals has a reduced capacity for delivering oxygen to tissues. The gene involved in sickle cell anemia encodes information for making part of the hemoglobin molecule, the oxygen-carrying protein that gives our red blood cells their color. In people with normal hemoglobin, these cells assume the shape of a biconcave disk, as seen in Figure 3-5, and they travel easily through arteries, veins, and capillaries, carrying oxygen to the tissues. The red blood cells of people with sickle cell disease appear normal if oxygen is plentiful, but when oxygen levels decline, as they do in blood that is exiting tissues through capillaries and veins, the hemoglobin in their red blood cells forms insoluble fibrous strands that distort the shape of the cells into long, thin sickles. Even slightly lower levels of oxygen are enough to cause sickling. These distorted cells cannot easily pass through the narrow capillaries of the tissues. Capillaries become clogged, and the tissues are starved for oxygen. These cells may rupture, causing severe anemia, great pain, and serious tissue damage. Without medical attention, sufferers may die.

We can analyze the inheritance of sickle cell anemia using the same approach that Mendel used for garden peas. We cannot, however, engineer crosses among human beings. Instead we must rely on family pedigrees to establish inheritance patterns. We

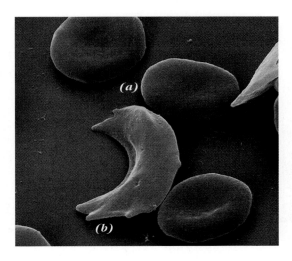

Figure 3-5. Red blood cells. (*a*) Cells containing normal hemoglobin are disk shaped with depressions on both sides. (*b*) Cells from persons suffering from sickle cell anemia become sickle shaped when the level of oxygen in the blood drops. (magnified 9,620 times)

will consider two alleles for hemoglobin: the allele for the normal hemoglobin protein, called *HbA*, and the allele for sickle cell anemia, called *HbS*. Individuals homozygous for normal hemoglobin, designated *HbA/HbA*, have no disease. Homozygotes for the sickle cell gene, *HbS/HbS*, are afflicted with the disease. In heterozygotes, *HbA/HbS*, about 1 percent of red blood cells exhibit the sickling trait. These individuals, called sickle cell **carriers**, show no symptoms of the disease under normal circumstances. Carriers lead normal lives as long as they avoid strenuous exercise, high altitudes, and other situations in which oxygen levels in their blood might get very low.

What would happen if a woman who carries two normal alleles for hemoglobin and a man who suffered from sickle cell anemia had children? The children born to this couple would be analogous to the F_1 hybrids from Mendel's first experimental crosses. Each child would receive one *HbA* allele from its mother and one *HbS* allele from its father. Each child would be heterozygous for the sickle cell gene. Each would be a sickle cell carrier.

Punnett Squares Predict Possible Genotypes

A clever way of envisioning how alleles are distributed during reproduction, called the Punnett square, was developed in the early 20th century by the British geneticist Reginald C. Punnett. For a cross involving a single trait such as sickle cell anemia, the alleles of the parents are arranged on the top and sides of a matrix as shown in Figure 3-6.

Each block of the square represents a possible combination of alleles, a genotype, that could occur in the F_1 generation. Each egg produced by the mother contains one

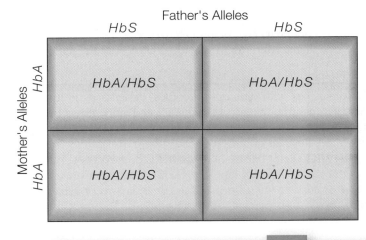

Figure 3-6. Punnett square illustrating possible genotypes of the offspring of a father suffering from sickle cell anemia and a normal mother. All children born to this couple are sickle cell carriers.

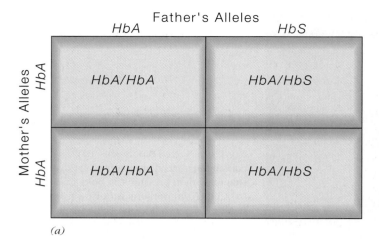

(a)

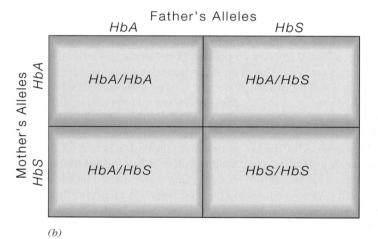

(b)

Figure 3-7. Punnett squares illustrating inheritance of the *Hb* gene. (*a*) The offspring of a normal mother and a father who is a carrier for sickle cell anemia have a 50 percent chance of being carriers themselves. (*b*) When both parents are sickle cell carriers, the offspring have a 25 percent chance of inheriting no alleles for the disease (*HbA/HbA*), a 25 percent chance of inheriting the disease (*HbS/HbS*), and a 50 percent chance of being sickle cell carriers.

of the maternal alleles. With regard to the gene for hemoglobin, all the possible eggs that this mother can produce with regard to the *Hb* gene are listed at the far left of the rows. Because this mother is homozygous and has only *HbA* alleles to contribute, all of her eggs contain the *HbA* allele. Likewise, the father is homozygous for the *HbS* allele, so all of his sperm contain the *HbS* allele. At the heads of the columns of the Punnett square, all of the possible types of sperm with regard to the *Hb* gene are listed. In each box, we insert the name of the allele at the head of that column and at the left of that row. For this match, regardless of which egg and which sperm join at conception, all the offspring will be heterozygous. This is shown by the *HbA/HbS* genotypes in each of the boxes of the square. This Punnett square clearly illustrates that all children in this family have a 100 percent chance of being heterozygous for the sickle cell trait, with one *HbA* and one *HbS* allele. All of them would be carriers of the allele for sickle cell anemia.

What happens when these children mature, marry, and have children of their own? We can envision several possible matches with different outcomes, depending on the genotypes of their spouses. First, let's assume that a boy of the above family grows up and marries a woman who is homozygous for the dominant normal allele (*HbA/HbA*). The possibilities for their children are illustrated on the Punnett square in Figure 3-7*a*.

The maternal eggs all carry the *HbA* alleles because the mother is homozygous. But the heterozygous father can produce two different kinds of sperm with regard to the gene for hemoglobin: either *HbA* or *HbS*. Of the four possible combinations of alleles that could occur in the children, two, or 50 percent, are heterozygous, and two, the other 50 percent, are homozygous for the dominant trait. Thus, *each* child from this union would have a 50 percent chance of inheriting one copy of the sickle cell allele.

Now consider the possibility that a man of the F_1 generation chose a wife who was also heterozygous for the sickle cell trait (Figure 3-7*b*). Of the four possible genotypes that could occur in their children, two are heterozygous (*HbA/HbS*). Each child in this family would have a 50 percent chance of being a carrier of the sickle cell trait. The two other possible genotypes include homozygous dominant (*HbA/HbA*), in which case the child would have neither the disease nor the allele for the disease, and homozygous recessive (*HbS/HbS*), in which case the child would be afflicted with sickle cell anemia. Each child in this family would have a 25 percent chance of each of the homozygous genotypes. Each child would have a 75 percent chance of leading a normal life and a 25 percent chance of suffering from the disease. Note that the ratio of genotypes exhibiting the two phenotypes, 3:1 normal/afflicted, corresponds exactly to the 3:1 dominant/recessive ratio that Mendel described in his experimental crosses. Were these parents to have lots and lots of children, we could reasonably predict that about 75 percent of them could lead normal or nearly normal lives and about 25 percent would suffer from sickle cell anemia.

3.1.4 Alleles Are Randomly Donated from Parents to Offspring

Using all seven of the traits he studied in peas, Mendel found that the F_2 generation averaged about three offspring exhibiting the dominant trait for each one exhibiting the recessive trait. Punnett squares illustrate that underlying the 3:1 ratio for phenotypes is a 1:2:1 ratio of genotypes—one homozygous dominant to two heterozygotes to one homozygous recessive. Mendel recognized that the 3:1 ratio of phenotypes from so many different crosses, and the 1:2:1 ratio of genotypes, implies that heterozygous parents are equally likely to donate either of their two different alleles to their offspring. Were this not the case—if, for example, only the dominant allele could be passed on—then 100 percent of the offspring would exhibit the dominant phenotype. Likewise, if the recessive allele were favored, offspring would all exhibit the recessive trait. The allele donated by a heterozygous parent is random. This simple but elegant insight is the best explanation of the experimental results—the 3:1 ratio of phenotypes and the 1:2:1 ratio of genotypes—from a cross between hybrids. This is called **Mendel's law of segregation**. Formally stated, it says that a parent contributes only one of its two alleles for a trait to each offspring, and if the parent is heterozygous for that trait, the particular allele that is donated to the offspring is random. Thus each embryo has exactly the same chance of receiving a particular allele from each parent. Assuming all combinations of alleles are equally likely to survive, the offspring of crosses between heterozygous parents will exhibit a 3:1 phenotypic ratio of dominant/recessive. If the phenotypic ratio resulting from a hybrid cross differs significantly from 3:1 (as we shall see in some special situations explained in Section 3.3), other forces must be operating.

3.1.5 Mendel's Factors Can Act Independently... Sometimes

Thus far we have considered the inheritance of a single characteristic, seed texture in peas or the gene for hemoglobin in humans, much as Mendel did at the beginning of his plant breeding experiments. But organisms are combinations of many traits, all of which are derived from the genes of their parents. Do Mendel's laws apply if two or more traits are considered simultaneously? When Mendel performed crosses in which he followed two traits at a time, he found that all of his original conclusions applied to

Figure 3-8. Female and male fruit flies of the genus *Drosophila*. Note that the male has a mutation that causes white eyes. The red color of the female's eyes is the wild type.

both traits. Remarkably, the two different traits appeared to be operating utterly independently of one another as they passed from parents to offspring. In his words "the relation of each pair of different characters in hybrid union is independent of the other differences in the two original parental stocks." This important observation is called **Mendel's law of independent assortment**. It states that the alleles of one gene are passed to offspring independently of the alleles of other genes.

There are exceptions to the law of independent assortment, but the exceptions do not invalidate Mendel's general conclusions. We shall see in Chapter 5 that the Mendelian factors, the genes, are carried on structures in cells called **chromosomes**. Each chromosome can carry many, many genes. When the genes for two different traits are found on the same chromosome, they have a tendency to travel together. But that does not minimize the importance of independent assortment. Genes that occur on different chromosomes *do* segregate independently, and even those that share a carrier chromosome may exhibit some degree of independence. It is well worth a closer look at independent assortment.

While Mendel preferred to work with the accommodating garden pea, we can just as easily describe independent assortment using another species that has played a starring role in the history of genetics: the fruit fly *Drosophila melanogaster* (Figure 3-8). Fruit flies are tiny, are easy to keep, and have short life cycles, all characteristics that make them good candidates for studying the transmission of genetic traits. Females lay hundreds of eggs in a lifetime. When many offspring are produced from a single cross, it is easy to calculate the ratios of different phenotypes that result from that cross. Hundreds of different traits have been experimentally bred into the many strains of flies that inhabit genetics laboratories throughout the world. Let's look at two of them simultaneously: body color and wing size.

The Dihybrid Cross

Most fruit flies that live in the wild have broad, straight wings and pale-colored bodies with dark transverse stripes. These traits are dominant to their alternatives, vestigial

Wild type
(V–E–)

Ebony
(V–ee)

Vestigial wings
(vvE–)

Figure 3-9. Wild type and mutant fruit flies. A wild type fly has at least one *V* allele for wing shape and at least one *E* allele for body color. An ebony fly must have two *e* alleles for body color. Likewise, a fly with vestigial wings must have two *v* alleles for wing shape.

(shriveled) wings and ebony body color (Figure 3-9). A trait that is usually found in organisms in their natural, or wild, state is called the **wild type**. Thus, broad wings and pale, striped bodies are the wild type forms of the traits for wing size and body color. Vestigial wings and ebony body color are recessive forms of the traits, called "mutant" forms.[1] Fruit fly genes are usually named for the mutant form; uppercase letters designate the dominant alleles, and lowercase letters designate the recessive alleles. Starting with parents that are homozygous for both traits, let's engineer a cross between a broad-winged *(VV)*, striped *(EE)* parent (wild type for both traits) and a vestigial-winged *(vv)*, ebony *(ee)* parent (mutant for both traits). Now our shorthand for the genotypes must include two sets of alleles, one for two different traits:

P_1 *VV EE*   ×   *vv ee*
(broad-winged, striped)   ×   (vestigial-winged, ebony)

Compare this with the first parental crosses that Mendel made between parents showing different forms of a single trait. All the F_1 of this cross will exhibit the dominant forms of both traits: broad wings and pale, striped bodies. Their genotype is heterozygous for both traits:

F_1 *Vv Ee*
(broad-winged/striped)

Now engineer a cross between the members of the F_1. Because the F_1 flies are hybrids (heterozygous) for both traits and we are following the fate of two traits simultaneously, this experimental cross is called a **dihybrid cross**:

F_1 *Vv Ee* × *Vv Ee*

Each new fly receives one allele for wing shape and one allele for body color from each parent, but recall that the law of segregation states that the actual allele that an offspring receives for each trait is random. It is just as likely that a fly will inherit a *V* allele as it is that it will inherit a *v* allele for wing size. The same goes for the *E* and *e* alleles for body color.

The law of independent assortment also applies. Hence, these alleles can be inherited by the offspring in any combination, as long as each new fly receives two alleles for wing shape and two alleles for body color. The possible genotypes that can result from this cross can be examined using a Punnett square, as shown in Figure 3-10.

These four alleles can combine to give a total of 16 genotypes, not all of them different, but each with equal likelihood. Of the 16 combinations, 9 of them are different. How many different phenotypes do these 16 genotypes represent? Recall that alleles designated with an uppercase letter are dominant to those designated with a lowercase letter. Figure 3-11 shows a Punnett square filled in with phenotypes instead of genotypes. From the nine different genotypes, a total of four different phenotypes can occur, and they occur *in a very specific ratio*: 9:3:3:1, or nine wild type for both traits to three wild type wing size with mutant body color, to three mutant wing size with wild type body color, to one mutant for both traits.

The 9:3:3:1 ratio is characteristic of the offspring of a dihybrid cross in which both traits follow Mendel's laws of segregation and independent assortment. Mendel did many dihybrid crosses choosing several different pairs from the seven traits he studied in pea plants, and in every case he found that the ratio of different phenotypes of the offspring approximated the predicted 9:3:3:1. For each pair of genes, the alleles segregated to the offspring independently of one another.

---

[1]In this particular example, the mutant forms are recessive to the wild type forms. It is important to note, however, that there are many examples of mutant forms that are actually dominant to the more common wild type alleles.

Male Gametes

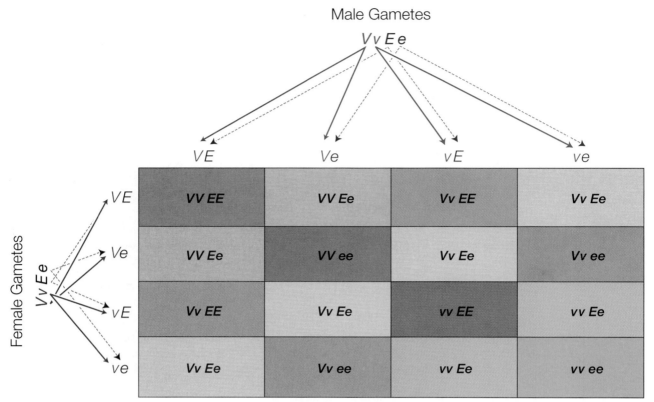

Figure 3-10. Genotypes of the F₂ generation of a dihybrid cross. Each parent fly is heterozygous for two traits, wing shape and body color. This kind of cross creates nine distinct genotypes.

Male Gametes

	VE	Ve	vE	ve
VE	Wild type	Wild type	Wild type	Wild type
Ve	Wild type	Broad wing ebony body	Wild type	Broad wing ebony body
vE	Wild type	Wild type	Vestigial wing striped body	Vestigial wing striped body
ve	Wild type	Broad wing ebony body	Vestigial wing striped body	Vestigial wing ebony body

Female Gametes

Figure 3-11. Phenotypes of the F₂ generation in a dihybrid cross. Four phenotypes result from this cross between flies heterozygous for both wing shape and body color. The ratio of phenotypes is nine wild type flies (broad wings, striped body), to three flies with vestigial wings and striped bodies, to three flies with broad wings and ebony bodies, to one fly with vestigial wings and an ebony body.

More Combinations What general rule about the ratio of genotypes can you infer from a dihybrid cross? What would happen if three traits were considered simultaneously in a trihybrid cross? Is there a mathematical rule that can predict the number of possible genotypes and phenotypes that can occur when any number of traits are considered simultaneously?

3.1.6 Chance Determines Which Alleles an Individual Inherits

A young couple anxiously awaits the arrival of their first child. Among the many things they wonder about before that child is born is whether their new baby will be healthy. One concern of prospective parents is cystic fibrosis, a genetic disease caused by a recessive allele of the *CF* gene. The recessive allele is carried by about 1 of every 25 white North Americans, making cystic fibrosis the most common genetic disorder among this group. It affects about 1 in 2,500 babies in this population. The disease is characterized by production of a thick, sticky mucus in the lungs that is difficult to propel from the airways (Figure 3-12). Victims suffer from chronic infections that progressively destroy the lungs. It is not surprising that the disease is a concern of prospective parents.

Because the allele is recessive, parents may be carriers of the disease-causing *CF* allele but show no signs of the disease. In other words, healthy parents may be heterozygous for the *CF* gene. Genetic testing can determine whether a person is a carrier of the recessive *CF* allele. If a couple knows from genetic tests that the mother is heterozygous for the *CF* gene and that the father does not carry the disease-causing allele, they can use a Punnett square to predict that the chances of having a child who is a carrier are equal to the chances of having a child who is not: 50 percent in each case. They can predict that if they had 10 children, about 5 of them would probably be *CF* carriers and 5 would be homozygous for the normal allele. But until their baby is tested for the *CF* allele, they cannot know for certain whether that child will be a carrier of cystic fibrosis. And even if they had 99 children, they could not predict whether or not child number 100 would carry the *CF* allele.

Donating alleles during reproduction is like tossing a coin. Both are random events. The best possible prediction of the outcome of a random event is nothing more than the probability that each of the possible outcomes will occur. To better understand segregation, independent assortment of alleles, and the different ratios of phenotypes

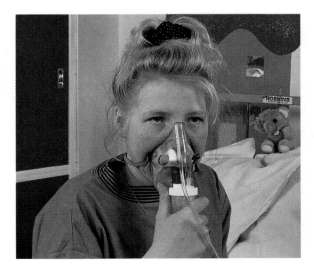

Figure 3.12. Cystic fibrosis treatment. In 1992, an experimental treatment was introduced in which the patient inhales a mist containing enzymes that thin the mucus that accumulates in the lungs.

and genotypes predicted from hybrid crosses, it is useful to know some facts about probability.

Probability

Every time a coin is tossed, there are two possible and equally likely outcomes: heads or tails. When a single die is tossed, there are six possible and equally likely outcomes: the numbers one through six. In any random event, the probability, or likelihood, that any single outcome will occur is equal to the number of times it can or actually does occur divided by the total number of possibilities. Thus the probability of getting heads in a single toss of a coin is $\frac{1}{2}$, or 0.5. Heads is one of two possible outcomes. Likewise, the probability of getting a specific number, say a four, on any given roll of the die is $\frac{1}{6}$, or 0.17. An event that always occurs has a probability of $\frac{1}{1}$, or 1.0. An impossible event has a probability of 0.

We can apply probability theory to the distribution of alleles at conception. Imagine a parent who is heterozygous for some hypothetical gene we will call *G*. Because the genotype is *Gg*, each gamete (egg or sperm) produced by that parent will contain one of the two alleles, either *G* or *g*, but not both. If neither allele is favored when gametes are produced, as we predict based on Mendel's law of segregation, the probability, *P*, that any one gamete will contain the *G* allele, and hence be passed on, is P(*G*) = $\frac{1}{2}$, or 0.5. Likewise, the probability that the *g* allele is passed on is P(*g*) = $\frac{1}{2}$, or 0.5. Notice that the probability that either the *G* or the *g* allele is passed on is equal to the sum of the probabilities of the two events:

$$P(G \text{ or } g) = 0.5 + 0.5 = 1.0$$

Either *G* or *g* is passed on by this parent, but not both. This is an example of a mutually exclusive event. It is analogous to tossing a coin. Either heads or tails can occur, but never both from the same toss of the coin. The equation above illustrates the **sum rule** that applies to mutually exclusive events. The combined probability of two or more mutually exclusive events occurring is equal to the sum of their individual probabilities.

Now let's consider two events that are not mutually exclusive, but instead are independent of one another and happen simultaneously. This is exactly the situation when the gametes of two heterozygous parents join at fertilization, each donating just one allele for each trait to a single offspring. For example, assume a *Gg* mother and a *Gg* father have one child. What is the probability of that child having a *gg* genotype? The probability of the mother's egg carrying a *g* allele is 0.5. The probability of the father's sperm containing a *g* allele is also 0.5. The allele that the mother contributes has no effect on the one that the father contributes; thus, we are describing two independent events happening simultaneously. To calculate the probability that any two specific outcomes, say a *g* from mother and a *g* from father, will result, we must multiply the probabilities of each of the events that happen individually:

or $\quad$ P(*g* from mother and *g* from father) = P(*g* from mother) $\times$ P(*g* from father)

$$P(gg \text{ genotype in offspring}) = 0.5 \times 0.5 = 0.25$$

This is known as the **product rule**. It states that the joint probability that both of two independent events will occur is the product of the individual probabilities of each. The joint probability in this case is equal to 0.25; in other words, there is a 25 percent chance that any given offspring of this union will be homozygous for the *g* allele. This is precisely what we saw when we applied the Punnett analysis to a hybrid cross.

Probability is a more formal way of determining the likelihood that an individual conception will yield a specific genotype. Keep in mind that probability cannot determine the actual outcome of any single event; it can only give the odds that a given outcome might occur. Thus, our expectant parents can make an educated guess about the likelihood of each of their children inheriting a gene for cystic fibrosis, but they cannot know for sure until their baby is tested.

The more times an event occurs—the more times we toss a coin, or the more babies are born to our heterozygous couple—the more likely it becomes that the ratio of different outcomes matches probabilities. Toss a coin two times or four times, and you may get two heads or four tails. But toss a coin 100 times, and the number of times you get heads will approach 50, approximately the probability of 0.5.

When Mendel crossed pea plants, he counted thousands of offspring. Because there were so many, the ratios of the different phenotypes were very close to the theoretical values that his laws of segregation and independent assortment predicted. They were close, but never exactly right on.

Half Is Enough The examples we have considered thus far apply to organisms, like humans, that spend most of their life history carrying two alleles for each gene. Only the gametes carry single alleles. Some organisms, however, such as algae and mosses, spend much of their lives carrying only a single copy of each gene. How do Mendel's laws apply to these creatures?

Piecing It Together

Mendel's principles, derived from his experiments with peas, are true of all sexually reproducing organisms—those that produce gametes that join at fertilization. Here, we summarize Mendel's conclusions about the inheritance of traits, as well as some of the insights that have emerged from his work.

1. Hereditary characters are passed from parent to offspring as units or particles. We refer to the basic unit of inheritance for a given trait as a gene. The different forms of a gene are called alleles. Mendel's factors correspond to alleles.

2. Individuals carry two alleles for every gene. The two alleles for a given trait may be identical, in which case the individual is said to be homozygous for that trait. Alternatively, the two alleles may differ, in which case the individual is said to be heterozygous for that trait.

3. Prior to reproduction, pairs of alleles are separated so that specialized reproductive cells called gametes contain only one allele from each pair. At fertilization, gametes fuse, each contributing one allele for each trait to the new offspring.

4. Some genes show dominance; that is, heterozygous individuals may express only one allele, the dominant allele, while the other, the recessive allele, is masked. Thus, the phenotype—the traits that are expressed in an individual—may not always reveal the genotype, or the full complement of alleles that the individual carries. Recessive alleles remain hidden in the genes.

5. Mendel's law of segregation states that heterozygous parents are equally likely to pass either of their two alleles on to their offspring. In other words, gametes combine at fertilization without regard to which alleles they carry. Because the alleles that an individual inherits are purely a matter of chance, the rules of probability can be used to determine the likelihood that any given allele is passed on. The Punnett square is a tool that illustrates the law of segregation.

6. Mendel's law of independent assortment applies when two or more genes are considered simultaneously. It states that the alleles of one gene are passed to offspring independently of the alleles for other genes. The Punnett square illustrates this law, as well.

3.2 WHY AREN'T MEMBERS OF THE SAME SPECIES IDENTICAL?

Fast Find 3.2 Sources of Variation

Like begets like. When pea plants are crossed, the resulting offspring are never roses or geraniums; they are always peas. When dogs are bred, the result is puppies, and when race horses are bred, inevitably they bear foals. Each species is a particular combination of genetic traits that characterize that species and make it different from all the others. But to anyone who has bred race horses for competition or dogs for show, it is abundantly clear that there are differences among members of the same species. These differences among individuals in natural populations are at the very core of Darwinian evolution. Where do these differences come from? What is the source of variety among natural populations?

3.2.1 Independent Assortment Is an Important Source of Variety

Almost every organism, every phenotype, is the result of perhaps thousands or tens of thousands of genes working together. When we considered just two of those genes in our dihybrid cross, body color and wing shape in the fruit fly, and ignored all the other genes of fruit flies, we saw that a single breeding pair of flies, both heterozygous for those two genes, could produce offspring with as many as four different phenotypes. There is a mathematical rule that expresses the number of possible different phenotypes that can result from a cross between heterozygotes for any number of traits:

$$\text{number of possible phenotypes} = 2^n$$

where n = the number of traits, or genes, considered. The 2 represents the two different forms of each trait; there are two alleles possible. For a monohybrid cross there can be $2^1 = 2$ possible phenotypes (for example, round or wrinkled seed shape; ebony or pale striped body color; broad or vestigial wings). For a dihybrid cross, n is equal to 2 because we are considering two different genes simultaneously, each occurring in two forms. The number of possible phenotypes is 2^2, or 4. This is exactly the number of phenotypes we saw with regard to body color and wing shape in fruit flies.

Now consider the number of possible phenotypes that could result if we considered many traits simultaneously, say, 100 traits or 1,000 traits, each of which could occur in two different forms. It would be a daunting task to list all of the combinations, but we can easily calculate how many different ones are possible:
For 100 traits, the number of possible different phenotypes = 2^{100}

$$= 1.26765 \times 10^{30}$$
$$= 1,267,650,000,000,000,000,000,000,000,000$$

For 1,000 traits, the number of possible different phenotypes = $2^{1,000}$

$$= 11 \times 10^{300}$$

or the number 11 followed by three hundred zeros. That is more phenotypes than there are stars in the heavens or grains of sand on all the beaches of the world. The astonishingly large number of new combinations of alleles that can occur each time a breeding pair produces offspring is an important source of variety in populations—exactly the kind of variety that is required for evolution by natural selection. This is one example of **genetic recombination**, or the production of new *combinations* of genes not found in either parent. We will see other ways in which genetic recombination can be achieved in Chapter 5.

In fact, not all of the genes in living organisms have different forms, or alleles. It has been estimated that, on average, about 30 percent of human genes are actually heterozygous in a way that influences phenotypes and thus are a source of variety. If we assume that humans have about 100,000 genes (this is an estimate; the exact number is unknown), then 30 percent of 100,000 would be 30,000 heterozygous genes. To calculate the number of possible human phenotypes this could produce, you would raise 2 to the power of 30,000. Chances are you will exceed the capacity of your calculator. Even if 70 percent of the genes in humans are homozygous and thus not a source of genetic recombination, there is still sufficient variety to produce a nearly infinite number of different phenotypes—variety upon which natural selection can operate. It is no wonder that, aside from identical twins who begin life with all the same alleles, no two people present the same phenotype.

In our discussion of heredity thus far, we have considered only two different forms, or alleles, for each gene. Mendel's peas were either smooth or wrinkled; the gene for one chain of the human hemoglobin protein was described as either normal or the sickle form. But whereas any individual can carry only two alleles for a given gene, there may be many different alleles for a trait in a population. Different combinations of these various alleles may produce more than two phenotypes for the trait. For example, there are three different alleles that determine blood type in human populations, namely *A, B,* and *O.* An individual can have only two of the three possible alleles, although a population of humans will have all three alleles represented. The possibilities for variety are endless.

Now imagine a certain *combination* of traits that bestows on its owner some competitive advantage. Imagine a desert plant, say a cactus, with an allele that creates a waxy coating on its surfaces. That trait alone would prevent water loss from evaporation and help the plant to survive in the hot, arid environment of the desert. But living plants also need a mechanism for taking up carbon dioxide gas for photosynthesis. A waxy cuticle might be a barrier to carbon dioxide uptake. If that same cactus had the alleles that permitted it to take up carbon dioxide only at unwaxed entry points, the combination of a waxy cuticle and special carbon dioxide uptake sites would be a powerful advantage. That cactus would fare much better than either one with only a waxy cuticle or one with specific gas exchange sites. Natural selection operates not on single traits, but on whole organisms, who are combinations of many, many traits.

 Sex Among Microbes Sexual reproduction allows members of the same species to create new combinations of genes by genetic recombination. But not all organisms reproduce sexually. Bacteria, for example, reproduce *asexually* by a process called *binary fission*. In binary fission, each bacterium makes a complete copy of all its genes, then splits, apportioning a complete copy of the genetic blueprint to each of the two new cells. Yet bacteria, too, exhibit genetic variety. Are there mechanisms by which organisms like bacteria that reproduce asexually can share genes?

3.2.2 Mutations Are Another Source of Genetic Variety

Occasionally, when large numbers of plants or animals are grown domestically, an individual is born with an entirely new characteristic never before seen in that group or any of its ancestors. This phenomenon, which occurs in natural populations as well as domestic ones, was well known to Darwin. He referred to these rare individuals as "sports of nature"; we call them **mutants**. A long pedigree of red roses, for example, bred for many years to produce only crimson blooms, may suddenly produce a pink rose. Or within a herd of sheep, bred for many years with long legs, may appear a single, short-legged lamb. Darwin was at a loss to explain such phenomena, although he recognized the role these "sports" have in providing variety upon which natural selection can act. (Indeed, knowing nothing of genetic recombination, Darwin wondered if mutations alone could provide all of the necessary variety for natural selection.) Mendel's laws do not explain the origin of these anomalies, although new features can often be propagated by careful breeding. Once present in a breeding population, a new trait obeys all of Mendel's laws as if the allele that causes it had been present all along. What, in genetic terms, is a mutation, and how do mutants fit into our Mendelian view of inheritance?

A **mutation** is the sudden appearance of a new allele. Although mutations can occur at any time, they become *heritable* mutations when genes are copied and partitioned into gametes during sexual reproduction. When males and females make sperm and eggs (the gametes that carry one allele for every gene into the next generation), they make copies of every single allele to be passed on. This process of gene replication has been studied extensively in the past several decades and is well understood, even at the level of the molecules involved; we'll focus more on that in Chapter 6. For now, suffice it to say that the copying process is pretty good, but not perfect. Occasionally a mistake slips in, and a new allele emerges as a result. Many times, errors in copying are fatal.

Imagine yourself as a maker of fine watches. Your timepieces are assembled from carefully designed parts that fit together to work harmoniously to keep time, much as the genes of a living thing work together to build a finely tuned organism. Each component of your watches is made by a different machine. Suddenly, a machine that makes a single part malfunctions, and the new versions of that part are somehow different from what they had been. Chances are the watches made with the "mutant" part will not keep time; after all, every part was carefully designed to fit with each other part. On rare occasions, however, the mutant part may have no effect on the functioning of your watches, or they may work even better than the original. If that is the case, you would not be anxious to repair your part-making machine. Your new watches would be at least as good, perhaps even better, than your old watches.

The history of life on Earth is a history of random mutations, most lost forever due to their deleterious effects, many simply carried along from generation to generation because they have no effect on the survival or reproduction of their recipients. But occasionally a copying error has produced an organism better suited to survive the rigors of life. The lucky recipients of beneficial mutations have gone on to produce many offspring and left many copies of that new allele in the next generation. Mutations are the ultimate source of new alleles. If errors never occurred during gene replication, all of life on Earth would resemble the first living thing (whatever that was—a topic addressed in Chapter 8). Despite the mostly deleterious effects of mutations, evolution would be impossible without them.

 Mutations That Kill Genes are copied during sexual reproduction, when gametes receive one copy of each allele, but they are also copied when organisms grow, making more cells that compose the living tissue. Can mutations occur in growing, nonreproductive tissues as well as reproductive tissues? What effect would mutations in nonreproductive tissues have on evolution and natural selection? How do mutations relate to cancer or other human diseases? Are you likely to be carrying any mutations in your genes?

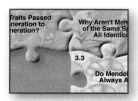

Piecing It Together

Variety is more than just the spice of life; it is the raw material upon which natural selection operates. Without it, there could be no evolution. Some of the genetic reasons for differences between individuals are summarized here.

1. The number of different phenotypes that can result from alleles coming together in new combinations as a result of sexual reproduction is essentially infinite. We can see this when we mathematically estimate the number of possible phenotypes arising from just a few traits.

2. Genetic traits can work together in combinations that enhance or hinder the survival and reproduction of their owners. In other words, having a certain trait may or may not provide an advantage, depending on the other traits that are present.

3. All genetic variety can ultimately be traced to mutation—the sudden appearance of new alleles in the genes of a population or species. Most mutations arise from errors that occur when the genetic information is copied.

4. Most mutations are deleterious or neutral. Because organisms are finely tuned combinations of traits, errors in the information encoding those traits are likely to do more harm than good. The occasional mutation that improves the functioning of an organism can be advantageous to its owner and may be passed on to future generations.

3.3 DO MENDEL'S LAWS ALWAYS APPLY?

Mendel's results were reported in 1865 as a series of lectures to the Brünn Society for the Study of Natural Science, and they were published in German in 1866 in a volume of proceedings of that society. Although the volume was widely distributed to libraries throughout Europe, few of Mendel's contemporaries recognized the significance of his work. The quantitative approach that Mendel used, counting the number of offspring showing each trait and calculating ratios, is a hallmark of modern science, but it was virtually unknown and certainly unappreciated in the latter part of the 19th century. Then in the early part of the 20th century, three botanists—Carl Correns in Germany, Hugo de Vries in the Netherlands, and Erich von Tschermak in Austria—who were also conducting experiments in plant hybridization, all rediscovered Mendel's original paper and realized its significance. Unfortunately, Mendel died in 1884, too soon to see his work assume its rightful place in the history of great discoveries.

3.3.1 Mendel's Laws Are Extended by Experimental Evidence

In 1899, Carl Correns was performing controlled crosses using a flowering plant called the four-o'clock. The blooms of the four-o'clock occur in three colors: white, red, and pink (Figure 3-13). When Correns crossed true-breeding red-flowered four o'clocks with true-breeding white-flowered plants, the hybrid offspring all had pink flowers—a shade intermediate between those of the parents. If Correns had ended his experiments there,

Figure 3-13. The flowers of the four-o'clock plant. It appears that the flower color trait of the four-o'clock exhibits blending in the F_1 generation. However, the blooms of the F_2 generation confirm that blending does not occur at the genetic level.

he, too, might have drawn the same erroneous conclusion that many of his predecessors, including Darwin, had drawn. He might have been fooled into thinking that the traits of offspring were blended averages of parental traits. But Correns went on to cross the pink-flowered plants of the F_1 with other pink F_1 plants and found that the F_2 plants exhibited all three traits: some were white, others were pink, and still others were red.

Dominance Relations

Fast Find 3.3 Incomplete Dominance

In Correns' experiment, the ratio of phenotypes in the F_2 was one white to two pink to one red—exactly the ratio of *genotypes* that results from a monohybrid cross (a cross between two individuals, both heterozygous for a single trait). Four o'clocks that were homozygous for the flower color gene were either red or white, depending on which allele was present. But when both the allele for red and the allele for white occurred in the same plant, neither the red allele nor the white allele masked the presence of the other. The alleles showed **incomplete dominance**, a condition in which all three genotypes are expressed. The phenotype of the heterozygote is intermediate between the phenotypes of the two homozygotes. The Punnett square for this cross, including both genotypes and phenotypes, illustrates this nicely (Figure 3-14). Because there is no truly dominant allele, we use the uppercase letter R for both alleles, distinguishing white and red with either a 1 or a 2, respectively.

Do these results invalidate Mendel's conclusions? At the level of the phenotype, the F_1 generation of the four o'clock appears to be showing blending, contrary to what Mendel found. But at the level of the genes, each allele remains intact as it is passed from parent to offspring. Each allele is a discrete entity, capable of showing its true colors when it is paired with the same allele in a homozygous plant. Mendel's laws are not invalidated.

What's Your Type? In the four o'clock, each allele is fully expressed in the homozygote and partially expressed in the heterozygote. But are there genes in which both alleles can be fully expressed in the same individual, neither one being dominant or even partially dominant? This occurs in the gene that codes for human blood groups. What is this relationship among alleles called? How are human blood group designations inherited?

Lethality

Mendel made much of the ratios of different phenotypes he found in the offspring of his crosses between heterozygotes. The ratio of dominant to recessive phenotypes, 3:1, gave him the idea that factors occur in pairs, that alleles show dominance, and that alleles segregate during reproduction. The 9:3:3:1 ratio of phenotypes that is character-

Male Alleles

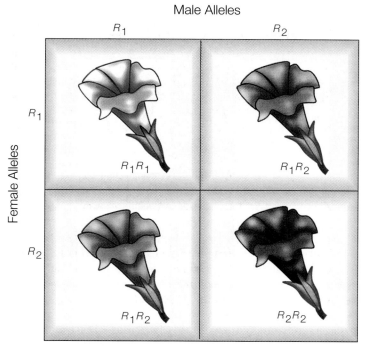

Figure 3-14. Punnett square illustrating possible offspring of a cross between a red-flowered and a white-flowered four-o'clock plant. The ratio of phenotypes is the same as the ratio of genotypes, namely 1:2:1 white/pink/red.

istic of a dihybrid cross between individuals heterozygous for two traits was the basis for his law of independent assortment. But what would happen to the ratio of phenotypes if certain combinations of alleles were deadly to the newly formed embryo? Because each unlucky embryo that receives a lethal combination of alleles dies, that phenotype would not be represented in the next generation at all. This condition, called **lethality**, was first documented soon after Mendel's groundbreaking paper was rediscovered in the early part of the 20th century. A dominant lethal allele is one that kills its recipient. A recessive lethal allele is only deadly when it is paired with another recessive lethal in a homozygote.

In 1904, the French geneticist Lucien Cuenot was performing experiments on the inheritance of coat color in mice. He found that the yellow coat color was dominant to the wild-type brownish color called agouti (Figure 3-15). When heterozygous

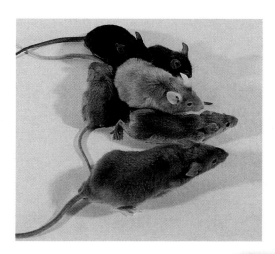

Figure 3-15. Mice with agouti (brown) and yellow fur.

Male Alleles

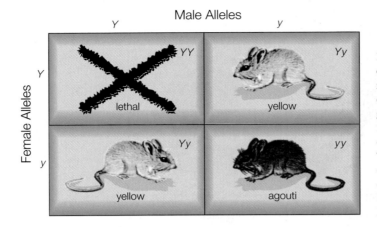

Figure 3-16. Inheritance of coat color in mice. The phenotypic ratio in the offspring of a cross between heterozygous yellow mice is closer to 2:1 than to the 3:1 ratio predicted by Mendelian genetics. The difference between the observed ratio and the predicted ratio is due to the recessive lethality of having two yellow *Y* alleles.

mice, yellow in color but each carrying an allele for the agouti color, were bred, the phenotypic ratio of offspring was 2:1 yellow/agouti, instead of the 3:1 dominant/recessive predicted by Mendel's laws. Furthermore, test crosses between yellow mice and homozygous agouti mice proved that all yellow mice were heterozygous. (How can a test cross prove that individuals showing the dominant trait are heterozygous and not homozygous? Check the CD to remind yourself of how a test cross works.) Cuenot could not find a single mouse that was homozygous for the dominant allele.

Where were all the homozygous yellow mice? Later experiments showed that a double dose of the yellow allele is a deadly combination. All such embryos die early in development, skewing the ratio of offspring away from 3:1 dominant/recessive and toward the observed 2:1 outcome (Figure 3-16).

The yellow allele in mice, although dominant in its effect on coat color, is an example of an allele that is recessive in its lethality. It may seem that recessive lethals would be the victims of natural selection, quickly eliminated from populations by their deleterious effects. But because the carriers of many recessive lethals can survive and reproduce normally, such alleles can remain hidden within the genes of a population for many generations, only killing their carriers when, by random chance, they are paired with similar alleles in deadly combination.

But what of dominant lethals, or alleles that kill when present in a single copy? Common sense might lead us to conclude that a dominant lethal allele could not possibly survive in a population. After all, any individual unfortunate enough to acquire even a single lethal dominant would surely die, eliminating the deadly allele from the population. That is certainly the case with many lethal dominants, but not all. Huntington's disease is an example of a deadly human disease caused by a dominant lethal allele. The disease is characterized by uncontrolled movements and mental deterioration, followed by death. Individuals need inherit only one copy of the Huntington's gene to exhibit the lethal phenotype. Because the onset of the disease occurs late in life, usually between the ages of 30 and 60, an afflicted person may already have passed on the allele to children before he or she is even aware of its presence. On average, about half of the offspring of an afflicted person inherit the disease. In this manner, a lethal allele can have a long history within a population.

One Gene Can Influence Two or More Traits

Just for a moment, let's go back to the Moravian garden and take a closer look at the wrinkled and smooth peas that inspired Mendel to derive his laws of inheritance. While he was focusing on the texture of the seed coat, he most likely was not aware that another trait of pea seeds was also affected by the coat texture gene. If he had

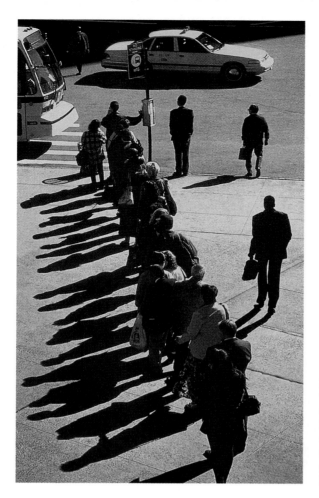

Figure 3-17. Human height is a trait that does not fall into discrete categories, but more closely resembles a continuum. Height is a result of several genes working together, as well as environmental factors such as nutritional level and childhood diseases.

examined smooth and wrinkled peas in a microscope, he would have seen that the tiny starch grains that provide food for the developing pea plant were different in these two varieties. The same gene that influences seed texture also affects starch grain shape and size. This single gene has multiple effects. The phenomenon whereby a single gene affects two or more traits is called **pleiotropy**.

Sometimes pleiotropy is explained by examining a gene's effect at an early stage of life. For example, cats that are either homozygous or heterozygous for a certain allele called *W* have pure white fur. These same cats are also deaf in one or both ears. What is the connection between the *W* allele, white fur, and deafness? The answer has to do with a particular kind of cell called the melanocyte. Melanocytes produce the pigmentation of the fur, and they also play a role in the inner ear of the animal where they contribute to the hair cells that sense sounds. When a cat inherits a *W* allele, its melanocytes fail to develop properly. Although deafness and white coat color appear to be unrelated phenomena, both phenotypic manifestations are attributed to this same failure. These are pleiotropic manifestations of a single allele.

Two or More Genes Can Influence a Single Trait

If all the students in your class formed a line with the shortest person at one end and the tallest person at the other, we would see a continuum of heights from short to tall (Figure 3-17). Even in this small subset of the population, human heights do not fall into discrete categories. Part of the variation can be attributed to factors such as nutritional status and childhood diseases, among others. But much of what determines an individual's height has an underlying genetic cause. Yet even with the powerful tools of modern molecular biology, no single gene for human height has been identified.[2] Human height is an example of a trait that is affected by many genes—it is a **polygenic** trait. Traits that are determined by a single gene with two alleles, or **monogenic** traits such as seed texture in peas or flower color in the four-o'clock plant, will occur as two or three distinct phenotypes. If as few as three different genes work together to determine a single trait, each represented by only two alleles, the distribution of phenotypes within a population becomes continuous. Other examples include weight, human skin color, coat color in many mammals, and even some traits that *do* form discrete phenotypes, such as the number of whiskers on the face of a mouse.

3.3.2 Mendel's Work Generated New Fields of Inquiry

At the turn of the 20th century, when Mendel's laws were gaining new adherents in the scientific community, it was not entirely clear how Darwin's theory and Mendel's laws complemented one another. Many who read Mendel's work believed that genetic traits must occur as discrete, all-or-none phenotypes, such as the purple versus white flower color in peas, or yellow versus agouti coat color in mice. The way in which polygenic traits can create a continuum of phenotypes was unknown. Darwin's champions pointed to the many traits that formed a continuum of phenotypes and insisted that such gradual differences between individuals could not be explained by Mendelian genetics, yet were necessary for natural selection. Many believed that the two theories were incompatible; contentious, even vitriolic disagreements developed. In the 1930s, the controversy was finally put to rest. A new field of biology emerged, called **population genetics**, which employed mathematics and statistics to prove that the variety required for natural selection could arise from Mendelian genetics. Instead of focusing on just one or a few genes occurring in individuals, the population geneticists used mathematical models to study the movements of many genes through entire populations over time. Population genetics is an exciting field that has continued to grow and provide answers to evolutionary questions ever since. We will take up the topic of population genetics again in Chapter 7.

Recall that, to Mendel, the hereditary "factors" were nothing more than theoretical entities. His theories did not require the ability to visualize genes moving from parent to gamete then to offspring. But the question remained, what exactly are these mysterious hereditary factors? And how are they copied, passed on, and expressed in their carriers? The answers to these questions have a fascinating history, leading up to the development of modern molecular biology. While Mendel was cultivating peas, others in Europe and elsewhere were learning that all living things are composed of fundamental units called cells. The hereditary factors, we will learn in Chapter 4, were found in cells.

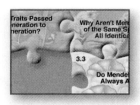

Piecing It Together

In the century that has elapsed since Mendel's work was rediscovered by Correns, de Vries, and Tschermak, new experimental evidence has provided the exceptions that prove the rules. Each new discovery can be explained in the context of the laws of inheritance as set forth so many years ago by this humble priest.

---

[2]The exception to this rule is achondroplasia, or dwarfism. In this genetic condition, a single dominant allele prevents its carrier from growing to normal adult heights.

1. Alleles occurring together in heterozygotes do not always exhibit true dominance, but instead may be incompletely dominant.

2. Some alleles are lethal. Most lethal alleles that survive in populations are recessive. Other lethals are dominant, killing their carriers with a single copy, such as the one that causes Huntington's disease.

3. Pleiotropic genes exert two or more effects on the phenotype.

4. Polygenic traits are influenced by two or more genes. These traits often do not occur as discrete phenotypes, but rather form continuous distributions of phenotypes in populations.

5. Mendelian genetics, originally thought to be incompatible with Darwinian evolution, was reconciled with natural selection in the 1930s. This synthesis of Darwinian evolution and Mendelian genetics gave rise to the modern science of population genetics.

Where Are We Now?

Some of the most exciting discoveries of recent years have come from the field of genetics. In the past few decades, we have identified the genes that cause many human genetic disorders, developed tests for detecting these disorders both prenatally and in prospective parents, and engineered therapies to minimize or eliminate many genetic diseases.

Among the most surprising recent discoveries in genetics is a class of genes that "remember" their parental origins—imprinted genes. Usually, once fertilization has taken place, the maternal and paternal alleles for most genes cannot be distinguished. Mendel's peas were just as round when the round allele came from the egg as when it came from the sperm. But for imprinted genes, alleles of maternal origin act differently from those of paternal origin.

Two Cambridge University researchers, Eric Keverne and Azim Surani, have discovered that, in mice, certain alleles from the mother contribute more to the development of the brain's reasoning portion, the cortex, and the same alleles from the father have a greater impact on the development of the brain region responsible for more primitive functions such as feeding, fighting, and reproduction. Paternal alleles also do the work of building the placenta and making the hormones responsible for the growth of the fetus. The alleles of certain genes that contribute to cortex development are "silenced" if they come from the sperm—they are chemically modified so that they are unable to function. Likewise, maternal alleles that contribute to growth hormones and placenta development are similarly silenced.

While neither researcher is willing to speculate on how human intelligence and behavior might be affected by imprinted genes, some human genetic disorders hint that we are not entirely dissimilar to the mouse. For example, children suffering from Prader-Willi syndrome have brain disorders that cause overeating, obesity, a placid personality, mild retardation, and a reduced sexual drive. These behaviors are under the control of the brain's primitive regions, the areas whose development can be traced to paternal alleles. Maternal alleles for these brain areas are normally silenced. Prader-Willi syndrome sufferers lack the paternal allele due to a deleterious mutation. Their abnormal phenotype implies that the maternal alleles for these brain areas cannot make up for the absence of the alleles from the father.

In addition, another human genetic disorder called Angelman syndrome occurs when a baby is born lacking a certain maternal allele that resides on chromosome number 15. Children with Angelman syndrome have defects in activities controlled by the higher brain centers. These defects include mental retardation, speech deficiencies, and jerky movements. The symptoms occur even though the chromosome 15 of paternal origin is present and intact. The abnormal phenotype of children with Angelman syndrome implies that the paternal alleles on chromosome 15 cannot substitute for the absence of those of maternal origin. While it is too soon to say whether human intelligence is maternal in origin, these exciting discoveries may give new insights into why we are the way we are.

REVIEW QUESTIONS

1. How did Mendel's approach to answering scientific questions differ from that of his contemporaries? How did his novel approach contribute to his success in describing how traits are inherited? What advantages did he gain by choosing to study the garden pea?

2. When Mendel crossed plants that bred true for round pea seeds with those that bred true for wrinkled seeds, what was the phenotype of the F_1 generation? What was the genotype of the F_1 generation?

3. What is a gamete? In terms of the number of alleles for each gene, what general statement can you make about how gametes differ from other cells in a mature pea plant?

4. In a cross between a pea plant that breeds true for purple flowers and one that breeds true for white flowers, the F_1 generation has all purple flowers. Draw a Punnett square showing how the alleles for flower color are combined in the F_1. Then draw a Punnett square showing the genotypes that can result in the F_2 generation.

5. The presence of freckles is a dominant trait in humans. What can you say about the genotype of a person with freckles? How could you find out for sure what that person's genotype is?

6. You are a genetic counselor, and a married couple comes to you for advice. The man's father suffers from sickle cell anemia. No one in the family history of the woman has ever had the disease. What would you tell this couple about the probability that their children will have the disease? What would you tell them about the probability that their children will be carriers of the disease?

7. What is an allele? How is an allele different from a gene? Give an example of a gene. Now give an example of two alleles for that gene.

8. In a dihybrid cross between individuals heterozygous for both traits, how many different genotypes are possible in the F_1 generation? Assuming that both traits show dominance, how many different phenotypes are possible?

9. How many different types of gametes can be formed by individuals with a genotype of *AABB*? With a genotype of *AaBb*?

10. Which of Mendel's laws is illustrated by a dihybrid cross? What does the law state?

11. How many different phenotypes can result from a cross between individuals heterozygous for five different traits? How are these different phenotypes important to the process of evolution by natural selection?

12. Assume a new allele, called *q*, arises by a mutation in the *Q* gene. A gamete that acquires the mutant *q* allele dies. Will this allele show up in the next generation? Illustrate your answer with a Punnett square.

13. In Mendel's crosses, one of the two forms of the traits he studied was dominant to the other. Are all traits inherited in this way? Support your answer with an example.

14. In genetic terms, what does it mean to be "true breeding"? Using the information in Section 3.3.1, could mice with yellow fur be true breeding? Why or why not?

15. Suppose Mendel had decided to study inheritance using the trait of human height. How would his methods have been different? Do you think he could have derived his principles of inheritance using this trait? Why or why not?

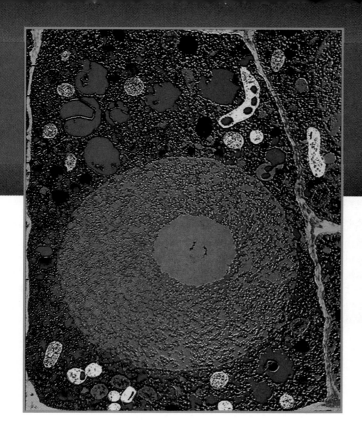

Chapter 4

Cells: What Are the Building Blocks of Life?

*Organisms are made up of cells much as societies
are made up of individual beings, and for much the same reasons.*
—Isaac Asimov, 1988

—Overview—

You might imagine that the torch ignited by Mendel would pass quickly. Surely his contemporaries in biology would see the central importance of the rules of inheritance, confirm their validity in other organisms, and carry the search forward. Such did not occur. Not until after the turn of the century—long after his death—was Mendel's work appreciated. There were several reasons. The monastery at Brunn was isolated from the mainstream of nineteenth-century science. Communications among scientists in the mid-1800s lacked the lightning speed of today. Those who did read Mendel's work may not have understood it. His use of mathematics in analyzing the results of his experiments was novel. His contemporaries may have gotten lost in his numbers.

Perhaps the most important reason Mendel's work was ignored was that biologists were already distracted by other new concepts that rocked the foundations of the field. In 1859, seven years before Mendel published his monograph, Darwin published *The Origin of Species*. If that were not controversy enough, another theory thrust itself into the limelight. This new theory, that all organisms are composed of tiny structures called cells, was as novel to nineteenth-century biology as was evolution.

New ideas in science are seldom readily accepted. They are often proposed before the evidence to support them is fully convincing. Scientists are skeptical and often slow to accept a new idea, especially without powerful supporting evidence. Skepticism is one of science's greatest strengths. It keeps science out of blind alleys and false path-

Chapter opening photo—Root tip cell from onion. This color-enhanced photograph was taken using a transmission electron microscope (magnified 800 times).

ways. But skepticism can also be one of science's greatest frustrations. Acceptance is often torturously slow and tedious. Such is the story of the **cell theory,** another of the foundation stones on which the house of biology stands.

4.1 WHAT IS THE CELLULAR NATURE OF LIFE?

When you look at an organism such as a tree or a human, what you see is a single being. What you cannot see is the vast community of cells that constitute that being. The **cell** is the fundamental unit of life: the smallest entity capable of exhibiting the characteristics of life. Cells acquire and use energy; they acquire and organize materials; they grow; they reproduce. Trees and humans are composed of trillions of cells working together, but this fact is not immediately obvious. To scientists of the nineteenth century, the discovery of the cellular basis of life was a surprise.

4.1.1 Microscopes Made Possible the Discovery of Cells

Although ancient thinkers had wondered about the composition of life, the existence of cells could not have been imagined until the development of the proper tool—the **microscope** (Figure 4-1). Before the invention of the microscope in the seventeenth century, anatomists had contented themselves with more and more detailed descriptions of what could be directly observed, and organs and tissues had been described in detail. But inventors in fields outside of biology were creating tools that would extend our vision and ultimately reveal cells.

In 1662, the Englishman Robert Hooke was appointed Curator of Experiments in the newly formed Royal Society of London. His duties included preparing demonstrations or experiments for each weekly meeting of the prestigious scientific society. Hooke was exceptionally clever and inventive. Among other devices, he built a primitive microscope. In Hooke's microscope, candlelight was reflected from a mirror onto an object of interest; then the image traveled through a series of lenses. Viewers peered into a tubelike instrument and saw a magnified image of the specimen. Hooke examined many things through his microscope, including thin pieces of cork (Figure 4-2). To his amazement, the cork appeared to be made of "a great many lit-

(a)

(b)

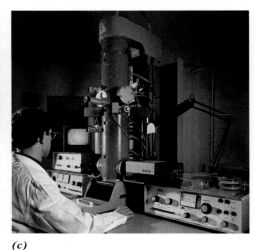
(c)

Figure 4-1. Microscopes have changed considerably in the past four hundred years. (*a*) Some antique microscopes. (*b*) A modern light microscope with attached video camera. (*c*) An electron microscope.

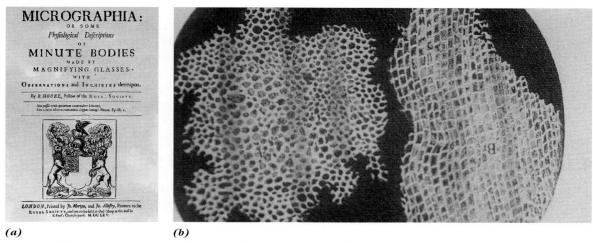

(a)   (b)

Figure 4-2. (*a*) Robert Hooke published drawings of his microscopic observations in a book called *Micrographia.* (*b*) Among the many things he observed were thin slices of cork. He named the tiny compartments "cells." This cork has been treated with a red stain; special stains and dyes create contrast (magnified 80 times).

Fast Find 4.1a Microscopes

tle Boxes." They reminded him of the small, plain rooms that monks occupied, called "cells." Hooke believed that this was an important observation, not because it explained the basic structure of life, but because it explained the properties of cork—why it floated, why it didn't soak up liquids, why it was easily crushed. Hooke published his observations in 1665 in a widely read book called *Micrographia.* The term "cell" persisted.

Nine years after Hooke published *Micrographia,* the Royal Society learned of the exceptional work of a Dutchman named Anton van Leeuwenhoek. Leeuwenhoek was a gifted craftsman with an inquiring mind. By the time the Royal Society contacted him, he had built dozens of microscopes of excellent quality, grinding his lenses to near perfection. Over the next 53 years, Leeuwenhoek maintained a regular correspondence with the Royal Society. He sent more than 200 letters, many with drawings. What he described delighted and intrigued not only the Royal Society, but all of Europe's elite community of scientists.

In 1675, Leeuwenhoek chanced to examine a drop of cloudy green water taken from a lake near his home. To his amazement, the water teemed with tiny creatures (Figure 4-3). He named them "animalcules" and described the tiny creatures in great detail in his correspondence with the Royal Society. Over the next few months animalcules showed up nearly everywhere Leeuwenhoek searched—in river water, seawater, rainwater flowing off his roof, even drinking water freshly drawn from his well.

Fast Find 4.1b Leeuwenhoek

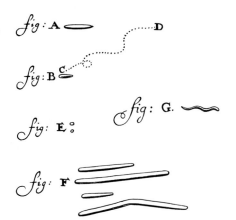

Figure 4-3. Leeuwenhoek attracted the attention of Europe's scientific elite with his drawings of "animalcules."

He found them in mud and in scrapings taken from between his teeth. Hooke, then secretary of the Royal Society, was asked to verify Leeuwenhoek's incredible reports. "Wonderful little animals" appeared nearly everywhere. Leeuwenhoek was the first to observe living, single-celled microorganisms—bacteria, algae, protists—all invisible to the unaided eye.

Leeuwenhoek took his secrets of fine lens making to his grave, and for many years after his death no one continued his work. Little of note was reported about cells for nearly 100 years.

4.1.2 The Cell Theory Was Proposed in the Early 1800s

Leeuwenhoek's discovery of creatures "a thousand times smaller than the eye of a big Louse" generated a great deal of excitement among scientists and laypeople alike. But these exotic single-celled creatures were only part of the story. Nearly a hundred years later, microscopes would yield even more remarkable discoveries about the cellular nature of life. By the early 1800s, careful observations would show that cells are characteristic of all living matter, and several scientists had made bold statements to that effect. For example, the French scientist R. J. H. Dutrochet wrote in 1824 that "All organic tissues . . . are actually only a cellular tissue variously modified." But the credit for formulating the cell theory, stated at the end of this section, is generally given to two German scientists, Matthias Schleiden and Theodor Schwann.

In 1837, Schleiden and Schwann met at a dinner party. Both had been studying cells. Schleiden, a lawyer turned botanist, was interested in the origins of plant cells. Where did they come from? He proposed and tested three hypotheses: (1) cells appear at the surfaces of previously formed tissues; (2) cells arise from preexisting cells; (3) cells appear spontaneously within the spaces between other cells. Because plant cells arose in no particular pattern throughout a tissue's mass, he rightly concluded that the second hypothesis was correct. All plant tissues were composed of cells, and the growth of tissues was simply the accumulation of new cells originating from other cells. He tried to describe how this might be possible. Unfortunately, he devised an erroneous process. He imagined that the nucleus of a new cell forms as matter spontaneously organizes within an existing cell. He envisioned that the new cell arises as a bubble on the surface of the nucleus, the large structure inside the cell. Other observers quickly found Schleiden's error, but not until he had published his observations in 1838.

Theodor Schwann was interested in animal tissues and had previously observed that the spinal cords of frog larvae had a cellular structure. This was an inspired observation because previously cells had been recognized on the basis of their square or ovoid shape. Nerve cells, however, are often long and thin with obscure membranes. After his discussions with Schleiden, Schwann too wondered if the growth of animal tissues arose from the accumulation of more and more cells from preexisting cells. He was able to show that a developing embryo starts out as a single cell which divides, resulting in few globular, generalized cells that grow into a much more complicated organism composed of many cellular tissues. Schleiden's hypothesis appeared to apply to animals as well as plants.

But do all animal tissues grow this way? Living tissues were clearly cellular, but what about teeth, hair, and feathers? These were not. Schwann looked at the tissues from whence these structures grew—the bed of a fingernail, the base of a growing tooth, the skin of birds from which feathers grew. In all cases, he found that the "primordial substance" was always a few generalized cells from which other cells grew and from which noncellular structures such as bone, teeth, hair, and feathers originated.

In 1839, Schwann published a book in which he reprinted Schleiden's paper and concluded with a chapter on the cell theory. "All organisms," he wrote, "and all their separate organs are composed of innumerable small particles of definite form. . . . There is one universal principle of development . . . and this principle is the development of cells."

Schwann and Schleiden brought 200 years of scattered observations together into a concise, testable, widely distributed statement, called the **cell theory.** The cell theory states that all organisms are composed of one or more cells. Therefore, cells are the fundamental units of life—the smallest units that show all the characteristics that define life. Later, as we shall see, the theory was modified to include the idea that cells do not arise spontaneously. All cells come from preexisting cells.

Schwann's many insights into the cellular nature of life earned him a place as one of biology's most important founders. In 1845, the Royal Society awarded him its highest honor, the Copley Medal.

Piecing It Together

After 200 years of fits and starts, a new biological theory had been stated. Schwann's book was widely read, easily understood, and its main assertions tested. Schleiden's errors were quickly exposed and discarded. Other tenets were verified and accepted. The real importance of Schwann's and Schleiden's work came in its precise statement of the cell theory. According to the cell theory:

1. All organisms are composed of cells.

2. The cell is the fundamental unit of life.

3. Cells arise from preexisting cells.

4.2   WHAT IS THE CHEMICAL NATURE OF CELLS?

In the 160 years since Schleiden and Schwann there has been a veritable explosion of knowledge about cell structure and function. Much of this was made possible by two important developments outside the field of biology.

First, microscopes were greatly improved. Light microscopes quickly reached the limit of their capabilities. Even with perfect glass lenses, detail from images seen with the light microscope is not improved at magnifications above about 1000 times their actual size. The invention in 1931 of the electron microscope overcame this limit. Soon, magnifications in excess of several hundreds of thousands were possible. By the 1980s, parts of cells approaching the size of individual molecules could be seen.

Second, chemists came to appreciate the dynamic complexity of cellular chemistry and developed creative ways of studying it. The molecules of living matter are among the most complex molecules known. Chemists and biochemists have come a long way in turning complexity into comprehension. Thousands of processes occur simultaneously and continuously. Chemical reactions control, transfer, store, and use energy. Raw materials continually pass into cells where they are stored, consumed, or used to make new products that are, in turn, stored, exported, traded, or ultimately taken apart. As a result the cell grows and divides. All of this highly regulated activity happens on a scale too small to be seen directly. Understanding how and what cells do to master all this activity has captivated biologists for nearly two centuries. It continues to do so.

Using Electrons to See Cells How do regular light microscopes make objects appear to be bigger? Why are electron microscopes better than light microscopes in magnifying objects?

4.2.1 The Building Blocks of Cells Are Complex Molecules

Fast Find Chemistry Review

Living matter is made of cells, but what are cells made of? It helps to know a bit of basic chemistry to answer this question. Before you read further, you may want to use your CD to review the **structure of atoms,** the **elements** that are important to life, and the differences between **atoms, molecules,** and **compounds.** It will also help to review **chemical bonds** and **functional groups**—topics that will help you to understand what comes next.

The chemistry of life is based on the chemistry of carbon. Carbon, and the many compounds that can be constructed with it, is so central to life that the term "organic" (which technically means "having to do with carbon") has come to mean "having to do with life." Carbon has the ability to form four **covalent bonds** with other atoms (Figure 4-4). A covalent bond is a strong attraction between two atoms that share electrons. An atom of carbon readily shares electrons with four other atoms, some or all of which can be other carbons. Thus complex carbon-containing compounds, called organic compounds, can make up a virtually limitless variety of linear or ringed chains of carbon backbones. Other atoms, groups of atoms, or even more carbons can branch from these carbon backbones. But despite the infinite variety of possible organic compounds, carbon tends to associate with itself and a few other elements in a finite number of ways throughout the living world. This is apparent when we examine the large carbon-containing compounds in cells.

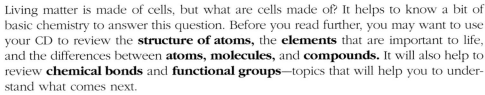

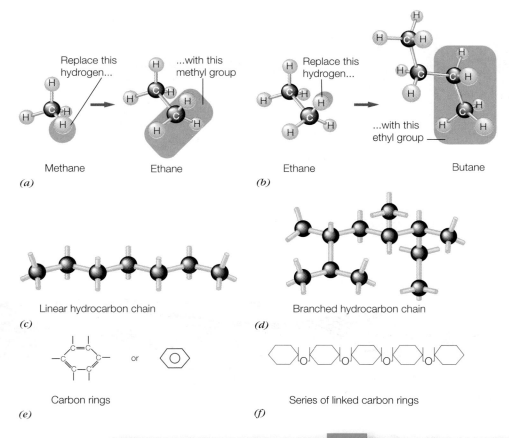

Figure 4-4. Carbon atoms form four covalent bonds. This feature is responsible for the many different organic molecules in cells.

(*a*) The simplest carbon-containing molecule is methane. Methyl groups occur on many organic compounds. Removing a hydrogen from methane creates a methyl group. Replacing a hydrogen in methane with a methyl group creates ethane.

(*b*) Similarly, removing a hydrogen from ethane creates an ethyl group. Replacing a hydrogen in ethane with an ethyl group creates butane.

(*c*) Hydrocarbon chains can be linear, or (*d*) branched.

(*e*) Carbon rings are also abundant in organic molecules.

(*f*) Chains of carbon-containing rings form the basic structure of starches, glycogen, and structural components of some cells.

**Fast Find
4.2b
Condensation
& Hydrolysis**

The large organic compounds found in cells fall into four broad classes, collectively called the **biological macromolecules.** They are **carbohydrates, lipids, proteins,** and **nucleic acids.** All four classes are built from smaller organic "building block" molecules. In the case of carbohydrates, the building blocks are **simple sugars.** Most lipids are built from **fatty acids** and other simple molecules. Proteins are linear arrangements of **amino acids.** Nucleic acids are linear arrays of **nucleotides.**

Carbohydrates

**Fast Find
4.2c
Carbohydrates**

Carbohydrates include simple sugars and all of the compounds made of sugar building blocks. A simple sugar, or **monosaccharide,** may have as few as three or as many as seven carbons bound in a linear array to form the backbone of the molecule. The most abundant monosaccharides have five or six carbons and are called **pentoses** and **hexoses** (Figure 4-5), respectively. Monosaccharides are quite soluble in water. When dissolved, pentoses and hexoses curl around and their ends combine to form a ring. Examples of monosaccharides found in cells include the pentoses **ribose** and **deoxyribose** and the hexoses **glucose, fructose,** and **galactose.**

Sucrose, or table sugar, is a **disaccharide** composed of two different monosaccharides, glucose and fructose, connected by a covalent bond. Maltose, or malt sugar, is a disaccharide made from two glucose units. Lactose, or milk sugar, is a disaccharide made up of one galactose unit and one glucose unit. The process need not stop at two, three, or even four. Hundreds, even thousands, of monosaccharides can be linked together when they are in ring form to make long **polysaccharides** (Figure 4-6). **Starch** is a polysaccharide made from thousands of repeating glucose units. Plants synthesize and store starch as a ready form of energy. It can be easily **hydrolyzed,** or broken down, into its glucose building blocks, and the glucose used for energy as the plant needs it. Animals make **glycogen,** another glucose-based polysaccharide, which can be stored in liver and muscles as a quick and ready source of chemical energy.

Cellulose is another polysaccharide with thousands of repeating glucose units. Cellulose is found in plants and is probably most familiar to you as cotton and paper, but it is an important structural component of many types of woody plants as well. It is so widespread in plants that it is the most abundant organic material on earth. The glucose units of cellulose are linked in such a way that they are unavailable to us as a source of energy. We lack the digestive machinery to break the bonds between the monosaccharides of cellulose and release the energy-rich glucose. Consequently, cellulose that we ingest passes through our system essentially unchanged, a property that makes it an excellent source of dietary fiber. Microorganisms that live in the digestive tracts of animals such as termites and ruminants are able to digest cellulose. Ruminants, such as cows, sheep, and deer, harbor an army of cellulose-digesting bacteria and protozoa in their stomachs, and they survive on the glucose that is made available to them by their friendly tenants.

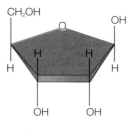

(a) Ribose (pentose)

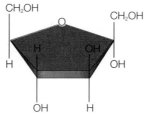

(b) Fructose (hexose)

Figure 4-5. Monosaccharides are simple sugars that typically have from three to six carbon atoms. The most abundant monosaccharides in cells are pentoses and hexoses, having either five or six carbons. These monosaccharides are illustrated in the ring form, a configuration that occurs frequently in sugars found in cells. (*a*) Pentose sugars, like this ribose, form part of the structure of nucleic acids. (*b*) Fructose, also called fruit sugar, is a six-carbon monosaccharide.

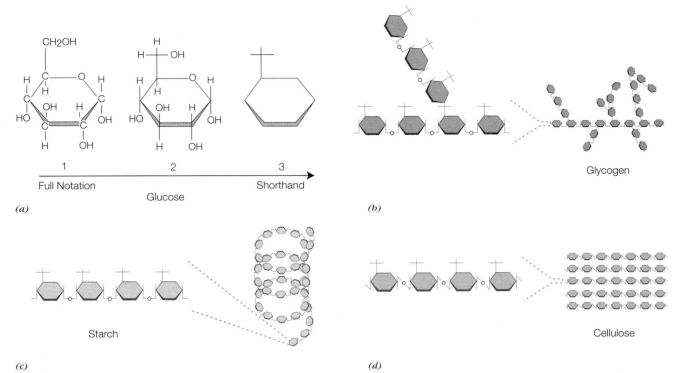

(a)

(b)

(c)

(d)

Figure 4-6. Carbohydrates have various roles in cells. (*a*) Glucose is a common hexose, shown here with all of its atoms labeled and in two shorthand drawings. (In 2, the carbon atom symbols are omitted; in 3, only the carbon atom locations are shown and the other atom symbols are omitted.) (*b*) Glycogen is a linear series of glucose units with branching strands; animals use glycogen to store chemical energy. (*c*) Starch is a slightly dif- ferent linear series of glucose that plants use to store chemical energy. (*d*) Cellulose is another linear series of glucose in which the covalent bonds between individual glucose units cannot be broken by the digestive machinery of most animals. Cellulose is a structural polysaccharide in plants, and also the main component in cotton and paper.

Lipids

**Fast Find
4.2e
Lipids**

The lipids, or fats, are a heterogeneous class of biological macromolecules that share a single property: They do not readily dissolve in water. Molecules or parts of molecules that are not water soluble are said to be **hydrophobic** ("water fear- ing"). Some lipids serve as energy reserves in the cell; others play important struc- tural roles in cell membranes; still others are hormones and other types of cellular messengers. Unlike the other biological macromolecules, lipids are not strictly strands of repeating building block subunits. The lipids can be subdivided into three major categories: the simple lipids that include **fatty acids**, the complex lipids that have fatty acids as part of a larger structure, and the steroids. First, we take a look at fatty acids.

Fatty acids are a group of hydrophobic molecules characterized by a long chain of carbon atoms with a single nongreasy, acidic molecular group, called a **carboxyl group** (-COOH), at one end (Figure 4-7*a*). There are about 100 naturally occurring fatty acids. They differ from one another in the number of carbons in the chain, and in the presence or absence of two bonds instead of a single bond between two or more carbons in the chain. Carbon atoms that are connected by means of a double bond do not bind to as many hydrogen atoms as those connected by means of a sin- gle covalent bond. **Saturated** fatty acids, abundant in tropical oils and many animal fats, have no double bonds connecting carbons in the chain; their carbons are fully saturated with hydrogen atoms. **Unsaturated** fatty acids, more common in plants, have varying numbers of carbon-to-carbon double bonds; their carbons are not carry-

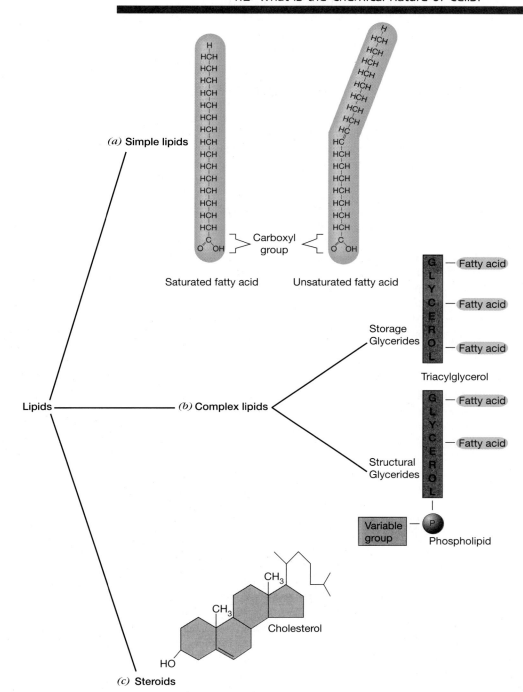

Figure 4-7. Different types of cellular lipids.

(*a*) Fatty acids are long hydrocarbon chains with a single carboxyl group at one end. Saturated fatty acids have no carbon-to-carbon double bonds. Unsaturated fatty acids may have one or several carbon-to-carbon double bonds.

(*b*) Complex lipids, including triacylglycerols and phospholipids, have fatty acids as part of their structure. Triacylglycerols are rich in chemical energy and are used by oilseed plants and animals to store energy. Phospholipids have both hydrophilic and hydrophobic ends. They form the basic structure of cell membranes.

(*c*) Steroids, such as this cholesterol molecule, are ringed hydrocarbons.

ing the full complement of hydrogen atoms. Monounsaturates, common in some vegetable oils, such as olive oil, have only one double bond. Polyunsaturates, abundant in other vegetable oils and fish oils, have two or more carbon-to-carbon double bonds in the chain.

Much has been made of saturated and unsaturated fats in our diets. The Surgeon General of the United States advises that we restrict our intake of saturated fats. Dangerous levels of serum **cholesterol** (Figure 4-7*c*) in some people have been correlated to a high intake of saturated fats. Saturated fats have high melting points relative to mono- and polyunsaturates. In other words, they remain solid at temperatures at which unsaturated fats are liquids. Picture the difference between shortening and vegetable oil at room temperature. Both are complex lipids containing fatty acids, but the lipids in shortening are rich in saturated fatty acids and those in vegetable oil have mostly mono- and polyunsaturated fatty acids.

One group of polyunsaturated fatty acids, called the omega-3 fatty acids, appear to be especially beneficial. A diet rich in these fatty acids appears to lower the incidence of a wide variety of ailments, including heart disease, hypertension, and inflammation. The omega-3 fatty acids are found in fish oils, among other things.

Fatty acids, though abundant in living cells, rarely occur on their own. They are usually covalently bound to a **glycerol** molecule, to form complex lipids called **glycerides** (Figure 4-7*b*). There are two major types of glycerides that differ in their structure and in the role they play in living cells. Storage glycerides have three fatty acids, one attached to each of the carbons of the glycerol molecule. They are very **hydrophobic**—utterly insoluble in water. The oilseed plants, such as soy and sunflower, and most animals (including humans) store chemical energy in this form.

Structural glycerides, called **phospholipids,** are also formed from a glycerol molecule bound to fatty acids. A different type of molecule branches from the third carbon of the glycerol backbone. This is always a nongreasy group—a **hydrophilic** (or "water loving") head group—and includes a phosphorus atom and some additional combination of water-soluble atoms. The presence of both hydrophobic and hydrophilic parts on the same molecule gives phospholipids unique chemical properties when they occur in watery environments such as cells.

**Fast Find
4.2f
Membranes**

Because hydrophobic parts of a molecule will always seek an environment without water, fatty acids tend to clump together, much like the oily droplets that float on the surface of soup. But the hydrophilic parts of a phospholipid seek water. Thus the most stable form that phospholipids will take in water is that of a bilayer sheet, shown in Figure 4-8, where fatty acid tails intermingle with each other, forming a greasy hydrophobic core, and hydrophilic, phosphate-containing head groups interact with water on either side. This stable bilayer is the basis for the structure of membranes in living cells. While cellular membranes contain proteins and carbohydrates in addition to phospholipids, basic membrane architecture is related to the dual nature of phospholipids. Membrane structure and function are discussed in more detail in Section 4.2.2, but take a look at Section 4.2 of your CD to get a "feel" for the three-dimensional structure of a cell membrane.

A third category of important lipids are those which have no fatty acids as part of their structure. These include, among others, **steroids** (Figure 4-7*c*). Cholesterol is perhaps the most notorious steroid, a well-known contributor to **atherosclerosis,** thickening of the arteries. But cholesterol is not always bad. First, along with the phospholipids, cholesterol is an important component of membranes in animal cells, where it helps maintain just the right consistency—pliability and resilience—for optimal membrane performance. In addition, the important steroid hormones, including the sex hormones testosterone, estrogen, and progesterone, are all made from cholesterol. These steroids share many structural features. Cholesterol gets its bad name from cases where it is overly abundant in our bloodstream.

Proteins

Of all the organic macromolecules, proteins are by far the most complex and varied, and the most ubiquitous in living cells. Proteins have multiple roles in cells. Some function as organic catalysts, called **enzymes,** helping to speed the rates of specific chemical reactions; others are responsible for cell movement or the movement of substances within cells; still others are structural materials forming parts of the scaffolding inside of cells, called the **cytoskeleton,** or contributing to cartilage, fingernails, hair, and other noncellular materials; and a vast array of different proteins have important roles as cellular regulators and messengers.

Scientists have known about proteins for centuries. In 1868, T. H. Huxley, Darwin's eloquent champion, wrote: "all forms of protoplasm [living matter] contain the four ele-

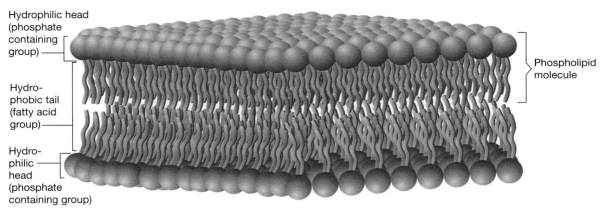

Hydrophilic head (phosphate containing group)

Hydro-phobic tail (fatty acid group)

Hydro-philic head (phosphate containing group)

Phospholipid molecule

Figure 4-8. Phospholipid bilayers are responsible for the basic architecture of cellular membranes. The hydrophilic head groups, shown here as tan balls, readily interact with the aqueous medium both inside and outside of cells. The hydrophobic fatty acid tails intermingle with each other, forming an oily core.

ments, carbon, hydrogen, oxygen, and nitrogen, in a very complex union . . . the nature of which has never been determined with exactness, [to which] the name of Protein has been applied." Thus the elements that constitute proteins were known, but how those elements combine to form the building blocks of proteins, called the **amino acids,** and how amino acids join together to form proteins were not fully understood until well into the twentieth century. The complexities of protein structure and their many different properties are still intensely studied today.

The amino acids that make up proteins are small, water-soluble compounds built around a central atom of carbon (Figure 4-9). Recall that carbon forms four covalent bonds. In amino acids, the central carbon bonds to a nitrogenous group called an **amino group** ($-NH_3^+$), a **carboxyl group** ($-COO^-$), and a single hydrogen atom (-H) at three of the four bonding sites. The structure that occupies the fourth position around the central carbon, called the **side group** or the **residue (R group),** is different in each amino acid. The 20 different naturally occurring amino acids are characterized by 20 different R groups. Huxley was quite correct in the four elements he assigned to proteins, but as you can see in Figure 4-9*b*, other atoms can be included in the R group. For example, the amino acid cysteine also has a sulfur atom in its R group.

Proteins are formed when amino acids are strung together, one after the other, connected by means of strong **peptide bonds.** Peptide bonds form between the carboxyl group of one amino acid and the amine group of the next, independent of the variable R group (Figure 4-9*c*). Thus any sequence of amino acids is possible—at least, theoretically. Consider the different structures you could build by linking together any number of the 20 different kinds of amino acids in any sequence. Although the possibilities are infinite, we estimate that cells make only a couple hundred thousand (give or take) different kinds of proteins, a large number to be sure, but still quite finite. (At this writing, about 50,000 different proteins have been identified. Many more are expected to be found when the human genome is fully sequenced, around the year 2003.)

The proteins in cells have precise and unique sequences of amino acids. Some large proteins have as many as 10,000 amino acids; others, small **peptides,** have just a few. The number and sequence of amino acids in a protein make up what is called its **primary structure** (Figure 4-10). Proteins with similar functions often have similar primary structures. Modern evolutionary biologists compare the primary structures of similar proteins to determine the degree of relatedness between species. For example, 98% of the sequence of amino acids of the protein hemoglobin in chimpanzees is the same as that in human hemoglobin. We are, indeed, close relatives of the chimpanzee.

But primary structure is only part of the story. Proteins curl, bend, and fold back on

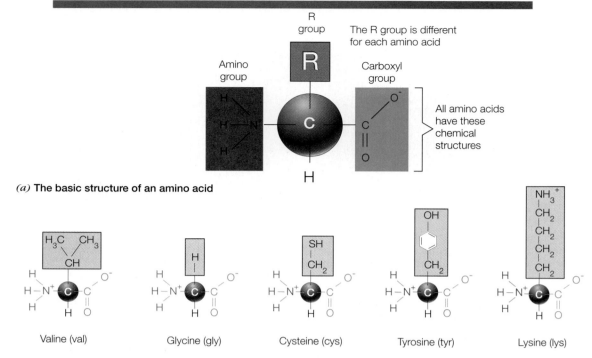

(a) The basic structure of an amino acid

Valine (val)   Glycine (gly)   Cysteine (cys)   Tyrosine (tyr)   Lysine (lys)

(b) Some examples of amino acids

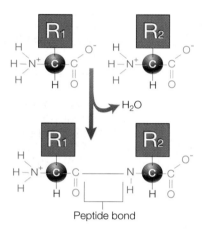

Peptide bond

(c) Joining amino acids to make proteins

Figure 4-9. Amino acids are the building blocks of proteins. (*a*) All amino acids have a central carbon, shown here in gray, to which is attached a single amino group, a single carboxyl group, and a hydrogen atom. The fourth bond of the central carbon is occupied by any one of twenty different molecular structures, some of which are also shown. (*b*) Different variable groups have different chemical properties which contribute to the structure and function of the protein. (*c*) When amino acids are joined to make proteins, one molecule of water is removed to make one peptide bond.

themselves in ways that are neither random nor insignificant. It is the three-dimensional forms of proteins that account for their many varied functions. These **higher order structures** (Figure 4-10) create the special shapes that make particular proteins well suited for certain jobs. Contractile proteins of muscle are long and fibrous. Enzymes are often globular (i.e., compact) with pockets for binding to their specific chemical reactants. The elucidation of the higher order structure of proteins and the role of each amino acid in determining protein structure and function continues to be an active area of biochemical research.

Nucleic Acids

Of the four classes of biological macromolecules, the last one to be discovered was the nucleic acids. Nucleic acids are chains of individual **nucleotides.** A nucleotide is a combination of three types of molecules: a ring-shaped five-carbon (pentose)

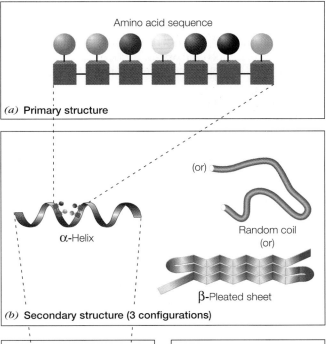

(a) Primary structure

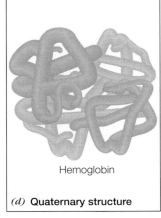

(b) Secondary structure (3 configurations)

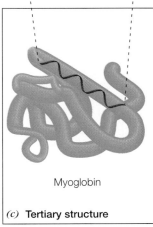

Myoglobin

(c) Tertiary structure

Hemoglobin

(d) Quaternary structure

Figure 4-10. Protein structure is considered at several levels.

(a) Primary structure is the number and sequence of different amino acids in the chain.

(b) Secondary structure is the basic shape that chains of adjacent amino acids assume in three-dimensional space. Some common secondary structures are helices, random coils, and pleated sheets. Any one protein may have regions that assume different secondary structures.

(c) Tertiary structure is the manner in which the helices, coils, and sheets of a whole protein are arranged. This figure shows a compact globular protein called myoglobin, but other tertiary structures may be long and fibrous or flat.

(d) Quaternary structure only applies to proteins that contain more than one chain of amino acids. It describes the manner in which different chains interact with each other. Hemoglobin, for example, has four chains of amino acids that interact in a specific manner to carry oxygen in the blood.

Fast Find Chemistry Review

sugar (either a ribose or a deoxyribose), a phosphate group, and a nitrogen-containing base. They are assembled as shown in Figure 4-11*a*. Any one of five different nitrogenous bases—adenine, guanine, cytosine, thymine, or uracil (abbreviated A, G, C, T, or U)—can be attached to the sugar. Nucleotides with a ribose sugar do not contain thymine, and those with a deoxyribose sugar do not contain uracil. Three of the nitrogenous bases, C, T, and U, are similar to one another and are members of the same family of molecules called the **pyrimidines.** The two others, G and A, are double-ringed structures and are members of the **purine** family of molecules. Nucleotides join together one after the other to form one of two kinds of nucleic acids, ribonucleic acid (RNA, with ribose sugar units) and deoxyribonucleic acid (DNA, with deoxyribose sugar units).

The bonds between nucleotides in nucleic acids link the phosphate of one nucleotide with the sugar of the next. Thus the backbone of a nucleic acid is a linear arrangement of sugar–phosphate–sugar–phosphate–sugar–phosphate–etc., with the nitrogenous bases oriented perpendicularly to the chain. There are two consequences to this arrangement. First, the formation of a nucleic acid is independent of the sequence of bases; any sequence is possible. Second, the nitrogenous bases are not occupied with the business of holding nucleotides together. They are free to participate in other activities, many of which will be discussed in later chapters.

Nucleotides and nucleic acids are central to life. Adenosine triphosphate (ATP), shown in Figure 4-12, is a ubiquitous small molecule made of a ribonucleotide bound to two additional phosphates (three phosphates altogether). The chemical bonds of ATP carry a small packet of energy used in numerous cellular activities. You will encounter ATP again in our discussion of cellular energetics and metabolism in Chapter 9. Another class of small molecules with nucleotide structures, called the nucleotide **coenzymes**, also have important roles in cellular energetics and metabolism. Two examples that will be discussed in Chapter 9 are NAD^+ and FAD. But it is the nucleotide polymers, the nucleic acids DNA and RNA, that contain and convey hereditary information—a primary function of nucleic acids in all cells. These critical roles are described in the next two chapters.

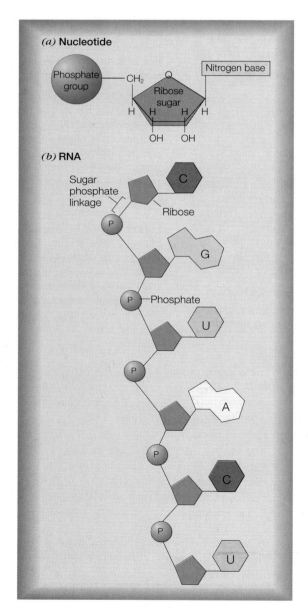

Figure 4-11. Nucleic acids are chains of individual nucleotides.

(*a*) Nucleotide building blocks of nucleic acids are, themselves, composed of three subunits: a five-carbon (ribose) sugar, a phosphate group, and a nitrogen-containing base.

(*b*) Individual nucleotides are linked together by covalent bonds between the phosphate group on one nucleotide and the ribose sugar on the adjacent nucleotide. This example is ribonucleic acid (RNA), because the ribose sugar has an oxygen atom on one of its carbons, and because the different bases include uracil. The other nucleic acid, DNA, lacks an oxygen atom on the sugar (making it a deoxyribose) and includes thymine as one of its nitrogen-containing bases in the place of uracil.

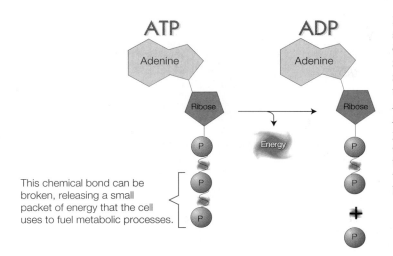

Figure 4-12. Adenosine triphosphate (ATP) is an important nucleotide in cellular metabolism. When the terminal phosphate on ATP is cleaved leaving ADP and a free phosphate, a small packet of energy is released that can be used by the cell to fuel metabolic processes.

Now let's see how cells make use of these macromolecules.

 Macromolecules on the Menu What are the contributions of carbohydrates, lipids, and proteins to our diets? Where do we get them and how do we use them?

4.2.2 Membranes Surround All Cells

**Fast Find
4.2g
Membrane
Structure**

All cells are enclosed in a cellular membrane called the **plasma membrane**, and many cells have extensive membrane systems within them. Despite the great variety of cell types, cell membranes are remarkably similar. A primary function of the plasma membrane is to separate the contents of the cell from its surroundings. At the same time, the plasma membrane provides a pathway for exchange between the cell and its surroundings. Materials and signals pass back and forth through plasma membranes. Obviously, activities involving cell membranes must be highly regulated to accomplish these potentially contradictory functions.

Cell membranes are made up of lipids (mostly phospholipids), proteins, and carbohydrates. Recall from the last section that phospholipids have special properties. One end, composed of two fatty acids, is hydrophobic; in other words, it is repulsed by water. The other end, which includes a phosphate group and other atoms, is hydrophilic; it is attracted to water. In a water-filled environment, phospholipid molecules spontaneously form a sheet two molecules thick—a **phospholipid bilayer**—with hydrophobic ends pointing inward and hydrophilic ends pointing outward. The lipid bilayer is a dynamic, fluid construction in which lipids can move in place as well as laterally in the plane of the bilayer.

Whereas the phospholipid bilayer provides the structural framework of the membrane and prevents water-soluble substances from entering and leaving the cell indiscriminately, proteins are responsible for the many specific functions of different membranes. Over 50 different proteins have been identified in the plasma membranes of red blood cells, and higher numbers may be found in other cell types. Some membrane proteins play a role in cellular identity and individuality; in other words, they act as recognition molecules. Others serve as enzymes, controlling chemical reactions occurring within or on either side of the membrane. In multicellular organisms, membrane proteins serve as couplers, connecting one cell to another. Still other proteins are intimately tied to the passage of materials into and out of the cell.

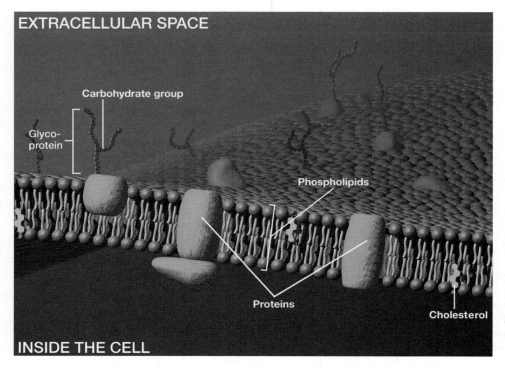

EXTRACELLULAR SPACE

Carbohydrate group

Glyco-
protein

Phospholipids

Proteins

Cholesterol

INSIDE THE CELL

Figure 4-13. The cell membrane is a fluid mosaic of phospholipids (orange) and proteins (green). Many membrane proteins and a few of the lipids have short carbohydrate chains extending into the extracellular space, making them glycoproteins and glycolipids, respectively.

Some membrane proteins rest on the surface of the phospholipid bilayer; others are deeply embedded in the phospholipid bilayer or penetrate it completely. Proteins, like lipids, move within the plane of the membrane, forming an active mosaic of particles within a lipid sheet. The term **fluid mosaic** is used to describe the dynamic nature of lipids and proteins in cell membranes (Figure 4-13).

Carbohydrates in cell membranes are typically attached to either proteins, in which case they are called **glycoproteins,** or lipids, called **glycolipids.** Because carbohydrates are hydrophilic, they extend outward from the surface of the membrane and not into the greasy membrane core. Additionally, carbohydrates nearly always extend from the membrane surface into the space outside of the cell, the **extracellular space,** and not into the cell's interior. Here they act in several capacities. Some membrane glycoproteins act as the "glue" that connects similar cells together to form a single tissue, such as those of the liver or skin. Membrane carbohydrates also function in cell identification. For example, the properties of blood cells that identify individuals as having a certain blood type, such as A, B, AB, or O, are due to the presence of specific membrane carbohydrates. Carbohydrates have also been implicated in some infectious diseases. For example, influenza viruses gain entry into cells by first binding to membrane glycolipids.

4.2.3 Materials Enter and Leave Cells through Their Membranes

Open a bottle of perfume in one corner of a quiet room. Make it a closed room, with no air currents, where people are sitting quietly. Even under these conditions, people sitting in the far corner of the room will eventually smell the perfume. Molecules of perfume move.

Diffusion and Osmosis

At temperatures characteristic of life, all molecules move. The more thermal energy they have, in other words, the hotter they are, the faster they move. Molecules of solids also move, but they are restricted by molecular bonds. Those of fluids—liquids and gases—are less constrained. As they move, fluid molecules bump into each other and rebound from place to place. Although their movements and collisions are random, over time these particles get pushed into areas where collisions are less and less frequent. Eventually, molecules become evenly distributed throughout all available space. Even then they continue to get pushed around, but for every one pushed up, another is pushed down. This is **diffusion**—the net movement of molecules from areas where they are highly concentrated to areas of low concentration by means of random collisions (Figure 4-14).

Cells use diffusion. Imagine a cell surrounded with potential nutrients dissolved in water. In situations like these, chemists use the words **solution** (one substance dissolved in another), **solvent** (the liquid portion), and **solute** (the particle dissolved in the solvent). If nutrient molecules diffuse from areas of higher concentration outside the cell to areas of lower concentration inside, the cell will obtain nutrients. The cell membrane, of course, may be a barrier, depending on the nature of the solute. Some smaller solutes and hydrophobic substances diffuse more readily through cell membranes than larger or more hydrophilic solutes.

Osmosis is a special type of diffusion involving the movement of water across a membrane. Water, too, moves from areas of high water concentration to areas of low water concentration, but predicting the direction in which water will move is not always easy. Imagine that the area on either side of the cell's membranes (inside the cell and outside the cell) has a definite number of spaces that can hold molecules. Now, consider a cell in pure water (one lacking solutes). Outside the cell, all of the spaces are filled with molecules of water. Inside the cell, most of the available spaces are filled with water, but some contain solutes—the proteins, carbohydrates, and other molecules that define the intracellular environment. Where is the water in highest concentration? In this example, the extracellular area has the highest water concentration. In pure water, therefore, water tends to diffuse *into* the cell. In the absence of mechanisms to compensate for the influx of water by osmosis, a cell in pure water will swell and may even burst. In this example, the cell's environment is said to be **hypotonic**

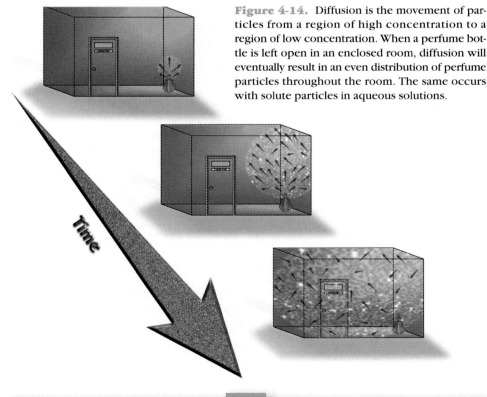

Figure 4-14. Diffusion is the movement of particles from a region of high concentration to a region of low concentration. When a perfume bottle is left open in an enclosed room, diffusion will eventually result in an even distribution of perfume particles throughout the room. The same occurs with solute particles in aqueous solutions.

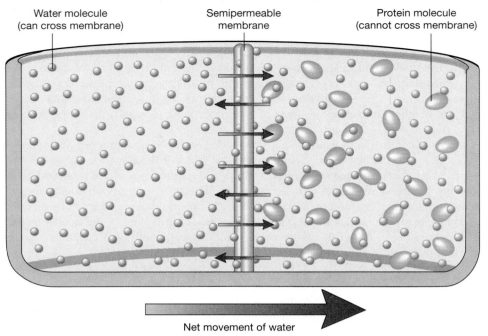

Water molecule
(can cross membrane)

Semipermeable
membrane

Protein molecule
(cannot cross membrane)

Net movement of water

Figure 4-15. Osmosis is the net diffusion of water across a selectively permeable membrane. In this illustration, green protein molecules cannot cross the membrane, but blue water molecules can. Water will move in both directions, but there is a net movement from the left compartment of the chamber to the right compartment. The left compartment is said to be hypotonic relative to the right compartment. Conversely, the right compartment is hypertonic relative to the left compartment.

to the cell's contents, that is, the cell's environment has a lower solute concentration than the cell's interior.

Now consider the same cell in heavily salted water. On both sides of the cell's membrane, spaces available to molecules are occupied by both solvent (water) and solutes, but inside the cell the concentration of water is higher than outside the cell. Where does the water flow? In this situation, the water diffuses from the intracellular compartment to the extracellular area, and the cell shrinks. The cell's environment is **hypertonic** relative to the cell's contents, that is, the cell's environment has a higher solute concentration than the cell's interior.

A cell placed in a solution with the same solute concentration on the outside as occurs on the inside will experience no net movement of water across the membrane— no osmosis. The cell's environment is **isotonic** to the cellular contents. In other words, the solute concentrations of the cell and its environment are equal. The cell neither swells nor shrinks under these conditions.[1] Notice that the *composition* of the solution can differ greatly on either side of the membrane. As long as the *number of solute particles* in the two solutions are the same, they are isotonic. Comparisons of tonicity (i.e., hypotonic, hypertonic, and isotonic) involve only the *number* of particles in aqueous solution, and not the *type* of particles (Figure 4-15).

Simple diffusion and osmosis are some of the most common and powerful means of exchange between cells and their environments. Diffusion, for example, is used to obtain oxygen, a necessary nutrient for most cells, and diffusion can rid the cell of excess water and metabolic wastes such as carbon dioxide. *Osmosis and simple diffusion cost the cell no energy; molecules move spontaneously.*

---

[1]The terms hypotonic, hypertonic, and isotonic always describe one solution relative to another and never just one solution by itself. Usually they are used to describe the extracellular compartment relative to the intracellular compartment.

Diffusion moves some solutes across cell membranes, but not all. When the cell membrane acts as an effective barrier, something more than simple diffusion is needed to transport other vital materials across cell membranes.

Membrane Channels and Facilitated Diffusion

Some large or hydrophilic substances traverse cell membranes through proteins called **channel proteins.** Channel proteins form holes through which atoms or molecules can diffuse. In particular, atoms or molecules that have an electrical charge (either positive or negative) called **ions,** cannot easily cross the membrane. They must move in and out of cells through ion channels, such as the one illustrated in Figure 4-16a. Some channels are permanently open; others can be opened or closed depending on signals originating from the cell or its surroundings. Most channel proteins are highly selective; that is, they will serve as pores for only certain atoms or molecules and exclude others. This type of movement through cell membranes is called **facilitated diffusion.**

As in simple diffusion and osmosis, facilitated diffusion is dependent on nothing more complicated than the random movement of molecules from areas of high concentration to areas of low. In other words, molecules move down a concentration gradient. Transport of solutes by these means does not cost the cells energy. For this reason, it is called **passive transport.** But not all substances are moved passively across cell membranes. Many substances must be moved in a direction that is opposite to their concentration gradient, in other words, from a region of low concentration to a region of higher concentration. For this type of movement, cells must expend energy, a process called **active transport.**

Active Transport and Other Energy-Requiring Kinds of Movement

Active transport is accomplished by a different class of membrane proteins, called **carrier proteins.** A carrier protein binds to the substance to be transported and, as the name implies, carries it to the other side of the cell membrane where it is released (Figure 4.17). At the same time, the protein carrier binds to a second energy-providing molecule and extracts the energy required to move the cargo uphill, against its concentration gradient. Anytime a substance is moved against its concentration gradient, energy is required.

In yet another type of transport, molecules and sometimes whole cells can be moved into the cell in bulk. Items to be transported come in contact with the cell mem-

**Fast Find
4.2k
Membrane
Transport**

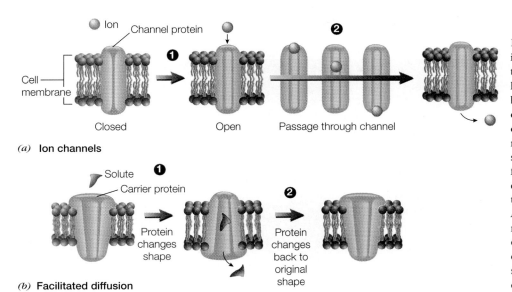

(a) **Ion channels**

(b) **Facilitated diffusion**

Figure 4-16. Membrane transport involves specific transport proteins that are embedded in the phospholipid bilayer. (a) Ion channels must be opened (step 1) before the ion can pass through (step 2). Channels can be opened or closed in response to various kinds of messages received by the cell. (b) In facilitated diffusion, substances enter a protein carrier and change the shape of the protein (step 1). After the solute is moved across the membrane, the protein resumes its original shape (step 2). For both ion channels and facilitated diffusion, substances move from high to low concentration. No energy is needed.

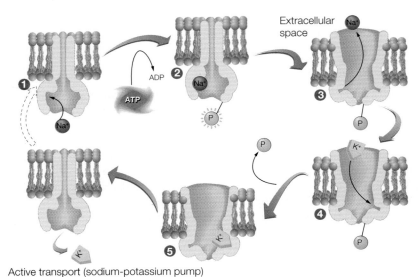

Extracellular space

ADP

ATP

Active transport (sodium-potassium pump)

Figure 4-17. Active transport moves substances from low to high concentration—against their concentration gradient. In this illustration, sodium (Na^+) outside of the cell attaches to the protein (step 1). The protein gets energy when the ATP donates a phosphate to the protein (step 2). This changes the shape of the protein and pushes the sodium across the membrane (step 3). Potassium (K^+) from outside the cell binds to the altered protein (step 4), and is expelled into the cell when the protein resumes its original shape (step 5).

brane, are engulfed by it, and end up in membrane-enclosed vesicles (bubbles) inside the cell. This is **endocytosis.** Similarly, **exocytosis** is used to export cellular products in bulk (see Figure 4-18).

The CD illustrates the relationships between passive and active methods of transport across cell membranes.

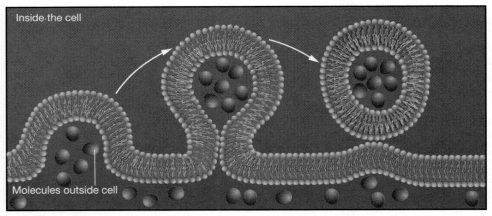

Inside the cell

Molecules outside cell

(a) **Endocytosis**

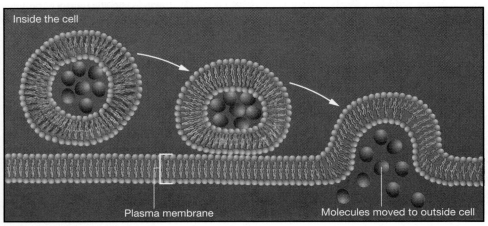

Inside the cell

Plasma membrane

Molecules moved to outside cell

(b) **Exocytosis**

Figure 4-18. Cells can transport substances in bulk across their membranes. (*a*) In endocytosis, the membrane engulfs the substances and the movement is into the cell. (*b*) In exocytosis, the movement is out of the cell.

Piecing It Together

1. The biological macromolecules include the following:

Classes	Building Blocks	Macromolecules	Examples
Carbohydrates	Monosaccharides	Disaccharides, polysaccharides	Glucose, starch, cellulose
Lipids	Vary, may include fatty acids and glycerol	Phospholipids, triacylglycerols	Animal fats, cholesterol
Proteins	Amino acids	Polypeptides, proteins	Enzymes, membrane-bound proteins
Nucleic acids	Nucleotides	Nucleic acids	DNA, RNA

2. Cell membranes surround all cells. Some cells have extensive intracellular membrane systems as well. Membrane structure is a fluid mosaic of phospholipids, proteins, and carbohydrates.

3. Materials traverse cell membranes from areas of high concentration to areas of low by passive means, including simple diffusion, osmosis, and facilitated diffusion. These methods involve no expenditure of energy by the cell.

4. Active transport, in which molecules are transported from areas of low concentration to areas of high, involves carrier proteins in the cell membrane and the expenditure of energy by the cell.

5. Cells can move substances in bulk, including solutes and bigger structures such as whole cells, across their membranes. Bulk uptake is called endocytosis; bulk movement out of the cell is called exocytosis.

4.3 WHAT TYPES OF CELLS ARE THERE?

Fast Find 4.3a Virtual Cell

Cells come in many shapes and sizes. Humans, for example, are made up of over 200 different types of cells (Figure 4-19). In some respects, each species of single-celled organism is a different type of cell. Cells vary greatly in shape. Many tend to be spherical, but others are elongated, box-shaped, or flat. Some have shapes that defy description; others change shape frequently. Cells vary in the kinds and complexity of parts found outside their membranes. Some have bare exteriors. Some have extensive cell walls, slimy coats, or elongated structures that push or pull them through their environments. Internally, some cells are rather simple; others are extensively compartmentalized with internal structures called **organelles.** In spite of their diversity, cells can be lumped into two broad types, prokaryotes and eukaryotes.

4.3.1 Prokaryotic Cells Have Few Internal Parts

More commonly known simply as bacteria, the **prokaryotes** include the true bacteria (Eubacteria) and the more primitive Archaebacteria. At first they appear to be rather simple (Figure 4-20). Compared to **eukaryotes,** they have many fewer internal parts. They are generally smaller, averaging only about one-tenth the size of typical eukaryotic cells.

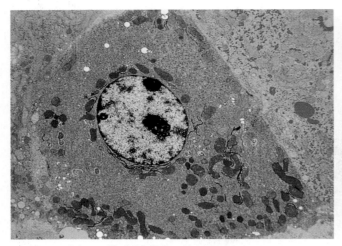

(a) Liver cell (magnified 3,048 times).

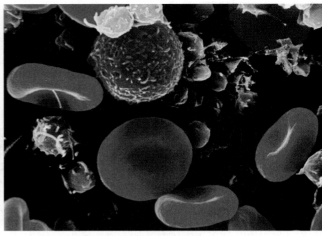

(b) Blood cells (magnified 4,445 times).

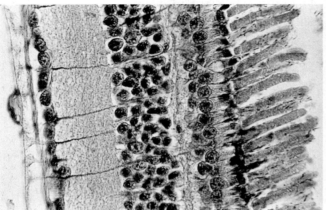

(c) Cells of the retina (magnified 720 times).

Figure 4-19. A sampler of the many different cell types found in human tissues. Colored stains enhance contrast.

(d) Human sperm cells (magnified 1,380 times).

The first impression that they are simple is misleading. Prokaryotes have been on the planet a long time. Their fossils date back at least 3.8 billion years. They have had more time to evolve and adjust to more of earth's environments than any other life form. As a result, they are found nearly everywhere—from the edge of space, and perhaps beyond, to the depths of earth and ocean. Every breath carries some bacterial spores. Nearly every surface holds them, too. Prokaryotes are found in nearly every drop of water. In drinking supplies and swimming pools there are only a few but in stagnant ponds they teem. In our intestines, some are welcome and helpful guests, where in exchange for space and nutrients they give us vitamins and help us absorb water. Prokaryotes are found naturally in soil, air, hot springs, acid pools, salad dressings, highly salty water, sauces, and festering sores. Some prokaryotes are our enemies, causing diseases such as tuberculosis, syphilis, and food poisoning. They cause disease in other living things also. Some prokaryotes are our allies, decomposing carcasses, recycling nutrients, tidying up ecosystems, putting oxygen in ponds and nitrates in soils.

Structurally, prokaryotes are relatively simple (Figure 4-21). A plasma membrane surrounds the cell. Often surrounding the plasma membrane is a thick **cell wall,** made not of cellulose, as in plant cells, but of other complex polypeptides and/or polysac-

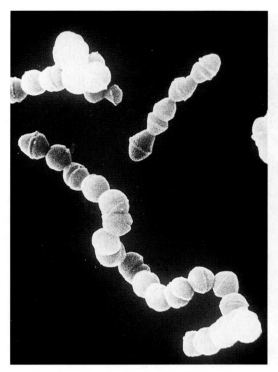

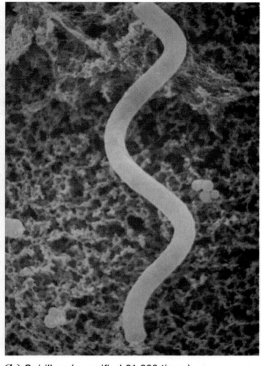

Figure 4-20. Prokaryotic cells come in many different forms.

(a) Streptococcus (magnified 85,100 times).

(b) Spirillum (magnified 21,300 times).

charides. Beyond the cell wall, some bacteria have a **capsule** or a slime layer. Cell walls, capsules, and slime layers help prokaryotes withstand hostile environments and attach to surfaces. Sometimes long fingerlike projections extend beyond the cell surface. These structures, called pili, allow some bacteria to attach to their hosts and occasionally to one another. Through their sex pili, cells exchange genetic information—a primitive form of sexual reproduction. One or more long, whiplike structures called **flagella** often extend beyond the cell wall, spin like propellers, and propel the cell through its medium.

What makes prokaryotes appear simple? Prokaryotic cells have virtually no internal parts or organization. They have no nucleus (the large central structure found in eukary-

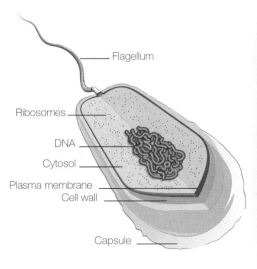

Flagellum

Ribosomes

DNA

Cytosol

Plasma membrane

Cell wall

Capsule

(a) Prokaryotic cell.

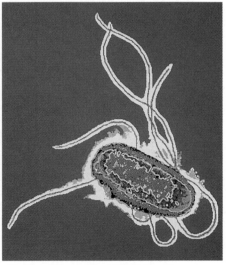

(b) E. coli (magnified 7,600 times).

Figure 4-21. (*a*) A generalized prokaryotic cell. (*b*) This color-enhanced electron micrograph of an *E. coli* cell clearly shows several flagella, but there is little obvious internal structure.

otic cells). One long chromosome, a continuous loop of coiled DNA, folds and twists within the **cytosol,** the cell's internal fluid. Few organelles are to be found. Numerous **ribosomes,** tiny structures made of RNA and protein, reside within the cytosol. Here, on the ribosomes, new proteins are constructed. Once formed and released from the ribosomes, proteins move out into a chaos of cytosol, free to find raw materials and trigger their own special reactions.

4.3.2 Eukaryotic Cells Have Numerous Internal Structures

Eukaryotic cells are the cells of all plants, animals, fungi, and protists. They come in a bewildering variety of types and forms (Figure 4-22). Some, the **protists,** are single-celled organisms. Protists are, for the most part, microscopic but larger than prokaryotes. There are exceptions. While some are scarcely larger than bacteria, other protists can just barely be seen without a microscope. Individual amoebas, for example, are about the size of the period at the end of this sentence.

The cells of all multicellular organisms are eukaryotic; in other words, only eukaryotic cells gave rise to multicellular organisms. In eukaryotes populations of several different types of cells are joined forming a single organism. Adult humans are composed of trillions of cells of over 200 different types. Rotifers are microscopic aquatic organisms smaller than some protists, but as multicellular organisms they are still composites of several types of eukaryotic cells.

Internally, eukaryotic cells are considerably more complex than prokaryotic cells. Eukaryotic cells have intracellular compartments enclosed in membranes that isolate and concentrate essential functions and processes for the cell. Membranes surrounding organelles are similar to plasma membranes, that is, they are fluid mosaics of phospholipid bilayers with proteins. The kinds of proteins that occur in intracellular membranes may differ from those of the membrane surrounding the cell.

Far from chaotic, the inside of the eukaryotic cell is a highly organized biological factory. Figure 4-23 illustrates some of the more prominent structures found in a eukaryotic cell. The **nucleus,** usually its most obvious organelle, is central to the cell's activities. Surrounded by a porous double membrane called the **nuclear envelope,** the nucleus also contains the cell's hereditary material, its **chromosomes.** Chromosomes act as a blueprint of instructions for making cellular proteins. Here is not one large closed chro-

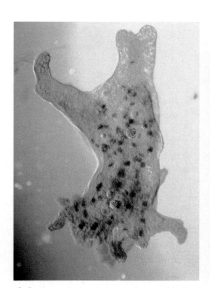
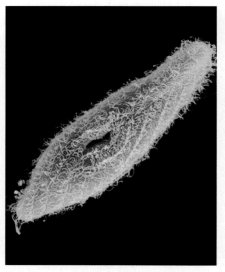
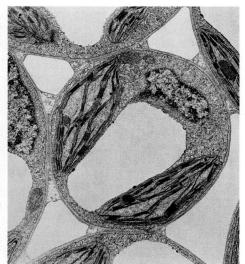

(a) *Amoeba pelomyxa* (magnified 100 times).

(b) *Paramecium* (magnified 3,400 times).

(c) Leaf cell (magnified 8,600 times).

Figure 4-22. Some examples of eukaryotic cells.

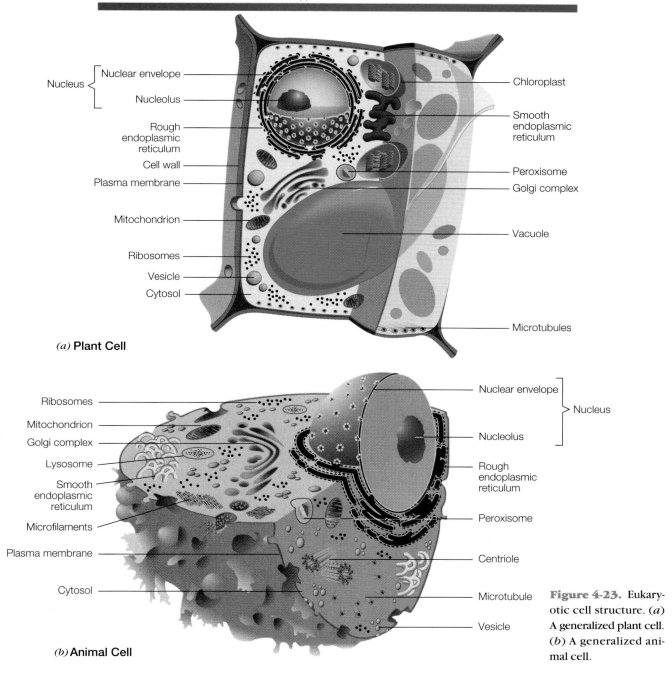

(a) **Plant Cell**

(b) **Animal Cell**

Figure 4-23. Eukaryotic cell structure. (*a*) A generalized plant cell. (*b*) A generalized animal cell.

mosome loop, as in prokaryotes, but a series of shorter, linear structures whose number varies, depending on species, from 2 to over 1000. Eukaryotic chromosomes are complexes of DNA as well as protein.

The nucleus also contains the **nucleolus,** a small, irregularly shaped structure of DNA and RNA where **ribosomes** are assembled.

In those cells that have them, namely, animal cells and some plant cells, **centrioles** are found outside of and adjacent to the nucleus. They play an important role in organizing parts of the cytoskeleton, discussed below.

Outside the nucleus, the cytoplasm is sprinkled with an array of structures. The largest of these is the **endoplasmic reticulum (ER),** a convoluted sheet of membrane enclosing a space within the cell that is separate from the cytoplasm. The twists and turns of the endoplasmic reticulum form numerous folds and pockets throughout the cell's interior. There are two kinds of ER. Relatively close to the nucleus, the

ER is covered with tiny bumps. Each bump is a ribosome, giving this region the appearance of sandpaper and hence its name, the **rough endoplasmic reticulum (RER).** Many cellular proteins are constructed on ribosomes that dot the RER. Other cellular proteins are made on ribosomes not affiliated with the ER but that float freely in the cytosol.

At other points further from the nucleus, the ER has no ribosomes and is called **smooth endoplasmic reticulum (SER).** The SER, too, encloses a separate compartment within the cell. Proteins embedded in the SER membrane synthesize other cellular products, such as lipids and carbohydrates. Tiny membrane-enclosed bubbles called **vesicles** leave the SER and join stacks of larger membrane-enclosed sacks that together form the **Golgi complex.** In the Golgi, cellular products—proteins, lipids, and carbohydrates—are modified, packaged, and prepared for their final destination. Many of these products are exported from the cell by exocytosis. Others are targeted to intracellular sites, such as specific organelles.

The endoplasmic reticulum connects the nuclear membrane, the Golgi complex, and cell membranes into a kind of supraorganelle, the **endomembranal system.** This is the cell factory's production line. Instructions for building proteins pass from the nucleus to the RER, where ribosomes serve as the protein construction site. Newly made proteins can move into the compartment of the RER, where they control the production of other products or are prepared for other destinations. Some travel from the compartment enclosed by ER to the Golgi complex, where they are packaged into vesicles and sent to other cellular locales.

Vesicles come in several types depending on what they contain or what occurs inside them. **Food vesicles** are formed, not by the Golgi complex, but by endocytosis. Food vesicles fuse with **lysosomes,** membrane-enclosed organelles that arise from the Golgi complex. Lysosomes contain digestive enzymes that break down proteins, carbohydrates, and lipids. Lysosomes are used to process cellular debris and worn out organelles. Similar looking structures, called **peroxisomes,** contain many different enzymes, including one that breaks down hydrogen peroxide—a toxic product of several cellular activities. Other vesicles store starch, carbohydrates, fats, atmospheric gases, inorganic salts, organic acids, water, or other materials.

All this cellular activity takes energy. Indeed, the acquisition and control of energy represents one of the cell's biggest challenges. Two prominent eukaryotic organelles, **mitochondria** and **chloroplasts,** deal specifically with energy. Nearly all eukaryotic cells contain mitochondria, but chloroplasts are found only in cells of plants and algae. In Chapter 9 we shall look in greater detail at how these organelles work. Here we will just summarize their functions: Both mitochondria and chloroplasts are elongated organelles, surrounded by two membranes. The innermost membrane of mitochondria is highly folded, a feature that increases the surface area on which important energy-producing reactions can take place. Chloroplasts, unlike mitochondria, have a third system of membranes, an internal membranous complex on which the reactions of photosynthesis occur.

External to the plasma membrane, many cells have surface structures. Plant and algal cells are surrounded by thick cell walls made of cellulose, a complex carbohydrate that not only protects the cell but gives it great strength.

Many protists have surface organelles that give them mobility. Some have flagella, superficially similar to those of prokaryotes, but with a very different internal structure. Other protists, the **ciliates,** are covered with hairlike structures called **cilia** that beat back and forth in unison producing motion. Structurally, cilia are identical to flagella, except they are shorter and more numerous. Some cells in multicellular organisms have similar surface organelles, such as the flagella of sperm and the cilia of several tissues, including the trachea, oviduct, and nasal passages. Wherever they occur, cilia and flagella are extensions of the cell that are enclosed in the plasma membrane. Their core is made up of microtubules, one of the protein filaments that also form part of the cytoskeleton described below.

In the 1980s, electron microscopes revealed that eukaryotic cells have an internal infrastructure, called the **cytoskeleton.** The cytoskeleton comprises long, thin protein filaments. They occur in three sizes: **microfilaments, intermediate filaments,** and **microtubules.** They give the cytoplasm rigidity, maintain shape, anchor and connect organelles, and move parts and products.

Some eukaryotic features are characteristic of certain cell types. Animal and protozoan cells have no thick cell walls and no chloroplasts. They are generally mobile. Plant and algal cells have thick cell walls of cellulose. They also have chloroplasts. Plants and algae can produce their own food (are **autotrophic**). Plant cells generally have a large conspicuous fluid-filled **vacuole** not present in algae. The cells of fungi have no chloroplasts and must rely on outside sources of organic material for food (are therefore **heterotrophic**) but do have thick cell walls.Our understanding of cells has come a long way since Hooke first reported tiny boxes in cork. Indeed, they are much more than mere structural units. Today, we recognize cells as the building blocks of life.

Piecing It Together

1. Prokaryotic cells are relatively simple, with few internal structures. They have no nucleus; a single, circular chromosome floats free in their cytoplasm. Externally, they may have cell walls, capsules, and flagella. They are the true bacteria and Archaebacteria.

2. Eukaryotic cells have numerous internal structures including the following:

   ◆ **Nucleus:** controls cellular activity and stores hereditary information;

   ◆ **Nucleolus:** organelle within the nucleus that controls assembly of ribosomes;

   ◆ **Endoplasmic reticulum:** extensive system of membranes that is either smooth or rough, where many proteins and other products are synthesized;

   ◆ **Ribosomes:** organelles, attached to the RER or floating free in the cytosol, that synthesize proteins;

   ◆ **Golgi complex:** packaging center for cell products;

   ◆ **Vesicles:** organelles that store and transport cellular products and raw materials;

   ◆ **Lysosomes:** vesicles that store digestive enzymes;

   ◆ **Peroxisomes:** vesicles containing many enzymes, one of which detoxifies hydrogen peroxide;

   ◆ **Mitochondrion:** cell's powerhouse where reactions occur that provide energy for the cell;

   ◆ **Chloroplasts:** organelles found in plants and algae where the reactions of photosynthesis occur;

   ◆ **Centrioles:** structures associated with the nucleus of animal and some plant cells that facilitate the organization and construction of the cytoskeleton;

   ◆ **Cytoskeleton:** system of microfilaments, intermediate filaments, and microtubules that provides internal cell structure.

3. Prokaryotes, fungi, algae, and plants have a cell wall that is external to the plasma membrane; cells of animals do not.

4.4 WHERE DO CELLS COME FROM?

Since antiquity, it had been generally believed that under the right circumstances life could generate spontaneously. It seemed obvious. In pond mud a variety of organisms appeared seasonally. Given time, milk soured and broth turned cloudy, teeming with life. Meat left several days in a warm place generated maggots. Later, maggots organized into flies. The feeding structures of barnacles found on the coasts of Europe resembled the body feathers of certain geese. These geese disappeared each spring only to reappear in fall. It was thought that barnacles spontaneously generated into barnacle geese. Elaborate recipes existed for the spontaneous generation of almost every living thing— except, of course, humans.

4.4.1 The Cell Theory Ran Contrary to the Theory of Spontaneous Generation

As we have seen, the nineteenth century was a period of great intellectual upheaval when many ancient beliefs were being questioned. New ideas about evolution and the cellular basis of life were proposed; conclusions were drawn from data carefully collected and analyzed. The theory of spontaneous generation, too, came under close scrutiny. Several years after Schwann proposed the first two tenets of the cell theory, another German biologist, Rudolf Virchow, proposed a principle that would both extend our understanding of cells and contribute to the debate on spontaneous generation.

"Where a cell exists," wrote Virchow, "there must have been a pre-existing cell, just as the animal arises only from an animal and the plant only from a plant. The principle is thus established, even though the strict proof has not yet been produced for every detail, that throughout the whole series of living forms, whether entire animals or plant organisms, or their component parts, there rules an eternal law of continuous development [reproduction]."

This seemingly simple statement was profound in its implications, especially within the context of the cell theory. Schwann had already alleged that tissues grow only through the accumulation of additional cells. Virchow extended this observation across generations of cells. If Virchow was correct, the cells of new organisms arise only from the cells of existing organisms. Spontaneous generation could not be possible. This insight, proposed in the absence of strict proof, required a leap of faith. But strong evidence is necessary to overturn long-held beliefs. Such evidence came from a French chemist named Louis Pasteur.

 One Big Family of Life   If Virchow is correct in stating that all cells come from preexisting cells, then all cells (and hence all organisms) must be related. Here is a profoundly important concept. What evidence supports the idea that all cells are related? *Hint:* Are there ways in which all cells are similar? Is this supporting evidence?

4.4.2 A Series of Brilliant Experiments Settled the Issue

Louis Pasteur was a fighter. Even as a young chemist he was not afraid of taking on conventional wisdom, even if that meant challenging some of the most respected and well-established names in science. In 1854, as a professor at Lille University in France, he became interested in fermentation. The father of one of his students ran a commercial distillery. Usually, the bubbling vats of sugar beets fermented into a drinkable, marketable brew. Sometimes the vats soured into a costly mess. Pasteur discovered that healthy vats invariably contained tiny globular organisms called yeast. Sick vats, on the other hand, contained rod-shaped microorganisms that he had never seen before. He concluded, correctly, that fermentation results from yeast producing alcohol, while the rod-shaped organisms produced a sour-tasting substance that we now know as lactic acid.

Fast Find 4.2b Pasteur's Experiment

Where had the rod-shaped organisms come from? The prevailing belief was that the organisms were generated spontaneously in the heady brew vats. But why only some of the vats? Pasteur reasoned that spores of the rod-shaped organisms must float through the air, sometimes landing in an open vat, where they took up residence and spoiled the brew. To prevent spoiling, Pasteur proposed to isolate the vats from the open air. It worked. Fermentation vats protected from the open air failed to grow rod-shaped organisms. No rods meant no lactic acid. Pasteur immediately saw the implications of his experiment for the theory of spontaneous generation.

Soon, Pasteur's theory was seriously challenged from a new direction. Felix Archemede Pouchet, another French scientist, presented a paper to the French Academy of Science describing an experiment once again "proving" spontaneous generation. First, he sealed a flask of boiling water. After the flask had cooled, he opened it and introduced into it hay and oxygen also previously exposed to temperatures high enough to kill any organisms. After resealing the flask, he waited. Within a few days, organisms appeared. Obviously, Pouchet mistakenly reasoned, they had not come from the air; they had been spontaneously generated.

Pasteur quickly saw the problem. It was easy to boil flasks and their contents, killing whatever lived within, but when the flasks were opened, room air was drawn into them, carrying with it microscopic organisms. Pasteur built an apparatus in which the air that was drawn into the flasks on cooling first passed over red-hot metal, killing any organisms. Nothing grew in the broth. Pasteur's detractors claimed that heating the air must have killed the elements requisite for life. The argument was still not settled.

The trick would be to devise a method whereby microbes in unheated air could be prevented from entering the flask as it cooled. Special flasks were built with **S**-shaped necks flowing off to one side (Figure 4-24). Nutrient-rich broth was boiled inside the flasks. As the flasks cooled, water condensed in the low point of the **S**-shaped curve. Microbes drawn in with the air were trapped in the moisture. The broth inside the flask did not sour. Only if some of the liquid in the neck were tipped into the broth or if the flask's neck were broken, exposing the broth to air, did microbes appear in the broth. It was a convincing demonstration that life did not appear spontaneously. But again, long-held beliefs were not quickly abandoned. For years, the argument raged.

Finally, the French Academy appointed a commission. A public debate was to be held and the issue settled once and for all. Pouchet failed to appear. In front of the commission and a large audience of scientists, business leaders, and even royalty, Pasteur produced sealed flasks filled with broth that for 4 years had not grown organisms. He darkened the hall and used a spotlight to expose dust particles floating in the air. These particles, he dramatically told his audience, included the spores of organisms. Because he protected his broth from spores, no organisms had grown in his flasks.

"Never will the doctrine of spontaneous generation recover from the mortal blow of this simple experiment," said Pasteur to thunderous applause. "There is no circumstance known in which it can be affirmed that microscopic beings come into the world without . . . parents similar to themselves."

Finally, the "strict proof" required by Virchow was at hand. Cells did not generate themselves spontaneously. Cells came only from preexisting cells. After a long gestation, another tenet of the cell theory became accepted.

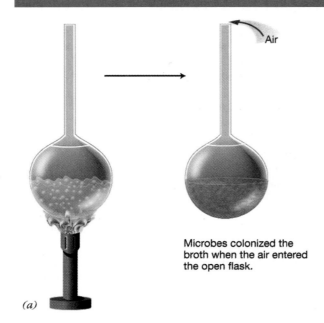

Figure 4-24. Pasteur's experiments in which he disproved spontaneous generation.

(*a*) When a nutrient-rich broth was heated in an open flask, all the microorganisms in the flask were killed, but after it had cooled microorganisms from the air entered through the neck and colonized the broth.

(*b*) If the neck of the flask were bent so that the microorganisms became trapped in the neck, the broth remained sterile. If the neck were broken, microorganisms entered and colonized the broth.

Air

Microbes colonized the broth when the air entered the open flask.

(*a*)

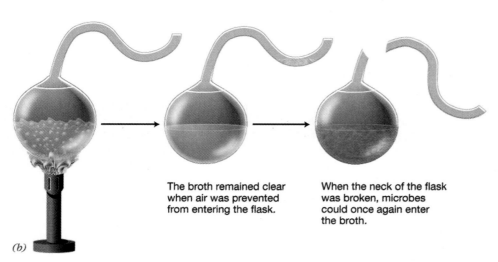

The broth remained clear when air was prevented from entering the flask.

When the neck of the flask was broken, microbes could once again enter the broth.

(*b*)

Pasteur and Beyond Although the emphasis of Pasteur's work was to dispel the theory of spontaneous generation, it led to important practical applications. What were they? *Hint:* Think pasteurization and the germ theory of disease.

4.4.3 A Brief Look at Viruses

Fast Find 4.4c Viruses

Almost lost in the discussions of spontaneous generation was the assertion of Pasteur and others that infectious diseases of both plants and animals are caused by bacteria—single-celled prokaryotic organisms. These observations were of profound importance to the fields of medicine and public health. One by one, the causative agents of various infectious diseases were isolated and identified as bacteria. Disturbingly, studies of some diseases of tobacco plants and cattle failed to produce any infectious bacteria. Indeed, sap of infected tobacco plants suffering from tobacco mosaic disease, and plasma of cattle with hoof-and-mouth disease, remained infectious after being passed through fil-

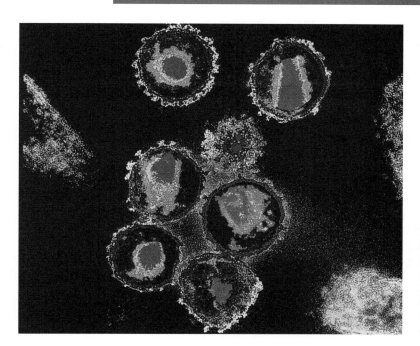

Figure 4-25. Human immunodeficiency virus (HIV). This virus is responsible for the worldwide AIDS crisis. Each viral particle is made of a core of RNA surrounded by a capsule of protein, enclosed in a lipid membrane (magnified 221,000 times).

**Fast Find
4.4c
Viruses**

ters whose pore sizes were tiny enough to exclude even the smallest bacteria. Obviously, these infectious diseases were caused by nonbacterial agents, far too small to be seen by conventional microscopes. Long before they were observed, these agents were given the name **viruses.**

In 1935, Wendell Stanley of the Rockefeller Institute discovered that tobacco mosaic virus could be crystallized, suggesting that it has a simple, regularly repeating structure—far too simple to be cellular in nature. He wrongly concluded that viruses were made solely of protein. Later work showed that they are structurally composed of nucleic acid cores (either DNA or RNA) surrounded by a protective protein coat. Some viruses, the AIDS-causing HIV (human immunodeficiency virus), for example, are also enclosed in a lipid membrane (Figure 4-25).

Are viruses living? Obviously, they are biological entities. They are made of biological molecules (nucleic acid, protein, and sometimes lipids). They reproduce and evolve. But are they organisms in the true sense? If so, they would have to be considered exceptions to the cell theory. The majority opinion among biologists is that viruses are not alive. Here are three main reasons why biologists do not classify viruses as living:

1. Viruses are acellular. One criterion for classifying organisms is how many cells they have. Viruses have none.

2. Viruses are totally dependent on other organisms. Not only must a virus enter a cell of its host to reproduce, it is utterly dependent on the host for energy, raw materials, and the machinery to make proteins.

3. When dried, some viruses form crystals—a characteristic shared with many obviously nonliving substances such as protein, salt, and sugar. No living cells form crystals when they are dried.

Cells Within Cells Consider the life history of the organism that causes malaria, *Plasmodium.* Although dependent on a host to reproduce, *Plasmodium* is still considered living. Why is *Plasmodium* considered living and viruses are not?

It is likely that viruses arose during the course of evolution as bits of nucleic acid, DNA or RNA, that escaped from the chromosomes of living organisms. These fragments contained genes that encoded proteins for the protective viral coat, as well as genes that enabled them to infect a cell and take over its cellular machinery. Two lines of evidence led biologists to this conclusion: First, there are examples of bits of nucleic acid that are not viruses but that leave chromosomes and take up residence at different places in the chromosomes, or even in different cells. These fragments are called **transposons.** Although transposons are not protected by a protein coat, it is likely that they resemble the early precursors of viruses. Second, the genes of many viruses closely resemble a few of the genes of living organisms—usually the organisms that the virus infects. Such similarities indicate that viral genes came from these similar cellular genes.

If the status of viruses as biological entities is debatable, their importance to living things is not. They cause serious diseases in most organisms and have complex, fascinating "life" histories. As an example, consider the life history of the virus that causes cold sores, named herpes simplex type 1. The virus invades the cells of the host organism, humans. It attaches by means of its protein coat. Next, its DNA core is inserted into the cell where it seeks the cell's ribosomes to use in synthesizing proteins, which are incorporated into new viruses. The cell, filled with new viruses, bursts, releasing the viruses to seek new host cells and start the cycle all over. Eventually, enough cells are damaged to produce the well-known cold sore.

Meanwhile, the host fights back. Subtle changes occur in susceptible cell membranes that interfere with the ability of the virus to enter the cell. First, the sore stops growing; then it heals. But the story is not ended here. Some viruses enter some of the host's nerve cells and go into a dormant stage. They may remain in this state for several years until something triggers them into activity. In the case of cold sores, the trigger appears to be an elevated body temperature or dry, cracked lips.

Another important viral disease of humans is AIDS, caused by a retrovirus—a virus with genetic information encoded in RNA and not DNA—called HIV. In retroviruses a protein coat surrounds an RNA core. HIV is also surrounded by a lipid membrane. Explore the World Wide Web for the latest information on this disease.

AIDS—A Virus Causes a Crisis Few diseases have raised such widespread concern as the sudden appearance of the AIDS epidemic in the 1980s. What are some of the reasons for this concern? What is the causative agent for the disease and how is it spread? Who are the at-risk populations for this disease, worldwide? How has the nature of at-risk groups changed over the last few years? What is the status of attempts to treat, slow the spread of, and cure this disease?

Piecing It Together

Today the idea that something as complex as a cell could arise spontaneously seems remote. But a century and a half ago, that was not the case. The third tenet of the cell theory, that cells arise only from preexisting cells, required experimental proof.

1. Louis Pasteur provided evidence that life could not arise spontaneously. Pasteur proved that a flask filled with rich broth could be kept free of microbes if it were boiled, then sealed in a way that prevented air from entering the flask.

2. Pasteur's work led to an understanding that many diseases are caused by bacteria.

3. Viruses are noncellular particles that cause many kinds of diseases. They contain nucleic acid—either DNA or RNA—and protein. Some, like the HIV virus that causes AIDS in humans, are surrounded by a lipid membrane. Most biologists do not consider viruses to be living.

Where Are We Now?

Cellular biology continues to be one of the most active areas of biological study and research. In 1977, Carl Woese (pronounced "woes") at the University of Illinois made a startling proposal. For over a decade, he had been comparing nucleotide sequences in ribosomal RNA of bacteria and found greater variation within the kingdom that includes the bacteria, called Kingdom Monera, than between Monera and cells of other kingdoms. Among the 60 bacterial types he studied, the greatest surprises came from a group known as the methanogens, primitive bacteria that produce methane. The structure of their ribosomes was so different that Woese concluded they are an entirely new cell type. He originally called them Archaebacteria but they have since been renamed Archaea. It was a revolutionary discovery that would change how biologists look at cells.

Imagine Woese's disappointment when his announcement was received with thunderous silence. At the time, biologists were accustomed to and content with the idea that life occurs in two broad cell types, prokaryotes and eukaryotes. What was needed was more evidence.

While many scientists chose to ignore Woese's discovery, not everyone felt that way. A group of biologists in Germany, headed by Otto Kandler, had previously discovered that the cell walls of methanogens had vastly different chemistry than those of other bacteria. They were much more willing to accept Woese's work and proceeded to garner evidence and support. Their efforts forced microbiologists worldwide to at least consider the possibility of a third cell type. Methanogens and other related bacteria were studied in detail, and their uniqueness was verified. Slowly the tide began to turn. In the summer of 1996, the tide became a torrent when the complete genome of the methanogen *Methanococcus jannaschii* was published by the Institute for Genomic Research in Rockville, Maryland. Investigators were able to identify 1,738 protein-coding genes in the methanogen's genome. The vast majority—62%—were utterly unique. Of the remaining genes, most were more closely related to the genes of eukaryotic cells than to those in bacteria. Woese's contentions seem to be vindicated, although acceptance is still far from universal.

REVIEW QUESTIONS

1. Why were the writings of Mendel ignored for nearly 50 years after his death?

2. Most of the major ideas in biology can trace their beginnings back to antiquity. Why not the cell theory?

3. What were the major contributions to biology of Anton van Leeuwenhoek?

4. Both Matthias Schleiden and Theodor Schwann were interested in cells. To what extent were their interests different? How were they compatible?

5. What are the major tenets of the cell theory?

6. Structurally, how are carbohydrates, lipids, proteins, and nucleic acids similar, and how are they different?

7. How are carbohydrates, lipids, proteins, and nucleic acids used by cells?

8. What are the functions of cell membranes?

9. By what processes do materials enter and leave cells? How do these processes work? Which of these processes requires the most cellular energy? Which the least?

10. What are some of the different functions of cellular proteins?

11. Sharks are isotonic to their environments, that is, the concentration of solutes inside their cells is the same as the concentration of solutes in their external environment. How will osmosis affect shark cells?

12. Too many or too few solutes creates serious problems for cells. What are these problems? *Hint:* Think of diffusion and osmosis. There are other problems as well.

13. What prevents hydrophilic atoms or molecules from entering or leaving cells? What cellular features overcome this barrier?

14. How are prokaryotic cells and eukaryotic cells similar? How are they different?

15. Describe the structure of a prokaryotic cell. What are the functions of its various parts (internal and external)?

16. Describe the most prominent structures of a eukaryotic cell. What are the functions of its various parts?

17. Describe how the inside of a eukaryotic cell is like a factory production line.

18. List two cellular organelles responsible for cell movement. How are they similar? How are they different?

19. How are animal cells similar to plant cells? How are they different?

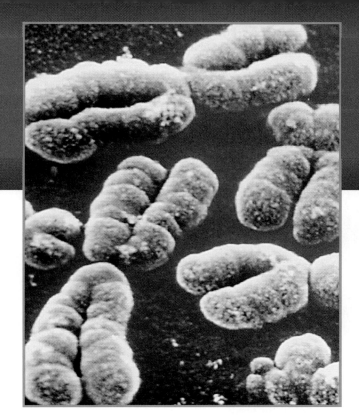

Chapter 5

DNA: Where Are the Genes Found?

If we consider that into these chromosomes are packed from
the beginning all that preordains, if not our fate and fortunes, at least
our bodily characteristics down to the color of our eyelashes, it becomes
a question whether the virtues of the nucleic acids may not rival those
of the amino acids in their importance. —J. B. Leathes, 1926

—Overview—

Consider a complex, many-celled creature such as yourself. You began life as a single cell formed by the fusion of an egg cell from your mother and a sperm cell from your father. This fertilized egg, called a **zygote**, then divided, making two cells, then four, then eight, and so on, until it had given rise to the trillions of cells that make up the complex adult that you are today. Even if you have stopped growing, most of your cells continue to divide. They will do so for as long as you live, replacing old, worn cells and repairing damaged tissue.

All of the instructions, or hereditary units, for building a human being were there at the start, when you were a single-celled zygote. Those instructions influenced what kind of organism you are, the color of your eyes, the way your nerve cells connect in your brain, the length of your legs—every physical trait that you possess. We call those hereditary units your **genes**. With each cycle of cell division, all of the genes are carefully copied and passed on to each new cell so that each of the trillions of cells of your body has its own complete set.

Cell theory brought new understanding to the nature of life, but in the late 1800s, the nature of the genes was not yet known. What are they made of? What structures, small enough to be contained in a single zygote, can harbor enough information to build an entire organism? How are they faithfully copied and passed on in full with each cell division? Genes, scientists reasoned, must be very tiny, indeed. The search for genes began under the microscope, with careful observations of cells in the process of dividing.

Chapter opening photo—Human chromosomes. This color-enhanced electron micrograph shows chromosomes in the condensed form just prior to cell division (magnified 2,280 times).

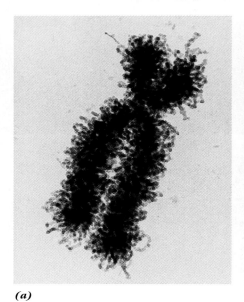

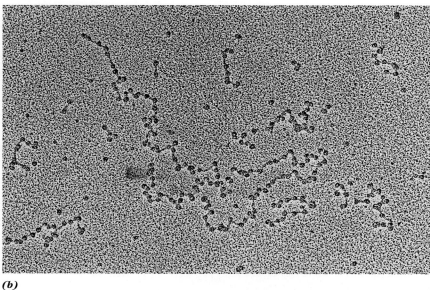

(a)　　　　　　　　　　　　　*(b)*

Figure 5-1. *(a)* A chromosome as it appears during cell division. The two chemical components of a chromosome, protein and DNA, cannot be distinguished in this electron micrograph. *(b)* In this electron micrograph of decondensed chromatin, proteins appear as tiny beads, and the DNA appears as a thin thread that surrounds and connects the beads.

5.1 WHAT CELLULAR STRUCTURE HOLDS THE GENETIC INFORMATION?

**Fast Find
5.1b
Genes,
Loci,
Alleles**

The earliest clues to the physical nature of the genes came in the 1880s from the German anatomist, Walther Flemming. Flemming had at his disposal some of the finest microscope lenses of the day. These oil immersion lenses could provide up to 1000-fold magnification while still maintaining a great deal of resolution. **Resolution** is the term that describes the amount of detail that can be seen in a magnified image. Flemming used a variety of chemical dyes that preferentially cling to certain cellular structures, staining them distinctive colors and improving the contrast between tiny cellular parts. Unfortunately, staining cells usually kills them. His observations were restricted to dead cells, killed in the act of dividing.

Flemming published highly elaborate drawings of dividing cells from salamander embryos. Included in his drawings were detailed renderings of threadlike structures that were all but invisible when cells were not dividing, and that became thick and sausage-like during division. Because of their affinity for colored dyes, he named these structures "chromatin." Today we use the term **chromosome** to describe the structures visible during cell division and the term **chromatin** for the decondensed, threadlike chromosomal material of the nondividing cell (Figure 5-1).

5.1.1 Chromosomes Are Apportioned to Daughter Cells During Mitosis

**Fast Find
5.1d
Mitosis**

Flemming was able to reconstruct the sequence of events during cell division, a process that he named mitosis. **Mitosis** is a form of cellular reproduction in which the parent cell divides, giving rise to two daughter cells that are genetically identical to each other and to the parent cell. Mitosis does not involve combining genetic information from two different parents; hence, it is a form of **asexual** cellular **reproduction**. It occurs when an organism grows, or when damaged or worn cells are replaced. Many unicellular organisms reproduce asexually by mitosis, as well.

Mitosis is a critical time in the life history of a cell. One cell is about to become two, and each new daughter cell will require a complete set of genes. We've learned a great deal about the details of mitosis since Flemming's time, but the behavior of the chromosomes that he described is still a good representation for eukaryotic cells. Take a look at a cell dividing by mitosis on your CD.

The Steps of Mitosis

Sometime prior to the onset of mitosis, the chromatin is copied in its entirety, a process called **replication** (described in detail in Chapter 6). Although this process is not visible using the light microscope, Flemming surmised that it must occur because at **prophase**, the start of mitosis (Figure 5-2), the chromatin begins to condense into a full set of rodlike chromosomes, each consisting of two identical rods, called **chromatids**, firmly connected at one point along their length. When Flemming compared the fine detail of the two rods of a duplex chromosome, he could see that each feature along the entire length of the rod had been exactly duplicated. Also during prophase, the nuclear envelope, the membrane that surrounds the chromosomes, disappears.

As mitosis proceeds, the chromosomes become even more compact. The two identical rods of a mitotic chromosome are called **sister chromatids**, and their point of attachment is called the **centromere**. There are four arms emanating from each centromere (see Figure 5-1a).

The doubled chromosomes move within the cell and align themselves in a flat plane at the equator (center) of the dividing cell, which is now in **metaphase**. Although metaphase is illustrated as a two-dimensional line of chromosomes (Figure 5-2), remember that cells are three-dimensional, and the chromosomes are actually in a plane extending through the cell.

Soon the centromeres of the duplex chromosomes divide, and the sister chromatids—now chromosomes in their own right—move apart in the stage called **anaphase**. The newly separated chromosomes are pulled toward opposite poles (ends) of the cell, and the distance between them increases. **Telophase** begins when the new chromosomes, no longer double, but single, rodlike structures, cluster at opposite poles, and their compact, sausagelike structure unravels, returning the chromatin to its nondividing, threadlike form. Two nuclear envelopes assemble, surrounding the two new sets of chromosomes. The cell nucleus has divided, and in the process, one full copy of each chromosome has been apportioned to each of the daughter cells.

Cellular Scaffolding Cell division involves a vast reorganization of the interior of the cell. In addition to the changes described here, such as disappearance of the nuclear envelope and the condensation and movement of the chromosomes, the cytoskeleton (described in Chapter 4) is also completely rebuilt. What are some of the changes that take place in the cytoskeleton during mitosis? What roles does the cytoskeleton play in mitosis? Which of the cytoskeletal elements are involved?

As the new chromosomes separate during anaphase and telophase, the cytoplasm of the dividing cell is divided into two compartments, each destined to become a new daughter cell with its own nucleus. This process, called **cytokinesis**, differs considerably between plant cells and animal cells. The differences between plant and animal cytokinesis can be attributed to the presence of a rigid cell wall in plants.

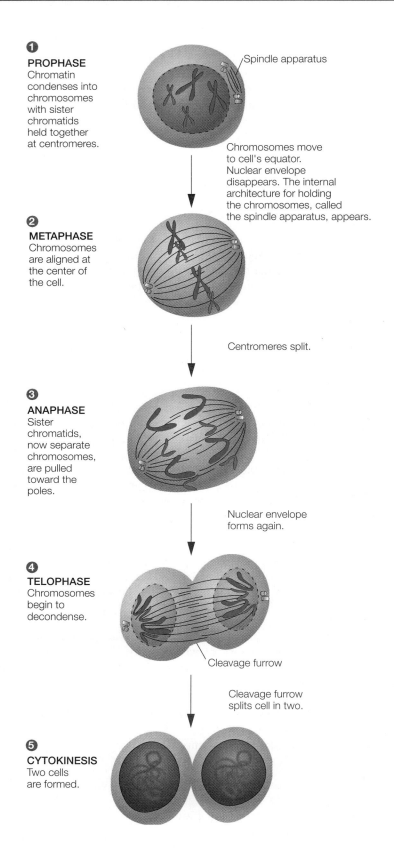

❶

PROPHASE
Chromatin
condenses into
chromosomes
with sister
chromatids
held together
at centromeres.

Spindle apparatus

Chromosomes move
to cell's equator.
Nuclear envelope
disappears. The internal
architecture for holding
the chromosomes, called
the spindle apparatus, appears.

❷

METAPHASE
Chromosomes
are aligned at
the center of
the cell.

Centromeres split.

❸

ANAPHASE
Sister
chromatids,
now separate
chromosomes,
are pulled
toward the
poles.

Nuclear envelope
forms again.

❹

TELOPHASE
Chromosomes
begin to
decondense.

Cleavage furrow

Cleavage furrow
splits cell in two.

❺

CYTOKINESIS
Two cells
are formed.

Figure 5-2. The stages of mitosis.

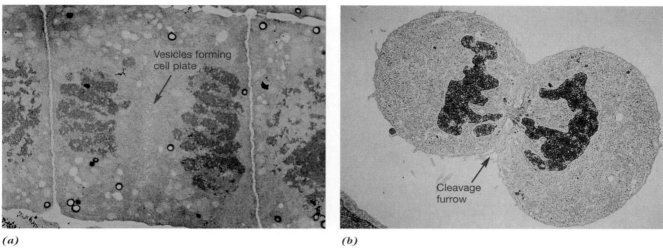

(a)   *(b)*

Figure 5-3. Cytokinesis in *(a)* plants (magnified 13,750 times) and *(b)* animals (magnified 17,325 times).

In cytokinesis in plant cells, membranous vesicles congregate at the equator of the dividing cell. These vesicles contain materials that will give rise to the new cellwall. The vesicles fuse, forming a **cell plate**, shown in Figure 5-3*a*. The vesicles of the cell plate then fuse with the plasma membrane, producing two daughter cells. In animal cells, a **cleavage furrow** forms around the periphery of the dividing cell (Figure 5-3*b*). The furrow becomes progressively deeper until it pinches the cell and its contents into two new cells.

No other cellular structures are as meticulously apportioned to the daughter cells during cell division as are the chromosomes. At the beginning of cell division, each chromosome is doubled and consists of two identical chromatids joined at the centromere. When the chromatids separate, each daughter cell receives one chromatid, now a chromosome, thereby obtaining a complete and exact copy of the chromosomes of the parent cell. Flemming's careful observations made chromosomes the prime candidates for carriers of the hereditary blueprint.

5.1.2 Chromosomes Come in Matched Pairs

With turn-of-the-century microscopes, chromosomes appeared as deeply stained, opaque structures. Nonetheless, careful observations revealed some fine structural detail. Chromosomes were seen to have distinct identities with certain individual features, such as size and shape. Chromosomes with the same features appeared in the nuclei of different cells from the same individual, as well as in cells from different individuals of the same species. Careful observations of several different species led cytologists to conclude that chromosome number is a characteristic of a species. Humans, for example, have 46 chromosomes (Figure 5-4). Our closest evolutionary relative, the chimpanzee, has 48 chromosomes. Mendel's peas have 14; fruit flies have 8; there are some species of fern that have over 1,000 chromosomes. While some species may or may not have the same number of chromosomes as other species, the members of any one species ordinarily share the same number of chromosomes.

Another important conclusion of cytologists was that the chromosomes of almost all mature plants and animals occur in **homologous pairs**, whose members are closely matched in size and shape. However, in many animal species, one pair of homologous chromosomes does not, in fact, appear to be exactly homologous. This is most often the case when cells are taken from males. In the males of some species, such as humans, one member of the odd pair is small and the other is large. Compare the homologous chromosomes of a human male and female illustrated in Figure 5-4. In

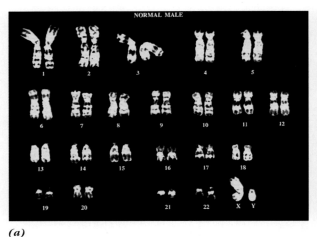
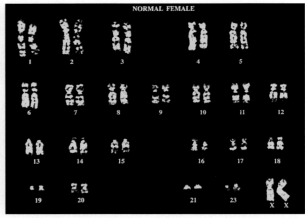

(a)   *(b)*

Figure 5-4. The 46 chromosomes of the human male *(a)* and female *(b)*. In these color-enhanced photomicrographs, homologous chromosomes are paired and numbered from 1 to 22. The 23rd pair of chromosomes, labeled XY in the male and XX in the female, are the sex chromosomes.

other animals, such as the grasshopper (on which these original experiments were done), the males have a single unpaired chromosome whereas the females have a corresponding homologous pair. These chromosomes, it was correctly reasoned, have something to do with determining the sex of the organism. They are called the **sex chromosomes**.

Ignoring the sex chromosomes for the moment (we will return to them in Section 5.2.1), we can see that the occurrence of chromosomes in pairs coincides perfectly with Mendel's description of hereditary factors, which also occur in pairs. By the early part of the 20th century, the discoveries in cytology were integrated with Mendel's genetics (described in Chapter 3) of nearly 40 years earlier to give rise to a new science called **cytogenetics**. The idea that Mendel's laws are the direct consequence of the organization and behavior of chromosomes during cell division was discovered independently by at least four different cytogeneticists in 1902 and 1903, but it was perhaps best articulated by a young American graduate student, Walter Sutton. Sutton's theory, the **chromosome theory of inheritance**, says that the Mendelian factors, the genes, are part and parcel of the chromosomes: *The two members of each homologous pair of chromosomes carry alleles for the same genes and, therefore, affect the same traits.* As we learned in Chapter 3, the two alleles, or forms, of a gene for any given trait can be identical, in which case the carrier is homozygous for that trait, or they can be different, as in a heterozygous individual. Consequently, the two chromosomes of a homologous pair need not be identical, although they carry the genetic information for the same traits.

Just as people reside at specific addresses in their home towns, alleles, too, reside at specific positions on chromosomes. The address of an allele, that is, its position on the chromosome, is called its **locus** (plural, *loci*). For example, the allele that causes sickle cell anemia *(HbS)* is located on the human chromosome designated chromosome number 11, near the tip of the short arm (the p arm) of the chromosome (Figure 5-5). Either an identical *HbS* allele or an alternative allele, perhaps the normal *HbA* allele, resides at the same position on the other copy of human chromosome 11, the other member of the homologous pair. The locus of the gene for hemoglobin is designated 11p15.5, where 11 is for the chromosome number, p for the short arm, and 15.5 for its exact position on the short arm. Likewise, the gene that is responsible for early-onset breast cancer is found at 17q21—about the midpoint of the long arm of human chromosome 17. Today we know the loci of thousands of genes on the 23 pairs of human chromosomes, and more are being found every day.

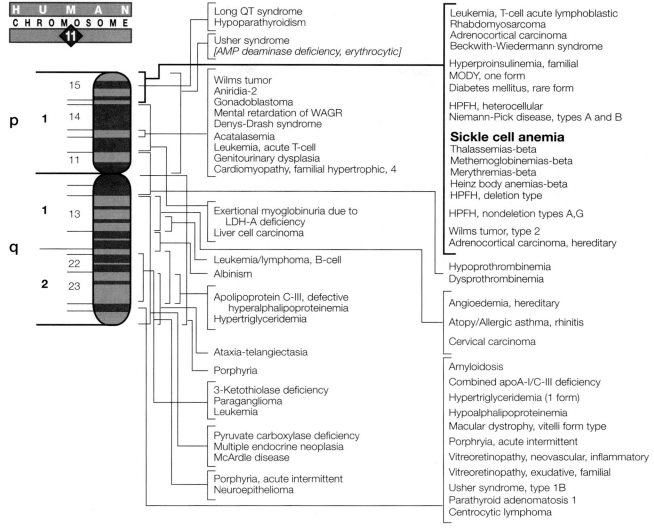

Figure 5-5. A map of the many genes that have thus far been found on human chromosome 11. Notice that all the traits that have been mapped to the chromosome are hereditary diseases. Sometimes many traits or genetic diseases map to a small area of a chromosome, as illustrated by brackets. Maps such as this one exist for all of the human chromosomes, and new genes are being located on chromosomes every day.

The chromosome theory of inheritance was first articulated in 1903, but it was not readily accepted by everyone at the time. There were important problems to resolve before the chromosome theory could gain wide acceptance. The first problem had to do with **sexual reproduction**, the type of reproduction in which the genetic information of a female is combined with that of a male. **Fertilization** is the fusion of gametes—an egg cell from the female and a sperm cell from the male—to form a new individual with a unique combination of genes (Figure 5-6). During fertilization, chromosomes of the male and female gametes combine. The number of chromosomes donated by the male and the female are equal, and the zygote ends up with exactly half of its chromosomes from its mother and half from its father. Thus, in each homologous pair, one chromosome is maternal (originating from the mother) and the other is paternal (originating from the father). In 1895, a prominent cytologist, E. B. Wilson, wrote: "The precise equivalence of the chromosomes contributed by the two sexes is a physical correlative of the fact that the two sexes play, on the whole, equal parts in hereditary transmission, and it seems to show that the chromosomal substance, the chromatin, is to be regarded as the physical basis of inheritance."

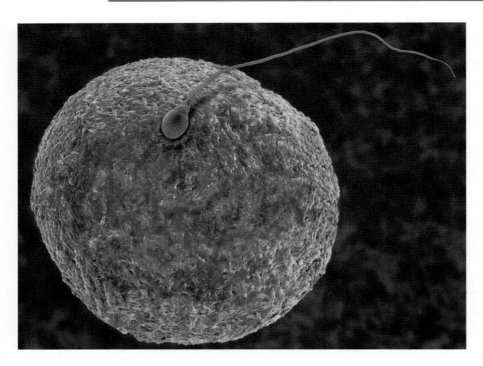

Figure 5-6. Human sperm fertilizing an egg. Although the egg is much larger than the sperm, both cells contribute the same number of chromosomes to the zygote (magnified 924 times).

If egg and sperm each donate an equal number of chromosomes to the new embryo, then in the absence of other mechanisms, the chromosome number of each succeeding generation would exactly double. It would take only a few generations for the number of chromosomes in an embryo to become impossibly large. Yet that doesn't happen; the chromosome number within a species stays constant from one generation to the next. There must be a mechanism for preventing chromosome doubling at fertilization, and that mechanism had to be addressed before biologists could entirely accept chromosomes as the heredity agents.

5.1.3 Meiosis Is Cell Division That Reduces the Chromosome Number

**Fast Find
5.1e
Meiosis**

Gametes are produced in the reproductive organs: the testes and ovaries in animals, the anthers and ovaries in flowering plants. As with all cells, gametes arise from parent cells by cell division. That division process, however, is not mitosis, but a different type of cell division called **meiosis**. Meiosis and mitosis have some similarities, but they differ in important ways that reflect the special role that gametes play in the life history of organisms (Figure 5-7). Whereas mitosis is asexual cellular reproduction, meiosis is a prelude to **sexual reproduction**. Meiosis results not only in multiplying the number of cells by cell division, but also in reducing the number of chromosomes in each daughter cell to exactly one-half of the number before meiosis. Unlike the daughter cells that result from mitosis, gametes receive only one member of each homologous pair, hence, only one allele for each gene. For this reason, meiosis is sometimes called **reductional division**. Cells with both members of each homologous pair are said to be **diploid**. In humans and many other organisms, all body cells *except* the gametes are diploid. Cells with only one member of each homologous pair are said to be **haploid**. In humans, only the gametes are haploid.

By the 1880s, meiosis had already been described in the ovaries of *Ascaris*, a parasitic roundworm. As is true of so many important discoveries in biology, choosing the right organism to study was instrumental in finding answers that could be generalized

MEIOSIS I

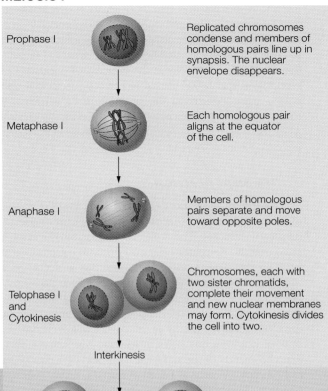

Prophase I
Replicated chromosomes condense and members of homologous pairs line up in synapsis. The nuclear envelope disappears.

Metaphase I
Each homologous pair aligns at the equator of the cell.

Anaphase I
Members of homologous pairs separate and move toward opposite poles.

Telophase I and Cytokinesis
Chromosomes, each with two sister chromatids, complete their movement and new nuclear membranes may form. Cytokinesis divides the cell into two.

Interkinesis

MITOSIS

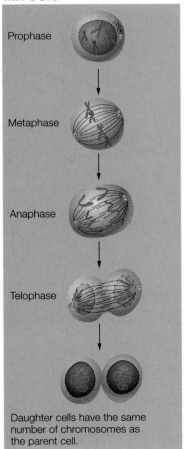

Prophase

Metaphase

Anaphase

Telophase

Daughter cells have the same number of chromosomes as the parent cell.

MEIOSIS II

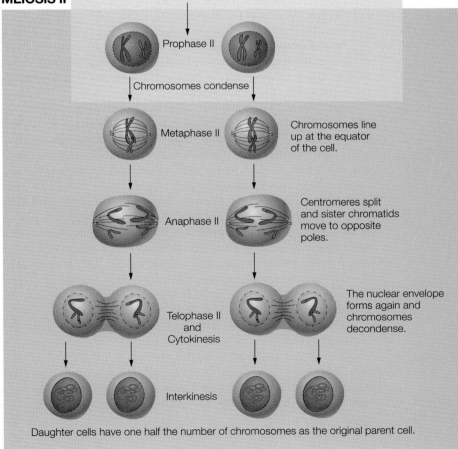

Prophase II
Chromosomes condense

Metaphase II
Chromosomes line up at the equator of the cell.

Anaphase II
Centromeres split and sister chromatids move to opposite poles.

Telophase II and Cytokinesis
The nuclear envelope forms again and chromosomes decondense.

Interkinesis

Daughter cells have one half the number of chromosomes as the original parent cell.

Figure 5-7. A comparison between the steps of mitosis and the steps of meiosis.

to the rest of the living world, and *Ascaris* proved to be the perfect organism for learning about meiosis and gamete formation. *Ascaris* is a large worm, almost a yard long, that lives in the digestive tract of its host and lays copious eggs that are passed to the outside with the host's feces. The eggs infect new hosts who ingest vegetation grown in contaminated soil. The body of the worm is largely occupied by reproductive tissue that is packed with cells in the process of meiosis. Oskar Hertwig, a German zoologist working in the 1880s, described two unusual cell divisions in *Ascaris* ovaries, both of which are necessary in the production of gametes. It is these two cell divisions that together constitute meiosis. Prior to meiosis, the chromosome number of *Ascaris* is four. After meiosis, each gamete has two chromosomes. Read about meiosis here, then take a closer look on the CD.

The Steps of Meiosis

Nearly all cells are capable of dividing by mitosis, but only some can divide by meiosis. In humans and other animals, only those cells destined to become gametes divide by meiosis. The steps resemble mitosis (Figure 5-7) with the important difference that the daughter cells end up with exactly half the number of chromosomes that were in the parent cell. In meiosis, this is accomplished with two cell divisions, one right after the other. Although there are two divisions, the chromosomes replicate only once, before the first cell division. In **prophase I**, the beginning of the first cell division of meiosis, the chromatin condenses into compact chromosomes and the nuclear envelope disappears, as in mitosis. Unlike chromosomes in mitosis, however, members of homologous pairs in meiotic prophase I become closely affiliated—so closely that chromatids of the different members of a pair become perfectly aligned. This joining of homologous pairs is called **synapsis**. As we will see in the next section, synapsis can result in the exchange of bits and pieces of chromosome between the two members of a homologous pair. The aligned pairs of replicated chromosomes move to the equator of the dividing cell in **metaphase I**, but unlike the events of mitosis, the members of homologous pairs move to the equator together. The pairs of duplex chromosomes moving together are called **tetrads**.

Anaphase I begins when members of homologous pairs of chromosomes separate from each other and move toward opposite poles of the cell. The separation of members of a pair of homologous chromosomes is independent of the separation of any other pair. In other words, members of homologous pairs undergo **independent assortment**. Can you recall where you have heard this phrase before? This event corresponds to Mendel's independent assortment of hereditary factors, providing further evidence that chromosomes contain the Mendelian factors. In **telophase I**, the chromosomes become clustered at opposite poles of the dividing cell, each cluster containing one member of each homologous pair. At this point, the chromosomes partially decondense. A nuclear envelope may or may not reform around each set of chromosomes, depending on the species. Cytokinesis divides the cytoplasm of the cells creating two daughter cells, each with half the number of chromosomes as the parent cell. There is a short-lived stage between the two meiotic divisions called **interkinesis**.

Already we see how the chromosome number is halved in cells destined to become reproductive cells. Each cell has only one member of each original homologous pair. But at interkinesis, the chromosomes are still in their replicated, duplex form. Meiosis is not complete until the duplexes separate and are distributed to two daughter cells. This occurs in the next stages of meiotic division, collectively called meiosis II. Both daughter cells resulting from meiosis I undergo the process of meiosis II.

During **prophase II**, partially unraveled chromosomes condense again in both of the cells. As the chromosomes move to the cell equator during **metaphase II**, their duplex structure is apparent, but there are no longer any homologous pairs, so chromosomes line up singly across the middle of the cell. The centromeres divide and the duplex chromosomes separate, with the newly formed single chromosomes moving to the opposite poles in **anaphase II**. As the cells move into **telophase II**, the chromosomes cluster at opposite poles, and the chromatin assumes its decondensed, interphase configuration. The nuclear envelopes assemble, surrounding the chromatin. Cytokinesis divides the cytoplasm, each with its new nucleus, into two haploid cells. Because meiosis II happens in both the daughter cells produced in meiosis I, a total of four new cells are produced from each cell entering meiosis.

Mitosis, meiosis, and cytokinesis are single steps in the **cell cycle**, a repetitive sequence of events that characterizes the life cycle of all cells. Cells that are not in the process of dividing are between mitotic divisions, a period called **interphase**. Flemming referred to nondividing cells as "resting cells," an unfortunate description because we now know that interphase cells are actively metabolizing and growing, among many other things. During interphase, the chromosomes are duplicated in preparation for the next round of division. As we will see in the next section, loss of control of the cell cycle has been implicated in many cancers, making the cell cycle and its regulation an area of intense research.

 Multiply and Divide Only diploid cells can undergo meiosis. But both diploid and haploid cells can undergo mitosis. Do human haploid cells (egg and sperm) divide by mitosis? Do the haploid cells of other species divide by mitosis? Which ones?

5.1.4 Cell Division Is Part of the Cell Cycle

As days warm up in the spring, trees leaf out. The leaves gather sunlight for a few months, then fall to the ground and decompose, providing the tree with nutrients to produce new leaves another spring. The young in all populations grow, produce offspring, age, and die, leaving the Earth to the next generation. Life is defined by cycles, and cells are no exception. Each cell of every living thing arises from a parent cell that preceded it. At any moment in time, every cell is at some point in its life cycle, called the **cell cycle** (Figure 5-8). The life of a cell begins when a parent cell divides, giving rise to two daughter cells. It ends either with its death or when the cell itself divides, giving rise to two new cells. We are still learning about the cell cycle and the complex mechanisms that have evolved to control just when and how a cell divides.

Based on what can be seen with a light microscope, there are two clearly visible phases of the cell cycle: the M phase and interphase. The M phase includes mitosis, itself a series of stages (prophase, metaphase, anaphase, and telophase), during which duplicated chromosomes are separated into two nuclei, and cytokinesis, during which the two nuclei are apportioned to two separate cells. For many years, interphase was believed to be a resting phase, because nothing terribly exciting can be seen happening under the light microscope. But there is more to interphase than meets the eye. For one thing, it is during this phase that the genetic material is copied in preparation for the next division. But that's not all.

A newly formed cell, entering interphase from cell division, is growing, taking in and processing nutrients and energy. The cell makes copies of many of its organelles (not including the nucleus) and gets bigger. This period of active metabolism is called the G_1 phase (for first gap). The new cell has a complete set of chromosomes. If it is a diploid cell, as are most of our cells, chromosomes are present in homologous pairs. But the members of the pairs, the individual chromosomes, exist singly, as very long strands. The length of time a cell remains in G_1 varies with the cell type.

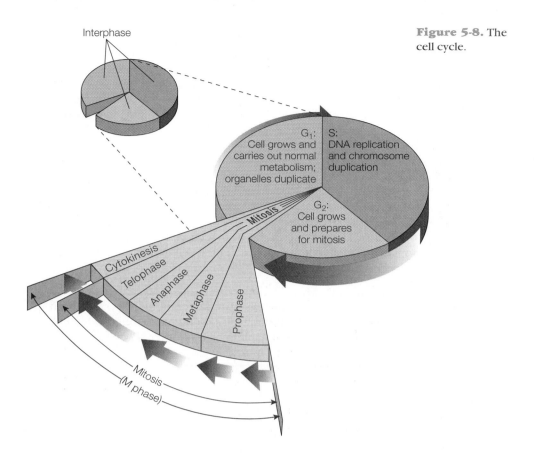

Figure 5-8. The cell cycle.

At some point, the cell receives a signal to proceed from G_1 into the next phase, the S (synthesis) phase of the cell cycle. At this time, cellular growth ceases. The most salient feature of the S phase is the **synthesis** of genetic material. This process is described in some detail in Section 6.2. For now, we will focus on the main events of the cell cycle. By the end of the S phase, the cell's chromosomes are doubled, each copied in its entirety. The two copies remain attached as sister chromatids, so that, under the microscope, the total *number* of chromosomes remains the same, but each one is doubled. (Recall that the doubled chromosomes do not separate until anaphase of mitosis or the second anaphase of meiosis.)

Synthesis of the genetic material marks a critical point in the cell cycle. At this time, the cell is committed to a round of division. Cells with duplicated chromosomes do not remain suspended at a point in the cell cycle, but proceed to the next phase, the G_2 phase (for second gap). The G_2 phase, during which the cell prepares for the upcoming M phase, separates the end of synthesis from the beginning of the next division.

The three parts of interphase, G_1, S, and G_2, typically occupy about 90 percent of the cell cycle. In human cells, M lasts about one to two hours. In rapidly dividing cells, such as those that line the digestive tract or those destined to circulate in the blood, the entire cycle may be anywhere from 16 to 24 hours long. The cell cycle of more slowly dividing human cells, such as those of a child's liver, may last several months. Certain cells cease dividing altogether. Some such cells are quiescent because they are very highly specialized and have lost the ability to divide, for example, the nerve cells in the central nervous system. Others are inhibited from dividing by some external factor. For example, the close proximity of other cells in the full-grown adult liver prevents liver cells from dividing. Cells that have ceased to divide are said to be locked into yet another phase of the cell cycle, called the G_0 phase. The G_0 phase does not occur in cells that continue to divide.

For the trillions of cells that make up a human being to work together as a unified whole, different cells must divide at different rates—some rapidly, others imperceptibly slowly or not at all. The same is true of all multicellular organisms. How is it that two cells of the same organism, sharing identical genetic information, can differ so vastly in their cell cycle? What controls the rate at which a cell divides?

5.1.5 The Cell Cycle Is Highly Regulated

Imagine what would happen if a cell, say a liver cell, lost all restraint over the rate at which it divided. With each turn of the cell cycle, the number of progeny arising from that single unregulated cell would double—first to two cells, then four, then eight and so on—until the number of cells became impossibly large. This is precisely what happens in cancer. Most cancers can be traced to a breakdown in the mechanism that controls the cell cycle in a single wayward cell. It is a heritable breakdown passed on to each of its daughter cells. If one such cell breaks free of the original cancerous mass and takes up residence and uncontrolled growth elsewhere in the body, the cancer has **metastasized**, meaning it has spread to wreak damage wherever it grows. Consequently, much research effort has been directed at discerning the complex mechanisms that regulate when and if a healthy cell divides, with the hope of learning how regulation fails in cancer. While the story is far from complete, much is known. What follows is a brief overview.

Control of the cell cycle is focused at two points, called "checkpoints." The first is just before the genetic material is synthesized, between G_1 and S, and the second is at the transition to division, between the G_2 phase and M. The passage of a cell through each of the two checkpoints is controlled by agents in the cytoplasm that trigger synthesis of the genetic material and the entry into mitosis (or meiosis), respectively (Figure 5-9). These regulating agents are proteins whose cytoplasmic concentrations rise and fall in a controlled manner at different points in the cycle. Their main job is to activate other proteins, including enzymes that initiate the synthesis of duplicated chromosomes at checkpoint one, and enzymes that stimulate the chromatin to condense (among other things) in preparation for cell division at checkpoint two.

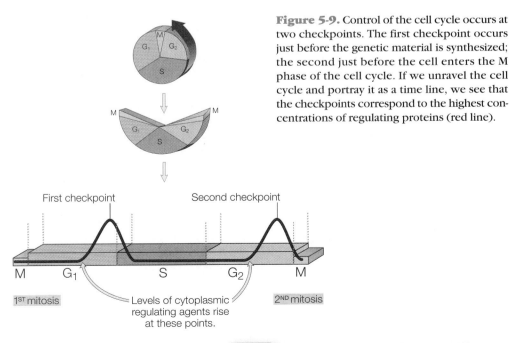

Figure 5-9. Control of the cell cycle occurs at two checkpoints. The first checkpoint occurs just before the genetic material is synthesized; the second just before the cell enters the M phase of the cell cycle. If we unravel the cell cycle and portray it as a time line, we see that the checkpoints correspond to the highest concentrations of regulating proteins (red line).

The regulating agents, in turn, can be controlled by stimuli that are external to the cell; for example, the proximity of other cells in the environment. If a liver cell is tightly packed with other liver cells, as in a healthy adult organ, their close proximity sends a message to the cell to cease production of the regulating agent. If the liver becomes damaged or if a piece is removed surgically, the cells surrounding the lesion respond with an increased production of regulating agent and proceed through the checkpoints to mitosis. In other words, more liver cells will grow, replacing those that are lost or damaged and repairing the lesion. Other external stimuli, such as the presence or absence of hormones, may accomplish the same task in other cell types.

The regulating agents in cells can also be controlled by factors inside the cell itself. For example, cells that have suffered damage from exposure to ultraviolet light or carcinogenic chemicals can often detect their own problems. Rather than reproducing and potentially passing the injury on to their progeny, damaged cells respond by producing proteins that inhibit the regulating agents. The cell cycle is thus halted at one of the checkpoints. The damaged cell may even commit suicide. One such protein, called p53, is so important in detecting damage and preventing an injured cell from entering mitosis that it has been dubbed the "guardian of the genome," a term that hints at its role in detecting damage to chromosomes.

Control of the cell cycle, you can see, is a multilayered operation. There are internal and external stimuli that activate or inhibit regulating agents, which in turn activate enzymes that make it more or less likely that a cell will divide. There are internal checkpoints and guardians monitoring the health of a cell. An error at one or more steps in this complex operation can be disastrous, giving rise to uncontrolled growth; indeed, some cancers have been traced to the failure of a single step.

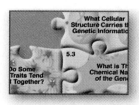

Piecing It Together

By the second decade of the 20th century, the new science of cytogenetics combined the study of cells with Mendelian genetics. We can summarize the important findings of cytogenetics as follows:

1. Genes reside on chromosomes in the cell nucleus. Chromosomes, and consequently genes, are carefully copied and apportioned to daughter cells during cell division. Each of the cells of an organism contains a complete set of chromosomes and, therefore, genes.

2. Mitosis is a form of asexual cell division in which each daughter cell gets an entire set of chromosomes. Organisms grow and repair damaged tissue by means of mitosis.

3. Chromosomes come in matched pairs, called homologous pairs. The two members of a homologous pair are similar in size and shape. During meiosis, or reductional division, the two members of each homologous pair are separated so that each daughter cell gets just one member of each pair and ends up with exactly half the chromosome number of the parent cell. This segregation of chromosomes corresponds to the segregation of factors reported by Mendel.

4. The way in which any one homologous pair is separated during meiosis and apportioned to the daughter cells has no effect on the way in which any other pair is separated. This independent assortment of chromosomes corresponds exactly to Mendel's law of independent assortment of factors.

5. The two alleles for a given gene occur at the same locus on each of the two chromosomes of a homologous pair. During meiosis, when homologous pairs segregate to the daughter cells, each gamete gets one allele for each gene. Genes on different chromosomes follow Mendel's law of independent assortment.

6. The cell cycle is a series of stages through which a cell passes in its lifetime. It begins when a cell is formed by the division of a parent cell, and ends when the cell dies or divides, giving rise to two daughter cells. The stages of the cell cycle include mitosis (or meiosis) and interphase. Interphase is further divided into the G_1, S, and the G_2 phases. Chromosomes are duplicated in the S phase.

7. The passage of a cell through the different phases of the cell cycle is regulated at two points, called checkpoints. The first checkpoint is at the end of the G_1 phase, just before the genetic material is duplicated. The second is in G_2, just before the onset of division. When regulation fails because of an error, the result can be uncontrolled growth and possibly cancer.

8. Proteins in the cytoplasm regulate the movement of the cell through the two checkpoints. When the concentrations of these regulators are high, the cell cycle progresses; when they are low, the cell remains suspended at one stage of the cycle, usually the G_1 phase. The cytoplasmic concentrations of the regulator proteins are, in turn, regulated by external and internal controls that monitor the need for a round of cell division.

5.2 WHY DO SOME GENETIC TRAITS TEND TO TRAVEL TOGETHER?

In the early part of the 20th century, understanding the mechanism of meiosis contributed to widespread acceptance of the chromosome theory of inheritance. Meiosis explained why the chromosome number does not double with each succeeding generation. Another problem for the chromosome theory, however, was the lack of correlation between the number of traits and the number of chromosomes in any species. For example, humans have thousands of traits, but only 23 homologous pairs of chromosomes. All organisms have many more traits than they have chromosomes, so clearly each chromosome must carry the information for many, many traits. During meiosis and fertilization, when chromosomes are segregated to gametes and combined again in a zygote, it might be reasonable to assume that all of the genes found on the same chromosome travel together; they may be apportioned to the gametes and end up in embryos in groups. This, it turns out, is only partially true. Despite their apparent integrity, homologous chromosomes can and do exchange bits and pieces of the genetic material before they are divided up among gametes during meiosis, but more on that later. For the time being, we will focus on the tendency for groups of traits to travel through the generations together on the same chromosome.

Recall from Chapter 3 that each of the seven traits that Mendel studied in pea plants was passed on to the F_1 generation independently of the other six. It just so happens that peas have exactly seven pairs of chromosomes. Peas certainly have more than seven traits, most of which Mendel never considered. According to cytogenetics, it is chromosomes, not individual traits (unless they happen to be on separate chromosomes), that follow Mendel's law of independent assortment. Some claim that Mendel was lucky, that if he had chosen traits that occur all on the same chromosome, he never would have concluded that traits are assorted independently. Sooner or later, however, the exceptions to the law of independent assortment had to emerge. That time was the early part of the 20th century, and it was the work of T. H. Morgan that demonstrated traits are sometimes linked.

5.2.1 Chromosomes, Not Genes, Follow the Law of Independent Assortment

Fast Find 5.2 Sex Linkage

Thomas Hunt Morgan began his career at Columbia College (now Columbia University) in New York City as an embryologist. His request for funds to study mammalian embryology was denied, so he turned his attention to a less expensive project, the genetics of the fruit fly *Drosophila melanogaster* (Figure 3-8 in Chapter 3). It turned out to be a fortuitous choice. The tiny fruit fly is easily collected, lives happily in half-pint milk bottles, and eats yeast that grows on mashed banana. Best of all, the fruit fly has a generation time of just two weeks. Many generations can be studied in a relatively short time. Fruit flies are perfectly suited for genetic studies. To this day, genetics departments in every major university have "fly rooms" filled with glass bottles housing many different strains of fruit flies for study, much like the famous fly room at Columbia in the early part of the 20th century. The fruit fly is one of the species whose entire genome has been sequenced as part of the massive Human Genome Project currently underway.

Although Morgan didn't know it when he began his studies, fruit flies have only eight chromosomes: three homologous pairs and one pair of "accessory" chromosomes. In the female fly, the accessory chromosomes are a homologous pair called XX; in the male, they are a nonhomologous pair called XY. The sex of a fruit fly is determined by which accessory chromosome is carried by the sperm involved in fertilization. When a female fruit fly produces eggs by meiosis, they all receive one X chromosome along with one member of each of the other three homologous pairs. When a male fruit fly produces sperm, they all receive one member of the three homologous pairs, but only half of the sperm get an X chromosome; the other half get a Y. This is illustrated in the Punnett square in Figure 5-10. If an X-carrying sperm fertilizes an egg, the fly embryo will be XX, a female. If the egg is fertilized by a Y-carrying sperm, the embryo that results is XY, a male. The X and Y chromosomes in the fly are the **sex chromosomes**. The sex of the individual is determined by which sex chromosomes are present. Humans have a similar system of sex determination involving X and Y chromosomes.

Figure 5-10. Punnett square illustrating how sex is determined in fruit flies. Females have a matched pair of X chromosomes, and males have one X and one Y chromosome. During meiosis, all of a female's eggs get an X chromosome, but only half of a male's sperm get an X. The other half get a Y. The sex of the offspring is determined by the sex chromosome that is contributed by the sperm.

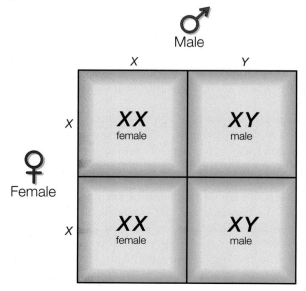

In 1910, an undergraduate named Calvin Bridges, hired to wash dishes in Morgan's laboratory, noticed an unusual white-eyed fly crawling on one of the bottles set aside for washing. Normal fruit flies, the **wild type**, have brilliant red eyes. Morgan carefully collected the lone white-eyed male and mated it with a wild-type, red-eyed female. He found that all 1,237 of the progeny had red eyes—a perfect Mendelian dominant... so far. But when males and females of the F_1 generation were bred, Morgan counted 2,459 red-eyed females, 1,011 red-eyed males, and 798 white-eyed males among the progeny. There were *no white-eyed females* in the F_2 generation. Perhaps, he reasoned, white eyes are incompatible with femaleness.

To test this hypothesis, he crossed a white-eyed male with one of its own daughters, a female carrying one allele for red eyes and one for white eyes—a **heterozygote**. This test cross produced 129 red-eyed females, 132 red-eyed males, 86 white-eyed males, and 88 white-eyed females! Morgan had disproved his own hypothesis. The viability of white-eyed flies is slightly less than that of red-eyed flies. Taking that into consideration, Morgan correctly interpreted these results to mean that equal numbers of red- and white-eyed flies were produced. Clearly, eye color and the factors that determine sex are connected, but given a suitable cross, either sex can have white eyes.

It took Morgan nearly a year to fully explain it, but by 1911, he was convinced that there was an exact correspondence between the inheritance pattern of a gene—the gene for white eyes—and the inheritance pattern of the X chromosome and the factors that determine sex. This demonstrated that the eye color gene is located on the X chromosome. This was the first experimental evidence that hereditary factors could travel together, linked on the same chromosome. Traits that occur on the X chromosome are called **sex linked**. To see how this works, visit Morgan's laboratory on the CD.

 Male, Female, and Beyond In fruit flies, humans, and many other animal species, females have two X chromosomes, and males have one X chromosome and one Y chromosome. In other kinds of organisms, however, different genetic mechanisms may determine the sex of the individual. What other mechanisms of sex determination can you find?

Sex linkage is just one example of a broader phenomenon, in which clusters of traits tend to be inherited in groups. Such groups are called **linkage groups**. By 1915, Morgan and his colleagues had studied the patterns of inheritance of more than 85 traits. They found that in *Drosophila* traits tend to be inherited in four groups corresponding to the four pairs of chromosomes. Humans have 23 linkage groups, pea plants have 7, and so on. The number of linkage groups corresponds to the haploid chromosome number of the species. This is just what one would predict if chromosomes carried the information of heredity and if members of homologous pairs each carried the alleles for the same group of traits. Before Morgan began his studies of *Drosophila* genetics in 1910, he had been an outspoken critic of the newly formulated chromosome theory of inheritance. Good scientists are skeptics who allow themselves to be persuaded only by experimental evidence. Morgan was indeed a superb scientist. By 1915, the results of his experiments had converted him into one of the most ardent supporters of chromosomal inheritance of the day. In 1933, T. H. Morgan was awarded a Nobel Prize for his contributions to our understanding of genetics.

Morgan's discovery of linkage groups, each associated with a chromosome pair, raised a new problem. While traits within a linkage group *tend* to be inherited together, there are many cases in which traits within a linkage group are inherited separately. For example, people with red hair tend to have freckles. But how many people do you know who have freckles but whose hair is brown or blond? A careful study of the exceptions to linkage provided new insights into the behavior of chromosomes during meiosis.

5.2.2 Chromosomes Can Exchange Parts During Meiosis

Several peculiar observations in Morgan's laboratory seemed to indicate that all the traits in a linkage group are not always inherited together. What's more, there was a clearly repeatable pattern in the exceptions to linked inheritance. For example, Morgan's group found that a set of three recessive genes—yellow body, white eyes, and miniature wings— formed a linkage group associated with the X chromosome in fruit flies. But about 1 percent of the time, the yellow-body trait was not inherited together with the white-eye trait. In nearly 34 percent of the offspring, the yellow-body trait did not travel with the miniature-wings trait. The percentage of offspring that showed such incomplete, or "broken," linkage was the same from generation to generation. Something regular and predictable happens when gametes are formed that results in new combinations of alleles, even when they occur on the same chromosome. To explain broken linkage, geneticists turned to the cytologists. Was there visible evidence that perhaps chromosomes exchanged parts during meiosis?

In 1909, the Belgian cytologist F. A. Janssens observed that, when chromosomes paired with their homologous mates during prophase I of meiosis, there were distinctive places where a chromatid on one member of a pair seemed connected to a chromatid of its homologous partner. Janssen called these points **chiasmata** (singular, *chiasma*), or points where there is a crossing-over between the four chromatids on the two members of a homologous pair (Figure 5-11*a*). He correctly reasoned that incomplete linkage could be explained if, in fact, chiasmata represented places where chromosomes break. Broken ends of chromosomes can rejoin, not in the original arrangement, but after switching places with corresponding broken ends on homologous chromatids (Figure 5-11*b*). Modern techniques in electron microscopy and molecular biology have proven that Janssens was correct. The phenomenon of broken and healed chromosomes during prophase I of meiosis is called crossing over

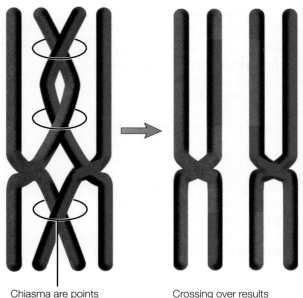

(b)

Figure 5-11. (*a*) Crossing over between chromatids of homologous pairs of chromosomes. (*b*) A chromosome in prophase of meiosis showing chiasmata (magnified 49 times).

Chiasma are points where homologous chromatids cross over.

Crossing over results in exchange of genetic information between members of homologous pairs.

(a)

because parts on one member of a homologous pair cross over to its homologous mate. Crossing over provides an important mechanism for creating new combinations of genes in gametes and, hence, in offspring. Crossing over should be added to the list of different ways in which alleles are recombined by sexual reproduction (see Chapter 3).

Since the discovery of crossing over, yet another way in which chromosomes exchange bits and pieces has been described. The chromosomes, it seems, are not as stable and unchanging as they appear to be.

5.2.3 Bits of Chromosomes Can Move

By the 1940s, the view of the chromosomes as stable arrays of alleles arranged in a linear sequence was firmly established among biologists. Into that milieu came a young loner, a thinker whose meticulous and revolutionary discoveries would not be appreciated for another 30 years. Barbara McClintock had already established her reputation in the 1930s as a leading cytogeneticist when she published a paper entitled "A Correlation of Cytological and Genetic Crossing Over in *Zea mays* [corn]." In that paper, she established once and for all that crossing over involves not just an exchange of bits of chromosomes, but of genetic *information* as well. Had that been McClintock's only contribution, she would have distinguished herself in the field of cytogenetics. But in the 1940s, her work took her in a new direction, one that was as startling as it was controversial. Certain genes, McClintock maintained, were capable of moving from one chromosomal address—one locus—to an entirely different one (Figure 5-12). She found evidence that certain genes "jump" from place to place among the chromosomes. The very idea of jumping genes was contrary to the prevailing view of chromosomes as static.

It wasn't until the 1960s that others found evidence of jumping genes, called **transposons**. Certain genes are more likely than others to leave their chromosomal address and take up residence elsewhere. There is ample evidence that many such pieces of chromosome have moved over the course of evolution. Transposons insert themselves

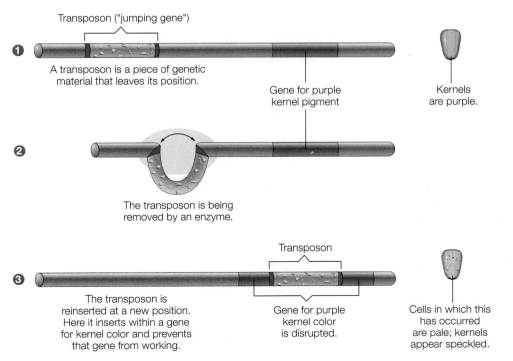

Transposon ("jumping gene")

❶ A transposon is a piece of genetic material that leaves its position.

Gene for purple kernel pigment

Kernels are purple.

❷ The transposon is being removed by an enzyme.

Transposon

❸ The transposon is reinserted at a new position. Here it inserts within a gene for kernel color and prevents that gene from working.

Gene for purple kernel color is disrupted.

Cells in which this has occurred are pale; kernels appear speckled.

Figure 5-12. Transposons move from one position to another, either on the same chromosome, as illustrated here, or on a different chromosome.

at random into new loci, sometimes disrupting other genes at their new address. Consequently, their effects can be devastating. We take up the topic of transposons again in Chapter 6 when we review the ways in which errors can be introduced into the genetic information by mutation.

Genetic Addresses Geneticists have discovered that genes located near the ends of chromosomes cross over more readily than genes that are located near the centromeres. This difference has been used to determine the position of various genes along the length of a chromosome. In other words, geneticists have mapped the loci (addresses) of various genes on chromosomes by determining the frequency with which alleles are exchanged between members of homologous pairs. What are some genes that have been mapped on human chromosomes? What are some modern techniques that are used to create genetic maps?

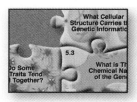

Piecing It Together

With the chromosomal theory of inheritance firmly entrenched in the early part of the 20th century, the focus of biological discovery shifted to meticulous study of chromosome behavior. A single laboratory, the fruit fly lab of T. H. Morgan and colleagues at Columbia University, established important tenets that formed the foundation of modern genetics for all organisms. These are some of Morgan's most important contributions:

1. Traits or characteristics appear to pass from parent to offspring in clusters called linkage groups. The number of different linkage groups in a given species corresponds to the number of homologous pairs of chromosomes in that species.

2. Linkage is not perfect. Traits that are normally inherited together do not always appear together in all offspring. Incomplete, or broken, linkage reflects the tendency for chromatids of homologous pairs to exchange parts during prophase I of meiosis, a phenomenon called crossing over. Crossing over can be seen using the microscope; the points where chromatids overlap are called chiasmata.

3. Some pieces of chromosomes are capable of moving from one locus to another. A transposon, a piece of chromosome that has moved, can insert anywhere at random. If a transposon disrupts a gene, it can interfere with the ability of that gene to express its trait.

5.3 WHAT IS THE CHEMICAL NATURE OF THE GENE?

Genes are present in all living cells. They are copied in their entirety every time a cell divides so that each new cell gets a complete set. They determine the color of fruit flies' eyes, the shapes and colors of seeds of peas, and virtually every other trait that living organisms possess. And they occupy a space small enough to fit into the nucleus of a single cell—a thousand times smaller than the period at the end of this sentence. What chemical substance could possibly account for all these properties? By the 1930s, biologists turned to the rapidly developing science of chemistry for answers to this question, and chemists and physicists turned their attention to biology in search of the secrets of life. Before you begin this section, you might want to review some of the basic chemistry that relates to living cells. Chapter 4 and the Chemistry Review on the CD have the chemistry you need to unravel the mysteries of the chemical nature of the gene.

5.3.1 Chromosomes Are Both Protein and Nucleic Acid

In the 1860s, the same decade that Mendel published his pioneering experiments with peas, a Swiss medical student named Frederich Miescher turned his attention from the practice of medicine to the study of the chemistry of life. His goal was to find the most fundamental constituents of life. Unfortunately, he didn't live long enough to learn just how close he came.

In the 19th century, aseptic technique was unknown. Surgical wounds nearly always became infected, producing copious pus. Using dead white blood cells collected from pus on bandages supplied by a nearby surgical clinic, Miescher searched for new human proteins. His experiments led to the discovery of an unknown substance that contained the elements carbon, nitrogen, oxygen, and hydrogen, as in proteins, but also phosphorus, an element not found in proteins. He determined that the new substance came from the nucleus of the cell and named it "nuclein." It was a large molecule, like protein, but unlike protein, it was acidic in character. One of his students later renamed the substance *nucleic acid*.

Soon after its discovery, nucleic acid was found to have some properties in common with chromatin. For example, both substances were resistant to breakdown in the presence of hydrochloric acid; both nucleic acid and chromatin swelled in the presence of salt solutions. Other cellular substances reacted differently to these treatments. By 1881, nucleic acid had been localized to the chromosomes. Furthermore, nucleic acid had been found in yeast, plants, and every other organism examined. By the 1890s, these observations led some to the prophetic conclusion that nucleic acid was the "idioplasm," the substance responsible for genetic inheritance. But it would be many years before these early conjectures would be proved correct and have widespread acceptance. Part of the reason was that protein seemed at the time to be a better candidate for the genetic material, and protein, too, was a component of the chromosomes.

In Chapter 4, we saw that protein is the most complex, varied, and ubiquitous substance in living cells. This complexity, combined with the many roles protein plays in cells and its presence in chromosomes, led most biologists to reject nucleic acid and accept protein as the substance of heredity. It would take until the middle of the 20th century to prove them wrong.

5.3.2 DNA as the Genetic Material

Pneumococcus is one of many disease-causing bacteria. It occurs in two distinct forms: the S form looks smooth and shiny when it is grown on the surface of an agar plate (agar is a nutrient-rich gelatin used as a substrate to grow bacteria in research laboratories). The R form appears rough and bumpy on agar plates. But the important difference is that the S form is highly virulent. A single, microscopic cell of S-form pneumococcus is lethal when injected into a mouse. The R form is harmless.

The reason for the differences is a smooth polysaccharide coat that envelops S-form bacteria and makes them indifferent to attack from the host's immune system. The R-form bacteria lack the polysaccharide coat, so they are easily conquered by the host's immune system. In 1928, Fred Griffith, a medical officer working for the

British Ministry of Health in London, noticed something strange and interesting about S and R strains of pneumococcus. When mice were injected with the R form, they showed no ill effects. When they were injected with the S form, they invariably died. If S bacteria were first killed by heating them, then injected into mice, the heat-killed S bacteria were as harmless as the living R bacteria. But if heat-killed S bacteria were combined with live R bacteria, the injected mice died just as if they had been injected with the virulent, living S bacteria (Figure 5-13). When those mice were autopsied, Griffith made an extraordinary discovery: the bacteria he isolated and grew from the corpses formed smooth, shiny colonies on agar plates. They had been transformed to the virulent S-form pneumococci. Visit Griffith's laboratory and try his experiments on the CD.

Griffith was a physician, not a geneticist, but his work attracted the attention of a dedicated group of geneticists working at the Rockefeller University in New York. They immediately recognized that the **transforming principle**, as it was ambiguously called, could be the genetic substance. A gene from the dead S-form bacteria had become incorporated into the hereditary blueprint of the live R-form bacteria. Oswald T. Avery and his colleagues, Colin MacLeod and MacLyn McCarty, set out to determine exactly what the transforming agent was. Their first discovery was that the transformation of R to S form using heat-killed S-form pneumococci could be accomplished in a test tube, without the unwitting help of the mice. From there, they proceeded by using a process of elimination. They broke apart the heat-killed S-form bacteria

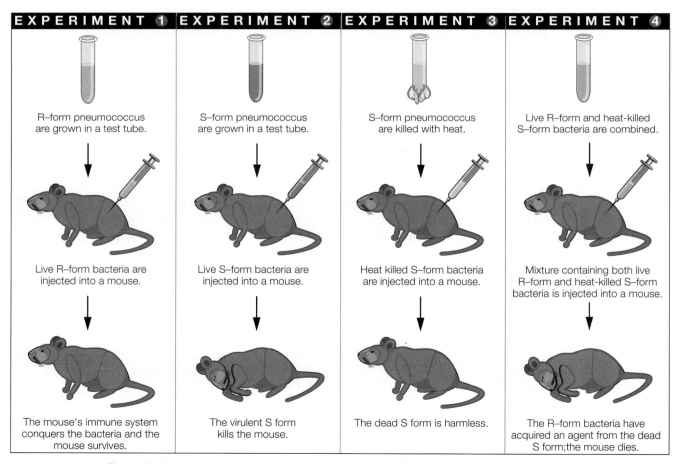

EXPERIMENT ①	**EXPERIMENT ②**	**EXPERIMENT ③**	**EXPERIMENT ④**
R–form pneumococcus are grown in a test tube.	S–form pneumococcus are grown in a test tube.	S–form pneumococcus are killed with heat.	Live R–form and heat-killed S–form bacteria are combined.
Live R–form bacteria are injected into a mouse.	Live S–form bacteria are injected into a mouse.	Heat killed S–form bacteria are injected into a mouse.	Mixture containing both live R–form and heat-killed S–form bacteria is injected into a mouse.
The mouse's immune system conquers the bacteria and the mouse survives.	The virulent S form kills the mouse.	The dead S form is harmless.	The R–form bacteria have acquired an agent from the dead S form; the mouse dies.

Figure 5-13. Griffith's experiments proving that a nonvirulent form of bacteria (the R form) can be converted to a virulent form (the S form) by a transforming agent. That agent was later identified as DNA, the genetic material.

by repeated cycles of freezing and thawing, releasing the cytoplasm of the virulent cells. Using enzymes that chop polysaccharides into their monosaccharide components, they removed the polysaccharide coats from the cell slush. The polysaccharide-free cytoplasm could still transform R-form bacteria into the S-form; the transforming agent remained active. Next they carved the proteins into their amino acid components with protease enzymes; this had no effect on the transforming agent. They extracted the heat-killed, broken S-form cells with alcohol, dissolving away the lipids but leaving other cellular components intact; still they saw no effect on the ability of the mixture to transform R-form cells. The transforming agent was not polysaccharide, protein, or lipid.

Fast Find 5.3c Avery's Experiment

Finally, Avery and his colleagues treated the mixture with DNase, an enzyme that breaks one of the nucleic acids, DNA, into its component nucleotides. In this final experiment, the transforming agent was destroyed, and transformation failed to occur. Using impeccable logic and careful experimentation, Avery, MacLeod, and McCarty had shown beyond reasonable doubt that the transforming agent, and hence the genetic material, is DNA.

5.3.3 Further Evidence for the Genetic Role of DNA

Fast Find 5.3d Hershey/ Chase Experiment

Before Avery, MacLeod, and McCarty published their landmark paper in 1944, most scientists believed that protein held the key to heredity. DNA was thought to be inert scaffolding, a boring molecule that positioned proteins in the chromosomes so that they, the proteins, could carry out the genetic work. The tide of opinion began to shift with Avery's elegant paper, but some resisters still searched for genes among proteins. The search ended once and for all with a 1952 publication by two Americans, Alfred D. Hershey and Martha Chase, working at the Cold Spring Harbor Laboratory on Long Island. They, like Griffith, were bacteriologists, but their studies used the common intestinal bacterium, *Escherichia coli*, as well as a virus that infects it, a virus called **bacteriophage**.

Viruses, as we saw in Chapter 4, are not considered living entities. They are tiny packets—smaller than the smallest cells—of nucleic acid and protein (and sometimes lipid). Incapable of living independently, they depend on the cells they infect for the raw materials and machinery to make more viruses. Despite their status as nonliving, their very simplicity has made them invaluable tools in studies of the most fundamental processes of life. Some bacteriophages (literally "bacteria eaters") are tadpole-shaped, submicroscopic viruses consisting of a protein capsule enclosing a strand of DNA. These viruses attach to the membrane of bacterial cells and inject their genetic material into the cell, leaving the rest of the particle behind (Figure 5-14). The viral genetic material then mingles with the host's genes and hijacks the cellular machinery to make hundreds, even thousands, of new virus particles, killing the bacterium in the process. Hershey and Chase reasoned that if they could somehow tag, or mark, the protein and the DNA in different ways, then follow the fate of the tagged material, they could determine which of the two components of the virus actually entered the bacterial cell. If tagged protein entered the cells, then the experiment would show that protein carries the genetic information. If, as they expected, tagged DNA entered the cell, then their experiment would demonstrate that DNA is the source of the genes.

How does one "tag" a molecule? One answer is **isotopes**. Recall from the Chemistry Review on the CD that isotopes are variants of elements that share all of the same chemical properties but differ in the number of neutrons in the atoms. Some isotopes, called *radioactive isotopes*, are unstable and emit rays or particles that can be detected with the proper instruments. Many of the elements that are abundant in biological molecules have isotopes that are radioactive. When these atoms are fed to living cells, the cells use them as they do their nonradioactive counterparts as building blocks for bio-

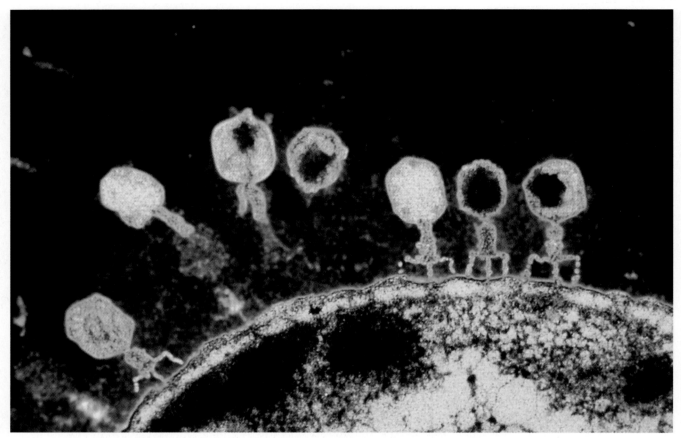

Figure 5-14. Bacteriophage infecting a bacterial cell (magnified 680,040 times). (A color-enhanced transmission electron micrograph.)

logical macromolecules, incorporating them into their cellular structures. Radioactive isotopes can be used to tag any kind of molecule in cells. The fate of the structures built with these isotopes can then be determined by following the path of radioactivity through the life of the cell.

To specifically tag either protein or DNA, Hershey and Chase had to use radioactive isotopes that are specific to only protein and only DNA. Phosphorus is an element that is abundant in DNA but doesn't occur in protein; sulfur is present in two of the amino acids, cysteine and methionine, and hence is found in proteins but is absent from DNA. Both elements have radioactive isotopes, called ^{32}P and ^{35}S, respec-

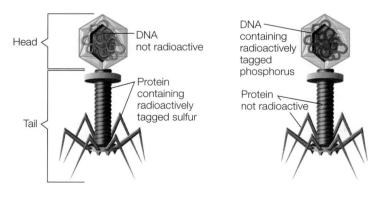

Figure 5-15. The bacteriophages that Hershey and Chase used in their experiments contained either protein with radioactive sulfur and unlabeled DNA, or DNA with radioactive phosphorus and unlabeled protein.

tively. In two separate experiments, Hershey and Chase grew *E. coli* in a medium rich with ^{32}P and another in medium rich with ^{35}S. These bacterial cultures were infected with bacteriophage, which readily incorporated the ^{32}P into their DNA in one case and the ^{35}S into their protein capsules in the other (Figure 5-15). The tagged viruses were isolated from their radioactive bacterial cultures and introduced into two fresh cultures of bacteria, devoid of radioactive isotopes. After about 20 minutes, the tagged viruses had found new bacterial hosts and had begun the process of infection all over again, injecting their genetic material into their new victims. The researchers separated the cells from the viral coats (using that sophisticated laboratory instrument, the kitchen blender!) and looked for the radioactive tags. Even after the viral coats had been sheared off the bacteria by the blender, the infected cells continued to produce new viruses. The ^{32}P, now part of the viral DNA, had found its way inside the *E. coli* cells, but the ^{35}S, part of the viral protein coat, remained outside (Figure 5-16). Try Hershey and Chase's experiment on the CD.

The Hershey-Chase experiment firmly established that DNA carries the genetic information. The combined experiments of Griffith, Avery and his associates, and Hershey and Chase marked the end of the search for the chemical that makes up Mendel's factors and the beginning of modern molecular biology. Once DNA was firmly established

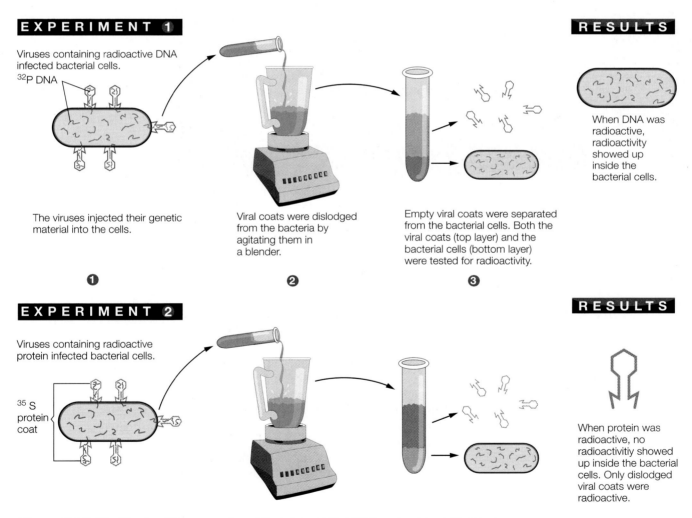

EXPERIMENT 1

RESULTS

Viruses containing radioactive DNA infected bacterial cells.

^{32}P DNA

The viruses injected their genetic material into the cells.

❶

Viral coats were dislodged from the bacteria by agitating them in a blender.

❷

Empty viral coats were separated from the bacterial cells. Both the viral coats (top layer) and the bacterial cells (bottom layer) were tested for radioactivity.

❸

When DNA was radioactive, radioactivity showed up inside the bacterial cells.

EXPERIMENT 2

RESULTS

Viruses containing radioactive protein infected bacterial cells.

^{35}S protein coat

When protein was radioactive, no radioactivitiy showed up inside the bacterial cells. Only dislodged viral coats were radioactive.

Figure 5-16. The Hershey-Chase experiment that showed that DNA and not protein enters the host cell when bacteriophage infect bacteria.

as the hereditary substance, scientists began to focus on the next big questions in biology: what is the structure of DNA, and how does it work?

Piecing It Together

With the experiments of Hershey and Chase, and many that followed, it was established beyond reasonable doubt that DNA is the genetic material, not just in viruses and bacteria, but in all living things. We can summarize our progress as follows:

1. Chromosomes in eukaryotic cells are made of both protein and nucleic acid. Both substances are found in all living cells.

2. For many years, scientists believed that only proteins were sufficiently complex to act as the hereditary material, but they were wrong. Experiments by Griffith using rough and smooth forms of pneumococcus bacteria proved that a chemical substance could transform one form to another. The transforming substance had to carry the genetic information.

3. Avery, MacLeod, and McCarty established that the transforming chemical was DNA.

4. Finally, Hershey and Chase proved that it is DNA that viruses inject into hosts, not protein. By 1952, DNA was firmly established as the hereditary material.

Where Are We Now?

In 1985, about a dozen leading molecular biologists from Europe and the United States gathered in Santa Cruz, California, to discuss the possibility of a new, world-wide effort in human genetics, the Human Genome Project (HGP). Before then, individual scientists and small groups of scientists had made remarkable strides in identifying human genes, almost all of which had a role in disease. For example, the gene that causes Huntington's disease was found on human chromosome number 4; the mutant beta-chain hemoglobin gene implicated in sickle cell anemia was found on chromosome 11; Duchenne's muscular dystrophy was traced to a mutant gene on the X chromosome. Even so, only a tiny fraction of the entire complement of human genes had been identified. A gene not implicated in a hereditary disease was nearly impossible to find.

By 1990, research agencies in the United States, Europe, and Japan had begun allocating funds to the HGP with the hope that it would identify and locate every one of the estimated 100,000 human genes. The project is an enormous international undertaking with several goals. The first goal is to create a map of traits—disease traits, physical traits, or any identifying characteristics—by assigning each trait to a particular chromosome and a particular position on that chromosome. Such genetic maps have been created for humans and other species since the beginning of the century. At first, using a combination of family pedigrees and inheritance patterns (Chapter 4), only a few genes could be assigned with any certainty to specific chromosomes. Now, human genetic maps contain the addresses of a several thousand genes, most of which have been identified because of their role in disease. A very recent example of a genetic map is shown in Figure 5-5. The hope is that every human gene will have a chromosomal address before the HGP is complete in 2003.

The second goal is to determine the nucleotide sequence of the entire human genome. Stretched end to end, the 6 billion DNA base pairs in a human cell are over a meter long. To be sequenced, the DNA is cut into smaller pieces, anywhere from 10,000 to 50,000 base pairs in length. Each piece, of which there may be several hundred thousand, is then separated from the others and amplified, or copied over and over, until there are several hundred thousand test tubes, each containing millions of copies of one relatively short fragment of human DNA. Test tubes containing copies of a fragment are assigned to different HGP laboratories around the world, where the base-by-base sequence of the fragments are determined. The technique used to cut the DNA generates a series of overlapping fragments. When the base sequence of each fragment is known, computers will use the overlapping portions to align the fragments in their proper order. When each of the several hundred thousand fragments is sequenced and properly aligned, this goal will be realized.

What can we learn by looking at the sequence of A, T, C, and G that constitute the human genome? The answer is a lot. There are techniques available today that can analyze the structure of genes and give a great deal of information about the proteins that they encode. For example, certain characteristic amino acid sequences occur in the proteins that recognize and bind DNA. Thus, some genes can be identified as encoding proteins that interact with DNA simply by the amino acid sequences they specify. The same is true for other classes of proteins, including messenger proteins and certain kinds of catalytic (enzyme) proteins. The ultimate hope is to learn the structure and function of every protein encoded by the human genome.

When the chromosomal address and nucleotide sequence of a gene is known, it is a fairly routine matter to develop a diagnostic test to determine the presence of mutations. The HGP may give us the ability to diagnose genetic diseases rapidly and cheaply. Such diseases may include cancer, Alzheimer's disease, mental illness, perhaps even obesity. Treatment could begin early, improving the chances for a good outcome for many human diseases.

Finally, a "library" of human genes can be kept in test tubes around the world. Researchers could obtain copies of a gene for further studies, or to use in manufacturing human proteins for therapeutic purposes.

But with new knowledge comes new power, and it is not always easy to decide how that power should be used. Critics fear that such detailed genetic information could be used to violate our privacy or our rights. Might people with genetic propensities for certain diseases, such as heart disease or Alzheimer's, be denied health insurance? Might employers demand genetic information from prospective employees? Even more frightening, might genetic information begin to define what is "normal" and what is "abnormal," or what is acceptable and what is not? Could detailed information about our genes be misused? Are the benefits of the Human Genome Project worth the risks? These are questions we all have to face. Perhaps the presentation of these issues on the *BioInquiry* web site will help you to define your own answers.

Beyond the Laboratory From the very start, scientists have been aware of the ethical dilemmas presented by the Human Genome Project. In addition to the ones mentioned above, what ethical questions are raised by our efforts to unravel the entire human genome? What steps have been taken to deal with these issues?

Search the web to find how much of the human genome has been characterized so far.

REVIEW QUESTIONS

1. What kinds of evidence did scientists use to establish that Mendel's "factors" were located on the chromosomes?

2. Review the steps of mitosis. At which step do the chromosomes separate? When does the chromatin condense into chromosomes? Why must mitosis be preceded by replication of the chromosomes?

3. How does cytokinesis differ in plant and animal cells? Why does it differ?

4. How do members of homologous pairs of chromosomes differ? How are they alike?

5. What is a gene's locus? Does the locus of a gene change after mitosis? After meiosis? Can you think of any situation in which a gene's locus might change?

6. Assume that you are a scientist working in the early part of the 20th century. You understand the events of fertilization, but you have no knowledge of meiosis. What argument would you use to convince your fellow scientists that meiosis must occur?

7. What is the difference between a haploid cell and a diploid cell? Are any of your cells diploid? Are any haploid?

8. Based on the description of the cell cycle in the text, propose a plausible mechanism that could account for some kinds of cancer.

9. How many linkage groups are there in humans? Explain your answer.

10. Hemophilia is a sex-linked trait in humans in which the blood lacks a certain factor involved in clot formation. Sufferers can bleed profusely from a simple cut. Most afflicted individuals are male. On what human chromosome might you find the gene for hemophilia? Why do more males suffer from the disease than females?

11. List the reasons why most scientists in the early part of the 20th century mistakenly believed that protein must contain the genetic information. Did Griffith's experiments infecting mice with R and S forms of pneumococcus prove that the genetic information was, in fact, in DNA? Why or why not?

12. Avery, MacLeod, and McCarty established that DNA was the transforming substance using a process of elimination. What did they eliminate and how?

13. Hershey and Chase used radioactive isotopes to provide additional evidence that DNA is the genetic material. What did they use radioactive isotopes for? Which isotopes did they use? What molecules did they label with each of the radioactive isotopes?

Molecular Biology: What Is DNA and How Does It Work?

The structure of DNA gave to the concept of the gene a physical and chemical meaning by which all its properties can be interpreted.
—Max Perutz, 1968

—Overview—

Old ideas die slowly, even when they are wrong. Such was the case with the chemical nature of the gene. The experiments of the late 1940s and early 1950s done by Avery, MacLeod, and McCarty and by Hershey and Chase had established DNA as the genetic material, but pockets of resistance to the new idea remained. While older, more set-in-their-ways researchers continued to argue that genes might be made of protein, two young scientists, James D. Watson and Francis Crick, embraced DNA as the genetic material and got a head start on what some have called the greatest discovery of the 20th century: the three-dimensional structure of DNA.

Far from being just an arcane detail, the three-dimensional structure of DNA was the key that opened whole new worlds of understanding about life on Earth. Once the structure of the genetic material was known, the way in which it encodes, expresses, and passes on the genetic information could be understood. From this understanding has come insights into the mysteries of life itself.

In this chapter we pick up the exciting story of discovery with Watson and Crick, two of the best-known names in biology. These brash young scientists, working at Cambridge University in England, solved the problem of how the genetic material is structured. Their approach was roundabout and unconventional—some even say it was unscrupulous. But their story is one of the most exciting in the history of biology.

Chapter opening photo—James D. Watson (left) and Francis H. C. Crick in 1953, with their model of the structure of DNA.

6.1 WHAT IS THE STRUCTURE OF DNA?

In the early 1950s, James D. Watson was a young American scientist visiting the world-famous Cavendish Laboratory in Cambridge, England. At the time, several well-known scientists at the Cavendish were studying the three-dimensional structure of proteins. Meanwhile, another group of top scientists just down the road at Kings College in London were studying the structure of DNA. The two groups often gave open lectures and shared their newest findings. It was at just such a lecture, given by Maurice Wilkins of the Kings laboratory, that the ambitious Watson became obsessed with DNA.

At the Cavendish, Watson shared an office with a graduate student named Francis Crick. Crick was in his mid-30s, older than most graduate students and anxious to make his mark in science. His career, like so many others at the time, had been sidetracked by the outbreak of war. Crick's Ph.D. project consisted of using a new technique called x-ray crystallography to study the structure of the red, oxygen-carrying protein, hemoglobin, found in red blood cells. Watson was at the Cavendish, ostensibly at least, to study the structure of a related protein, myoglobin, another red protein found in muscle cells. No sooner had Watson introduced himself to his new office mate than the two began dreaming of solving the structure of DNA. There would be no better way for two ambitious young scientists to make their mark in the world of science.

6.1.1 DNA's Structure Must Be Compatible with Its Four Roles

The genetic material plays four roles in cells, and Watson and Crick understood that the structure of DNA must be compatible with all four.

1. DNA Makes Copies of Itself

Before a cell reproduces, by either mitosis or meiosis, the chromosomes are duplicated in their entirety, and a copy of each chromosome is distributed to each daughter cell. In this manner, the hereditary information passes from generation to generation. The prevailing view at the time was that each cell must have a mold, or a template, that could be used to stamp out copies of the hereditary information. The starting assumption was that DNA had to act as a template for making more DNA.

2. DNA Encodes Information

In addition, the structure of DNA must be consistent with the ability to encode information that gives rise to discernible traits. All organisms have heritable traits, but they differ even among individuals of the same species; therefore, different DNAs, regardless of the source, must have some features in common and some differences, or else we would all look exactly the same.

3. DNA Controls Cells and Tells Them What to Do

Its not enough to simply carry information. That information must also be put to use controlling what a cell does and what it becomes. Genetic information is expressed by means of chemical processes that are well understood today but were completely unknown at the time. Consequently, the structure of DNA had to be compatible with processes that Watson and Crick knew must exist, but for which, in the 1950s, there was no explanation.

4. DNA Changes by Mutation

The structure of the DNA molecule must be compatible with the ability to change as a result of mutation. It cannot be so fixed that minor changes that result in the appearance of new genetic information render it nonfunctional. There must be some inherent flexibility built into the structure.

Enumerating the roles of the genetic material enabled scientists to recognize a plausible structure when they saw one, but it shed little light on what that structure might be. Watson and Crick needed a clue—a bit of experimental evidence—to get them started. In one of the most unusual episodes in the history of scientific discovery, Watson and Crick discovered the structure of the genetic material, yet neither one of them ever did a single experiment on DNA.

6.1.2 DNA Is a Double Helix

**Fast Find
6.1a
DNA
Structure**

By the time Watson and Crick began brainstorming, the chemical components of the nucleotides, the building blocks of DNA, had been known for many years. Recall from Chapter 4 that each nucleotide has three components: a five-carbon sugar, a nitrogen-containing base, and a phosphate group (Figure 6-1). The sugar and the phosphate are alike in all the DNA nucleotides, but the nitrogenous base can be any one of four different chemical structures, called **adenine**, **guanine**, **thymine**, and **cytosine**.

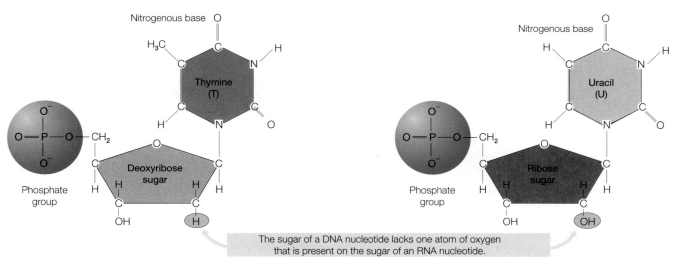

(a) **DNA nucleotide**

(b) **RNA nucleotide**

The sugar of a DNA nucleotide lacks one atom of oxygen that is present on the sugar of an RNA nucleotide.

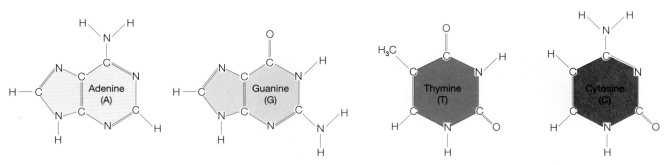

(c) **The four nitrogenous bases that occur in DNA**

Figure 6-1. The chemical structure of the nucleotides. *(a)* A DNA nucleotide with thymine as the nitrogenous base. *(b)* An RNA nucleotide, shown with uracil as the nitrogenous base.

(c) The four nitrogenous bases that occur in DNA. Except for thymine, they also occur in RNA.

Watson and Crick began with the bold assumption that, because different DNAs were all long chains (polymers) of nucleotides, the chemical groups responsible for connecting them together must be those that were most alike in all nucleotides. Thus, the nucleotides must be held together by the five-carbon sugar and the phosphate group—the two portions of the nucleotide that were always the same in all DNA. The variable portions, the nitrogenous bases, were not assigned a role in connecting nucleotides together. That meant, theoretically at least, that any sequence of nitrogenous bases could occur without changing the basic structure of the molecule. These assumptions, which turned out to be quite true, were consistent with the second role of DNA as a repository of information. The sugar-phosphate backbone provides for similarities between DNAs, whereas the sequence of nitrogenous bases in DNA accounts for the differences and plays the role of encoding information. But what of the other roles of DNA?

Just before Watson arrived at the Cavendish, a remarkable discovery about the three-dimensional structure of *proteins* had been made by the brilliant American biochemist, Linus Pauling, at the California Institute of Technology—a discovery that won Pauling the first of his two Nobel Prizes. Pauling began by examining photographs of proteins made by x-ray crystallography. Using this technique, a tiny bit of the sample material—protein, for example, or DNA—is bombarded with x rays. The atoms in the sample reflect the x rays in quite characteristic ways. Photographic film captures the reflected x rays and provides an image of their path. To most of us, such a photograph might appear as splotchy dark spots and light areas, but to a trained x-ray crystallographer, those spots and areas reveal how the atoms in the sample are arranged (Figure 6-2). They give clues to the three-dimensional structure of the sample molecule.

Pauling's x-ray photographs indicated that some proteins have a regular, repeating structure. He made paper cutouts folded to resemble the individual amino acid building blocks of proteins. He then assembled the pieces, much like a jigsaw puzzle, into a protein model that was consistent with the arrangement of atoms predicted by x-ray photographs. The result was an elegant, twisted helix winding around an imaginary axis or an elongated spiral. Because he was using x-ray photographs of a protein called alpha keratin (a component of hair and fingernails), Pauling named the structure an alpha helix.

The alpha helix had taken the biochemical community by storm, and talk of helices was everywhere. Could DNA be another of nature's helices? Could a string of sugar-phosphate molecules with dangling nitrogenous bases form an elongated spiral? If so, could that feature explain the other roles of the genetic material? A good x-ray photograph of DNA was needed to answer this question. But no one at the Cavendish could provide one. Watson and Crick turned to the Kings College laboratory in London.

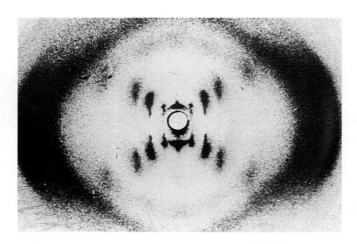

Figure 6-2. Photograph of DNA using x-ray crystallography. The central cross-shaped pattern indicates that DNA is a helix.

The laboratory at Kings College had just been joined by a topnotch x-ray crystallographer, Rosalind Franklin. Franklin had given a public lecture on her x-ray findings using DNA. But Franklin was a thorough and meticulous scientist. Her work was at that time unfinished, and she was unwilling to jump to conclusions about DNA's structure until she had enough experimental data to support her conclusions. If Watson could just get his hands on one of Franklin's x-ray photographs of DNA, he was sure he could at least determine if DNA was a helix. That would be a good start.

Maurice Wilkins at Kings College was also working on the structure of DNA. None of the three men, according to Watson's account in his book, *The Double Helix*, treated Franklin as a colleague. They obtained her x-ray photographs without her knowledge by means that some might consider devious,[1] and with the x-ray photographs of Franklin and Wilkins, they confirmed that DNA was a helix. But what kind of helix? How many chains of nucleotides composed the spiral? How were the chains arranged? And how did the helical structure account for the varied roles of the genetic material?

In 1950, a curious feature of DNA had been published by Erwin Chargaff, a biochemist at Columbia University. Regardless of the source of DNA, the relative amounts of the four nitrogenous bases seemed to conform to a rule: the amount of adenine always equals the amount of thymine, and the amount of cytosine always equals the amount of guanine. The amount of A + T together, however, is independent of the amount of C + G. No one, least of all Chargaff, understood if this regularity was (in Chargaff's words) "merely accidental, or whether it is an expression of certain structural principles." The A = T and C = G rule became known as **Chargaff's ratios**, and they would be another clue in solving the structure of the molecule. Whatever the structure of DNA, it had to account for Chargaff's ratios.

What if, pondered Watson and Crick, DNA were not a single helix as in protein but consisted of more than one chain of intertwined nucleotides? And what if, to satisfy Chargaff's ratios, every adenine base on one chain were somehow paired with a thymine on the other nucleotide chain, one-to-one, and likewise a cytosine was paired one-to-one with a guanine? In this manner, the two chains of nucleotides would not have the identical sequence of nitrogenous bases, but the sequence of nucleotides on one of the chains would be the exact complement of the sequence on the other chain. If the genetic information resided in the base sequence, both chains would contain the same, albeit complementary, instructions.

Using the model-building approach that Linus Pauling had pioneered, Watson and Crick had the machine shop at the Cavendish build a set of model molecules from metal. The repeating sugar-phosphate backbones of two model nucleic acids were assembled first—side by side like the rails of a ladder—then twisted into the exact helix predicted by the x-ray photographs. Then, projecting inward from each sugar, the nitrogenous bases—an adenine on one chain paired with a thymine projecting from the parallel sugar on the other; a cytosine paired with a guanine—formed the rungs between the rails (Figure 6-3). Thus paired, the bases of the model DNA satisfied Chargaff's ratios: every A paired with a T; every C paired with a G. In this way, one DNA strand is complementary to the other; the sequence of bases on one strand fits perfectly with the sequence on the other, although the two strands are not identical.

The chemical bonds connecting the nucleotides in a chain are strong covalent bonds, the type in which atoms share electrons. The bonds joining an adenine with a thymine and a cytosine with a guanine are relatively weak hydrogen bonds. (Review

---

[1]Several interesting books concerning the relationship between Franklin, Wilkins, Watson, and Crick have been published, and their accounts differ. It is interesting to read how different characters in this scientific drama have differing memories of how the individuals involved interacted.

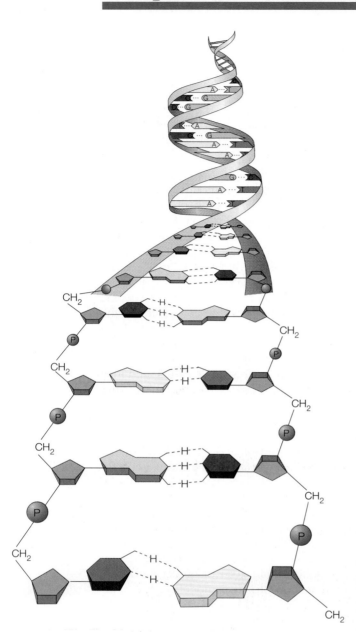

Figure 6-3. The double helix of DNA. The rails of the twisted ladder are held together by covalent bonds between sugars and phosphates. The rungs connecting the two rails are held together by weaker hydrogen bonds between adenine and thymine, or cytosine and guanine. The sequence of nitrogenous bases encodes the genetic information.

these different bonds in the Chemistry Review section on the CD.) Nonetheless, two long chains with many such connections would form a stable double strand. The pieces of the clunky metal DNA molecule, hurriedly cut from sheet metal in the Cavendish shop, fit together like pieces of an elegant puzzle. The model double helix of DNA took shape, rising in a gently twisted ladder from floor to ceiling of the tiny office shared by Watson and Crick, revealing to its discoverers what had been one of life's best kept secrets.[2]

Not only was the double helix an elegant composition, too beautiful *not* to exist, according to James Watson, but the structure was compatible with all four roles of the

---

[2]Watson and Crick were awarded the Nobel Prize in 1962 for discovering the structure of DNA. Wilkins, as head of the Kings College laboratory in London, shared the prize. Rosalind Franklin, however, died of cancer in 1958. Because the Nobel Prize is never awarded posthumously, Rosalind Franklin was deprived of her just reward by dying at the young age of 37.

genetic material. For the remainder of this chapter, we will take a closer look at those diverse roles and the manner in which the double helix of DNA acts as the cell's exquisite taskmaster.

Forwards and Backwards The nucleotides that constitute each of the individual strands of the double helix of DNA are connected by means of covalent bonds between the phosphate of one nucleotide and the sugar of the next. What consequence does this fact have for the end-to-end symmetry of each of the individual strands? Are the ends of the strands exactly alike? What does it mean that the two strands of nucleotides in DNA are "antiparallel"? What are the 5' (pronounced "five-prime") and 3' (three-prime) ends of DNA?

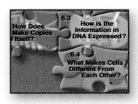

Piecing It Together

Most scientific discoveries are a combination of many years of hard work, meticulous experiments, and brilliant insight. The discovery of the three-dimensional structure of DNA was different. Watson and Crick made one of the most important discoveries of the 20th century without doing a single experiment. Their brilliant insight came from borrowed data, brainstorming, and model building. Here's what they found:

1. The three-dimensional structure of the genetic material, DNA, is a double helix. Two strands of nucleotides coil around each other, forming a twisted ladder in which the side rails are made of deoxyribose sugars alternating with phosphate groups, and the rungs are made of complementary pairs of nitrogenous bases.

2. The genetic information resides in the sequence of nitrogenous bases forming the rungs of the twisted DNA ladder. An adenine base on one strand of the double helix always corresponds to a thymine on the other strand; a cytosine on one strand corresponds to a guanine on the other strand. In this way, both strands of the double helix represent the same information but in complementary form.

3. This double helix is compatible with the four things that the genetic material must do: copy itself, encode information, express that information, and change over time by mutation.

6.2 HOW DOES DNA MAKE COPIES OF ITSELF?

One of the most striking properties distinguishing life from the inanimate is the ability to copy information and pass it from generation to generation. Replication is the process by which DNA copies itself, a process that must take place prior to each round of cell division.

6.2.1 DNA Replication Precedes Cell Division

"It has not escaped our notice," wrote Watson and Crick in their 1953 article describing the double helix, "that the specific pairing we have postulated immediately suggests a possible copying mechanism for the genetic material." By the time Watson and Crick's one-page article appeared in the British journal *Nature*, even skeptics agreed that the genetic information is encoded in the sequence of nucleotide base pairs constituting the rungs of the DNA ladder. DNA must be copied, that is, the correct sequence of every matched pair of nucleotides must be duplicated, and the copies packaged into the separate nuclei of daughter cells at cell division. The process by which this occurs is DNA **replication**.

Recall that the strands of DNA are held together by weak hydrogen bonds between complementary nitrogenous bases, T bonded with A and C with G. According to the Watson-Crick proposal, DNA replication begins when the weak bonds connecting the two parental strands break and the two strands of the molecule begin to physically separate, much like the halves of a zipper. Once separated, the exposed nitrogenous bases of the parent strands attract new nucleotide mates, a T and A paired together and C and G paired together. In this manner, each parental strand acts as a blueprint on which a new partner is assembled (Figure 6-4). As every exposed base on the parental strand becomes paired with a complementary nucleotide, the newly assembled nucleotides are linked together to form a complete strand. The result is two doubled-stranded daughter helices, with half of each daughter composed of one of the parental (old) strands, and the other half a newly synthesized strand. Because the new double helix is half parent and half new, this mechanism is called semiconservative replication.

Watson and Crick proposed this elegant mechanism for DNA replication solely on its logic, with no scientific evidence to support it. Later experiments provided the evidence that proved them correct. The most convincing of this evidence was published in 1958 by two American researchers working at the California Institute of Technology, Matthew Meselson and Franklin Stahl. What kind of evidence could be used to decipher

**Fast Find
6.2
DNA
Replication**

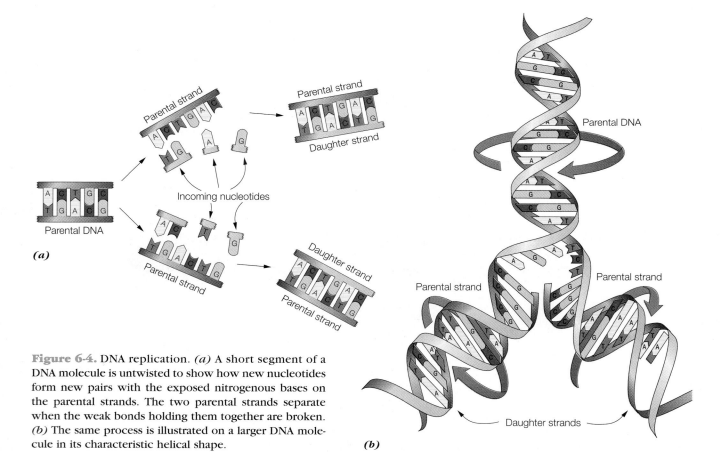

Figure 6-4. DNA replication. *(a)* A short segment of a DNA molecule is untwisted to show how new nucleotides form new pairs with the exposed nitrogenous bases on the parental strands. The two parental strands separate when the weak bonds holding them together are broken. *(b)* The same process is illustrated on a larger DNA molecule in its characteristic helical shape.

the mechanism of a process too small to be observed directly, even with the strongest microscope? How could Meselson and Stahl get inside a cell to reveal the mechanism of DNA replication?

Meselson and Stahl needed a model organism, a creature that could be manipulated and studied easily in the lab. Bacteria seemed the perfect choice. Bacteria have DNA, as do all organisms, and their generation time, that is, their cell cycle, is a short 20 minutes. With each generation, the bacterial chromosome is replicated in full and distributed to two daughter cells. While bacterial cell division is not the same as mitosis (explore the World Wide Web to learn the differences between mitosis in eukaryotes and bacterial cell division), the general mechanism of DNA replication is the same in both prokaryotes and eukaryotes.

The researchers needed a way of distinguishing between newly synthesized DNA strands and the parental strands that served as the template. For this, they turned to isotopes. Recall from Chapter 5 (Section 5.3.3) that isotopes are variants of elements that share all of the same chemical properties but differ slightly in the number of neutrons. Once inside a living creature, different isotopes of the same element are treated the same; organisms that are fed different isotopes cannot distinguish between them. But researchers, using a bit of ingenuity and some common laboratory equipment, can. For Meselson and Stahl, the trick was to grow bacteria whose parental DNA contained one kind of isotope and whose newly synthesized DNA contained a different isotope. If replication is semiconservative, as Watson and Crick proposed, one strand of each daughter helix should contain one isotope, and the other strand should contain the other isotope.

The element nitrogen is a good candidate. It is abundant in DNA and available as several different isotopes. Meselson and Stahl chose a heavy isotope of nitrogen, called ^{15}N. The common isotope of nitrogen, usually found in DNA, has an atomic weight of 14; that is, its nucleus contains seven protons and seven neutrons (atomic weights, isotopes, and related topics, are reviewed in the Chemistry Review section of the CD). The ^{15}N isotope has one more neutron than ^{14}N, a feature that alters none of its chemical properties but makes any compounds that incorporate it just a bit heavier than those containing only common nitrogen.

The researchers grew bacteria for many generations in a medium containing only heavy nitrogen. After several generations, virtually all of the nitrogenous bases in the DNA of these cells contained only heavy nitrogen. Then, all at once, the cultures were washed free of the old medium and placed in a new medium containing only light nitrogen. Any new DNA synthesized after the wash would contain the light isotope. Bacteria were harvested from the fresh growth medium at intervals, and the DNA was separated from the other cellular components.

A clever way of separating compounds based upon their different densities is to place them in a solution that forms a density gradient in a test tube. Sugar solutions, for example, can be very dense and viscous, as in syrup, or no denser than the water in which they are mixed, as happens when you sweeten your tea. If a test tube is filled first with a layer of syrup followed by several layers of progressively less dense solution, the result is a density gradient—thick and syrupy at the bottom and light and watery at the top. When compounds such as DNA are added to the top of the gradient they are pulled downward by gravity until they reach a position in the gradient where the density is equal to their own, or their equilibrium point.[3] DNA containing ^{15}N sinks low; DNA containing only ^{14}N stays near the top. Gravity is slow, however, so to achieve the separation rapidly, the test tube is spun in a centrifuge at high speed, pulling DNA

[3]While density gradients of sugar are often used to separate other cellular components, researchers working with DNA use a compound called cesium chloride (CsCl) instead of sugar. Not only is the range of densities of DNA closer to that of CsCl than that of sugar, but a density gradient of CsCl forms by itself when uniform solutions of it are spun in a high-speed centrifuge.

**Fast Find
6.3f
DNA vs RNA**

down with a force that is over 100,000 times that of gravity alone.

When parental DNA, assembled from nitrogenous bases containing only heavy nitrogen, was separated from cells and spun in a density gradient, all of the DNA reached its equilibrium point near the bottom of the test tube (Figure 6-5). After many generations of growth on only light nitrogen, the bacterial DNA settled at a high point in the density gradient, near the top of the test tube. But after a single generation grown on light nitrogen, the bacteria contained double helixes made up of one heavy

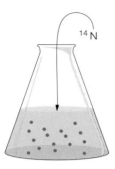

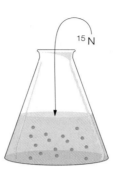

These bacteria were grown for many generations in a medium containing only "light" nitrogen, ^{14}N.

These bacteria were grown for many generations in a medium containing only "heavy" nitrogen, ^{15}N.

These bacteria were grown for many generations in a medium containing only ^{15}N, then switched for a single generation to a medium containing only ^{14}N.

In all three experiments, the DNA was extracted from the bacterial cells and centrifuged in a solution of cesium chloride (CsCl). The CsCl formed a density gradient, heaviest at the bottom of the tube and lightest at the top. DNA localized at a position within the test tube at which the density of CsCl was equal to the density of DNA.

DNA containing only "light" nitrogen formed a band near the top of the tube.

DNA containing only "heavy" nitrogen formed a band near the bottom of the tube.

DNA containing one strand of "light" nitrogen and one strand of "heavy" nitrogen formed a band near the center of the tube. This result of Meselson and Stahl helped prove that DNA replication is semiconservative.

Figure 6-5. Meselson and Stahl's experiment. The results demonstrated that DNA replication is semiconservative.

strand and one light strand. The hybrid heavy–light DNA reached its equilibrium position at a point exactly halfway between the all heavy and all light DNA, in the middle of the tube.

Meselson and Stahl's experiment demonstrated that Watson and Crick, once again relying only on their wits, had correctly guessed the mechanism of DNA replication. DNA replication is semiconservative, with each daughter DNA molecule containing one strand from the parent and one new strand. The process is fairly accurate; only occasionally are nucleotides mismatched in replication. The new strand is almost always a perfect complement of the parent strand that served as its blueprint. There are infrequent errors that slip by the DNA replicating machinery, but elaborate repair mechanisms involving enzymes have evolved to detect and correct errors, so that very few errors are actually passed on to daughter cells. In Section 6.5, we will look at how errors in replication result in mutations that can be passed on, but for now, let's examine another of the roles of the genetic material: the expression of the genetic information. How does the sequence of nucleotides in DNA translate into physical characteristics or traits? What is the language of the genes?

The Second Generation Meselson and Stahl repeated the density gradient experiment after two generations, too. Now that you know that DNA replication is semiconservative, can you predict what they found? Why was it necessary to allow the bacteria to undergo two rounds of division to firmly establish semiconservative replication? What information did the second generation bacteria provide that was not apparent after just one generation?

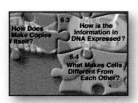

Piecing It Together

The genetic material, by definition, is passed on from generation to generation. This means that each time a cell divides, its entire genetic complement must be fully and faithfully copied. This process is called DNA replication.

1. DNA replication begins when the two strands of the DNA molecule separate, and the exposed bases on each strand bind to complementary nucleotides.

2. Meselson and Stahl proved this by growing bacteria in media containing different isotopes of nitrogen, so that parental DNA contained heavy nitrogen and newly synthesized DNA contained light nitrogen. After one generation, the DNA had one strand containing heavy nitrogen and one containing light nitrogen.

3. This mechanism of replication, in which the two parental strands of DNA separate and each strand acts as a template for the synthesis of its new partner strand, is said to be semiconservative.

6.3 HOW IS THE INFORMATION IN DNA EXPRESSED?

One of the keys to what makes us different—not only different from each other but different from other forms of life—is in the proteins our cells produce. The key to the different proteins expressed by each species is in its **genome**, or all of the nucleotide bases that constitute the entire complement of genes. The genetic information is encoded in DNA.

6.3.1 DNA Encodes the Information to Make Proteins

In 1902, an English physician named Archibald E. Garrod treated an infant for a strange and rare malady. The baby's diapers were stained a dark reddish black. Garrod recognized the condition as a rare disease called alkaptonuria. The urine of alkaptonuria patients contains alkapton bodies, chemicals that turn black on exposure to air.

At that time, Mendelian genetics, which had recently been rediscovered by plant breeders, was working its way into the consciousness of medical doctors. Garrod knew that the baby's parents were first cousins; he wondered if perhaps alkaptonuria could be a recessive genetic disorder, or an inherited disease only apparent in individuals that receive two alleles for the disorder. Because members of the same family are more likely than others to share the same alleles, recessive genetic diseases are much more common in children of closely related parents.

Taking it a step further, Garrod and his colleagues proposed an explanation for the baby's condition, one that was eerily prophetic. They proposed that their infant patient was missing a "certain ferment," in other words, an enzyme protein, that the body normally uses to break down alkapton bodies before they ever enter the urine. If the baby was deficient in such an enzyme, then alkapton bodies would be excreted intact, and the baby's diapers would turn black. The inability to make the enzyme was an inherited trait. For the first time, a connection between genes and proteins was made, at a time when little was known about either.

Garrod and his colleagues were far ahead of their time. The link between genes and proteins was all but forgotten as the interests of geneticists for the next 30 years centered on the patterns of inheritance, and those of biochemists centered on enzymes and how they work. But as more was learned about how genes act, it became obvious that the sequence of nucleotide bases in DNA somehow relates to the types of proteins present in cells. To understand how this works, it is necessary to review some features of proteins. Chapter 4 describes proteins in some detail, and you can review their structure and properties on the CD.

Proteins are polymers, chains of amino acid subunits connected end to end like beads on a string, that fold and twist into characteristic three-dimensional structures. Amino acids differ from one another in the molecular composition of their side groups, called **R groups**. The differences in R groups give amino acids their characteristic properties, and proteins get their characteristic properties from the types and arrangements of their amino acid components. The sequence of amino acids in proteins, also called the **primary structure**, gives each protein its own specific structure and function.

But what determines the primary structure of the proteins? The answer is the genes. The sequence of nucleotides of DNA determines the sequences of amino acids in proteins. The process is indirect; after all, DNA is part of the chromosomes, large and immobile and confined to the nucleus of the cell, whereas proteins are everywhere, within cells and even surrounding them. DNA codes for the construction of proteins through an intermediary, a related polymer of nucleotides called **ribonucleic acid**, or RNA.

6.3.2 RNA Acts as an Intermediary

Chromosomes are gigantic structures made of DNA intertwined with proteins. They sit at the heart of a cell like a colossal set of encyclopedias sitting in a library. The sequences of A, T, C, and G contain the information for building all of the proteins of the cell—indeed for building the entire organism. When you use the encyclopedias in a library, you pull one off the shelf and read it right there, perhaps making a few notes to take with you. To carry the entire set around with you would be unwieldy, not to mention against the rules. So it is with DNA. Their large size makes it impractical to move them to the different places in cells where the proteins they encode are needed. Instead, the bits of DNA that encode a needed protein are copied into a complementary sequence of nucleotides in RNA, and the smaller, much more mobile RNA then travels to the parts of the cell where its sequence can be decoded into protein.

There are two separate processes involved in decoding the DNA. The first process, in which a portion of the DNA is used as a template, or blueprint, to make RNA, is called **transcription**. The second process, in which RNA serves as a template for the sequence of amino acids in a protein, is called **translation**. We will discuss both transcription and translation in the next few pages, but first a few words about RNA.

Differences between DNA and RNA

**Fast Find
6.3f
DNA vs RNA**

The first difference between DNA and RNA is that they contain different sugars. DNA contains a sugar called deoxyribose, whereas RNA contains ribose. Each carbon in the ribose of RNA is chemically bound to an oxygen atom, but in deoxyribose, one of the carbons is missing the oxygen; hence their names, ribonucleic acid and *deoxy*ribonucleic acid. Another difference is in the nitrogenous bases in the two nucleic acids. DNA nucleotides include adenine, thymine, cytosine, and guanine; in RNA, a base called uracil substitutes for thymine. This means that where an adenine is paired with a thymine in DNA, that same adenine is paired with a uracil when DNA is used as a template to make RNA.

These seemingly slight chemical differences make a world of difference in how the two nucleotides behave. DNA is most stable as a double helix, or two twisted chains of nucleotides connected by hydrogen bonds between complementary base pairs. RNA most often exists as a single strand of nucleotides, although it may occasionally form short-lived associations with complementary sequences on other strands. Some RNAs fold back upon themselves and form pairings within the same molecule that give them distinctive shapes. These shapes, as we will see, are important to the function of RNA.

Three additional differences between DNA and RNA are their size, their mobility, and their life span. The DNA in cells exists as a small number of large, immobile molecules that change little during the life of the cell; RNAs are small and highly mobile, traveling easily from nucleus to cytoplasm, where they carry out their functions. They are broken down into their nucleotide components soon after their job is done.

All RNAs arise by means of transcription from DNA in the nucleus, but not all RNAs are decoded into protein. Some are part of the machinery that translates other RNAs into amino acid sequences. We can categorize RNAs into three classes based on their different functions: messenger RNA, transfer RNA, and ribosomal RNA. Each class includes different versions of that type of RNA, as you will see. Only one class, messenger RNA, is a template for protein synthesis.

Messenger RNA

As its name suggests, **messenger RNA**, or **mRNA**, carries genetic information from the DNA in the nucleus into the cytoplasm, where it can be translated into protein. At any given time, there are as many different mRNAs in a cell as there are different proteins being made, although different proteins are made at different times. Consequently, the mRNA population of a cell changes over time. In prokaryotic cells (bacteria), a single mRNA molecule may carry enough information to make

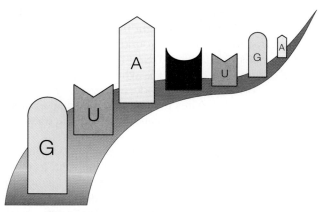

Figure 6-6. Messenger RNA carries the genetic information from the DNA in the nucleus to the cytoplasm, where it is translated into a sequence of amino acids.

several different proteins. In eukaryotes, however, a single mRNA usually codes for a single chain of amino acids. The characteristic shape of mRNA is a long, unfolded chain (Figure 6-6).

More mRNA is transcribed than either of the other two types of RNA, but if you measured the amounts of all three types, mRNA would be the least abundant in the cell because mRNA is the least stable of the three classes of RNA. A typical mRNA molecule

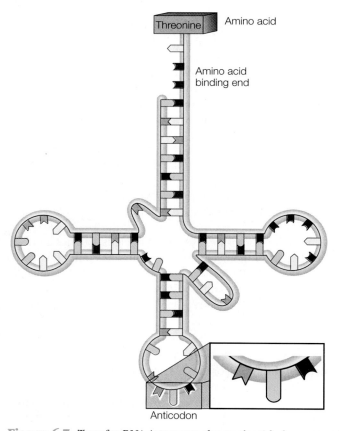

Figure 6-7. Transfer RNA interprets the nucleotide language of mRNA into the amino acid language of protein. Its characteristic t-shaped structure includes both an anticodon region and an amino acid binding region.

lasts on the order of minutes or hours before it is broken down to its nucleotide building blocks. This rapid turnover of mRNA is one way in which the cell regulates how much of any given protein is synthesized.

Transfer RNA

Transfer RNA, or **tRNA**, is the "interpreter" molecule. It brings amino acids to the sites where mRNA is translated into protein. To accomplish this, tRNAs must recognize both the mRNA template and also specific amino acids. For now, we will examine the characteristic structure of tRNA.

Transfer RNAs fold back on themselves to form a distinctive shape that looks almost like a folded lowercase letter t (Figure 6-7). This three-dimensional shape is held together by the tendency for nitrogenous bases within a single tRNA to form complementary pairs: A pairing with U and C pairing with G. In this configuration, there are three unpaired bases at the bottom of the t, called the **anticodon region**, and a loose end at the top. The loose end binds to an amino acid. Each tRNA recognizes and binds to only one of the 20 different kinds of amino acid. This specificity is crucial to the correct interpretation of the genetic message. The anticodon region, with three unpaired nitrogenous bases, associates with three complementary bases on the messenger RNA during translation. In this manner, tRNAs carrying specific amino acids recognize certain parts of the messenger RNA.

Transfer RNAs are relatively long-lived in the cell, lasting from hours to days before they are broken down and their nucleotides recycled.

Ribosomal RNA

Over 80 percent of the RNA in most eukaryotic cells is **ribosomal RNA, or rRNA** (Figure 6-8*a*). Several different ribosomal RNAs, along with many proteins, join together in the cytoplasm to form ribosomes, organelles large enough to be visible in the electron microscope. It is on the ribosomes that translation occurs. Ribosomes are composed of two subunits, a large subunit and a small subunit, each made of rRNA and protein (Figure 6-8*b*). The two subunits come together at the start of translation and separate when the protein is completed.

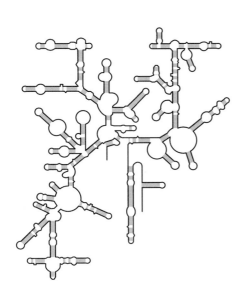

(a) A single rRNA molecule

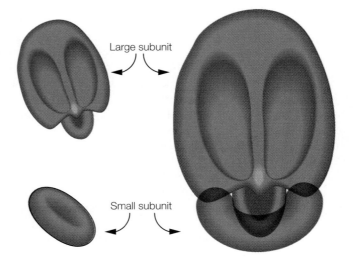

(b) A diagrammatic representation of the two subunits of the ribosome, each made of rRNA and proteins

Figure 6-8. *(a)* Ribosomal RNA has a complex folded structure that aids in its many roles as part of the structure of the ribosome. *(b)* Several rRNAs and many proteins join to make the subunits of the ribosome, shown here in a diagrammatic representation. The subunits of the ribosome join when it is participating in translation.

As shown in Figure 6-8a, ribosomal RNAs, like tRNAs, fold back on themselves in highly ordered, complex patterns, reminiscent of the elaborate higher order structure of proteins described in Chapter 4. Like proteins, rRNAs require these shapes to perform their cellular duties. The job of accelerating chemical reactions in cells is almost exclusively the role of proteins. The only other chemical substance in cells with that ability is RNA; rRNA accelerates some of the chemical reactions of translation.

All three classes of RNA come together in the translation machinery in the cell's cytoplasm. However, the pathway from gene to protein begins not in the cytoplasm, but in the nucleus, with the production of RNA by means of transcription. How are RNAs assembled?

6.3.3 RNA Is Synthesized in Transcription

Fast Find 6.3g Transcription Translation

Imagine you are in the business of making RNA, and you are building an RNA factory. What raw materials are needed for your budding business? The list might look like this: (1) the ribonucleotides A, U, C, and G, which are the building blocks of RNA; (2) a template, or blueprint, of the final product; (3) fuel to drive the assembly line linking ribonucleotides; and finally, (4) equipment to accomplish the actual assembly of the final product. Transcription utilizes the cellular equivalents of all of these. A plentiful supply of nucleotide building blocks (1) is produced by metabolic reactions in the cytoplasm or supplied by the diet. DNA serves as the template (2) that determines the order in which the ribonucleotides are assembled. The ribonucleotides themselves are present in the cytoplasm as nucleotide triphosphates, a form that can supply all of the necessary energy (3) to drive the synthetic process. (We will learn more about how nucleotide triphosphates act as a source of cellular fuel in Chapter 9.) Finally, the equipment (4) that makes it all happen is an enzyme called **RNA polymerase**. Now, where should the assembly begin?

The DNA template is large, and only short stretches are transcribed at any one time. Though there are no obvious starting points, it is nonetheless critical that transcription begin and end at specific places on the DNA. We need to know where on the DNA to attach our nucleotide-linking RNA polymerase so that we begin at exactly the right point. Certain sequences within the DNA, called **promoter sequences**, signal the RNA polymerase to attach to the template at that point, and to begin transcribing some predetermined number of nucleotides down the line. Not all promoters are created equal; strong promoters are better at attracting RNA polymerase than weak promoters. Genes preceded by strong promoters are transcribed often; the proteins they encode are required by the cell in large quantities. Other proteins are not in such high demand. Consequently, their genes may have weaker promoters.

Now our machinery is in order and ready to roll. The assembly equipment, the RNA polymerase, is situated at an appropriate starting point and it is time to begin linking together ribonucleotides to make RNA. Beginning in the small region of the DNA double helix where the enzyme sits, the two strands of DNA separate, thus leaving some nitrogenous bases unpaired. The exposed bases on the DNA attract complementary ribonucleotide partners. Exposed adenine bases on the DNA pair with uracil ribonucleotides; exposed thymines on DNA pair with adenine ribonucleotides. Likewise with cytosine and guanine on DNA, which attract guanine and cytosine ribonucleotides. When the ribonucleotides are in position on the template, hydrogen-bonded to their complementary mates, the RNA polymerase chops off the two extra phosphates of the first ribonucleotide, absorbing the energy provided by the broken chemical bond and using it to link the two ribonucleotides together. Without letting go, the polymerase moves down a step and connects the next ribonucleotide. The process is repeated until enough DNA has been transcribed to make an appropriately functional RNA (Figure 6-9). How does the RNA

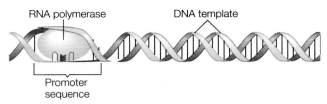

RNA polymerase — DNA template

Promoter sequence

❶ The RNA polymerase binds to the DNA at specific promoter sequences. The two strands of DNA are separated at a site near the promoter.

Growing RNA

❷ Complementary RNA nucleotides bind to only one of the exposed DNA strands. RNA polymerase joins the RNA nucleotides together to form a chain.

❸ The RNA polymerase moves along the DNA, separating the strands and joining matched RNA nucleotides as it goes.

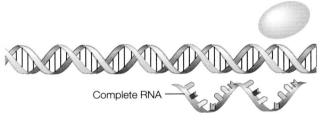

Complete RNA

❹ At the end of the coding sequence, the RNA polymerase and the newly synthesized RNA leave the DNA. The double helix closes again.

Figure 6-9. In transcription, the sequence of nucleotides in DNA is used to construct a molecule of RNA.

polymerase know where to stop? The answer is not entirely clear, but scientists know that certain nucleotide sequences can change the shape of the newly synthesized RNA, or the way it folds and bends. Such shape changes may be the signal to stop transcription.

Regardless of the exact nature of our finished product, mRNA, rRNA, or tRNA, the next step in the flow of information from DNA to protein takes us from the nucleus of the cell to the cytoplasm. Here, the sequence of ribonucleotides in mRNA is decoded in the process known as translation.

6.3.4 Proteins Are Synthesized in Translation

In transcription, information was converted from one nucleic acid to another, from DNA to RNA. The language of the two, while not identical, is quite similar. Consequently, the transcription machinery is relatively simple: only one kind of enzyme, some energy, and some raw materials. Accessory proteins (not described above) help to fine-tune the process. The next step, in which proteins are assembled from an mRNA template, is more complex. This complexity is reflected in the far more elaborate machinery of translation.

What materials and equipment are required for translation? As before, we need raw materials (this time in the form of amino acids), energy to drive the synthesis, a template to determine the amino acid sequence, and the machinery to make it all happen. In translation, we are changing molecular languages, from a nucleotide sequence to an amino

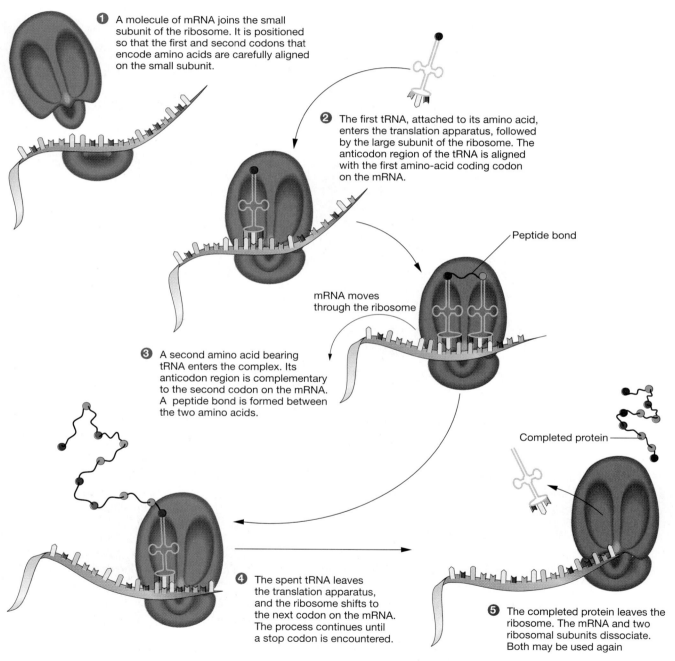

❶ A molecule of mRNA joins the small subunit of the ribosome. It is positioned so that the first and second codons that encode amino acids are carefully aligned on the small subunit.

❷ The first tRNA, attached to its amino acid, enters the translation apparatus, followed by the large subunit of the ribosome. The anticodon region of the tRNA is aligned with the first amino-acid coding codon on the mRNA.

Peptide bond

mRNA moves through the ribosome

❸ A second amino acid bearing tRNA enters the complex. Its anticodon region is complementary to the second codon on the mRNA. A peptide bond is formed between the two amino acids.

Completed protein

❹ The spent tRNA leaves the translation apparatus, and the ribosome shifts to the next codon on the mRNA. The process continues until a stop codon is encountered.

❺ The completed protein leaves the ribosome. The mRNA and two ribosomal subunits dissociate. Both may be used again

Figure 6-10. Translation converts the mRNA nucleotide sequence into an amino acid sequence. The linear mRNA is held in place on the ribosomal subunits, while complementary tRNAs are brought in. There are two tRNA binding sites on the ribosome. When both are occupied, the rRNA catalyzes the formation of a peptide bond between the amino acids on the tRNAs. The process continues until a stop codon is reached on mRNA, at which point the complex comes apart.

acid sequence. This means we need a reliable interpreter, a molecule that can read nucleotides and decode them into amino acids. This is the job of the various tRNAs. In addition, we need a stable platform on which to translate—a ribosome—to anchor all of the participants in just the right positions.

The first step in translation is to assemble all of the participants in exactly the right places (Figure 6-10). To begin, a long strand of mRNA joins with the small subunit of the ribosome.

For reasons that will become obvious in the next section, it helps to think of the ribonucleotides in mRNA as sets of three, with each threesome called a **codon**. Recall from our discussion above that tRNAs contain an **anticodon** region of three unpaired ribonucleotide bases at the bottom of the t shape. During translation, each codon pairs with a complementary anticodon on a tRNA.

At this point, a tRNA connected at its loose end to its particular amino acid joins the growing translation apparatus. This position can be filled only by those specific tRNAs with anticodon regions complementary to the codon on the mRNA. The tRNA with its attached amino acid is bound to the mRNA by the same weak hydrogen bonds that link the two strands of DNA in the nucleus: C with G and, in the case of the two RNAs, A with U (Figure 6-11). You can visualize this process on the CD.

Next, the large subunit of the ribosome enters, simultaneously hugging the smaller subunit and the tRNA bound to the mRNA. As it does so, it creates a pocket surrounding the next codon sequence on the mRNA. In comes a second tRNA, also carrying its appropriate amino acid on its loose end. This new tRNA fits into the pocket and aligns its anticodon end with the next three ribonucleotides on the mRNA (Figure 6-10). Not just any tRNA can slide into the pocket; only one with a complementary anticodon will fit.

At this point, the translation machinery is ready to begin making protein. Let's review what we have. The translation complex consists of the two-subunit ribosome, two tRNAs, each carrying its specific amino acid, and a strand of mRNA, all carefully positioned. The next step is carried out by one of the rRNAs that is part of the large ribosomal subunit—an rRNA with the ability to catalyze a chemical reaction.

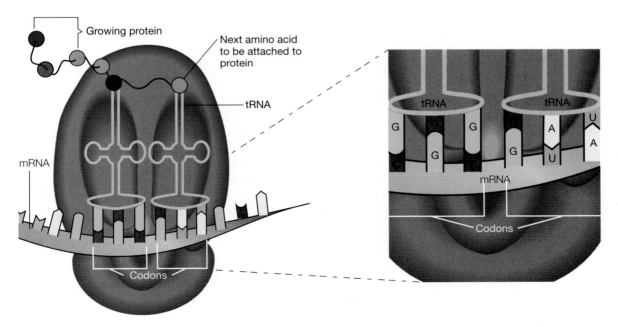

Figure 6-11. Transfer RNAs carrying amino acids are held in place during translation by hydrogen bonds between their anticodon regions and the codons on mRNA.

The amino acid on the first tRNA is released from its tRNA carrier and is joined to the amino acid on the second tRNA by a peptide bond. The result is the beginning of a protein, albeit a small, two–amino acid chain (small proteins are often called peptides). At this point, the peptide is still attached to the second tRNA, which is still associated with the mRNA at the codon adjacent to the codon where translation began. As the peptide grows by the addition of new amino acids, it remains attached to the newest tRNA in the complex until the amino acid chain is released at the end of translation.

After a tRNA gives up its amino acid to the growing protein, it separates from the mRNA and moves back into the cytoplasm, where it can pick up another amino acid on its loose end. Meanwhile, the ribosome shifts position on the mRNA exactly three ribonucleotides, or one codon, downstream. Figure 6-11 illustrates the two binding sites on the ribosome for tRNAs. With each shift of the ribosome, a spent tRNA is ejected and a position for a new, amino acid–charged tRNA is opened. As the position is filled with the correct tRNA, a new peptide bond is formed, elongating the growing protein by one amino acid. This process is translation in action.

Eventually, the ribosome encounters a codon on the mRNA for which there are no tRNAs with complementary anticodons. This codon is the signal to stop. Termination is complete when the translation apparatus, with the help of accessory proteins called release factors, dissociates into its original parts, and the new protein is released to take up its cellular role.

On the CD, you can see a few amino acids of hemoglobin, the oxygen-carrying protein of the blood, assembled in translation. Human hemoglobin consists of four separate chains of amino acids. Two of the chains each have 141 amino acids, and the other two each have 146 amino acids. Needless to say, seeing all four chains translated, with their total of 574 amino acids, would be tedious. Watch the synthesis of just the first six amino acids of one chain to learn the principles of translation.

The language of the genes is written in sequences of nitrogenous bases with a four-letter genetic alphabet, A, C, T, and G. The language of proteins is written in the sequences of the 20 different amino acids. Transcription and translation interpret the genetic letters into amino acid words, each word spelled with only three nucleotide letters. But how are the words spelled? What codons correspond to which amino acids? What is the genetic code?

6.3.5 Three RNA Nucleotides Code for One Amino Acid

How many different one-letter words can be made using only four different letters? Clearly the answer is only four. Using the four letters two at a time, you can make 4 x 4, or 16, different two-letter words. This is still not enough for each of the 20 amino acids to have a unique nucleotide code. But if the four nucleotide letters of DNA are taken three at a time, there are 4 x 4 x 4, or 64, different combinations. Even before the mechanisms of transcription and translation were understood, this kind of reasoning was used to propose that each amino acid in a protein must be encoded by at least three sequential nucleotides in a gene: a three-nucleotide genetic code. But it wasn't until Marshall Nirenberg and Heinrich Matthaei developed a technique for cracking the code that any of the codons could be assigned to specific amino acids. That happened in the early 1960s.

Nirenberg and Matthaei's big breakthrough was the ability to synthesize an artificial genetic message—a strand of mRNA with exactly the sequence of ribonucleotides that they specified. They began with a simple sequence: a long strand of mRNA made of nothing but uracil. When this artificial message was mixed in a test tube with all of the other ingredients necessary for translation, the result was a very simple protein: a long chain made up of molecules of the amino acid phenylalanine. Nirenberg and Matthaei correctly concluded that the triplet codon UUU on mRNA corresponds to the amino acid phenylalanine. Over the next four years, Nirenberg and Matthaei and many other researchers who had joined the effort used this technique to crack the entire genetic code. By 1966, the correlation between every codon and every amino acid was known. The genetic code, shown in Figure 6-12, is universal. It applies to humans and to all other living things. The universality of the genetic code is one very powerful piece of evidence that all organisms on Earth share a common evolutionary ancestry.

Look again at Figure 6-12. Do any features of this code strike you as unusual? First, as you might have guessed, most of the amino acids have at least two, and many have four, triplets that code for them. After all, there are 64 different three-letter codons and only 20 different amino acids. Furthermore, there is an obvious similarity in the different codons that specify the same amino acid. Often, within a set of triplets that codes for a single amino acid, codons differ from one another only in the last nucleotide. This feature has consequences for the integrity of the genetic message. Mistakes that occur during DNA replication sometimes involve mismatched nucleotide bases (as discussed in Section 6.5.2). If a mistake occurs involving the last nucleotide of a codon, there is a good chance the codon will still encode the same amino acid so that the

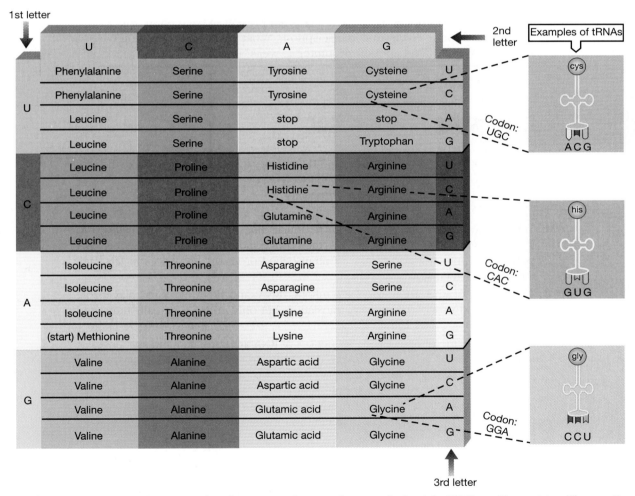

Figure 6-12. The genetic code. To use this chart to translate the codon UGC, for example, find the first letter (U) in the column on the far left. Follow the U row toward the right until you reach the column headed by the second letter (G) at the top. Then find the amino acid that matches the third letter (C) in the column on the far right. UGC specifies cysteine. The genetic code is usually written in the language of mRNA and not DNA or tRNA. The triplets correspond to codons that are found on mRNA. Some examples of complementary tRNAs are shown at the far right.

error would have no consequence for the cell. Also, mistakes that occur when tRNAs are fitted on the ribosome during translation usually involve mismatches at the last position of the codon–anticodon pairing. Many such mistakes will have no effect on the amino acid that is incorporated into the growing protein. The redundant aspect of the genetic code acts as a "safeguard" against errors that can occur during DNA replication and also during translation.

Like the genetic code, the processes we have discussed in this section are true of cells in general, not just one kind of cell. The greatest differences in transcription and translation are those between prokaryotic and eukaryotic cells, but the basic processes achieve the same goals in generally the same way. These processes alone, however, say little about how a cell knows which genes to transcribe and translate into protein. In the next section, we will look at ways in which gene expression (transcription and translation) is regulated.

Cookbook Expression In recent decades, scientists have developed quick and easy methods of achieving both transcription and translation in a test tube, in the absence of intact cells. What ingredients are required to reconstitute these complex pathways? How do reconstituted prokaryotic systems differ from reconstituted eukaryotic systems? What other accessory factors, besides those mentioned in the text, are often present in bacterial transcription? What about eukaryotic transcription? What makes one promoter a "strong" promoter and another promoter a "weak" promoter?

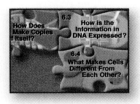

Piecing It Together

Proteins make up the organism, and DNA carries the information to make the proteins. The process is an indirect one involving two steps, transcription and translation, connected by an RNA intermediary.

1. Transcription is the process in which ribonucleotide triphosphates are matched to their complementary partners on DNA, then connected by means of an enzyme, RNA polymerase, into strands of RNA. Nontranscribed promoter sequences in DNA tell the RNA polymerase where to begin.

2. Three types of RNA are transcribed: mRNA, tRNA, and rRNA. All three play a role in the second step of the process, translation, but only mRNA is actually translated into protein.

3. In translation, the ribonucleotide sequences of RNA are translated into sequences of amino acids in proteins. The process occurs on structures in the cytoplasm called ribosomes. A strand of mRNA is held in place on the ribosome while two molecules of tRNA, each carrying a specific amino acid, are brought in. Three bases on each tRNA, called the anticodon, are complementary to a three-base sequence on the mRNA called the codon. In this manner, only certain tRNAs, carrying their specific amino acids, align on the mRNA.

4. A peptide bond is made when the two amino acids on adjacent tRNAs on the ribosome are joined. One tRNA exits the ribosome and another is brought in. With each new addition, the ribosome moves down the mRNA by three nucleotides, adding amino acids to the growing peptide chain until it reaches a stop codon, when it disassociates and releases the completed protein.

5. The genetic code is the set of three-ribonucleotide sequences (codons) that specify the 20 different amino acids, including sequences that specify where to begin

translation and where to end it. Most amino acids have two or more codons that code for their insertion into protein, a feature that protects the integrity of the genetic code.

6.4 WHAT MAKES CELLS DIFFERENT FROM EACH OTHER?

During a typical person's lifetime, he or she may manufacture as many as 100,000 different proteins. But at any given time, only about 5,000 different proteins can be found in any one cell. Some are made all the time and by nearly all the cells of the body. The enzymes that supply the body with energy are constantly being transcribed and translated. Other proteins are found in only a single cell type. For example, cells destined to become red blood cells are the only ones that make hemoglobin. Indeed, hemoglobin makes up over 95 percent of the protein produced by these cells. How do these cells know to make hemoglobin and not the many other proteins encoded in their genes? What accounts for differences in gene expression?

6.4.1 Prokaryotes Regulate Genetic Expression Mostly at Transcription

For several years after DNA had been established as the genetic material, little was known about the mechanisms by which genes are turned "on" and "off." Then, in the late 1940s, two researchers at the Pasteur Institute in Paris, François Jacob and Jacques Monod, made an important breakthrough. The two were studying bacterial cells of the species *Escherichia coli* when they made the following remarkable discovery.

An *E. coli* cell living in a laboratory flask uses the simple sugars provided by the rich medium in which it lives for energy. If its main food, glucose, is in limited supply, then before long the medium becomes depleted of glucose. Another sugar, lactose, may still be plentiful, but at first the bacterial cell lacks the proper enzymes to use lactose for food. Jacob and Monod discovered that within about an hour the problem is solved. The bacterial cell makes all of the enzymes required to digest lactose, and then uses lactose in place of glucose.

The genes that encode enzymes for digesting lactose are highly regulated in bacteria. They are transcribed and translated only if lactose is the sole source of food. In other words, the presence of lactose, and the absence of glucose, induces the synthesis of the lactose-digesting enzymes. Jacob and Monod had discovered the first good experimental system for learning about how genes are regulated. They set about using a combination of bacterial genetics and biochemistry to uncover the basic cellular mechanisms that regulate gene expression in these cells (Figure 6-13). Their experiments are now considered classics in the field of genetic regulation, and Jacob and Monod were awarded a Nobel Prize in 1965 for their groundbreaking discoveries.

Jacob and Monod found that the DNA of *E. coli* includes not only the genes for enzymes involved in lactose digestion, but also other DNA sequences that play a critical role in regulating when and if the protein-coding genes are expressed. The entire suite of genes constitutes a regulatory unit of DNA called an operon. An **operon** is a cluster of genes, including protein-coding genes and all of the regulatory DNA involved in their expression. The operon involved in lactose digestion was named the ***lac* operon**. It is an example of an inducible operon because the presence of a key substance in the environment causes otherwise quiescent genes to be transcribed and translated. Explore the *BioInquiry* web site to find out how the *lac* operon works. By using environmental cues to determine which genes are transcribed, the bacterial cell can tailor its cellular

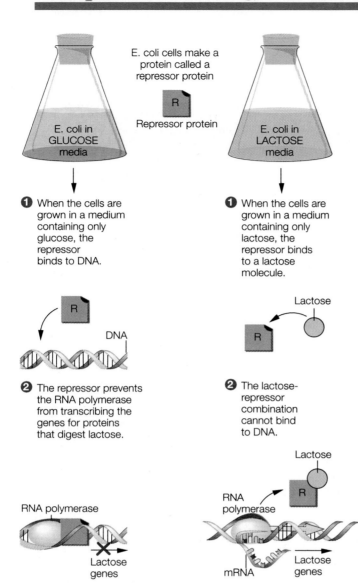

E. coli cells make a protein called a repressor protein

R

Repressor protein

E. coli in GLUCOSE media

E. coli in LACTOSE media

1 When the cells are grown in a medium containing only glucose, the repressor binds to DNA.

1 When the cells are grown in a medium containing only lactose, the repressor binds to a lactose molecule.

R

DNA

Lactose

R

2 The repressor prevents the RNA polymerase from transcribing the genes for proteins that digest lactose.

2 The lactose-repressor combination cannot bind to DNA.

RNA polymerase

Lactose genes

3 Transcription of the lactose genes is blocked by the repressor.

Lactose

R

RNA polymerase

mRNA

Lactose genes

3 The lactose genes are transcribed by the RNA polymerase.

Figure 6-13. Regulation of gene expression by means of operons. Jacob and Monod found that genes for proteins that digest lactose are only expressed when lactose is the only sugar present in the growth medium. Without lactose, the genes cannot be transcribed because a repressor protein blocks the RNA polymerase. With lactose, the repressor binds to a lactose molecule and cannot bind to DNA. The RNA polymerase can then transcribe the genes.

proteins to the prevailing conditions. This efficient mechanism saves energy and precious raw materials that can be devoted to other bacterial tasks, for example, making more bacteria.

Operons were soon shown to be widespread among the prokaryotes. Genetic regulation in eukaryotes, however, is far more complex, and a complete understanding of eukaryotic gene regulation is still a distant goal. Nonetheless, a few generalizations can lend insight to the main themes of eukaryotic genetic regulation.

6.4.2 Eukaryotes Regulate Genetic Expression at Many Different Levels

In prokaryotes, such as *E. coli*, the most important step in the regulation of gene expression is transcription. The same is true in eukaryotic cells, but that is where the similarity ends. In the past decade, a large number of regulatory proteins, called **transcription factors**, have been identified in eukaryotic cells. (A few transcription factors have also been found in prokaryotes.) Transcription factors recognize and bind to specific sequences in DNA called **regulatory sequences**. These sequences lie outside of the protein-coding regions. The genes of eukaryotic cells have many such sequences. Transcription factors act by increasing or decreasing the rate at which specific protein-coding genes are transcribed. With few exceptions, the exact manner by which they accomplish this task is poorly understood.

Eukaryotes exert control over the kinds of proteins that appear in cells at points other than transcription. For example, not all eukaryotic messenger RNAs are translated equally. Some mRNAs may be translated many times, so that each copy of a particular mRNA can give rise to many copies of the protein it encodes. Other mRNAs may be translated only a few times before they are degraded into their ribonucleotide building blocks. The key to the longevity, and hence productivity, of an mRNA lies in the sequences of ribonucleotides at its very end. Messenger RNAs with long chains of adenine bases at the end have longer life spans than those with just a few or none at all. But there is still more to learn about the mechanisms controlling the number of adenines at the ends of mRNA molecules.

Our picture of how eukaryotic cells regulate gene expression still has gaps. The control of embryonic development from fertilized egg to complex, integrated multicellular organism involves regulated gene expression in many different cell types. It remains, for the most part, a mystery. These are surely the questions that will generate some of the most exciting research efforts of the next several decades.

Genes Behind the Scenes Aside from the protein-coding genes, what other kinds of DNA sequences are found in the *lac* operon? How do these sequences regulate the expression of the protein-coding genes?

The *lac* operon is an inducible operon. What is a repressible operon? Can you give an example of a repressible operon?

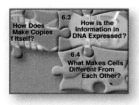

Piecing It Together

Two cells that share the same genetic information may express such diverse collections of proteins that their differences appear to outweigh their similarities. Even the same cell can change the proteins it makes from one moment to the next. These processes are only partially understood, particularly in eukaryotic cells. Here is some of what is known:

1. Some genes are always transcribed and translated; others can be turned on and off by signals from the environment. Gene expression is highly regulated.

2. In prokaryotes, much gene regulation occurs by means of operons, which control whether or not a DNA sequence (gene) is transcribed. An operon is a group of genes that includes one or more protein-encoding genes and several regulatory sequences.

3. Eukaryotes do not have operons, but their genes are nonetheless highly regulated. Transcription factors are regulatory proteins that have been identified in both prokaryotic and eukaryotic cells, but they are most widespread in eukaryotes. They recognize and bind to regulatory sequences, regions of DNA that lie outside of the protein-encoding sequences.

6.5 HOW DOES DNA CHANGE OVER TIME?

DNA makes copies of itself. DNA encodes information. DNA controls cells and tells them what to do. DNA changes by mutation. These are the four roles of the genetic material—the starting point for Watson and Crick and all who have searched for the secrets of heredity since them. Some of the most remarkable advances in the past 50 years of science have come from our growing understanding of these four phenomena. We have discussed in this chapter how DNA accomplishes the first three of these tasks, but it is the fourth task, the accumulation of alterations in the genetic material, that brings our story back to its beginnings. Without change, there can be no evolution. And without evolution, there would be no life.

We have mentioned *mutation* several times, but only in the vaguest terms. In Chapter 3, for example, we defined mutation as the sudden appearance of a new allele. Now, armed with a new understanding of DNA and how it works, we are ready for a new look at mutations. What are mutations in terms of chromosomes, genes, and DNA? How do they happen? We will see that there are, in fact, several kinds. Some involve a single nitrogenous base in DNA; others involve whole chromosomes, even entire sets of chromosomes. We shall begin with the macromutations, or those that involve whole chromosomes or chromosome sets, and work our way down to micromutations, or those that occur in one or only a few DNA nucleotides.

6.5.1 Some Mutations Involve Whole Chromosomes

Polyploidy and Aneuploidy

A pollen-producing cell on the stamen of a flower is ready to undergo meiosis, that is, to divide so that the number of chromosomes in each daughter cell is one half that of the meiotic mother cell (Chapter 5). But something goes wrong. At the end of the first meiotic division, instead of separating, the homologous pairs of chromosomes stay together. One gamete gets all of the chromosomes, and the other gets none and dies. It is a rare event, but it happens, and in the tens of millions of years that flowering plants have been around, even this rare event has happened many times. The outcome of events like this depend on the fate of the chromosome-heavy pollen grain. If it finds its way to a plant ovary and fertilizes an ovum in which the same failure of meiotic separation has occurred, the result will be the seed of a new plant with *double* the usual number of chromosomes, that is, four sets of haploid chromosomes, or two complete sets of homologous pairs.

The plant in the scenario above is said to be **polyploid**—it has more than two haploid sets of chromosomes.[4] Polyploidy arises as a genetic accident, but the consequences for the polyploid individual, especially among plants, can be quite advantageous. Polyploid plants are generally bigger, more hardy, and produce more seeds than their diploid relatives. Natural selection favors these traits; so much so that it is estimated that nearly 50 percent of flowering plants are polyploids, including many of commercial value. For example, wheat has six haploid sets of chromosomes; strawberries have eight. While polyploidy is most common among plants, there are a few animal species that are polyploids, including some amphibians, fishes, a few species of

---

[4]Because polyploidy and other changes in chromosome number do not result in the formation of new alleles, just more alleles packaged in the same organism, one could argue that these are not mutations in the strictest sense, but a form of genetic recombination. However, it seems appropriate to discuss these large chromosomal changes here because they are most definitely alterations in the sum total of genetic material.

beetles, and earthworms. We do not know how many times gametes with extra sets of chromosomes have been produced and failed to survive—probably many times. However, when a polyploid gamete has successfully fertilized another polyploid gamete and given rise to viable offspring, the result has been a new species. Polyploidy is one way in which new species can arise.

Some changes in chromosome number involve a single chromosome or single homologous pair, either a pair that becomes a threesome or a pair that loses one member. The mechanism is similar to that described for polyploidy. Sometimes during meiosis, members of a homologous pair of chromosomes fail to separate from one another, a phenomenon called **nondisjunction**. One gamete ends up with both members of a pair, and the other gets neither member. If a gamete that results from nondisjunction joins another gamete in fertilization, the result is an **aneuploid**, or an individual whose chromosome number is either greater or less than normal. Most human aneuploidies are lethal before birth and result in spontaneous abortion. There are a few aneuploidies, however, that come to full term. Most cases of Down syndrome, for example, result from a condition in which the individual has an extra copy of chromosome number 21 (Figure 6-14). This type of nondisjunction occurs about once in every 750 live births. Down syndrome children are characterized by mental retardation, a distinctive facial appearance, and an outgoing disposition. Other human aneuploidies, especially involving the sex chromosomes X and Y, are responsible for a substantial proportion of congenital abnormalities.

Transposable Genetic Elements

DNA molecules consist of a pair of stable chains, each comprising deoxyribonucleotides linked end to end. While we have described how the two halves of the double helix separate during DNA replication and transcription, it has been useful to think of each half of the DNA duplex as remaining intact from generation to generation, connected by strong covalent bonds between the phosphates and sugars that form the backbone of each chain. This is how DNA was viewed by most molecular biologists working in the 1950s and early 1960s. Now, thanks to the work of Barbara McClintock, we also know of important exceptions.

In the late 1940s, McClintock found that certain mutations in the genes of corn, or maize, involve DNA sequences that move from place to place, either on the same chromosome or on a different chromosome altogether. She called this genetic rearrangement **transposition**, and the bits of DNA that were on the move she referred to as **transposable genetic elements**, later called **transposons**. In Chapter 5 (Section 5.2.3), we saw that transposons can break away from their starting position in the DNA and become spliced into a new position on the DNA, called the *target* DNA. Although the sequences of transposon DNA are not random (only certain DNA sequences can move around), the target sites are thought to be random; a transposon can land anywhere. Transposition is one way in which new combinations of genes are created. Here we see that transposition can introduce errors in the genetic material, as well.

Imagine that you are writing an essay using your word processor. Now imagine that you have chosen a few phrases or sentences and cut them from your text with the cut-and-paste function of your computer program. If you insert these phrases randomly into the essay, the result might be a silly, but meaningless, sentence. Alternatively, the phrases could land within key sentences and undermine the meaning of your essay. Similarly, if a transposon just happens to land in an important part of the DNA, say, in the middle of a gene, the result could be harmful. It is estimated that about one in every 500 mutations in humans results from the insertion of a transposon into a gene or into the DNA sequences that control the expression of a gene. Several forms of hemophilia, the human genetic disease in which afflicted people lack one

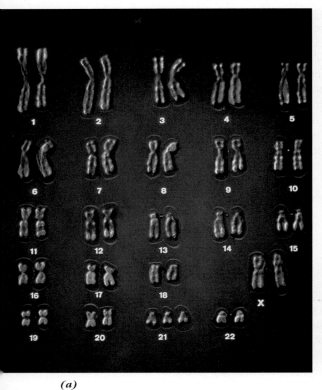

(a)

(b)

Figure 6-14. *(a)* Chromosomes of a child with Down syndrome. Note that there are three copies of chromosome number 21. *(b)* Children with Down syndrome have distinctive facial features and pleasant dispositions.

or more proteins involved in blood clotting, are due to transposons inserted into genes encoding blood-clotting proteins.

 Inside-out Genes Other kinds of chromosomal mutations are **deletions**, which occur when parts of a chromosome are spontaneously deleted, or **inversions**, which occur when a piece of a chromosome is broken and then reincorporated into a chromosome in a reversed order. What are some examples of deletions and inversions that have been documented in humans? What genetic abnormalities have been identified as arising from these kinds of mutations?

6.5.2 Some Mutations Involve Single DNA Bases

Thus far we have considered mutations that involve large chunks of DNA—pieces of chromosomes, whole chromosomes, and even entire sets of chromosomes. But if we look very closely, we see that the genetic material can change at the level of one or just a few deoxyribonucleotide bases. Such micromutations are termed

point mutations. An excellent example of a point mutation occurs in sickle cell anemia. We saw in Chapter 3 that sickle cell anemia is an inherited disease that results when a person carries two copies of a mutant allele for the oxygen-carrying protein hemoglobin. How is the gene for hemoglobin mutated in people with sickle cell anemia?

Hemoglobin is a protein composed of not one, but four intertwined chains of amino acids. The four chains occur as pairs: two identical **alpha** (α) **chains**, and two identical **beta** (β) **chains**. The alpha chains of normal and sickle cell hemoglobin are identical. But the amino acid sequence of the beta chains differs between normal and sickle cell hemoglobin by a single amino acid. The amino acid glutamic acid that occurs at the sixth position in a normal beta chain is replaced with a valine in the sickle cell beta chain. This tiny difference can be sufficient to condemn its bearer to a lifelong affliction with sickle cell disease.

The beta chain of hemoglobin is encoded by a gene called the human β-globin gene (*HbA*). Using the genetic code shown in Figure 6-12, we can compare the codon for glutamic acid with that for valine. The two codons that instruct the ribosome to insert a glutamic acid into protein are GAA and GAG. (Recall that the genetic code is written in terms of mRNA, not the DNA of the actual gene. These two mRNA codons are complementary to the DNA sequences CTT and CTC.) Two of the four codons for valine are GUA and GUG.

Consider what would happen if, at some time in the course of human evolution, an error occurred in DNA replication such that the middle base of the triplet encoding the sixth amino acid in the β-globin gene of one individual was accidentally changed from T to A (Figure 6-15). A new allele would emerge: the *HbS* allele. A molecule of β-hemoglobin transcribed and translated from this allele would carry the erroneous valine at position six in place of glutamic acid. Because the error occurred in only one of the two alleles for the β-globin gene, there would be no immediate consequence. Over the course of many generations, the allele could spread throughout the population unhindered. Its consequences are revealed only when two carriers each donate an *HbS* allele to an offspring.

The substitution of an A for a T in the β-globin gene is an example of a point mutation that results in the exchange of one amino acid for another in the final gene product. We can envision several other kinds of point mutations, each with different consequences. Imagine what would happen if a point mutation involved the deletion or insertion of a single deoxyribonucleotide. During translation, genes are read almost like sentences of three-letter words, each set of three ribonucleotides constituting a single word. The addition or deletion of a single "letter" in one word would shift the reading frame of the entire sequence. The following sentence makes a good analogy:

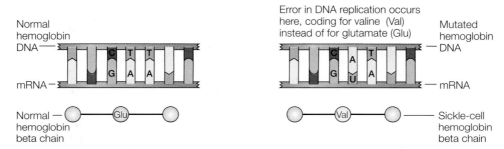

Figure 6-15. The point mutation in sickle cell anemia. A mutation involving a single nucleotide base at a critical position in DNA results in the substitution of an A for a T. When this DNA is transcribed and translated, the final protein, the beta chain of hemoglobin, has a valine where a glutamic acid occurs in the normal beta chain of hemoglobin.

J O E   A T E   H O T   D O G

Even without punctuation, you can make sense of this simple sentence. Consider a point substitution, as we saw in the case of the β-globin gene:

J O E   T T E   H O T   D O G

The sentence loses some of its meaning, but you can still discern at least part of its meaning. Now look at the consequences of a deletion:

J E A   T E H   O T D   O G

The combinations of three letters that compose the words in this sentence are frame shifted. Consequently, none of the words following the mutation are recognizable. Such mutations, called **frame-shift mutations**, are especially deleterious.

Try constructing a sentence in which a letter is added at a random place. Addition mutations are deleterious in the same way that deletions are: they shift the reading frame of the message.

Neutral Mutations

Most of the examples of mutations that we have seen so far are harmful. An individual who inherits these mutations is worse off than one who does not. That is because the DNA suffers damage at some critical place, either where a gene is encoded or where the expression of a gene is regulated. There are, however, many mutations that have little or no impact on their recipients. For example, eukaryotic cells have long stretches of DNA that are noncoding and whose functions are not fully understood. It is estimated that only about 5 percent of the DNA in human cells actually codes for protein. We know that some noncoding regions are regulatory sequences that control which genes are transcribed and translated; other regions are transcribed into the nontranslated RNAs, tRNA and rRNA; still others appear to travel through the generations as hitchhikers, or sequences that are replicated and passed on from generation to generation with no known function. Most mutations in these sequences have no known effect on their recipients and are said to be **neutral mutations**.

Throughout the course of evolution, many of the mutations that have accumulated in populations are ones that have produced individuals that are better suited for survival, as was the case with the polyploid flowering plants. These are the mutations that have played significant roles in evolution. Any change, even a small one, that confers some survival advantage on its recipient is likely to be preserved and spread throughout the species. The study of these mutations and the way that they move between and among groups of interbreeding individuals constitutes the topic of the next chapter. In Chapter 7, we will leave the hereditary molecule per se and watch as genes travel through populations and time. This is the topic of **population genetics**.

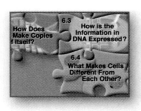

Piecing It Together

The history of life on Earth is a chronicle of changes in DNA, or mutations, passed from one generation to the next. As scientists learn more about the structure and function of genes, the definition of mutation changes to accommodate this ever-expanding knowledge. There are several ways in which mutations might arise:

1. Polyploidy and aneuploidy are conditions in which organisms have inherited a different number of chromosomes than their parents had. Both conditions result from errors during meiosis in which members of homologous pairs of chromosomes failed to segregate into separate gametes. Such mutations may involve one or a few chromosomes, as in aneuploidy, or an entire set of chromosomes, resulting in polyploidy. An error of this type is the result of nondisjunction.

2. Transposable genetic elements, or transposons, are pieces of DNA that move from place to place within the genome. While only certain DNA sequences have the ability to move, any part of the DNA can act as an insertion point. If transposons are inserted into genes or the sequences regulating genes, they can alter the genetic message. Transposons can thus cause mutations.

3. Point mutations occur at the level of one or a few nucleotides. Substitution mutations arise when one deoxyribonucleotide base is accidentally substituted for another, usually during DNA replication. If the substitution falls within the protein-encoding sequences of a gene, the result may be the incorporation of a different amino acid into the protein it encodes. Addition and deletion point mutations shift the reading frame of the genetic material. The effects of additions and deletions are almost always harmful.

Where Are We Now?

Remarkable advances in the ability to manipulate genes and DNA in the laboratory have caused nothing short of a revolution in the life sciences. All areas of biology have been affected, including basic research aimed at understanding fundamental cellular processes, medicine, agriculture, and environmental science. DNA technology has made its way into the courtroom, where it is having a profound influence on criminal cases. Imagine the power of tools that make it possible to identify individuals with utter confidence based on a single hair or a speck of blood. New techniques enable biologists to identify the genes for specific proteins, cut or copy those genes from their source, insert them into another organism (even another species), and control their expression in their new host. DNA technology has provided all this and more.

The power of DNA technology comes from understanding a few things about genes—all described in this chapter—and the laboratory techniques for manipulating them. Here we will look at some of those techniques and how they are being used in DNA technology today.

Molecular Scissors and Molecular Paste

Restriction enzymes, also called **restriction endonucleases**, are proteins made by bacteria as part of their natural defense mechanisms. These enzymes evolved as a mechanism for ridding the bacteria of unwanted DNA, perhaps from viruses or other invaders. The enzymes cut unwanted DNA into fragments that can be digested and removed by the bacterial cell. But restriction enzymes owe their recent fame to the ways in which they have been used to manipulate all kinds of DNA in the laboratory. They serve as "molecular scissors" and are one of the most important tools of the molecular biologist.

Each of the several hundred restriction enzymes that have been identified and isolated from various bacteria recognizes a specific sequence of nucleotides on DNA, called its **recognition site**. A restriction enzyme cuts both strands of the double helix at its recognition site. Thus, when a restriction enzyme is present, it makes a cut wherever that particular sequence occurs in DNA. One of the first restriction enzymes to be discovered was isolated from the bacterium *Escherichia coli*. It was named *Eco*R1, for the first restriction enzyme found in *E. coli*. *Eco*R1 recognizes the sequence

It cuts this sequence in a manner that leaves some unpaired bases on both strands of the DNA, like this:

The unpaired bases of DNA are called "sticky ends," a term that describes their tendency to find other similar unpaired sequences and form hydrogen bonds with them. Imagine that DNA from a bacterium and a human were both cut with *Eco*R1 in the same test tube. Some of the sticky ends of the bacterial DNA would find complementary sticky ends from other bacterial DNA molecules, but others would find partners from among the human DNA strands with complementary sticky ends. Once such partnerships have formed by hydrogen bonding, the breaks within strands can be healed using another bacterial enzyme, **DNA ligase**, which serves as "molecular paste." (This enzyme plays a role in DNA replication and heals naturally occurring breaks in DNA.) A DNA molecule formed from the DNAs of different organisms is called **recombinant DNA**. In this example the recombinant DNA would be part bacterial and part human.

Made-to-Order Cells

Prior to the advent of DNA technology, human proteins, such as insulin for the treatment of diabetes and growth hormone for treatment of growth abnormalities, were prohibitively expensive or not available at all. Sufferers of these maladies were treated with proteins taken from cows or pigs brought to slaughter. These animal proteins were inefficient and costly to prepare. Now there are cultures of bacteria expressing the human genes for these proteins in quantity, cheaply and accurately. Recombinant DNA technology has revolutionized the pharmaceutical industry. But how are recombinant DNA molecules introduced into bacterial cells?

Bacteria have a single loop of DNA called a **bacterial chromosome**. But they can acquire additional small pieces of DNA from outside the cell. For example, under the right conditions, fragments of DNA in the fluid surrounding the bacterial cell can enter the cell in a process called **transformation**. Recall the experiments by Frederick Griffith, discussed in Chapter 5, in which harmless pneumococcus bacteria of the R strain took up DNA from the medium that *transformed* them into virulent bacteria of the S strain. The conditions for transformation can be easily mimicked in the laboratory simply by changing the salt concentration of the medium in which the bacteria are living. Recombinant DNA molecules are synthesized and introduced into the medium as **plasmids**, or small circular fragments of DNA that contain the gene of interest, for example, the human insulin gene. Bacteria that take up the plasmid by transformation produce the protein encoded on the plasmid's DNA. In this manner, strains of bacteria capable of synthesizing great quantities of human insulin have been developed.

Plasmids in the bacterial medium act as **vectors**, or vehicles for carrying foreign DNA into a suitable host bacterium. Vectors must contain not only the gene of interest, but also sequences that permit the vector to be replicated inside the host and passed on to daughter cells when the bacterium divides. Another vector that has been used successfully is the bacterial virus called *bacteriophage lambda*, or simply phage lambda. Phage lambda infects its bacterial host by injecting its DNA into the host cell where it becomes integrated into the bacterial chromosome. A recombinant DNA molecule containing phage lambda DNA and a gene of interest has all of the molecular instructions for integrating the gene into the bacterial chromosome. Once integrated into the host's DNA, the gene is replicated when the cell divides, and it is also transcribed and translated.

Gene Therapy

Recombinant DNA technology has been remarkably successful at producing human proteins from bacteria. But could we insert a healthy gene directly into the cells of a human patient so that the cells could make their own protein instead of relying on a bacterial protein factory? Many molecular biologists think that **gene therapy**, or the treatment of human genetic disorders by introducing healthy genes into human cells, will soon be commonplace.

The principles of human gene therapy are the same as those of recombinant bacterial cells. A recombinant molecule containing a healthy gene is synthesized in the laboratory using restriction enzymes and DNA ligase. Human gene therapy, however, has thus far met with only limited success. One problem has been finding an appropriate vector to carry the gene to the target cells within the human body. Human viruses, pathogens whose life cycle includes injecting viral DNA into a human host cell, have the advantage of delivering DNA effectively, but the viruses must be attenuated, or made nonvirulent, before they can be used. Attenuation, too, is accomplished using DNA technology. The viral genes that cause disease symptoms are usually removed from the viral DNA before they are used as vectors for gene therapy. The *BioInquiry* web site can direct you to some of the latest developments in human gene therapy.

DNA in the Courtroom

While all humans have many DNA sequences in common, there are places in the genome in which each person's DNA is unique. This simple fact has provided a powerful mechanism for identifying individuals who have been involved in a crime. Inevitably, there are traces of DNA left at the scene of a crime, usually from a few hairs or drops of blood, semen, or other body fluids. When the DNA from a person's cells is exposed to restriction enzymes, the numbers and lengths of different fragments that result are unique to that person, much as a fingerprint is a unique feature of each person. DNA fragments are separated according to their size using a technique called **gel electrophoresis** (Figure 6-16). The pattern of DNA fragments from specimens at a crime scene is compared with patterns of DNA fragments taken from suspects, so that the source of the DNA at the crime scene can be identified with certainty.

Often, however, the amount of DNA left at the scene of a crime is tiny. Analysis by restriction enzymes and gel electrophoresis requires more than just a trace. In 1983, a clever technique for making large amounts of DNA from a small sample was developed; this technique is called the **polymerase chain reaction**, or **PCR**. Figure 6-17 describes how this technique can amplify just one strand of DNA to millions of copies in a short time. The polymerase chain reaction has widespread application for many aspects of DNA technology, not just forensics.

This is just a sample of the ways in which DNA technology is being used in medicine and society. The *BioInquiry* web site shows how these techniques are being used to develop better crops, clean up pollution, and even improve some of our leisure activities.

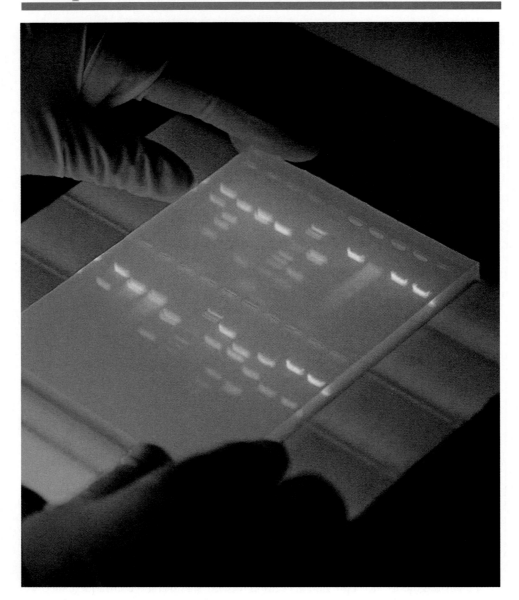

Figure 6-16. Fragments of DNA are separated using a technique called gel electrophoresis. A mixture of DNA fragments of different sizes is introduced at one end of a jellylike slab, called an agarose gel, which is placed in an electrical field. Small fragments of DNA move quickly from the negative end of the field to the positive end; larger fragments move more slowly. The result, shown in this photo, is a series of bands, each one representing DNA of a different size. When DNA from certain parts of the human genome is used, the banding pattern is unique for each individual.

Figure 6-17. The polymerase chain reaction, also called PCR, is a procedure in which tiny amounts of DNA are duplicated many times over to yield quantities large enough to be analyzed. Each strand of DNA that is produced is exactly like the original strand.

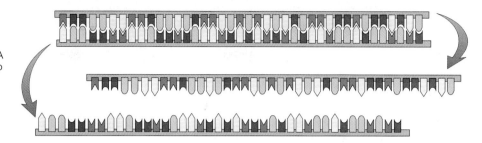

1 The tiny bit of starting DNA is "melted" so that the two strands separate.

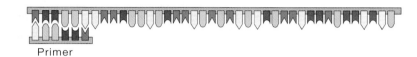

2 Primers are short DNA sequences that are complementary to the ends of the DNA. Primers are added to the mixture upon which new DNA strands can be made. The primers line up with their complementary bases on the separated strands.

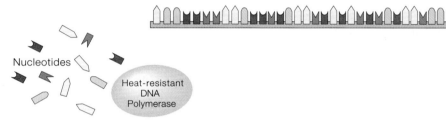

Primer

Primer

Nucleotides

Heat-resistant DNA Polymerase

3 A heat-resistant DNA polymerase and nucleotides are part of the mixture. This enzyme adds nucleotides to the primers, using the original DNA as a template. This creates exact copies of the original DNA.

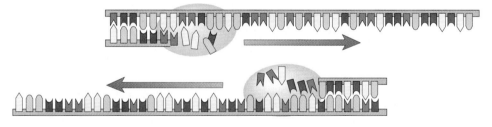

4 This process is repeated many times. Each cycle doubles the amount of DNA, until the original DNA has been amplified into many identical copies.

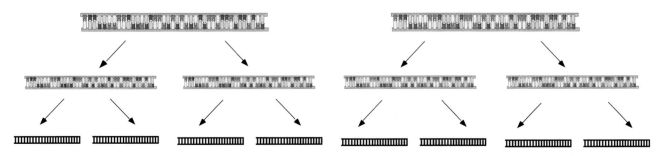

REVIEW QUESTIONS

1. What evidence did Watson and Crick use to determine that DNA is a helix? What other cellular macromolecule can assume a helical shape?

2. What are Chargaff's ratios? Why were they important in discovering the structure of DNA?

3. Name the four roles that DNA must play in cells. How is the structure of DNA consistent with these four roles?

4. What cellular components are necessary to replicate DNA? Where does the energy come from to link nucleotides together during DNA replication?

5. Where in the cell does DNA replication take place? Where does transcription take place? Where does translation take place?

6. Before Meselson and Stahl proved that DNA replication is semiconservative, what other mechanisms had to be considered? How did Meselson and Stahl's experiments eliminate those other possible mechanisms?

7. The flow of information in cells is from DNA to RNA to proteins. Why do you think this two-step process evolved instead of a simpler one-step process, say, from DNA directly to proteins?

8. What are the differences in molecular structure between DNA and RNA? What are the differences in three-dimensional structure? How do the cellular roles of these two nucleic acids compare?

9. How do the three different kinds of RNA differ from each other, both in their three-dimensional structure and in the cellular tasks they perform?

10. What is the smallest possible number of different tRNAs that must exist in cells?

11. What would happen if the different tRNAs in cells could bind to just any amino acid? How does the specificity of tRNA for particular amino acids maintain the integrity of the genetic information?

12. What happens if a ribosome encounters a codon on mRNA that does not correspond to any of the amino acid–linked tRNAs?

13. Name two kinds of nucleotide sequences that do not code for amino acids in proteins. How are they important to cells?

14. Why do cells express (transcribe and translate) only some of the proteins encoded in the genetic material? Name some ways in which gene expression is regulated.

15. What kind of mutation has occurred in the DNA of people with sickle cell anemia?

16. Frame-shift mutations occur when a nucleotide is accidentally omitted or added to DNA. Would a frame-shift mutation be more harmful if it occurred at the start of a gene or near the end of a gene? Why?

Chapter 7

Population Genetics: How Do Genes Move through Time?

The importance of the great principle of selection mainly lies in the power of selecting scarcely appreciable differences... which can be accumulated until the result is made manifest to the eyes of every beholder. —Charles Darwin, 1858

—Overview—

Prior to 1858, the study of life was limited to descriptions of different species with little thought as to how they came to be or how they were related. But the revolutionary ideas spelled out so brilliantly by Darwin in *The Origin of Species* provided the framework for understanding the connections among living forms. We saw in Chapter 2 that Darwin's description of evolution by natural selection made two important assumptions: (1) traits are passed from parents to offspring, and (2) there are important differences between individuals, even of the same species. In Chapters 3 through 6 we followed the pathways of discovery that gave insight into both inheritance and variation. But it may not be immediately obvious to you how Darwinian evolution and Mendelian genetics are connected. It certainly was not obvious to the biologists of the early 20th century. Yet it is the marriage between these two central paradigms that makes modern biology intelligible.

The relationship between evolution and genetics had rocky beginnings. In this chapter, we will explore these beginnings, and we will see how the reconciliation of Mendelism and Darwinism in the 1930s gave rise to a new field of study called population genetics. **Population genetics** is the study of how genes of entire populations change over time. In other words, it describes how groups of organisms evolve. This integration of evolution and genetics was not in itself a scientific "revolution" in the sense that Darwin's and Mendel's theories were, wherein an entirely new idea is presented for the first time.

Chapter opening photo—A field of tulips showing the vast color differences that can occur between individuals of the same species.

Rather it was a *new synthesis* of existing ideas, a synthesis that enabled biologists to look at both evolution and genetics with new eyes. By combining Mendelism and Darwinism, population genetics has given rise to new and powerful tools for understanding both of its parent ideas.

We begin by looking at an idea that is central to both theories: variation. Darwin noticed that the individuals in natural populations may each show slightly different forms of a given trait or characteristic. He recognized this variation as the raw material of evolution. Without it, there would be nothing upon which natural selection could act. Mendel, too, used individual variation to derive his laws of inheritance, but he focused on traits that show only two distinct forms. Is there a connection between the smooth, continuous variation observed by Darwin, and the discrete, nonoverlapping variation observed by Mendel? How do we reconcile the Darwinian and Mendelian views of variation?

7.1 HOW DO WE CHARACTERIZE VARIATION?

Nearly 80 years after Darwin published *The Origin of Species* and about 30 years after the rediscovery of Mendel's laws, an important and influential book appeared by the Russian-American geneticist Theodosius Dobzhansky. Dobzhansky's 1937 book, called *Genetics and the Origin of Species*, marked the end of over a half-century of bickering between two schools of biological thought that had previously dominated ideas about life on Earth. The book brought together the "naturalist" and the "experimentalist" views of biology. The naturalist school found Darwin's ideas most consistent with their observations but rejected Mendel's ideas; the experimentalist school agreed with the principles of Mendelian genetics but rejected natural selection as the driving force of evolution. Their biggest disagreement concerned the nature of variation.

Both of these groups had important and accurate insights into the nature of life, but both were guilty of relying on erroneous assumptions that prevented each of them from seeing the truth in the ideas of the other. In the early 20th century, legendary animosities arose between the leading personalities in each of the schools. It took the appearance of new young thinkers, such as Dobzhansky and others, to bring the warring factions together and show how evolution by natural selection and the laws of heredity were two parts of the same exciting story of life.

7.1.1 Variation Can Be Smooth or Discontinuous

To the naturalists, Darwin's evolution by natural selection made perfect sense. These were the field scientists and the paleontologists, scientists who spent their time outdoors observing populations of organisms in their natural settings. These biologists had firsthand knowledge of the variation among individuals that was so central to Darwinian evolution. In particular, they saw that most traits, when examined within a population of individuals of the same species, exhibited a continuum of forms—a few individuals at either end of a spectrum and many others showing all manner of intermediate forms. Mendel, on the other hand, had described individuals with either one or another discrete form of a trait. Darwin and the naturalists saw that plants in their natural settings could be tall, short, and all sizes in between. Flowers came in a spectrum of different shades, not just a pair of nonoverlapping alternatives. Mendel's carefully cultivated pea plants were either tall or short; flowers were either purple or white, but none were in-between. Naturalists believed that the ability to survive and reproduce might depend on having traits that fall within some certain range of the spectrum. They recognized that the environment in which an organism lives might change over time, making a different range of the same trait more advantageous. In this manner, traits within populations change or evolve as features of the environment change. This is evolution by natural selection as described by Darwin: slow, gradual changes in populations of organisms eventually give rise to new species. But these observations seemed contradictory to the laws of heredity spelled out by Mendel. The naturalists stubbornly (and wrongly) refused to believe that Mendel's factors had any role in evolution.

The experimentalists, on the other hand, rejected the naturalists' view of evolution, choosing instead to focus on the sudden changes that occur as a result of mutations. Their views coincided well with Mendel's description of traits, which occurred in only two or a few distinct forms, such as white flowers and purple flowers, or red-eyed and white-eyed fruit flies. New forms of traits resulted from mutations of existing alleles. In the view of the experimentalists, evolution progressed by leaps and bounds, driven by these sudden, random mutations. From their point of view, evolution could not possibly be a gradual process. New species arose when one or more mutations resulted in an individual unlike any before it. The experimentalists worked in the laboratory, focusing on individual genes. Genes, they knew from Mendel, occurred as discrete entities. They stubbornly (and wrongly) refused to believe that the continuous variation in traits that occurs in natural populations had any relation to genes or to evolution. Natural selection, they mistakenly reasoned, might play a minor role in weeding out deleterious mutations, but it was surely not important in evolution and the origin of new species.

The first breakthrough in the stalemate between the naturalists and the experimentalists came in 1909 from the work of a Swedish plant breeder, Herman Nilsson-Ehle. Working with wheat kernels, Nilsson-Ehle proved that traits that appear in populations as a continuous spectrum of forms have a genetic basis, that the smooth variation within natural populations described by Darwin and the naturalists fits quite nicely into the genetic framework described by Mendel and the experimentalists.

Some wheat kernels are white; others are deep red (Figure 7-1). Most are shades in between, from pale pink to light red. The color trait in wheat kernels appears to exhibit continuous variation. But when Nilsson-Ehle made careful crosses, much like Mendel had, he noticed that a cross between a true-breeding red-kerneled plant and a true-breeding white-kerneled plant produced all light red kernels in the F1, which was intermediate between the shades of the two parent kernels. Recall from Chapter 3 that this is just what happens when genetic traits exhibit incomplete dominance: neither of the parental forms of a trait is completely dominant over the other form; hence the offspring appear intermediate between the parents.

When these light red seeds of the F_1 generation were grown to mature wheat plants and crossed with each other, the second (F_2) generation seeds had exactly seven categories of color: dark red, moderately dark red, red, light red, pink, light pink, and white.

(a)   *(b)*

Figure 7-1. Wheat kernels vary in color from white (*a*) to dark red (*b*). Kernel color is a polygenic trait determined by alleles at three different genetic loci.

Nilsson-Ehle repeated these crosses many times, and each time the ratio of the seven colors of F_2 wheat kernels remained the same: 1 dark red to 6 moderately dark red to 15 red to 20 light red to 15 pink to 6 light pink to 1 white. Such highly predictable results could not possibly be due solely to chance.

A genetic principle does, indeed, determine the spectrum of colors that occurs in wheat kernels. We will look closely at the genetic basis for continuous traits in Section 7.1.3, but first let's examine a useful graphical tool for illustrating variation in populations, called the **frequency diagram**. The frequency diagram also helps show how Nilsson-Ehle discovered the genetic basis for smooth, continuous variation.

7.1.2 Frequency Diagrams Illustrate Variation

Fast Find 7.1a Discontinuous Traits

A frequency diagram is a simple graph. The horizontal axis, or x axis, shows the range of different forms that a trait can exhibit within a population, for example, white at one end and dark red at the other. The vertical axis, or y axis, shows the number of individuals (or some relative number, such as percent) in the population that exhibit each form of the trait. We can begin by examining a simple frequency diagram for a Mendelian-type discontinuous trait. Let's take a familiar example from our discussion of incomplete dominance in Chapter 3. Recall that the flower of the four-o'clock plant is white, red, or pink. Flower color is determined by a single gene that has two different forms, or alleles. Plants that have two alleles for white flower color (R_1R_1) appear white; those with two alleles for red flower color (R_2R_2) appear red. If a plant is heterozygous for the flower color gene (R_1R_2), then the flower's phenotype is pink. When heterozygous plants are carefully crossed, the ratio of white/pink/red in the next generation is 1:2:1. (If you are having trouble remembering the difference between phenotype and genotype or how to perform crosses between heterozygotes, refer to Chapter 3. You can also practice crosses between four o'clocks on the CD.) These ratios can be illustrated by a frequency diagram as shown in Figure 7-2. Because there are only three forms of the trait, the frequency diagram has only three bars.

Now let's consider an example of the other extreme, human height. People come in all different sizes. If we tried to plot a frequency diagram of the heights of everyone

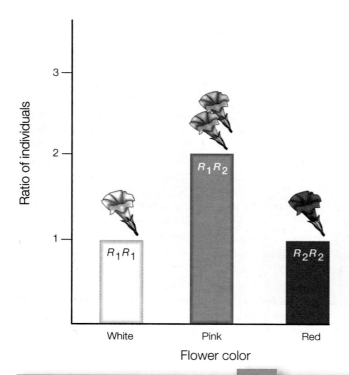

Figure 7-2. The results of a cross between a white-flowered and a red-flowered four-o'clock plant can be described with a frequency diagram. The ratio of offspring in this cross is 1:2:1 white/pink/red flowered plants.

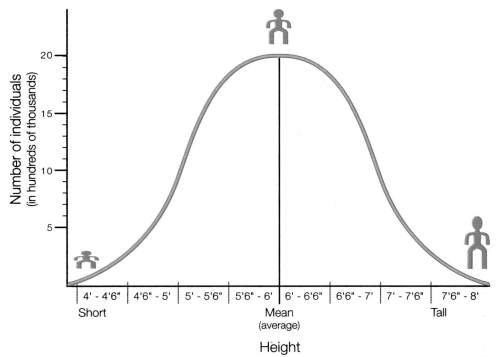

Figure 7-3. A frequency diagram showing that the range of human heights produces a bell-shaped curve. Human height is a polygenic trait.

in North America, there would be so many phenotypes that the categories would blend into one another forming a smooth, bell-shaped curve (Figure 7-3). This is perfectly continuous variation of the type Darwin and the naturalists reported in natural populations. There are some very tall people and some very short people, but most individuals fall somewhere in the middle. The highest point on a perfect bell curve represents the mean, or average, form. We calculate the average height by summing all the heights of all the people, then dividing that sum by the number of people we measured. The result represents the center of distribution of the trait.

Nilsson-Ehle's wheat kernels in the F_2 generation ranged from white (no red color) to dark red (maximum red color), with five different intermediate forms. For the phenotypes of wheat kernels of the F_2 generation, the frequency diagram would look somewhat intermediate between a perfectly discontinuous trait such as flower color in the four o'clock, and a perfectly continuous trait such as human height (Figure 7-4).

Now imagine that the wheat kernels, formerly cultivated under carefully controlled conditions in Nilsson-Ehle's laboratory, were, instead, growing in the wild. Some wheat seedlings might get a bit more sunlight than others; some might get a bit more water or nutrients from the soil. These environmental factors influence the amount of color that each seed develops, creating variation within each of the seven distinct color categories. Each category might show its own bell curve frequency diagram. Plotting them together, we get a frequency diagram as shown in Figure 7-5. Under these conditions, there is considerable overlap between the different forms, and the distinctions between them are blurred. What had been seven clearly delineated colors in the laboratory becomes one continuous spectrum of color variation in nature.

Frequency diagrams are helpful tools for visualizing how, given the added influence of environmental factors, a series of discretely different forms can appear as a smooth spectrum. But what of the genetic basis of continuous variation? Why did Nilsson-Ehle's wheat kernels exhibit seven different forms whereas Mendel's peas showed only two? Can genes be responsible for seven different forms—a nearly continuous spectrum—of the same trait? The answer (the experimentalists notwithstanding) is a resounding yes.

**Fast Find
7.1b
Continuous
Traits**

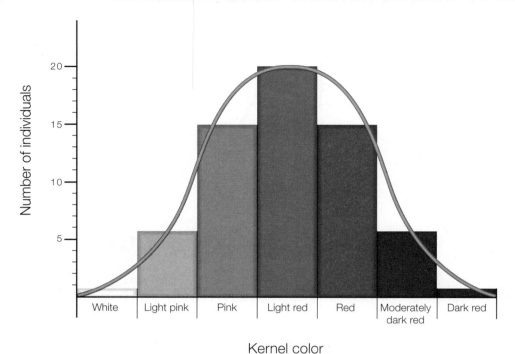
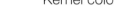

Figure 7-4. A frequency diagram showing the range of colors in the F_2 generation of a cross between a wheat plant with white kernels and one with dark red kernels.

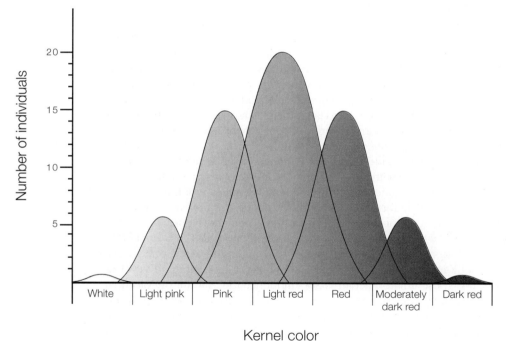

Figure 7-5. A frequency diagram showing kernel color in the F_2 progeny of a cross between a wheat plant with white kernels and one with dark red kernels. In this diagram, each discrete color category is influenced by nongenetic factors, producing a series of seven overlapping bell-shaped curves.

7.1.3 Continuous Variation Is Determined by Two or More Genes

Many traits of evolutionary importance, including height, weight, and growth rate, cannot be traced to a single gene, but in fact are due to the simultaneous expression of two or more genes all influencing the same trait. Such traits are called **polygenic**, or **quantitative**, traits. Each single gene has a specific chromosomal address, called a **locus** (plural, *loci*), which is a fixed position on the chromosome that is occupied by one of

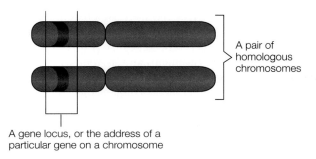

A gene locus, or the address of a
particular gene on a chromosome

A pair of
homologous
chromosomes

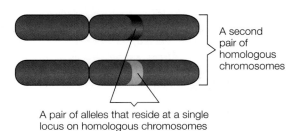

A second
pair of
homologous
chromosomes

A pair of alleles that reside at a single
locus on homologous chromosomes

Figure 7-6. Two pairs of homologous chromosomes are shown with a pair of alleles at each of two loci. The alleles at a given locus may be the same on both chromosomes, as in the first pair, or they may be different, as shown in the second pair.

the alleles, or forms, of that gene. Polygenic traits are influenced by two or more genes residing at different loci on the same or on different chromosomes. The relationship between alleles, genes, and loci is shown in Figure 7-6; these topics were introduced in Chapter 5 of the text and CD.

For a **monogenic** trait, one determined by a single gene locus, an individual can carry at most two different alleles, one on each member of a homologous pair of chromosomes. (The corresponding positions on homologous chromosomes are considered a single locus.) Consequently, there are only three possible combinations of alleles: two different homozygous genotypes and one heterozygous genotype. But what would happen if more than one locus, say two loci, affected a single trait? Now each individual carries two alleles at each of the two loci, a total of four alleles in all influencing the same trait. If three different loci influence the same trait, then an individual would contain six alleles for that trait, and so on. In wheat, three different genes contribute equally to the color of the kernel. When all six alleles at the three loci code for red pigment, the kernel is dark red; when none of the six alleles encode red pigment, it is white. If a wheat kernel inherits just one allele for red pigment, its color is light pink; if it inherits two alleles for red pigment, it is pink, and so on. The actual color of a kernel, its phenotype, is the sum of the individual effects of each of the six alleles. In this case, the greater the number of dominant alleles, the darker the color.

As the number of genes determining a trait increases, the variation within a population changes from discrete categories to a smooth continuum. Add the effect of environmental influence on the expression of each allele and the frequency diagram is transformed from a few nonoverlapping bars to a smooth bell curve. Given the effects of the environment, a trait that is influenced by as few as two genes can show a smooth spectrum of forms within a population—exactly what the naturalists knew to be the case for many traits in natural populations.

No one knows how many traits are polygenic and thus exhibit continuous variation and how many are monogenic, in humans or in any other species. We do know that most traits having to do with size, shape, and form are polygenic. Even Mendel's peas have many traits that are polygenic, although he carefully chose not to study those. The difference between the two kinds of traits, however, has nothing to do with their mode of inheritance. Both are passed from generation to generation in a Mendelian manner, as discrete particles or factors that we call genes. This most fundamental and important insight was the first step in reconciling the Darwinian and Mendelian views, by applying genetics to the study of evolution by natural selection. The next step was to define evolution not in terms of the appearance and disappearance of different species through the ages, or even in terms of changes in the traits expressed within a species, but as changes in the genetic composition of a population over time. How do changes in genes, or more precisely, alleles, accomplish the gradual differences that characterize the evolutionary process in living things? What part does natural selection play in increasing or decreasing the abundance of different alleles?

For Whom the Bell Curves Can you think of other human traits in addition to height that are continuous? What do we currently know about the genes that influence those traits? Are there ways in which the application of population genetics to the study of humans might be controversial? How?

Piecing It Together

The brilliant insights of both Charles Darwin and Gregor Mendel are at the root of modern biology. But even after natural selection and genetics had become part of the consciousness of the field, it took several decades for their interconnectedness to be appreciated. The relationship of evolution and genetics begins with the following ideas:

1. Some traits in living things, like those described by Mendel, show discontinuous variation; that is, they come in only a few discrete forms. Others show continuous variation; that is, they come in a spectrum of different forms, such as human height. Regardless of whether traits are discontinuous or continuous, all are encoded in the genes of living organisms, and all are inherited in the same manner.

2. Traits that appear to exhibit continuous variation are usually encoded by two or more genes found at two or more different loci. Such traits are said to be polygenic, or quantitative. On a genetic level, quantitative traits are determined by the sum of all the alleles that influence that trait.

3. The environment can have a profound influence on the expression of some genetic traits. Consequently, part of the variation we see may be attributable to nongenetic factors.

7.2 HOW DO POPULATIONS DIFFER?

Brachydactyly is a human trait in which the terminal bones of the fingers and toes do not grow to their normal length. Consequently, affected individuals are characterized by stubby digits (Figure 7-7). The cause of this condition is a single dominant gene inherited in a strict Mendelian fashion. People with only one allele for brachydactyly exhibit the trait. We know from Mendel's rules of inheritance that if two people, both heterozygous for the trait (and hence both exhibiting the trait), marry, the odds are 75 percent, or three out of four, that a child from that union will have at least one allele for the trait. Yet fewer than 0.1 percent of humans are brachydactylous. This fact puzzled geneticists early in the 20th century. If a trait is dominant, they reasoned, Mendel's laws predict that three-fourths of all people should exhibit the trait. Can you see the flaw in their reasoning? Before we can predict how many individuals in a population

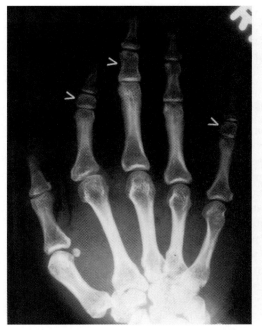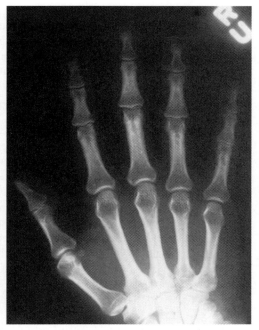

Figure 7-7. X-rays of the hand of a person with brachydactyly (left) and a normal hand (right).

will exhibit a certain trait, we first must know what proportion of reproductively active individuals carry the allele for the trait.

Population genetics takes Mendelian genetics one step further. Understanding the genetics of entire populations, and therefore the manner in which they evolve, requires knowledge of *all* the alleles in the population, not just those in a single mating pair. This task was made much simpler when, in the first half of the 20th century, some straightforward mathematical tools were devised for calculating the occurrence, or frequency, of different alleles in natural populations.

7.2.1 Populations Are Collections of Alleles

**Fast Find
7.2
Genes and
Populations**

What is a **population**? A traditional, and intuitively correct, definition might be this: a group of interbreeding organisms of the same species that exist together in both time and space. That is how we have used the term thus far in our discussion. The early population geneticists realized, however, that to learn how populations evolve, we must visualize not a group of *individuals*, but a group of *alleles*. All of the alleles found in a population are collectively called the **gene pool**. Though we know that alleles cannot exist without the individuals that carry them, it is useful to ignore the individuals for the time being. Imagine the alleles of a population randomly assorting and individually segregating as predicted by Mendel. Later researchers branded this kind of visualization as "beanbag genetics." The alleles are like beans in a beanbag, and the entire beanbag full of beans is the population's gene pool. There are, as you can imagine, limitations to this view. Not least among these limitations is it ignores the fact that the beans can influence each other in important ways.

One of the strengths of this view of populations is we can imagine all of the alleles for all the different traits at once, the entire gene pool, or alternatively we can focus on some subset, such as all the alleles for a single trait. For example, we might focus on all the alleles for one of the amino acid chains in the human hemoglobin protein. To test our beanbag approach, we will do an analysis of the alleles that encode the hemoglobin beta chain in a human population. One type of allele is the normal allele, *HbA*; another is the allele that encodes a slightly altered form of hemoglobin beta chain, *HbS*. We will consider a population in which there are just these two types of alleles. An individual carrying two *HbS* alleles suffers from sickle cell anemia, a debilitating and painful blood disorder. We have used this example several times before, in our discussion of Mendelian inheritance in Chapter 3 and our discussion of how genes encode proteins in Chapter 6. Now we will see how the *HbA* and *HbS* alleles behave in human populations.

7.2.2 Alleles Occur at Certain Frequencies

We begin by envisioning a human population as a collection of alleles. In this case, we are concerned only with the alleles that encode the hemoglobin beta chain. Although any one individual has only two alleles for the beta chain, we are considering an entire population. Consequently, our beanbag contains many more than just two beans. It contains twice as many beans as there are people in the population. Even so, there are only two kinds of beans in the bag. There are beans representing the *HbA* allele and beans representing the *HbS* allele. Our first question is how many of each kind of allele are represented in the population? In other words, what proportion of the total number of hemoglobin beta chain alleles in the population are *HbA* and what proportion are *HbS*? We cannot count the alleles directly, but we do have a few tools for calculating the proportion of each kind of allele based on things we can count.

Let's assess what we already know about our population of alleles. Because we are considering only two different kinds, we already know that the proportion of *HbA* plus the proportion of *HbS* must add up to 100 percent, or all of the hemoglobin beta chain alleles in the population. Stated in mathematical terms, we can assign the letter p as the proportion of *HbA* alleles and q as the proportion of *HbS* alleles. Therefore,

$$p + q = 1.0$$

In this equation, we used the number 1.0 to represent 100 percent. The letters p and q are proportions—fractions of the whole that must add up to 1.0—also called the **allelic frequencies**.

We also know what proportion of the population carries two *HbS* alleles. These are the people afflicted with sickle cell anemia. We can simply count them. But we do not know what fraction of the population is carrying an *HbS* allele masked by the (nearly) dominant *HbA* form.[1] Recall from our discussion in Chapter 3 that the *HbA* allele is nearly dominant to the *HbS* allele. Although people carrying one *HbS* allele and one *HbA* allele can suffer sickle-cell anemia symptoms under conditions of oxygen stress, most of the time they exhibit no symptoms at all. Therefore, we will not always be able to identify the carriers of a single *HbS* allele. Is there a way to use the information we have about phenotypes within a population to calculate the frequency or proportion of the two alleles, p and q? The answer, though not obvious, is yes.

7.2.3 The Hardy-Weinberg Principle Relates Genotypes and Allelic Frequencies

Working independently, an English mathematician named Godfrey H. Hardy and a German physician named Wilhelm Weinberg developed a principle for calculating allelic frequencies, p and q, based on phenotypes. Hardy and Weinberg assumed that populations are very large and that no individuals enter or leave the population. They also assumed that individuals mate at random to produce the next generation. This means that the frequency of any allele in the population is the same as the frequency

---

[1] There are certain genetic tests that can determine if an individual is heterozygous for the hemoglobin beta chain, that is, if the individual carries the *HbS* allele. But for analyzing whole populations (not just individuals), it is not necessary to genetically screen every individual for heterozygosity.

of that allele in the haploid gametes, the egg and sperm. This is the same as saying that the probability that any given egg or sperm that carries the *HbA* allele is p and the probability that any egg or sperm carries the *HbS* allele is q. Using the *HbA* and *HbS* alleles as an example, the probability that an *HbA*-carrying egg will be fertilized by an *HbA*-carrying sperm is $p \times p$, or p^2 (see Chapter 3, Section 3.1.6 to remind yourself why). Thus the proportion of *HbA/HbA* individuals in the populations is equal to p^2. Likewise, the probability that an *HbS*-carrying egg will be fertilized by an *HbS*-carrying sperm is q^2; consequently, the proportion of *HbS/HbS* individuals is q^2. Finally, there are two different ways in which fertilization can result in a heterozygote: either an *HbA*-carrying egg can join with an *HbS*-carrying sperm, or vice versa, an *HbS*-carrying egg can join with an *HbA*-carrying sperm. The probability of either of those events happening is $p \times q$, but because these two different scenarios for conceiving a heterozygous individual are mutually exclusive (and equally likely) in any one conception, the sum rule applies (Chapter 3, Section 3.1.6). To calculate the total probability of a heterozygote, we must add $(p \times q) + (q \times p)$; simplified, the probability that a person has the *HbA/HbS* genotype is $2pq$.

When we sum the probabilities of all the different possible genotypes, we account for the entire population. Therefore,

$$p^2 + 2pq + q^2 = 1.0$$

In words, this equation states that the proportion of homozygous *HbA/HbA* individuals, plus the proportion of heterozygous *HbA/HbS* individuals, plus the proportion of homozygous *HbS/HbS* individuals account for all the members of the population. Each term in the equation represents one of the three possible genotypes that humans can have. The first term, p^2, stands for the homozygous normals; the second term, $2pq$, stands for the heterozygous people, and the third term, q^2, stands for the homozygous people with sickle cell anemia. If you've studied algebra, you may recognize this equation as an expansion of an algebraic binomial expression, $(p + q)^2 = 1.0$.

This relationship between genotypes and allelic frequencies is called the Hardy-Weinberg principle after the two mathematicians that derived it in 1908. Stated in words, the Hardy-Weinberg principle says that the frequencies of genotypes for a gene with two different alleles are a binomial function of the allelic frequencies. In addition, the Hardy-Weinberg principle says that, barring outside influences such as immigration, emigration, or selection, the allelic frequencies of a population will remain the same from one generation to the next. Regardless of how the alleles are paired during sexual reproduction, p and q do not change.

Populations in which p and q do not change are said to be in **genetic equilibrium**. But Hardy and Weinberg made some rather sweeping assumptions regarding the way in which other factors influence p and q when they derived their equation. Let's review those assumptions.

1. *Populations are very large*. Many natural populations are large enough that this assumption approximates reality. Very small populations may experience drastic changes in allelic frequencies due only to chance. Such small populations, therefore, are not in genetic equilibrium.

2. *Individuals mate at random*. As humans, we are careful in our choice of mate, often (but not always) choosing mates that resemble ourselves in many ways. This, too, influences genetic equilibrium, as is most obvious in tightly inbred social groups such as religious sects or highly isolated communities. Even so, many species of plants and animals in nature do exhibit random, or nearly random, mating patterns.

3. *Populations do not gain or lose individuals by immigration and emigration*. When individuals come into or leave a population, the alleles they carry come and go with them. The extent to which this happens in natural populations varies considerably. When immigration and emigration occur frequently in a population, that population is not in genetic equilibrium.

4. *Natural selection is not occurring.* This, we know from Darwin (see Chapter 2), is an assumption that does not always reflect reality. Natural selection may lower the frequency of harmful alleles and raise the frequency of beneficial alleles in any naturally occurring group. As we will soon see, alleles that are subject to changes due to natural selection are not in genetic equilibrium.

5. *Mutation is not occurring at a high enough rate to influence genetic variation.* Although all genetic variation originates with mutations (different alleles could not exist unless a mutation had occurred at some time in history), mutation is really a quite rare event. For this reason, mutations cause only negligible changes in allelic frequencies from one generation to the next. Mutation is generally ignored as a factor influencing genetic variation over just a few generations.

How, you might ask, can the Hardy-Weinberg principle teach us anything about populations if its assumptions are not always true? The power of the Hardy-Weinberg principle is that it allows us to calculate what would happen if natural selection were not occurring, then compare what does happen in the real world. In this way, the Hardy-Weinberg principle can be used to learn about all populations; not just those in perfect genetic equilibrium.

At this point, we have gathered the information we need to answer our original question regarding the hemoglobin allele. What proportion of the total hemoglobin alleles in the population are *HbA* and what proportion are *HbS*? Assume we have a population in which 2 percent of the people suffer from sickle cell anemia. This is the same as saying that a proportion of the population equal to 0.02 has two *HbS* alleles (*HbS/HbS*=q^2). This relatively high level of afflicted individuals is characteristic of many human populations in sub-Saharan Africa and the Near East. Using this information and the Hardy-Weinberg principle, we can calculate the proportion of *HbS* alleles in this population:

$$q^2 = 0.02$$
$$q = \sqrt{0.02}$$
$$q = 0.14$$

Because we know that $p + q = 1.0$, we can use that equation to calculate p:

$$p + 0.14 = 1.0$$
$$p = 1.0 - 0.14$$
$$p = 0.86$$

About 14 percent of the alleles for hemoglobin in this population are the sickle cell *HbS* allele, and about 86 percent are the normal *HbA* allele.

The Hardy-Weinberg principle also allows us to calculate the proportion of individuals in the population that have each of the three possible genotypes. As an example, we can calculate the proportion of people that are heterozygous at the hemoglobin locus. Recall that the term $2pq$ in the Hardy-Weinberg equation refers to the proportion of heterozygotes. Therefore,

$$\text{Proportion of heterozygotes} = 2pq$$
$$= 2\,(0.86)\,(0.14)$$
$$= 0.24 \text{ or } 24 \text{ percent}$$

Remarkably, 24 percent of the individuals in this population are heterozygous at the hemoglobin locus. This is a rather large percentage considering the devastating disease that the allele causes in the homozygous condition. Why hasn't natural selection eliminated the *HbS* allele? For reasons we will see in the next section, natural selection has acted to keep the frequency of the *HbS* allele high, although based on previous discussions you might have predicted otherwise.

We began Section 7.2 with a riddle faced by the early population geneticists: if Mendel's laws predict that on average 75 percent of offspring resulting from a cross between heterozygotes will exhibit the dominant trait, why do so few people exhibit the dominant trait of brachydactyly? The answer, of course, is that when a single cross between heterozygotes is considered, the allelic frequencies in that pair, p and q, are both equal to 0.5. Within a population, however, the allelic frequency of the brachydactyly allele is nowhere near 0.5. It is, in fact, quite low–less than 0.001. Although every individual who inherits the brachydactyly allele will have the brachydactyly phenotype, very few will inherit the allele.

With the tools for calculating allelic frequencies of populations in hand, we can begin to look at the genetic basis of evolution and the origin of species. Speciation begins when the allelic frequencies of two or more populations start to diverge. Natural selection is the most important mechanism driving speciation, but there are others, too.

What Are the Odds? Would the Hardy-Weinberg principle be useful in estimating the frequency of other human genetic disorders? Give some examples. Choose a genetic disorder that occurs in humans and use information from the World Wide Web to determine its occurrence. Now use the Hardy-Weinberg principle to estimate the frequency of the allele that causes that disease.

Piecing It Together

By visualizing populations as groups of alleles instead of groups of individuals, geneticists have been able to develop powerful models for studying the genetic basis of evolution and change. Here are some of the concepts they use:

1. Within a population, every different form of a gene—every allele—composes some fraction of all the different forms of the gene for that trait. The proportion of a given allele among all the alleles for that locus is called its allelic frequency. For alleles that occur in only two forms, the allelic frequencies are designated p and q. Allelic frequencies always add up to one.

2. The Hardy-Weinberg principle states that the frequencies of two alleles in a population will remain the same from generation to generation unless their occurrence is altered by immigration, emigration, nonrandom mating, natural selection, or random chance due to small population size.

3. The Hardy-Weinberg principle is demonstrated mathematically by the Hardy-Weinberg equation, which relates the frequencies of two alleles to the genotypes that can result from the same two alleles. The Hardy-Weinberg equation is

$$p^2 + 2pq + q^2 = 1.0$$

where p is the frequency of one allele; q is the frequency of the alternative allele; and p^2, $2pq$, and q^2 are the proportions of the population that are homozygous for the first allele, heterozygous, and homozygous for the second allele, respectively. The number 1.0 represents the entire population.

7.3 HOW CAN ALLELIC FREQUENCIES CHANGE OVER TIME?

A sparrow flew overhead, scanning the tree trunks for a tasty moth to eat. The year was 1848, just before the start of the industrial revolution, and the place was Manchester, England, the site of one of the first textile mills. The trunks of the trees were covered with healthy lichens, giving the bark a mottled appearance, patchy brown and white in color.

Figure 7-8. Speckled and carbonaria forms of peppered moths on a tree trunk.

The occasional black-colored moths, the carbonaria form of the peppered moth, were easy to spot on the lighter background (Figure 7-8), and the sparrow took advantage of this, swooping down to devour the carbonaria moths, but oblivious to the much more common speckled forms that were camouflaged by their cryptic coloration. To be a carbonaria moth in preindustrial England was a bit of bad luck. A carbonaria moth had only a small chance of surviving the flocks of hungry sparrows. More often than not, the carbonaria moth was the victim of natural selection.

Let's revisit Manchester in the year 1900. The Manchester mills have been belching out black smoke for 50 years, and the lichens have long since died from the pollution. A pale, speckled moth has little chance of hiding on tree trunks black with soot, but a carbonaria form is well hidden on this dark surface. Now nearly every moth in the population is the carbonaria form; the speckled form is rare. The pressures of natural selection have shifted, and the moth population has responded.

7.3.1 Natural Selection Changes Allelic Frequencies

**Fast Find
7.3
Change in
Allelic
Frequencies**

Industrial melanism is the name given to the rapid shift in color that occurred in populations of the peppered moth, *Biston betularia*, during the last half of the 19th century in Manchester, England. It is one of the most spectacular and well-documented examples of natural selection ever recorded. The color of the moths is due to a pair of alleles at a single locus in which the carbonaria form, *M*, is dominant to the speckled form, *m*. It has been estimated that before 1848, the *M* allele accounted for only about 10 percent of the body color alleles in the Manchester population. The other 90 percent of alleles were the recessive *m* allele. We can use the Hardy-Weinberg principle to follow the shift in allelic frequencies due to natural selection—evolution in action.

First let's take a look at what would happen to a population when *no* allele is favored by selection. We can begin with the preindustrial population of peppered moths. We will assign p to represent the proportion or frequency of the *M* allele, which we defined earlier as 10 percent, or 0.10. We will assign q as the frequency of the *m* allele; therefore $q = 0.90$. Using the Hardy-Weinberg equation, we calculate the pro-

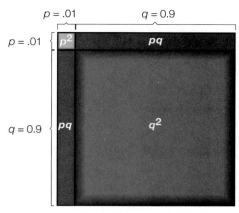

(a) Before industrial melanism

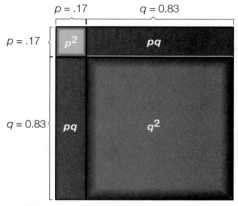

(b) After industrial melanism

Figure 7-9. The Hardy-Weinberg proportions of genotypes in a population can be illustrated using a square. The lengths across the top of the square and down the side of the square are divided into proportions p and q, representing the frequencies of alleles M and m in populations of peppered moth. The areas represent the proportions of the different genotypes in the population. In (a), the proportions represent the situation before industrial melanism. In (b), the proportions represent the situation after natural selection has increased the frequency of the M allele. Note that the areas of both the p^2 and $p \times q$ boxes are larger in (b), and the area of the q^2 box is smaller.

portion of the population that is homozygous for each of the two alleles, and also the proportion of heterozygotes:

$p = 0.10$ This is the frequency of the M allele.
$p^2 = 0.01$ This is the proportion of individuals with the MM genotype.
$q = 0.90$ This is the frequency of the m allele.
$q^2 = 0.81$ This is the proportion of individuals with the mm genotype.
$2pq = 2(0.10)(0.90)$
$2pq = 0.18$ This is the proportion of individuals with the Mm genotype.

Therefore, 1 percent of the population is homozygous for the carbonaria form, 81 percent is homozygous for the speckled form, and 18 percent is heterozygous (Figure 7-9a). Remember that the M allele is dominant to the m allele, so the heterozygotes have the carbonaria phenotype. That means that 19 percent of the moths exhibit the carbonaria form.

If every moth, regardless of its phenotype, is equally likely to survive and reproduce (in other words, natural selection is not occurring), then neither the allelic frequencies nor genotype frequencies will change from generation to generation. The proportion of each allele in the gametes is equal to its proportion in the gene pool, and all gametes have an equal chance of combining with all other gametes of the opposite sex. This is a population in genetic equilibrium. But what if one phenotype were more likely to be eaten by a sparrow?

As the air in Manchester became filled with soot and the lichens on the tree trunks died, the speckled form of peppered moths no longer had the advantage of camouflage. They became easy prey for the sparrows, and their numbers declined. Let's assume that after industrialization, about half of the speckled forms were eaten in each generation, but none of the carbonaria forms were eaten. For every 100 moths in the parental generation, 19 were black and 81 were speckled. Of the 81 that were speckled, let's estimate that 40 were eaten. That leaves only 60 moths to reproduce, 41 of which are speckled. Speckled forms, which constituted 81 percent of the population before industrialization, account for only about 68 percent of the population after a single generation of predation. (In other words, $^{41}/_{60} \times 100 = 68$ percent.)

Using the Hardy-Weinberg equation and the numbers of speckled and carbonaria form of moth, we can also calculate how the allelic frequencies of the M and m alleles have changed. There are 41 moths with the mm genotype and 18 moths with the Mm genotype. Therefore, there are $41 \times 2 = 82$ m alleles among the remaining speckled forms and 18 m alleles among the carbonaria forms, making $82 + 18 = 100$ m alleles in the population. But with only 60 moths remaining in the population (and each moth with 2 alleles for color), there are $60 \times 2 = 120$ alleles altogether. Therefore, the m allele accounts for 100 of the total of 120 alleles, or about 83 percent. The allelic frequency of the m allele has dropped from $q = 0.90$ to $q = 0.83$ (Figure 7-9b).

As a result of selection pressure from hungry sparrows, the proportion of the two alleles in the moth population has shifted. The m allele, which formerly comprised 90 percent of the alleles for color, now only accounts for about 83 percent; the M allele, formerly at 10 percent is now up to nearly 17 percent. It is easy to see how over the course of several generations of natural selection, the relative numbers of the two different alleles can change dramatically. Using the Hardy-Weinberg equation, can you calculate the proportions of each genotype in the next generation?

Depending upon features of the environment, natural selection may favor one or the other allele, m or M, for body color in the peppered moth. The frequency of the unfavored allele declines. Why, then, hasn't natural selection reduced the frequency of the hemoglobin allele that causes sickle cell anemia in human populations? The case with the sickle cell allele is also an example of natural selection at work, but in this case, certain environments make it advantageous to carry two different alleles. In some regions of the world, people who are heterozygous for the human hemoglobin gene are at a distinct advantage over either of the homozygotes.

Heterozygote Advantage

On the African continent, particularly near the equatorial belt in central Africa, malaria is endemic (Figure 7-10). Children are exposed to the disease year-round, and they suffer repeated infections during their early years. It has been estimated that as many as 2 million children die each year of malaria in this region. If they survive, these children

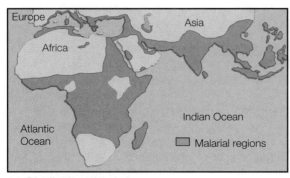

(a) Distribution of malaria

(b) Distribution of sickle-cell anemia allele (HbS)

Figure 7-10. These maps illustrate the relationship between the distribution of malaria and the *HbS* allele in Africa and Asia. Individuals that are heterozygous for sickle cell anemia (*HbA/HbS*) are best able to survive malaria.

acquire considerable immunity to malaria, increasing the likelihood that they will ward off further infections. But what factors help children survive early infections?

One well-established factor is the presence of the *HbS* allele. The tendency for some red blood cells in sickle cell carriers (heterozygotes) to assume the sickle shape makes them particularly resistant to penetration by the parasite that causes malaria. Heterozygous children are better able to survive the disease and, at the same time, acquire immunity that prevents further infections. Meanwhile, those who are homozygous for the normal hemoglobin allele, *HbA/HbA*, are more susceptible to malaria. Most who are homozygous for the sickle cell allele, *HbS/HbS*, do not survive long enough to have children. The best genotype to have in equatorial Africa is *HbA/HbS*; this is called the **heterozygote advantage**. Natural selection can sometimes act to reduce or eliminate detrimental alleles; alternatively, natural selection can maintain genetic variation by favoring heterozygous individuals.

Both of these examples, the peppered moth in Manchester and the sickle cell carriers in Africa, represent a kind of evolution called **microevolution**, or changes in allelic frequencies within a population. Microevolution can occur in a relatively short time, such as over just one or a few generations, as in the example of peppered moths. Microevolution is reversible. In the past 35 years, the Manchester mills have begun taking measures to reduce pollution. The lichens on tree trunks are making a comeback, and the speckled form of moth is on the rise. Macroevolution, on the other hand, describes the large-scale changes that lead to the origin of higher taxa, or categories of organisms such as orders, families, and phyla (see Chapter 8). With some exceptions, most evidence indicates that the higher taxa that result from macroevolution are formed by the accumulation of many small changes in gene frequencies at the population level; that is, microevolution gives rise to macroevolution.

Microevolution proceeds by natural selection, whereby individuals varying in one or more traits differ in their ability to survive and reproduce. As a result, the frequencies of different alleles shift over one or more generations. Often selection favors one extreme form or another at one end of the spectrum of variation, as in the case of the carbonaria form of peppered moths in postindustrial Manchester. But that is not always the case. Sometimes it is most advantageous to be average, or right in the middle of the frequency diagram. In other cases, natural selection can favor both ends of the variation spectrum at the expense of the middle.

7.3.2 Selection Can Favor Any Part of the Range of Phenotypes

A green and brown mottled frog sits silently on the river's edge, blending with the mud and vegetation of the bank. The frog's tongue flicks out with speed and precision to capture a passing damselfly. The long tongue ensures that this animal has a good chance of securing a ready food supply. In a population of frogs with different tongue lengths, those with longer tongues have an advantage: they can capture more insects from farther away.

Directional Selection

Natural selection has favored long tongues in frogs. This type of selection is called **directional selection**, in which the mean value for a trait shifts in a particular direction, in the frog case, toward greater tongue length. This is illustrated in the frequency diagram in Figure 7-11. Alleles that increase the length of the tongue are favored, and therefore are present in successive generations at higher frequencies than those for shorter tongues. However, there are also factors that prevent a frog's tongue from becoming infinitely long—the ability of the animal to close its mouth, for example. In directional selection, the forms of traits at one end of the range of variation become more common with each

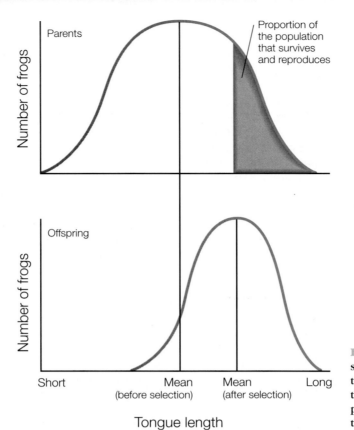

Figure 7-11. Directional selection is illustrated by frog tongue length. After selection, the mean tongue length in a population of frogs is longer than it was before selection.

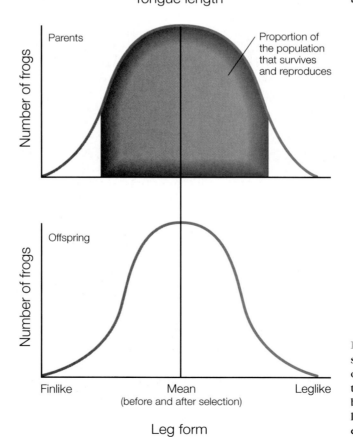

Figure 7-12. Stabilizing selection maintains the form of frog legs. After selection, the most common phenotype becomes even more prevalent. The phenotypes at the extremes are fewer.

succeeding generation. But selection does not always have to favor one end of the spectrum or the other. Sometimes the most advantageous place is right in the middle.

Stabilizing Selection

As an amphibian, the frog must be able to move in two different environments, water and land. Frog legs are a perfect compromise for moving in these two environments: the webbed feet are adapted for swimming, and the long muscular legs are adapted for jumping. Frogs with more finlike legs would be less effective on the land, and those with smaller, unwebbed feet would be weaker swimmers. Selection acts to maintain this compromise. The most common form is the one most likely to survive and reproduce. This is an example of **stabilizing selection** (Figure 7-12). In unchanging environments, stabilizing selection is the rule. Organisms at either end of a spectrum of traits are likely to be less well adapted to a stable environment and thus less successful at passing their alleles on to the next generation.

The gene pools of large populations usually retain enough genetic variation to produce some members at the two extremes, even after many generations of stabilizing selection. This is because quantitative traits, or those traits that exhibit bell-shaped frequency distributions, are the result of many alleles functioning at many different genetic loci. The individuals who breed most successfully, the ones near the mean of the bell-shaped curve, are likely to be heterozygous at some of these loci. When alleles are figuratively plucked from the beanbag and recombined in each new generation, combinations of similar alleles that were not coupled in either parent can result in the offspring. By chance, some members of each new generation will have a particular combination of alleles that places their phenotypes at the extremes of the curve.

This variation is, in itself, an adaptive trait of populations. In spite of many years of selection against the carbonaria form of moth prior to industrialization, the population retained enough variation to produce a few carbonaria forms in each generation. When the environment changed and the tree trunks became black with soot, it was the variant form, the previously unadapted form, that prevailed. Populations that lack variety are likely to become extinct when features of their environment change.

Disruptive Selection

There are some situations in which it is most advantageous to be at either end of a spectrum of forms, but not so advantageous to be in the middle. Suppose, for example, that a population of birds shows a great deal of variation in beak length. Suppose further that the normal food supply for this population, say, a certain fruit or insect, becomes scarce. Perhaps another species of bird invades the territory and competes for the limited food supply. Birds with small beaks are adept at exploiting a new and different food source, beetles and worms from small crevices in fallen trees, for example. Large beaks, on the other hand, are strong and can crack the seeds of large fruits or crush the shells of snails. But the birds with beaks intermediate in size are not very good at either. Selection would favor birds with small beaks and birds with large beaks, but it would work against the most common form, birds with intermediate-sized beaks. After several generations of selection, we would expect the two extremes to become more common and the intermediate forms to begin to decline. This is an example of **disruptive**, or **diversifying, selection**. Disruptive selection works against individuals near the middle of the distribution and favors the individuals at both ends (Figure 7-13). Eventually, as a result of selective changes, one population splits into two subpopulations, an event that may lead to the formation of two separate species.

The three types of selection we have just seen, directional, stabilizing, and disruptive selection, all result in some alleles becoming more common in the population because their bearers are better able to survive and reproduce. Sometimes alleles may become more common, however, simply as a matter of chance.

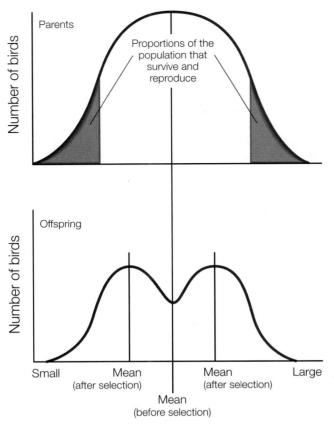

Figure 7-13. Disruptive selection favors extremes over intermediates. After selection, the most common phenotype becomes less prevalent, while phenotypes at both ends of the spectrum become more prevalent.

7.3.3 Some Changes in Allelic Frequency Are Random

The snakes on the Ohio shore of Lake Erie are good swimmers. Boaters and fishermen have seen an occasional snake swimming the mile or so out into the lake to the nearest island. These islands have small populations of the same species of snakes found on the Ohio shoreline, but the island forms are evenly colored whereas most of the mainland forms have banded skin patterns. Perhaps the banded pattern gives the mainland snakes some survival advantage—maybe the ability to blend with the grassy shore—that the island forms do not share. That would be an example of directional selection. Or, alternatively, perhaps the colonization of the islands by plain, unbanded forms was simply a matter of chance. If only a few mainland snakes survived the long swim to the islands, and if, purely by chance, those few that made it were variants of the main population with faint bands or no bands at all, the differences between the two populations, and the differences between their skin pattern alleles, would be a matter of pure chance.

Genetic Drift

When a population is small, chance can be an important factor in determining the composition of the gene pool. Without the influence of selection, flux (immigration and emigration), or mutation, passing alleles from parent to offspring is a bit like blindly drawing beans from the beanbag. If you draw many beans, the numbers of variations you end up with will quite likely reflect the proportions of different beans in the bag. If half the beans in the bag are red and half are green, approximately half of the many beans you choose will be red. But if you draw only a few beans, you may draw only red beans, or mostly green ones. There is a good chance that your small sample will not reflect the bean population at large. Likewise, when only a few randomly chosen alleles from a population successfully make it into the next generation, the likelihood that they will not be representative of the parental gene pool is great (Figure 7-14).

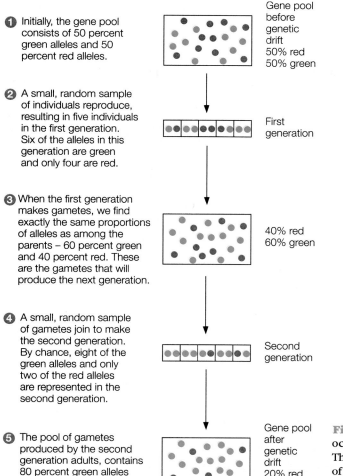

1 Initially, the gene pool consists of 50 percent green alleles and 50 percent red alleles.

Gene pool before genetic drift
50% red
50% green

2 A small, random sample of individuals reproduce, resulting in five individuals in the first generation. Six of the alleles in this generation are green and only four are red.

First generation

3 When the first generation makes gametes, we find exactly the same proportions of alleles as among the parents – 60 percent green and 40 percent red. These are the gametes that will produce the next generation.

40% red
60% green

4 A small, random sample of gametes join to make the second generation. By chance, eight of the green alleles and only two of the red alleles are represented in the second generation.

Second generation

5 The pool of gametes produced by the second generation adults, contains 80 percent green alleles and 20 percent red alleles.

Gene pool after genetic drift
20% red
80% green

Figure 7-14. Genetic drift occurs in small populations. The change in the frequencies of alleles within a gene pool is due to chance.

Such chance events, called genetic drift, can significantly alter the frequency of different alleles in the next generation. **Genetic drift** is the alteration in allelic frequencies that results from chance variation in survival and/or reproductive success.

Because genetic drift results when small samples of alleles are taken from a larger pool, the complete variety of alleles in the parental pool is almost never fully represented in the sample. For this reason, genetic drift invariably leads to a loss of genetic variation. Because the microevolutionary changes that result from genetic drift are independent of natural selection, it is often referred to as **neutral selection**. Unlike natural selection, whereby alleles are preserved or lost due to the advantages or disadvantages they confer upon their owners, the alleles that are perpetuated or lost due to genetic drift are random.

Founder Effect

In the southern Atlantic Ocean, about 1,500 miles off the western coast of Africa, lies a remote volcanic island called Tristan da Cunha. Its population of about 275 people is descended from a group of 15 English settlers that arrived on the island during the 19th century. A recent study by scientists from the University of Toronto revealed that an astonishing 57 percent of the islanders suffer from asthma or asthmalike symptoms, compared with about 10 percent of the population of England. Digging through historical records of the island, the researchers found that at least two of the original settlers were asthmatic. Although the exact causes of asthma have not been established with certainty, these researchers concluded that alleles carried by the founders of the island population have made their descendants particularly susceptible to this debilitating lung disease.

When the settlers arrived at Tristan da Cunha, the alleles they carried became the basis for a new gene pool, one that differed from the parent population in England. This microevolutionary phenomenon is a form of genetic drift, called the **founder effect**. The founder effect occurs when there are differences in allelic composition of a gene pool due to the initiation of a population by a small number of founder individuals.

The founder effect is a primary factor in the abnormally high incidence of other genetic diseases in small, closed human populations. Certain isolated religious groups, for example, have a high incidence of rare and deleterious diseases. The Amish of Lancaster County, Pennsylvania (Figure 7-15), have a high incidence of a recessive genetic disorder called Ellis–van Creveld syndrome, in which those affected have short limbs, malformed hearts, and six fingers on each hand. About 1 percent of the population suffers from the syndrome. The Pennsylvania Amish society was established by a small number of founders who settled in Lancaster County from Holland, many of whom trace their ancestry to a single man, Samuel King, in the last century. It is safe to assume that either Samuel King or his wife was a carrier of the allele for Ellis–van Creveld syndrome.

Bottlenecks

In the late 19th century, the northern elephant seal (Figure 7-16) was heavily hunted. So many seals were killed that the population dwindled to fewer than 20 individuals. Since that time, conservation efforts have resulted in an increase in the north-

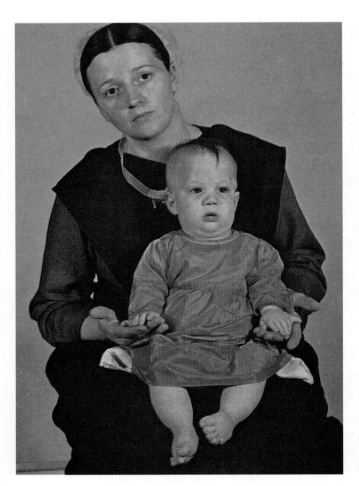

Figure 7-15. The Amish of Pennsylvania have a high incidence of Ellis–van Creveld syndrome. This Amish child has six fingers.

Figure 7-16. The northern elephant seal was hunted to near extinction.

ern elephant seal population; now there are over 50,000 elephant seals. When a population undergoes a temporary decline from which the survivors of all future generations are derived, genetic drift plays a role in determining the composition of the gene pool. This phenomenon is called a **bottleneck**.

Bottlenecks, as with all types of genetic drift, reduce the genetic variation in a population. The full range of genetic variation of the original population is never fully represented in the few surviving individuals. This is invariably detrimental, and often it is catastrophic. Many populations that lack genetic variation perish in the face of even minor changes in environmental conditions. As human activities reduce the number of natural habitats and more species become endangered, more and more evolutionary bottlenecks result. Even heroic attempts to preserve endangered species in zoos and wildlife preserves cannot bring back the genetic variation that characterized the original population. **Conservation genetics** is a subdiscipline of biology that attempts to maintain genetic variation by careful programs of breeding and artificial selection.

Mutation

In the early part of the century, before genetics and evolution were reconciled into the modern science of population genetics, the biologists of the experimentalist school attributed all evolution to the sudden and random appearance of new alleles by means of mutation. In some ways, the early experimentalists were mistaken. We now know that natural selection and genetic drift are both important factors in establishing the frequency of different alleles, and that new species arise from existing species when the allelic frequencies of different populations diverge sufficiently to prevent them from interbreeding. But in other ways, the experimentalists were right. Mutation is a random process that plays a crucial role in evolution by creating new alleles. Without it, there would be no genetic variation upon which selection or genetic drift could act.

In the ideal population that Hardy and Weinberg envisioned when they derived their principle of genetic equilibrium, it was assumed that mutation did not occur. But in real populations, mutations arise with infrequent, but measurable, regularity. For example, in a maternity hospital in Copenhagen, Denmark, there are records of nearly

95,000 births. Of these, ten newborns exhibited achondroplasia, a form of dwarfism caused by a dominant allele. Two of these babies each had an affected parent. Thus eight newborns appeared to be new mutants. Keeping in mind that for each genetic locus there are two alleles, the mutation rate for this gene is

$$\text{mutation rate} = 8 \text{ mutations}/(2)(95,000) \text{ alleles}$$
$$= 4.2 \times 10^{-5} \text{ mutations per allele}$$

or about one mutation for every 23,800 alleles. The rates at which genes mutate differ for different genes and also for different species, but this order of magnitude, somewhere around 10^{-5} mutations per allele for each generation, is a reasonably characteristic mutation rate.

Mutations introduce new alleles into the gene pool. Based on the effects of the mutant gene on the recipient, three somewhat arbitrary categories of mutations with three different fates have been established. First, there are mutations whose effects are lethal or near lethal. In diploid species, lethal alleles may hide as recessives, masked by their non-lethal, dominant counterparts for several generations. But selection acts to keep the frequency of such alleles low or to eliminate them completely. Second, and probably most commonly, mutations may be selectively neutral or nearly neutral. These mutations may spread throughout the gene pool, or they may be eliminated by random means such as genetic drift. Most of the considerable genetic variation in natural populations is probably perpetuated in this neutral manner. Finally, there are a small number of mutations that are advantageous and are thus selected for within a population. Like the deleterious lethals, if such advantageous alleles are hidden by dominant counterparts, they too may languish in the gene pool until or unless they are paired with each other in a homozygote where their benefits can be fully expressed. The ability of mutation alone to effect large changes in allelic frequency is negligible, but combined with natural selection and the random forces of genetic drift, mutation becomes the very stuff of evolution.

Gene Flow

In Chapter 2 we saw that new species arise when subpopulations become reproductively isolated from each other. Sometimes ecological or geographical barriers or differences in behavior prevent two groups from mating, effectively creating two gene pools from one. As the level of genetic connectedness (the ability to interbreed) between two subpopulations declines, yet another of the Hardy-Weinberg assumptions is violated, that of random breeding. Two separate gene pools are free to undergo changes in genetic composition by natural selection or genetic drift. But oftentimes individuals are able to move from one population to another, introducing the genetic elements that characterize their parent population to the new population through mating and reproduction. This movement between populations is called gene flow. **Gene flow** results in the introduction of alleles into a population by means of immigration or the loss of alleles from a population by means of emigration. Often the effect of gene flow coupled with natural selection can be significant. Such was the case with mosquitoes and DDT.

After World War II, the insecticide DDT was used liberally to control populations of various insect pests, including the malaria-carrying mosquito. While this approach kept the numbers of mosquitoes down for a while, it wasn't long before resistant strains began to appear. A mutant mosquito, high in levels of a single enzyme called esterase, appeared by random mutation in one population. Esterase breaks down DDT into harmless metabolites. Just after World War II, most mosquitoes had levels of esterase that were too low to detoxify the insecticide. Before long, this highly successful mosquito produced many offspring, some of which also produced excess esterase (Figure 7-17). These progeny, and their progeny as well, infiltrated the mosquito populations of the world, carrying with them the alleles that enabled them to produce copious esterase. The high esterase alleles flowed from population to population by immigration and emigration. These alleles

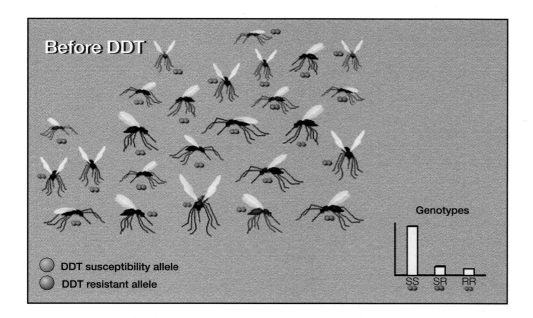

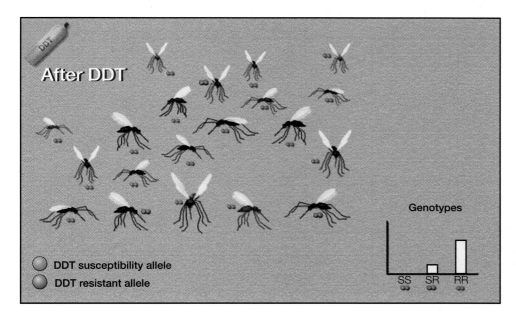

Figure 7-17. The frequency of the DDT-resistance allele in malaria-carrying mosquitoes increased dramatically as a result of selection. Mosquitoes carrying the DDT-resistant allele infiltrated mosquito populations worldwide.

conferred an obvious advantage, spreading rapidly in new generations. Now most mosquito populations are DDT resistant. As a consequence, malaria is on the rise in many Asian and African countries. Effective pest control programs must take into account the principles of population genetics.

Hardy-Weinberg Revisited

At this point you may be wondering whether the assumptions made by Hardy and Weinberg are ever met in the real world. And if not, what is the use of the Hardy-Weinberg principle? Recall that the principle states that in a large, randomly breeding population lacking any outside influences (such as genetic drift, gene flow, natural selection, or mutation), genetic equilibrium is reached; the frequency of different alleles does not change from generation to generation. We refer to such populations as being in genetic, or Hardy-Weinberg, equilibrium. But these stringent restrictions are almost never met in nature. Most populations are subject to outside influences, and all populations are finite.

Nonetheless, for many populations, the effects of size and outside influences are small enough to be negligible. While it is rare to reach perfect Hardy-Weinberg equilibrium, many natural populations approximate such a state, particularly in stable environments. But the real value of the Hardy-Weinberg principle lies not in what it says about populations that conform, but in what we can learn about populations that do not conform to the principle. The power of the Hardy-Weinberg principle lies in learning exactly how real populations differ from genetic equilibrium, then uncovering the factors that are responsible for the differences.

Most of this chapter has been concerned with the processes of microevolution, or the manner in which allelic frequencies change over time and distance within a species. When conditions are such that the changes are vast and irreversible, the end result can be a new species entirely, with a gene pool sufficiently different from the parent population that members of the two groups can no longer interbreed. New species have changed and evolved over time, giving rise to new orders, new classes, new phyla, and new kingdoms. Over the course of 3.5 billion years, since the dawn of life, an astounding array of living forms has populated the Earth. In the next chapter we will take a look at some of the different forms of life, and the trends that have characterized the evolution of life on Earth.

Disappearing Cat The cheetah is an example of a species that has experienced a genetic bottleneck due to near extinction. Why did the cheetah nearly become extinct? What conservation efforts have been made to save the cheetah from extinction?

Piecing It Together

Genetic equilibrium is the condition in which the frequencies of alleles in a population do not change over time. No evolutionary change occurs in a population in genetic equilibrium. Population geneticists have used comparisons between real populations that experience change and theoretical populations in genetic equilibrium to learn about the factors that cause evolution.

1. Natural selection is one of the most powerful of the factors that change the genetic composition of populations. When individuals with a particular genetic complement are more likely to survive and reproduce than others, the gene pool of succeeding generations reflects the alleles of the successful individuals. This is the genetic basis for evolution by natural selection.

2. Natural selection can favor any part of a range of phenotypes. It can favor one of the extremes, in which case selection is directional. It can favor the middle of the range, in which case selection is stabilizing, that is, maintaining the status quo. There are some cases in which both extremes are favored over the middle of the range, this is disruptive selection.

3. The genetic composition of a population can change due to random chance. Such changes are called genetic drift. They are most evident in populations that are small, where chance is more likely to play a role in determining the alleles that get passed on to future generations.

Where Are We Now?

The principles of population genetics have been useful in understanding problems of human genetic disease, including the origins of some disorders and the risks to members of families, ethnic groups, or populations. Recently researchers have been applying these principles to the study of two genes implicated in breast cancer, *BRCA1* and *BRCA2*. The most common alleles for these genes encode normal proteins that play an important role in regulating cell division. Certain less common alleles, called mutant alleles, encode proteins that do not function properly. Women who inherit mutant alleles for either gene are at high risk for developing both breast and ovarian cancers. Mutant *BRCA1* and *BRCA2* alleles account for about 6 to 10 percent of breast cancer cases in most populations.

Researchers around the world are interested in several questions about mutant alleles of the *BRCA* genes. First, how many different alleles exist that have been implicated in cancer? Although any one person carries only two alleles for a gene, an entire population may harbor many different alleles. So far, over 30 different alleles of the *BRCA1* gene and more than 15 alleles for the *BRCA2* gene have been identified in populations worldwide, although many of them are found in only one or a few countries or populations. Researchers expect that more alleles will be identified as more of the world's populations are studied.

Second, where and when did mutant alleles originate, and how have they dispersed from one population to another? Human population structure, including patterns of migration, natural selection, and marriage traditions, all influence the manner in which alleles occur in populations. For example, one mutant allele of the *BRCA1* gene, called the *185delAG* allele,[2] is found in about 1 percent (frequency = 0.01) of people of Ashkenazi Jewish heritage. The same *185delAG* allele has also been identified among Iranian and Iraqi families. The occurrence of the identical allele in different populations suggests that the original mutation probably occurred before the populations became separated. History tells us that before the year 70 B.C.E., Jewish life centered around the Temple in Jerusalem. In 70 B.C.E., the Romans occupied Jerusalem, destroying the Temple. Many Jews fled to Mesopotamia, an area that now encompasses Iraq. The distribution of the *185delAG* allele among descendents of both Jews and Mesopotamians suggests that the mutation probably originated before that time, making the mutation at least 2,000 years old.

Third, researchers are interested in the risks associated with different mutant alleles and what steps can be taken to minimize those risks. The *185delAG* allele, for example, is present at approximately equal frequencies in both the Iranian and Iraqi population and the Ashkenazi Jewish population (about 1 percent), but the incidences of both breast and ovarian cancer are higher among the Ashkenazi women than either Iranian or Iraqi women. Why do some individuals who inherit a mutant allele develop cancer whereas others do not? Studies are now addressing how other factors, such as environmental, cultural, or even other genetic differences, can account for the discrepancy. Population genetics is a starting point for all of these questions.

---

[2]Mutant alleles are named for the way in which the DNA has been altered when compared with normal alleles. In this case, the mutant allele (a deletion) is missing two nucleotides, A and G, at a position in the DNA that corresponds with the 185th base pair.

REVIEW QUESTIONS

1. Having traveled back in time to the 19th century, you are debating an early naturalist. Your opponent is arguing that traits in natural populations exhibit a continuum of forms and therefore cannot be attributed to genes. Explain to your opponent why he is mistaken.

2. What is the difference between a polygenic trait and a monogenic trait? Give an example of each.

3. What is a genetic locus? What is the relationship between a locus and an allele?

4. What is a gene pool? How does a gene pool differ from a population?

5. In the Hardy-Weinberg equation, what do the letters p and q represent? What do the mathematical terms p^2, $2pq$, and q^2 each represent?

6. Describe in words what it means when a population is in genetic equilibrium. Now describe genetic equilibrium using mathematical terms.

7. What assumptions did Hardy and Weinberg make when they derived their mathematical principle describing genetic equilibrium? What factors may prevent a population from reaching genetic equilibrium?

8. You are studying a population of caterpillars in which some of the individuals are woolly and some are bare. The woolly trait is caused by the recessive w allele of the woolly gene. The dominant W allele causes caterpillars to be bare. Sixteen percent of the population is woolly. What is the frequency of the w allele? What is the frequency of the W allele? What proportion of the population is heterozygous at the woolly locus?

9. What is the heterozygote advantage? Give an example.

10. Distinguish between microevolution and macroevolution. How are they related?

11. What kind of selection—directional, stabilizing, or disruptive—is illustrated by industrial melanism of the English peppered moth in the last half of the 19th century? Explain.

12. Why is genetic drift more likely to occur in small populations than in large populations?

13. The founder effect and genetic bottlenecks are two examples of changes in gene frequencies that occur by chance. Explain how these two mechanisms are similar and how they differ.

14. What is gene flow? Is gene flow more likely to occur between two populations on distant islands or two populations that overlap in time and space? Why?

15. How has the Hardy-Weinberg principle been used to learn about populations that are not in genetic equilibrium?

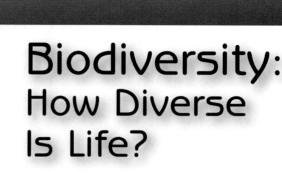

Chapter 8

Biodiversity: How Diverse Is Life?

Classifications both reflect and direct our thinking. The way we order represents the way we think. —Stephen Jay Gould (1983)

—Overview—

Life on Earth is amazingly diverse. In the last 250 years, biologists have described roughly 1.5 million different kinds of organisms, and they have only scratched the surface of the diversity that exists.

In the 1950s, biologists in the tropics discovered, in the tops of trees hundreds of feet high, a rich assemblage of living things. Because these organisms only rarely and accidentally come near ground, they were largely unknown. Borrowing equipment and techniques from mountain climbers and those who build tall buildings, biologists took to the treetops and found thousands of new species. Their work goes on. When completely cataloged, these and other tropical ecosystems may well add 15 million additional species to Earth's catalog of diversity.

The deep oceans, once thought to be largely biological wastelands, are also proving to be surprisingly diverse. In the 1970s, rich biological communities were discovered associated with geologic vents on the ocean floor. These vents spew hot, mineral-laden water that provides energy and nutrients to microbes. Feeding on the microbes were found hitherto unknown species of fish, crabs, clams, giant worms, and others. Away from the vents, oceanic bottom muds don't exactly teem with life, but even there, extensive and unique life forms were found. The same surprises awaited biologists who investigated the tops of sea mounts and the thin transition layer between ice and water under the pack ice of the Arctic and Antarctic Oceans.

More recently, in the crawlspaces between soil particles, a rich assemblage of organisms has been discovered. A little too large to be studied by microscopists and a little too small to be easily noticed by those without a microscope, the so-called *mesofauna*

Chapter opening photo—A rich assemblage of organisms lives deep in the world's oceans associated with hydrothermal vents.

have quietly gone about their business largely unnoticed. Now alerted to their existence, biologists are finding them everywhere. Apparently, their numbers and diversity are extensive. Likewise, new organisms are being found thousands of feet deep in rock cracks and in muds and soils beneath lakes. Their life cycles and the extent of their diversity are still being discovered.

How many species are there on Earth? We don't know. Educated guesses now run in excess of 100 million species, and biologists will continue to discover new species for a long time to come. Such diversity is both a blessing and a curse for biology. Each species represents a unique solution to the challenges of life, and coming to understand the breadth of such solutions is exciting, challenging, awe inspiring. The curse of such diversity comes in describing, identifying, classifying, and naming these species. What is needed is an extensive data management system. Ideally, it would arrange species in some logical order, indicate how each is related to others, assist in identifying them, be sufficiently robust to handle at least 100 million separate objects (species), and be "user friendly." Biologists who are not deeply interested in classification should be able to get information into and out of the system with minimal training.

Fortunately, biologists have indeed devised such a system, and it comprises two branches of biology. **Taxonomy** is the science of describing, classifying, and organizing organisms according to their similarities and differences. **Systematics** is a more interpretive approach; it is the study of evolutionary relationships among organisms. The first step in classification, accomplished by taxonomists, is to group species according to shared characteristics. These indicate ancestry. That is, the more similar two species are, the more closely related they are. But all organisms, no matter how dissimilar, share some characteristics. If taken far enough, the study of similarities and differences and of evolutionary relationships shed light on the very origins of life. First, we will discuss taxonomy and systematics, and then turn to the story of the origins of life.

Discovering Vertebrates Are biologists still finding any new terrestrial vertebrates, that is, amphibians, reptiles, birds, and mammals? If so, where are they being found?

8.1 HOW DO BIOLOGISTS KEEP TRACK OF SO MANY SPECIES?

Supermarkets face problems similar to biologists, namely, how to organize thousands of brand names and food types so that they can be easily found by shoppers and easily stocked, maintained, and inventoried by employees. Suppose you're shopping for a can of peas. One way the store might arrange its food is in alphabetical order. On a shelf somewhere in the middle of the store you would find "Peas, canned," "Peas, dried," and "Peas, fresh." No problem for the shopper, but next to impossible for the grocer.

Instead, supermarkets are arranged in a series of sections, "Dairy products," "Meat," "Produce," "Package goods," "Gourmet foods," and so forth. To find "Canned peas," you go first to "Package goods," then you find "Canned goods," then "Canned vegetables," and finally "Canned peas." Notice that the organizational system is based on similarities and differences. All package goods share certain characteristics that distinguish them from produce. Notice, too, the system is *hierarchical*. "Chicken" is a subset of "Meats" and is itself subdivided into "Whole chickens" and "Chicken parts," which are further subdivided into "Breasts," "Drumsticks," and so on. Finally, notice that the system is somewhat arbitrary. Canned garbanzo beans might be considered canned vegetables in one store and a gourmet item in another (Figure 8-1).

Whereas grocers deal with a few thousand items, biologists deal with roughly a million and a half and anticipate dealing with many more. Their system of organization is

Figure 8-1. Modern supermarkets are organized so that it is relatively easy to find a particular food item among racks and stacks of thousands of similar items. The classification system used by biologists can bring similar order to and keep track of millions of species.

also based on comparisons and hierarchical groupings, and it is somewhat arbitrary. Not surprisingly, the biological system of classification took several hundred years to develop.

8.1.1 Prior to Darwin, Classification Was Concerned with Describing "Natural Order"

As with so many other areas of biology, taxonomy can be traced back to Aristotle. He arranged objects, including animals, into groups through a series of "either–or" comparisons. Either an object is living or nonliving; animal or plant; "blooded" or "non-blooded;" with feet or without; many footed or four footed, and so on. This resulted in a listing of animals by what was perceived to be a natural order leading progressively from the simplest animals to the most complex. His principal student, Theophrastus, used similar "either–or" groupings to classify plants as trees or shrubs, subshrubs, or herbs. Their system of classification, such as it was, persisted with little change for nearly 1,500 years.

The 14th century saw a vast increase in exploration, as mariners, particularly from Europe, plied the world's oceans. Many expeditions included naturalists, who described and collected specimens and brought them back to Europe. For several hundred years, museum collections grew and, as they did, so did the need to organize, classify, and name organisms.

Throughout these early days of classification, the dominant philosophical view was that species were fixed, or unchanging, and could be arranged in some sort of natural order. The job of the taxonomist was to describe species as accurately as possible. This would serve, first, to assist in accurate identifications and then, perhaps more importantly, to reveal life's natural order. Each species was seen as a separate type of life, perfectly conceived. Species descriptions sought to list each species' idealized characteristics, which became the *archetype* for that species. The perfect or archetypal rose was seen to have one set of characteristics, while the archetypal dog another. Individual variations were largely ignored.

The archetypes that shared a particular set of characteristics could be arranged together in groups. At first there was little agreement as to which characteristics should be chosen for grouping or how the groups should be arranged. By the beginning of the 18th century there were dozens of separate classification systems; some focused on plants, while others focused on particular groups of animals. Organizationally, they shared little in common and were largely incompatible. Biological classification was on the verge of chaos.

Into this chaos came the Swedish botanist Carl von Linné, who lived from 1707 to 1778. He was one of biology's most enigmatic and complicated characters. He was interested in numerology, especially taken by the numbers 5, 12, and, 365. He was highly religious and later in life became a mystic. He had great literary powers and enjoyed great fame during his life. He traveled widely, but spoke only Swedish and Latin. He was also a pragmatist who wrote meticulously detailed descriptions of plants and animals. His obsession with classification knew few bounds. Not only did he classify plants, he developed an elaborate classification of botanists (plant physiologists, "botanophiles," collectors, and so on).

His contributions were great. By the strength of his personality and sheer volume of work, he brought simplicity and consensus to the chaos of the field and in the process described over 8,000 species of plants and animals (Figure 8-2). To each he assigned two Latinized names, unique to each species. (Latinized names were so important to him that he changed his own name to Carolus Linnaeus.) The first name was the **genus** name; closely related forms could share this name and thus be grouped together. The second name was the **species** name, which distinguished between closely related forms. For example, in the genus *Canis*, he assigned to the domestic dog the scientific name *Canis*

(a)

(b)

(c)

Figure 8-2. Carolus Linnaeus named over 8,000 organisms, including *(a)* the domestic dog, which he named *Canis familiaris; (b)* the club moss, which he named *Lycopodium salago;* and *(c)* the red fox, which he named *Vulpes vulpes.*

familiaris and to the gray wolf *Canis lupus*. This system of **binomial nomenclature** developed by Linnaeus continues in the naming of species today.

To deal with so many species, Linnaeus saw the need to organize them into higher taxonomic categories. Related genera (plural of genus) were combined into categories he called "orders," which were similarly combined into "classes." The highest taxonomic category was kingdom, and there were two: Plant and Animal. He did this as a matter of necessity. "An order," he wrote, "is a subdivision of classes needed to avoid placing together more genera than the mind can easily follow." Whereas he felt his descriptions of species and genera represented "natural order," he saw orders and classes to be largely artificial.

The Linnaean system of classification was widely accepted into mainstream biology. Buffon, a contemporary of Linnaeus; Lamarck, one of Buffon's students; and Cuvier (refer back to Chapter 2 and the web site for more extensive discussions of some of their other contributions to biology) who accepted and refined the Linnaean system, concentrating on animals. Others extended Linnaeus' work with plants. Thus, the task of describing Earth's "perfect" species and life's natural order continued.

8.1.2 After Darwin and Mendel, Emphasis Shifted to the Ancestry of Species

Darwin's theory of evolution through natural selection stimulated a complete reevaluation of the meaning of biological classification. Prior to Darwin, classifiers saw themselves as describing fixed, immutable, perfect entities. They placed no special significance on the fact that species share characteristics. That species do was simply a convenience that allowed them to be grouped. Darwin not only viewed species differently, he provided an explanation of why certain characteristics are shared. Rather than perfectly immutable entities that could be arranged into a "natural order," he saw species as constantly, albeit slowly, changing. The fact that they share characteristics was to be expected, because, he realized, all species, at some point in their past, share common ancestors. Species that share the most characteristics are the most closely related. Thus, dogs are closely related to wolves, and neither is too far removed from their common ancestor. Dogs and cats are more distantly related; their common ancestor is further back in evolutionary history.

Darwin's theory had little effect on the everyday work of the taxonomist. The archetype was replaced with the **type specimen**, which was either the first specimen collected or a representative specimen of a given species. A detailed description of the type specimen defined each species' essential characteristics against which subsequent specimens could be compared. However, individual differences between specimens were to be expected after Darwin's theory became known. They were understood to be the result of natural selection.

Mendelian and population genetics affected biological classification by stressing the importance of populations. Species were no longer seen as "types of life," but rather as sets of individuals that could interbreed. The job of the taxonomists became to describe those characteristics that differentiated one distinct *population* of a species from another. Type specimens are still collected and described, but any given characteristic within a population is expected to range through a predictable set of values. For example, the weight of an adult male wolf is not expected to match some idealized value, but to range between 20 and 80 kilograms (22 to 175 pounds).

8.1.3 Today, Modern Tools Are Yielding Direct Evidence of Evolutionary Relationships between Species

Fast Find 8.1 Classification

The number one objective of modern systematists is to describe the evolutionary relationships between species. In Chapter 2, we defined species as "a group of organisms that can interbreed in nature." Determining whether or not individuals interbreed is often difficult and is sometimes impossible under natural conditions. Traditionally, taxono-

Figure 8-3. Although ichthyosaurs resemble dolphins and sharks (see Figure 2-12), they are not considered to be closely related. Ichthyosaurs had no mammary glands and were therefore not mammals. Their skeletons were made of bone, not cartilage, as in sharks and their relatives. Because of the structure of their skeletons, ichthyosaurs are classified as reptiles.

mists defined species by comparing their structures, forms, and other relatively easily observed characteristics. They assumed that if individuals are different they cannot or will not interbreed, but even this is not as straightforward as it might sound. First, natural selection is not an evenhanded process. Different populations change at different rates. At what point two populations have differentiated enough to be considered separate species is sometimes a difficult judgment call. Natural selection does not make life easy for the taxonomist.

Inferring evolutionary relationships between species is often more difficult still. Sometimes, distantly related species evolve similar characteristics when faced with similar environmental challenges (**convergent evolution** was discussed in Chapter 2). Compare dolphins (mammals), sharks (primitive fish), and ichthyosaurs (extinct marine reptiles) (Figure 8-3). If they share so many characteristics, why aren't they more closely related?

Under other circumstances, closely related species may differentiate (**divergent evolution** was also discussed in Chapter 2). At first glance, hummingbirds seem to have little in common with ostriches. What makes us think they are related?

In spite of these and other difficulties, taxonomists and systematists regularly identify new species and infer relationships between them. Given the complications, it is perhaps not surprising that adjustments are sometimes necessary. Not too many years ago, snow geese (*Chen hyperborea*) and blue geese (*Chen caerulescens*) were thought to be separate species, and although inbreeding was known to occur, the offspring were assumed to be sterile. When this was seen not to be so, snow geese and blue geese were combined into one species, *Chen caerulescens*. Blue forms and white forms are now seen as **color phases** of the species, that is, individuals with different appearances. Red foxes of Europe (*Vulpes vulpes*) were once thought to be separate from red foxes of North America (*Vulpes fulva*). How could they possibly interbreed, separated by an ocean? Recent historic evidence suggests red foxes were released in the southeastern United States from Europe around 1750. Today, we recognize only one species of red fox on both continents, *Vulpes vulpes*. Determining relationships through comparison of structures and form is seldom easy.

Recently, new technologies in molecular biochemistry are making systematics more straightforward.[1] Relationships between species can be worked out by comparing their proteins, RNAs, and DNAs. Early efforts compared various species' cytochrome *c*, an enzyme present in most cells. First, samples of cytochrome *c* were isolated from cell samples of a variety of species. Next, the order of their amino acids was determined. Mutations change the amino acid sequence of proteins and are thought to occur at relatively

---

[1]For a review of biochemistry, see Sections 4.2 and 6.3 of the text and the Chemistry Review section of the CD.

constant rates. Closely related species that have recently separated from their common ancestor should have accrued few mutations, and the structures of their cytochrome c should be similar. By the same reasoning, more distantly related species should show greater variations in their cytochrome c.

Similar analyses compare species' mitochondrial DNA (mtDNA), ribosomal RNA (rRNA), and nuclear DNAs. For the most part, these techniques have yielded few surprises. Traditional taxonomists and systematists did their work carefully, and the relationships they worked out have passed yet another test of validity. These chemical tests are most valuable in settling on going controversies. For example, evolutionists have long contended that the first land vertebrates, early amphibians that are related to today's salamanders and frogs, evolved from fish. But which group of fish (Figure 8-4)? To

(a) Early amphibian

?   ?

(b) Lobe-fin fish   *(c)* Lungfish

Figure 8-4. Who were the ancestors of early amphibians? Because of similarities in their skeletons, it appears that amphibians evolved from bony fish. But from which group of fish? The pattern of bones in the amphibian's limbs closely resembles the arrangement in the lobe-finned fish's fins. Therefore, they were thought to be the ancestors to amphibians. But recent studies suggest otherwise. Comparisons of RNA of these three animal groups suggest that the more likely ancestors of early amphibians are lungfish.

answer, we might rephrase the question, "Which group of fish shares the most characteristics with primitive tetrapods (four-legged vertebrates)?" Two distantly related candidates emerge. Lungfish have lungs, which the first tetrapods surely required, so perhaps they were the ancestors to tetrapods. But the lobe-finned fish, of which the *Coelacanth* is an example, has fins with a tetrapod-type skeleton. Over the years, most specialists have favored lobe-finned fish as the tetrapod ancestor, but recent comparison of rRNAs suggests otherwise. The sequence of nucleotides in lungfish more closely resembles that in amphibians than it does that of lobe-finned fish. Is the question settled? Not yet. Remember, biologists are always skeptical. Some would like to see more comparisons of protein and nuclear DNA structure. But the scale of probability currently tips a bit toward lungfish.

Classifying so many different organisms is a daunting and sometimes esoteric task. Exceptions abound and make life for the taxonomist interesting. An ancient Chinese proverb has it that naming things is the first step to wisdom. Try being a taxonomist, and as you encounter organisms day to day, classify them and try to look for their similarities and differences.

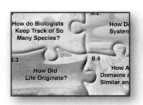

Piecing It Together

1. The purposes of biological classification are threefold:
   - **To assist in species identification.**
   - **To assign formal, consistent scientific names to species.**
   - **To describe ancestral relationships between species.**

2. Traditionally, species have been described and relationships proposed through detailed descriptions, stressing structure and form. Recently developed biochemical techniques allow comparisons of amino acid sequences in proteins and nucleotide sequences of mtDNA, rRNA, and nuclear DNA.

3. Biology's classification system is both hierarchical—based on groups within groups—and comparative—based on the characteristics groups share. Groups and shared characteristics are more than conveniences; often they show evolutionary relatedness.

8.2 HOW DOES THE SYSTEM WORK?

Stated simply, the work of the taxonomist is to place organisms into a series of hierarchical groups called **taxa** (plural of taxon). The broadest taxonomic groups contain the most organisms and are themselves broken up into less-inclusive groups, which are subdivided still further until the level of individual species is reached. All organisms within a particular taxon share certain characteristics. Until recently, the choice of which characteristics to use for taxonomic comparisons was somewhat arbitrary and often involved analyses of homologous structures (discussed in Chapter 2). As new species evolve, they inherit and modify the characteristics of their ancestors. As we shall see, interpreting relationships between organisms, the work of the systematist, is far from arbitrary.

8.2.1 At First, There Were Thought to Be Only Two Kingdoms; Now There Are at Least Six

Fast Find 8.2a The Kingdom of Life

From the time of Aristotle to the middle of the 20th century, biological classifiers first divided organisms into either of two **kingdoms**: every organism was either plant or animal. These were the largest of the taxonomic groups.

To be classified a plant, an organism had to possess certain characteristics. Plants tend to be *sessile;* that is, they don't move around much. Structurally, they are made of leaves that are generally green, stems, and roots. By the middle of the 19th century it was known that plant cells are surrounded by thick walls. Later, the significance of green was realized: plants are photosynthetic and produce their own food. Animals possess other characteristics. They are generally more responsive; that is, they move around more than plants. Their cells have no walls, and they don't produce their own food. The first step of the taxonomists—placing organisms into kingdoms—seemed straightforward and easy.

There were some exceptions. What about fungi? They are generally sessile, with thick-walled cells, but they are not photosynthetic. To be grouped together, organisms don't need to possess *all* characteristics of a given group; where necessary, shared characteristics are prioritized. In the case of fungi, thickness of cell walls was thought to be more diagnostic than the mode of nutrition. The predominance of evidence suggested fungi were plants. Bacteria, too, were included as plants. Other exceptions were not so easily dealt with. A microscopic, one-celled organism named *Euglena* was particularly worrisome. It has a long flagella. It has no cell wall. During summer, *Euglena* has well-defined chloroplasts, gathers at the surface of ponds, and produces its own food. But with light less available in winter, particularly under ice, *Euglena* loses its chloroplasts, descends into bottom muds, and moves from place to place decomposing dead material. In summer, *Euglena* is clearly a plant; in winter, clearly an animal. Which is it, really? Not even prioritized characteristics help much here.

In the late 1960s, a radical solution was proposed: create more kingdoms to accommodate the exceptions. Kingdom Protista would include not only *Euglena*, but all similar single-celled organisms. All protists have **eukaryotic** cells, that is, cells with organelles. Some are **photosynthetic**; some are not. Some have thick cell walls, some do not. Kingdom Monera was proposed to accommodate bacteria, or single-celled **prokaryotes**. (If you've forgotten the characteristics of prokaryotes, review Section 4.3.) A third new kingdom included fungi. Biological classification entered into a five-kingdom phase: Monera, Protista, Fungi, Plantae, and Animalia (Figure 8-5).

Still, there were exceptions. What should be done with green algae? Most are single-celled, but many are not. Indeed, multicellularity evolved in this group, although it may have evolved independently in the seaweeds known as brown algae. Today, there are no single-celled brown algae, but fossilized forms are known. Obviously related to the brown algae are the red algae, another kind of seaweed. Presumably, all red algae are and were multicellular (no single-celled fossil forms are known). Are these groups protists or plants? There is not much agreement among biologists. Some consider them primitive plants, whereas others—and we will follow this convention—lump them with the protists.

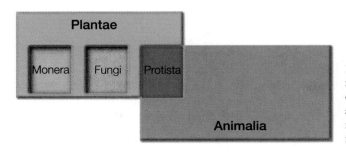

Figure 8-5. How many kingdoms are there? Formerly, all organisms were classified as either plant or animal. These became the first kingdoms. Then, as the differences between organisms became more understood, organisms were reclassified into five kingdoms.

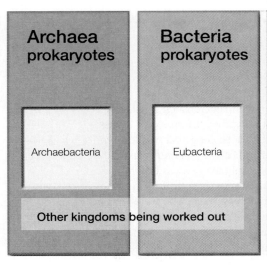
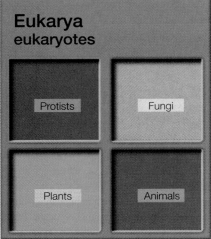

Figure 8-6. The recently proposed addition of domains, a taxonomic category larger and more inclusive than kingdoms, represents a major change to the system of classification used by biologists. The basis of distinction between domains is the structure and chemical composition of cells. Within these three domains there are now six kingdoms.

In Chapter 4, we discussed the possibilities of a third type of cell in addition to prokaryotes and eukaryotes, the Archaebacteria. The existence of a third cell type creates special problems for taxonomy. Where do the new cells fit into biological classification? Today, we recognize a new taxonomic category larger than kingdoms, the **domain**. There are three domains (Figure 8-6). The Domain Archaea includes the newly discovered cell types. The Domain Bacteria includes the other members of the old Kingdom Monera. Each of these domains has one Kingdom, namely Archaebacteria within Domain Archaea and Eubacteria within Domain Bacteria. The Domain Eukarya includes all of the kingdoms and organisms comprised of eukaryotic cells. Notice that within the Eukarya are several kingdoms, namely protists, fungi, plants, and animals. Will there be additional kingdoms within Archaea and Bacteria? Probably. Because of the inherent difficulties in studying them, bacterial classification has lagged far behind that of other organisms. As that changes in the next few years, our understanding of bacterial diversity is bound to expand.

A third cell type creates even more problems for systematists. What are the evolutionary relationships between the domains? Much more data are needed before this question can be answered with certainty. At least some leading questions can be proposed: What does it mean that Archaea share more genes with Eukarya than with Bacteria? Which domain evolved first?

So, how many kingdoms are there, really? The answer is; As many as we need to have. Remember that biological classification is a work in progress. Furthermore, it is work that has only just begun.

8.2.2 Within Each Kingdom, There Are Many Species, Requiring Additional Categories

Groups within groups and shared characteristics—these are the hallmarks of the modern biological classification system. Let's trace a path through this system to see how it works.

All members of the Domain Eukarya have one characteristic in common: their cells are eukaryotic (Figure 8-7). Further, this characteristic excludes all members of other domains; that is, no members of Bacteria or Archaea possess this trait.

Within the Domain Eukarya are literally millions of different species. Some are single-celled, whereas others are multicelled. Some have cells with thick walls, whereas others have cells with no walls. These two criteria go a long way to differentiate the eukaryotes into kingdoms. Multicellular organisms with cells that lack walls belong in the animal kingdom. Biologists prefer the name Kingdom Animalia, following Linnaeus' lead in Latinizing taxon names.

(a)

(b)

(c)

Figure 8-7. In spite of their obvious and great differences, all of these organisms are placed in the same domain, Eukarya. Organisms within this domain are composed of one or more eukaryotic cells, that is, cells with internal membranes and organelles. *(a)* Diatom (magnified 9,200 times); *(b)* fungi; *(c)* roundworm (magnified 156 times); and *(d)* Fairy tern.

(d)

Now think of all the different kinds of animals. The variety is still nearly bewildering, but they can be grouped. Each group is a subdivision of the next larger taxon (Figure 8-8). For example, kingdoms are subdivided into groups called **phyla** (botanists prefer to use the taxon **divisions** to group plants at this level). Simple organisms with tentacles such as jellyfish form one phylum. Those with no legs at all such as worms form several phyla. Many different kinds of animals have legs; numbers vary from two to dozens. Let's follow the classification of organisms with legs and backbones; they belong to the Phylum Chordata.

Within the Phylum Chordata there are many kinds of animals with backbones, still "too many for the mind to grasp," as Linnaeus might say. More taxonomic categories are needed, and in the modern system of classification, phyla are subdivided into **classes**. Phylum Chordata is subdivided into several classes of fish and several classes that, for the most part, live on land. One group of terrestrial, backboned animals are warm-blooded, have hair as a body covering, and possess mammary glands with which females feed young. These are the mammals, Class Mammalia.

Hairy Whales? Given the previously mentioned, why are whales considered mammals?

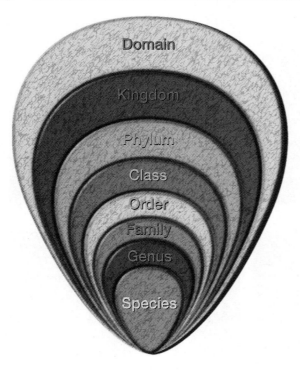

Figure 8-8. The system of classification used by biologists is a series of groups within groups based on similarities and differences.

Classes are subdivided into **orders**, of which Class Mammalia has several. One includes all the monkeys, apes, and their relatives—Order Primate. What do you suppose are the characteristics that would differentiate primates from all other mammals?

Orders are subdivided into **families**. Within the primate order, monkeys can be separated from the so-called great apes. The great apes belong to the Family Hominidae. This family includes gibbons, orangutans, gorillas, chimpanzees, and humans, each of which constitutes one or more **genera**, the taxon into which families are divided. Humans belong to Genus *Homo*.

In times past there were several human ancestors, known only from fossils. These are the species included in the Genus *Homo*. Today, there is only one example of this genus, modern humans, *Homo sapiens*. In Latin this means "self, the wise".

The complete classification of humans is as follows:

> Domain Eukarya (eukaryotic cells)
> Kingdom Animalia (multicellular; no cell walls)
> Phylum Chordata (backboned animals)
> Class Mammalia (warm-blooded, hair, mammary glands)
> Order Primate (opposable thumbs)
> Family Hominidae (no tails)
> Genus *Homo* ("self")
> Species *sapiens* ("wise")

Every species has a similar set of taxa to which it has been assigned, based on its similarities and differences with other organisms.

In particularly numerous groups, especially insects and certain plants, additional taxa are sometimes required. These are formed using prefixes "sub-" (below) and "supra-" (above). Thus, between kingdom and phylum, taxonomists could, if necessary, have subkingdoms and supraphyla. These refinements are available at all taxonomic levels, wherever needed.

Figure 8-9. The shape and structure of antennae are used to separate butterflies from moths. *(a)* The antennae of butterflies end in knobs; *(b)* those of moths either have no knobs or are feather shaped.

(a)   *(b)*

When writing scientific names, biologists use certain conventions. Note that all names are Latinized. Usually, not all taxa are listed, but genus and species often are (and always should be in biological writings). Genus is capitalized; species is not. Scientific names are italicized or underlined to distinguish them from other kinds of terms. In scientific writing, the second reference to an organism within a paragraph or so can be abbreviated. For example, the bacterium *Escherischia coli* is often referred to as *E. coli.* (but not the first time it is referred to in writing).

The biological system of classification is extremely useful to biologists. It can accommodate any number of species. A newly discovered organism can be fit into the system by noting what characteristics it shares with other organisms. For example, if it has eukaryotic cells, it belongs in the Domain Eukarya. If it is a multicellular eukaryote, has cells with walls, and is photosynthetic, it is a plant.

The system is somewhat arbitrary in the selection of characteristics used for comparison. Ideally, characteristics are chosen that easily separate groups, irrespective of importance to the organism. For instance, knobs on the end of antennae separate butterflies from moths. A characteristic with seemingly no overwhelming importance to the organism is of extreme importance to the taxonomist (Figure 8-9).

The system is also focused on interpretation. Theoretically, taxa indicate evolutionary relationships. Organisms that share the same genus are quite closely related. Those that only share domains are distantly related. Furthermore, relationships denote evolutionary history. Whatever point in history two organisms stop sharing taxa is the point at which they stop sharing ancestors. *Panthera onca* (jaguar) and *Panthera leo* (African lion) share a common ancestor that lived a few million years ago. These two cats (Domain Eukarya) share a common ancestor with *E. coli* (Domain Bacteria) dating back to billions of years ago.

Inherent in the statement "at some level, all organisms share a common ancestor" is one obvious exception. What about the first organisms? The question directs our attention to the origins of life itself.

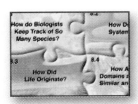

Piecing It Together

1. The biological system of classification catalogs, organizes, characterizes, and names the millions of organisms that constitute life on Earth. Basically, taxonomists place organisms into various classification categories, called taxa, based on their similarities and differences with other organisms. Systematists describe the evolutionary relationships between organisms and taxa. All organisms are placed in the following categories:

Domain (largest, most inclusive taxon)

Kingdom (subdivisions of domains)

Phylum or Division (subdivisions of kingdoms)

Class (subdivisions of phyla or divisions)

Order (subdivisions of classes)

Family (subdivisions of orders)

Genus (subdivisions of families)

Species (distinct type of life)

2. Today, biologists recognize three domains
- ◆ **Archaea (ancient bacteria with prokaryotic cells)**
- ◆ **Bacteria (true bacteria, also with prokaryotic cells)**
- ◆ **Eukarya (organisms with eukaryotic cells)**

3. The Domain Eukarya currently has four kingdoms: Protista (single-celled and simple multicelled eukaryotes), Fungi (single- and multicelled, eukaryotic, nonphotosynthetic organisms with thick-walled cells), Plantae (single- and multicellular, eukaryotic, photosynthetic organisms with thick-walled cells), and Animalia (multicellular, eukaryotic, nonphotosynthetic organisms with cells that have no walls). The old Kingdom Monera (single-celled, prokaryotic organisms) has recently been split into the other two Domains, Archaea and Bacteria. The number and descriptions of kingdoms within these domains are currently in a state of flux.

8.3 HOW DID LIFE ORIGINATE?

**Fast Find
8.3a
Origin of
Life**

Today there are millions of species living in virtually all of Earth's nooks and crannies, but we can imagine that at one time in its far distant past there was no life on Earth. Where did life come from? Let's attempt to answer this most basic of questions.

Interest in life's origins is probably as old as humanity. Indeed, ancient Greek and Roman scholars may have first proposed spontaneous generation to answer the question about the origin of life. It's a question that plagued Darwin. "Life," he answered, "may have originated in a warm, little pond" (Figure 8-10). In 1859, when he first published *The Origin of Species*, the implications of evolution on the idea of spontaneous generation may not have been obvious. The idea, inherent in the cell theory, that all life comes from preexisting life had not yet been proposed. Darwin ducked the question and moved on to other arguments.

Other scientists, too, ducked the question. Science does best with questions dealing with measurable, repeatable phenomena that can be tested, experimented on, and analyzed. How can one do experiments on the origin of life? It was best to leave such questions in the hands of philosophers.

Other scientists were less reluctant. Besides, the question does not go away. If biologists study life, then its origins will continue to be inherently interesting. Definitive answers may never be possible. We may never say, "This is the way life evolved on Earth." But we may garner enough circumstantial evidence to at least say, "It may have happened like this." Let this be our goal.

8.3.1 Early Speculations on the Origins of Life Lacked Experimental Evidence

In the 1920s and 1930s, the Russian biochemist A. I. Oparin and the Scottish biochemist J. B. S. Haldane wrote a series of papers that tweaked contemporary interest in life's

Figure 8-10. Darwin imagined that life might have originated in some small, warm pond (1). Other biologists have proposed (2) hydrothermal vents at the bottom of the ocean, (3) heat-stressed ponds near ancient volcanoes, (4) clay beds in estuaries or bays, (5) tidal pools, (6) within bubbles of foam formed by ocean waves, and (7) even asteroids.

origins. They reasoned that Earth's early atmosphere was considerably different from today's atmosphere. Where is the evidence? Astronomers at the time studied the atmospheres of other planets in the solar system and found little free oxygen. Why should the early Earth be different? Also, rocks that were on Earth's surface 3 billion or so years ago contain free iron. In today's atmosphere, free iron quickly reacts with oxygen to form rust (iron oxide). Absence of rust in those ancient rocks suggests Earth's ancient atmosphere had no free oxygen.

What gases would be present in the early atmosphere? Oparin and Haldane speculated on an abundance of methane; ammonia; nitrogen; water vapor; and, perhaps, free hydrogen. They also envisioned a variety of energy sources present on primitive Earth. Earthquakes and lightning would have been more common than today. No free oxygen would mean no ozone layer in the outer atmosphere to keep the sun's ultraviolet radiation from reaching Earth's surface. All those energy sources working on all those atmospheric chemicals would have stimulated chemical reactions. In particular, they speculated, amino acids, the building blocks of proteins that are the building blocks of cells, would arise—dare we say it—spontaneously.

Initially, Oparin and Haldane's speculations were not well received. Would the idea of spontaneous generation never go away? Besides, where is the evidence? How can there possibly be evidence? Skeptics reasoned that the formation of amino acids would have taken millions, perhaps billions, of years to occur. Untestable hypotheses in the absence of evidence are nothing more than idle speculation, little more than science fiction.

8.3.2 Early Experiments Spontaneously Produced Organic Compounds

The speculations could well have died then, except for an ingenious experiment conducted in 1952. Harold Urey of the University of Chicago and Stanley Miller, a graduate student, built an apparatus that modeled Oparin and Haldane's atmosphere (Figure 8-11). They used electric sparks to simulate lightning in a simulated atmosphere of methane, ammonia, hydrogen sulfide, and water vapor. Water in a flask simulated an ocean. The water's evaporation and condensation simulated the water cycle. Amazingly, in less than

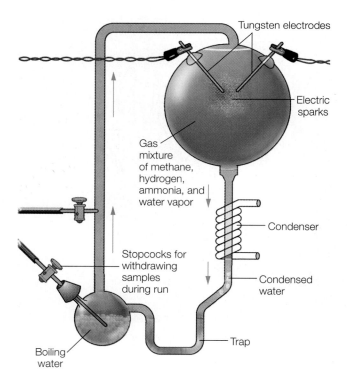

Tungsten electrodes

Electric sparks

Gas mixture of methane, hydrogen, ammonia, and water vapor

Condenser

Stopcocks for withdrawing samples during run

Condensed water

Trap

Boiling water

Figure 8-11. The apparatus used by Miller and Urey. Conditions inside the apparatus duplicated conditions of Earth's early atmosphere. Gases contained no free oxygen, but were rich in methane, ammonia, and hydrogen. Water was heated in the flask at the lower left. Water vapor flowed through the tubes and past the sparking electrode, representative of lightning, which supplied an energy source for the reacting chemicals. The water condenser turned water vapor into droplets that flowed into the trap and eventually back into the flask. In a surprisingly short period of time, organic compounds, including amino acids, gathered in the trap.

a week, their water turned cloudy. Amino acids had formed!

Others repeated the experiments with similar results. By varying the composition of gases and other conditions, other carbon compounds appeared, including carbohydrates, lipids, the components of DNA and RNA, and other amino acids. Not only was it possible to synthesize the building blocks of life in the proposed ancient atmosphere, the syntheses took place easily and quickly.

Could Oparin and Haldane have been correct? More circumstantial evidence accumulated. Astronomers found simple organic compounds in meteorites. Again, scientists were initially skeptical. Some derided the findings. Maybe the meteor hit a bird on its trip to Earth. A more reasonable explanation was that, as they streak toward Earth, meteors may pick up bacteria and other spores. Careful analyses of several meteorites found the questionable compounds on the inside. Other instruments remotely sensed methane deep in open space. Carbon-based compounds, once thought only to be organic, are apparently common throughout the universe.

Then the pendulum of technical opinion swung in the other direction. Astronomers and geologists became convinced that Earth's initial atmosphere could not possibly have been as Oparin and Haldane had imagined. It was much more likely to have been composed mainly of carbon dioxide, nitrogen, and water vapor. In this atmosphere, still lacking free oxygen, organic compounds would be much less likely to form. Speculations continued: perhaps meteors or comets brought Earth its first precursor carbon compounds. More circumstantial evidence accumulated: fossils of ancient bacteria that are 3.5 billion years old turned up in western Australia. At least 11 different kinds of fossils were found there, suggesting that life must have evolved quickly, in something less than a billion years.

What is the current state of conventional wisdom with respect to the origin of life? To paraphrase Rudolph Virchow, "even though the strict proof has not yet been produced in every detail" (see Chapter 2), it must have been something like this: Almost as soon as Earth cooled to the point where oceans formed, precursor carbon compounds appeared. They were either transported in by comets and meteors, were formed in an atmosphere different from today's, or both. The first steps in the spontaneous generation of life had been taken.

8.3.3 The Next Step Was to Move Beyond Isolated Carbon-Based Compounds to Cells

Just as a pile of bricks is a far cry from a building, the presence of precursor carbon compounds is a far cry from life. How could even an ocean of simple carbon compounds spontaneously organize themselves into living cells? Some of the most complicated chemical reactions must have occurred *spontaneously*. In today's world, such reactions are invariably controlled by enzymes. Among other functions, enzymes bring together amino acids, combine them, and produce proteins. But enzymes are themselves proteins. Producing proteins in the absence of proteins is biologically impossible, at least in today's world.

But perhaps in the ancient world this was not so. Speculators offer us several scenarios. (1) When ocean tidal pools evaporate, salts and other impurities left behind become highly concentrated. In ancient oceans, evaporation would have concentrated the amino acids, making it much more likely that they would combine and form proteins. (2) Bubbles are common in today's ocean and would have also been common in an ocean of carbon compounds. Powerful electrostatic forces inside bubbles would attract amino acids, pulling them into close proximity where again they might interact. Furthermore, when bubbles burst, they spew contents into the atmosphere where other, important chemical reactions could occur. (3) It is also possible that iron pyrite crystals, also known as "fool's gold," and clay crystals, both common on Earth's surface, could similarly attract and concentrate amino acids. Over perhaps hundreds of millions of years, any, all, or similar processes could have transformed an ocean of simple carbon compounds into an ocean of more complicated organic compounds—proteins, carbohydrates, nucleotides, and phospholipids.

Even in this early ocean, a kind of natural selection would tend to favor certain molecules over others. Some of these compounds would have been more stable than others. These would tend to persist. Less stable compounds would tend to disintegrate. Over time, the variety of chemical compounds in the ancient ocean would decrease as more stable compounds evolved and persisted.

The next step in the evolution of life depended on a peculiar property of some of those chemical compounds, the phospholipids. As has been mentioned previously, one end of these rather large molecules is attracted to water while the other end is repulsed by it. If you swirl a mixture of phospholipids in water, they arrange themselves into tiny bubbles covered with a two-molecule-thick skin of phospholipids whose water-seeking ends point out and whose water-avoiding ends point in. As we have seen, today's cells are surrounded with a cell membrane two molecules thick composed largely of phospholipids. Tiny bubbles in an ancient ocean are certainly not yet cells. But their outer coverings begin to resemble cell membranes.

Inside ancient bubbles, perhaps chemicals were concentrated. The contents of one bubble would be considerably different from the contents of another. Their existence was transitory; bubbles exist for only a few moments, then burst. Some, by pure chance, might contain proteins and other chemicals that would stabilize the phospholipid membranes. These bubbles would tend to persist, again favored by natural selection. Some, by pure chance, might contain proteins that could break down complicated chemicals into simpler ones, releasing energy. Bubbles in ancient seas might fuse together, mixing contents, mixing capabilities. Over time, the more stable and complicated of these fortuitous combinations of chemical bubbles would persist and thrive. Eventually, they would reach a level of complexity that could be called **protocells** (Figure 8-12).

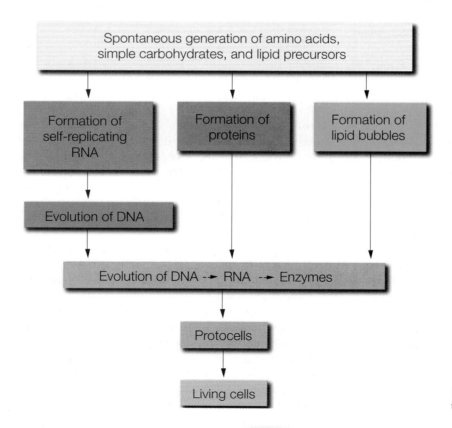

Figure 8-12. Key events in the chemical evolution of life.

Protocells are not yet living organisms. One crucial characteristic of life is still lacking: the ability to reproduce. For years, scientists assumed that cells could not reproduce until the chemical DNA evolved. In today's cells, DNA is the hereditary material passed from one cell generation to another. DNA controls the production of proteins through a much simpler intermediary, RNA. DNA is one of the most complicated chemical compounds known to science. How might it have evolved?

There is a more basic question: Is DNA absolutely essential? Certain viruses, called retroviruses, may provide a clue. Retroviruses contain no DNA. Their "hereditary material" is only RNA. Once inside a host cell, each viral particle contains sufficient RNA to synthesize a new generation of viral particles. Could RNA have done the same in protocells? In 1993, scientists at the Scripps Institute in La Jolla, California, chanced upon a small molecule of synthetic RNA with remarkable properties. Within an hour of its synthesis it began to make copies of itself, and the copies made more copies. Then the copies began to change—evolve, if you will—acquiring new chemical characteristics.

Is the molecule alive? Not yet, because its existence depends on a steady supply of preformed proteins. But in an ancient ocean filled with organic compounds, protocells containing RNA with similar properties, plus RNA that could synthesize enzymes capable of breaking down other organic compounds, plus RNA that could synthesize enzymes capable of building and maintaining cell membranes, might well qualify as the first cells. Later, DNA could have evolved as a method of conveniently and safely storing the vital chemical information contained in the RNA of the cell.

Is this how life and the first cells evolved on Earth? Possibly, or in some process similar to this. Dozens of laboratories around the world are conducting experiments, testing hypotheses that may someday synthesize life. It is an active area of research.

 Synthesizing Cells What is the current state of attempts to synthesize cells in a laboratory? If scientists were successful in synthesizing "life in a test tube," what would be the moral, legal, and philosophical implications?

8.3.4 These First Cells Evolved into the Different Cell Types We See Today

However it may have happened, life appeared on Earth and changed forever conditions of the oceans. These first cells would have been totally dependent on the ocean's preformed carbon compounds for nutrients. They would have had no predators, nothing to control their populations. What probably happened next is easy enough to duplicate in the laboratory. Introduce a few bacteria into a beaker of nutrient broth and what happens? The numbers of bacteria soar until nutrients are exhausted. Then the population crashes. In the ancient ocean, the first cells would have done the same. They would have been totally dependent on the ocean's nutrients for energy. It would have taken time, of course—the ocean is a huge beaker—but eventually, the first cells would exhaust the ancient ocean of its organic soup.

Furthermore, the very existence of these **heterotrophic** cells—cells incapable of producing their own food—would make the spontaneous generation of new organic compounds impossible. Any simple carbon compound floating around in the environment would be quickly gobbled up by a living cell.

With increasing numbers of cells and decreasing nutrients in the early ocean, early life may have faced its first crucial test. Any cell at this time with the capacity to trap and store energy would have had a tremendous advantage. One process early cells seized upon was to trap the energy of sunlight and store it in relatively simple chemicals they produced. Chemicals called pigments absorb light energy, and those related to chlorophyll may have been available (discussed more fully in Chapter 9). Cells that can produce chemicals that store energy are called **autotrophs**, and their numbers soared.

Similar cells, called **chemoautotrophs**, evolved the ability to do the same thing using energy contained in certain inorganic chemicals found in or near the ocean.

Autotrophs proliferated as the carbon-based soup played out. Heterotrophs had to switch from dependence on the organic soup to dependence on organic autotrophs. Forevermore, heterotrophs would limit the numbers of autotrophs, while autotrophs, by their presence or absence, would regulate numbers of heterotrophs. The capacity for balanced ecosystems had evolved.

These are the organisms that left fossils 3.5 billion years ago in the rocks of western Australia (Figure 8-13). A billion years later, some autotrophs evolved methods that improved their ability to trap and store the energy of sunlight, and modern photosynthesis evolved. These new autotrophs were so much more efficient at storing energy that their numbers must have soared. One of the byproducts of photosynthesis is free oxygen, which most organisms of the ancient world found intolerable. In the oceans, processes that built first simple and then complex organic compounds removed much of the carbon dioxide from the Earth's atmosphere. Now latter-day autotrophs removed most of the rest and replaced it with oxygen. Their excess oxygen changed forever the chemical nature of Earth's atmosphere. The atmosphere we know today had evolved.

Other improvements in cellular efficiency also evolved. The first cells were necessarily simple in structure with few internal parts, what we call prokaryotes. Somewhere between 2 billion and 1.5 billion years ago a new kind of cell appeared in the fossil record. These cells, the eukaryotes, had membranes on the inside producing complex systems of parts and pockets in which particular kinds of chemical reactions could be isolated from others (more information about prokaryotic and eukaryotic cells can be found in Chapter 4 and on the CD).

Cellular life could then evolve into what we know today. Organic soups and protocells were gone forever, incompatible with the new conditions of the ocean and atmosphere. Cells perhaps identical to the first oxygen-intolerant heterotrophs are still present as the Archaebacteria. They are confined to deep muds, inside carcasses, within intestinal tracts of more complicated organisms, and in other places free of oxygen. Chemoautotrophs are with us still, found near volcanic vents deep in the ocean, in hot sulfur springs, and similar environments. Today's blue-green algae, or cyanobacteria, must at least be similar to the early prokaryotic autotrophs. They still trap the energy of the sun and make it available to other organisms.

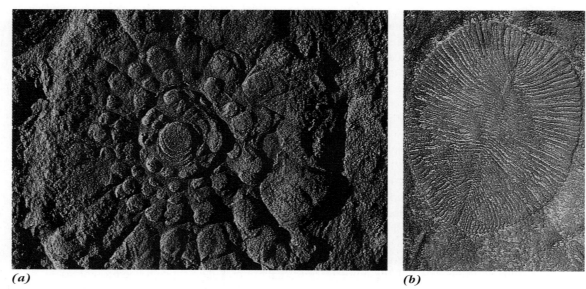

(a)   *(b)*

Figure 8-13. These fossils from South Australia are some of the most ancient multicellular animals ever found. *(a) Mawsonia springgi; (b) Dickinsonia cosata.*

A great deal of evolution occurred among the eukaryotic cells. Today, some are single celled and some are multicelled. Some are autotrophs, and some are heterotrophs. Some are plants, and some are animals. All are composed of cells with similar parts and capabilities.

8.3.5 Within Four Billion Years, Life as We Know It Evolved

However it may have happened, life appeared on Earth and evolved into the myriad forms we see today. Geologists and paleontologists have worked out a rough time line describing when the major milestones of life's evolution, discussed in the last section, occurred (Figure 8-14). They tell us that Earth first appeared around 4.6 billion years ago and was initially much too hot to support complex chemicals of life. Earth must have cooled considerably before life could appear. The oldest known fossil cells found in rocks 3.5 billion years old are far too complex to have been the first cells. Therefore,

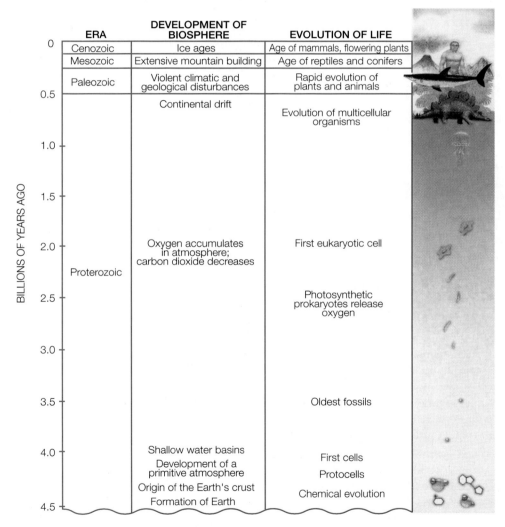

ERA	DEVELOPMENT OF BIOSPHERE	EVOLUTION OF LIFE
Cenozoic	Ice ages	Age of mammals, flowering plants
Mesozoic	Extensive mountain building	Age of reptiles and conifers
Paleozoic	Violent climatic and geological disturbances	Rapid evolution of plants and animals
	Continental drift	Evolution of multicellular organisms
Proterozoic	Oxygen accumulates in atmosphere; carbon dioxide decreases	First eukaryotic cell
		Photosynthetic prokaryotes release oxygen
		Oldest fossils
	Shallow water basins	First cells
	Development of a primitive atmosphere	Protocells
	Origin of the Earth's crust	Chemical evolution
	Formation of Earth	

BILLIONS OF YEARS AGO: 0, 0.5, 1.0, 1.5, 2.0, 2.5, 3.0, 3.5, 4.0, 4.5

Figure 8-14. Earth came into existence approximately 4.6 billion years ago. Life appeared less than 1 billion years later. Those vast spans of time are divided into four great eras based on the structure of rocks and the fossils found within them.

life must have first appeared within this span of time, sometime between 4.6 billion and 3.5 billion years ago. As a rough estimate, let us say that those first most primitive cells must have appeared around 4 billion years ago. For the next 2 billion years, they ruled Earth as dominant forms of life. During this vast span of time, evolution and genetic processes changed and shaped populations to tolerate every known aquatic and damp environment on Earth.

About 2.5 billion years ago, photosynthetic cells formed and changed Earth's atmosphere from one dominated by carbon dioxide to one dominated by oxygen. (How? What happened to the carbon dioxide? Where did the oxygen come from?)

About 2 billion years ago, primitive eukaryotic cells evolved and were a more efficient cell type that allowed further evolutionary possibilities.

Somewhere around 1 billion years ago, multicellular organisms appeared. Groups of cells living together were nothing new; some of the most primitive prokaryotes form colonies of single cells that simply fail to separate after cell division. Multicelluarity evolved whenever some colonial cells specialized, for example, concentrating on movement, food digestion, or reproduction, and other cells became dependent on them for those functions.

Around 600 million years ago, some multicellular organisms evolved hard parts—shells, skeletons, and the like—and the number and diversity of fossils increased. By 500 million years ago, exploitation of terrestrial Earth was in full swing as plants, animals, fungi, and even some algae and bacteria became increasingly less dependent on watery environments for wetness and buoyancy. Life proliferated everywhere. Some species have persisted, little changed from their first appearance. Many others became extinct, replaced by new forms better adapted to environmental conditions. Earth has changed and so has life. Overall, diversity has increased.

Piecing It Together

1. Earth's early atmosphere was considerably different from today's, containing much more carbon dioxide and nearly no free oxygen.

2. Experiments conducted since the early 1950s have confirmed that simple carbon-based compounds, including amino acids, spontaneously form under conditions typical of Earth's early atmosphere. Astronomers commonly observe carbon-based compounds throughout the universe.

3. Over time, naturally occurring chemical processes combined simple carbon-based compounds into more complex ones, including proteins, carbohydrates, lipids, phospholipids, and nucleotides (RNA and DNA), leading to protocells.

4. At some time prior to 3.6 billion years ago, the first primitive, heterotrophic cells evolved. Relatively soon after, primitive autotrophic cells evolved. These cells began to remove carbon dioxide from Earth's atmosphere.

5. The oldest fossils of primitive cells date to 3.5 billion years ago.

6. About 2.5 billion years ago, autotrophic cells evolved whose photosynthesis produced free oxygen.

7. Consumption of carbon dioxide and release of free oxygen by early autotrophs changed Earth's atmosphere into one resembling today's.

8. Around 1.8 billion years ago, eukaryotic cells evolved.

9. Around 1 billion years ago, multicellular organisms evolved.

10. The last billion years have seen an overall steady increase in Earth's biodiversity.

8.4 HOW DIVERSE IS LIFE ON EARTH?

It is beyond the scope of this book (or, indeed, any other single volume) to review all of Earth's biodiversity. Instead, we will hit some high points. To date, taxonomists have described and classified roughly 1.5 million species into three domains, six kingdoms, and dozens of other taxa. New species are being added to the lists daily. Let's discuss the major characteristics of the main groups.

8.4.1 A Brief Look at the Archaea and Bacteria Domains

Fast Find 8.4b Archaebacteria

The diversity of the Archaea and Bacteria domains should not be surprising. They are directly related to the oldest organisms on Earth and have had more time to evolve and differentiate than all others. They thrive nearly everywhere, from the depths of the ocean and deep in the Earth to the upper atmosphere; they are on surfaces of everything you touch, in every body of water, and anywhere it is the least bit damp. Some depend on oxygen, others are indifferent to its presence, and still others find oxygen toxic. Some are highly tolerant of extreme heat, extreme acidity, extreme salinity, and combinations of these harsh environments. Others are no more tolerant of harsh conditions than you are. They are incredibly numerous. Indeed, if you are healthy, the number of individual bacteria in your intestines right now exceeds the number of humans that have ever lived.

Members of both these domains have several characteristics in common (Figure 8-15). All are single-celled organisms. Their cells contain no nuclei and few other organelles except ribosomes. Their DNA is contained in a single, twisted, and circular chromosome that floats free in their cytoplasm. All have relatively thick cell walls made of some substance other than cellulose (unlike plants). When environmental conditions become intolerable, many form spores: they lose most of their cytoplasm; shrink in size; surround themselves with thick cell walls; and become metabolically inactive, sometimes for centuries, until conditions improve. So light are they as spores that they virtually float in air, move vast distances, and occur everywhere.

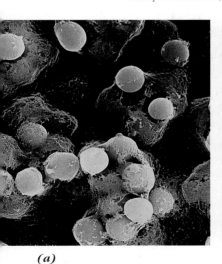
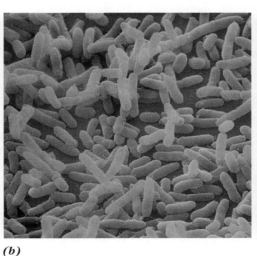
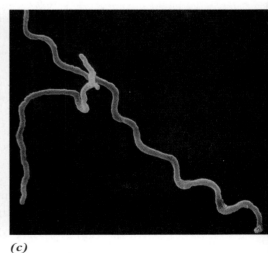

(a)   *(b)*   *(c)*

Figure 8-15. Bacteria are relatively simple organisms. Their single cells have few internal structures and only a single, albeit large, chromosome. Of all life now found on Earth, bacteria are thought to most closely resemble life's earliest cells. *(a) Streptococcus pyrogens,* the causative agent of strep throat (magnified 2,500 times); *(b) Escherichia coli,* a normal inhabitant of human intestines (magnified 3,000 times); and *(c) Borrelia burgdorferi,* the causative agent of Lyme disease (magnified 720 times).

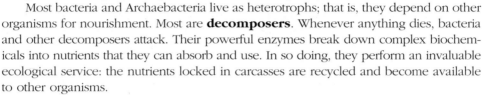

Reproduction is mainly through simple cell division in a process called **binary fission**. However, rudimentary sexual reproduction also occurs. In a process called **conjugation**, two cells form a cytoplasmic bridge, called a **pilus**, through which they pass at least a portion of their chromosome. Some of these microbes routinely pick up stray genes from their environment as a result of decomposition and feeding. Most are readily digested, but occasionally, these genes and the characteristics they control become incorporated into the genome of microbes. Characteristics that evolved in one organism pass directly into another. This blurs the distinction between species.

Within their environments, some are virtually immobile, whereas others come equipped with one or more **flagella** that whip back and forth, pushing them rapidly along. Still others glide, using mechanisms that are poorly understood.

Fast Find 8.4c Eubacteria

Most bacteria and Archaebacteria live as heterotrophs; that is, they depend on other organisms for nourishment. Most are **decomposers**. Whenever anything dies, bacteria and other decomposers attack. Their powerful enzymes break down complex biochemicals into nutrients that they can absorb and use. In so doing, they perform an invaluable ecological service: the nutrients locked in carcasses are recycled and become available to other organisms.

The world of decomposers is highly competitive. Individuals lucky enough to find a carcass must do their work quickly and defend their find from interlopers. Some produce and excrete powerful poisons that only they can tolerate. *Clostridium botulinum* produces the most powerful poison known. Humans know it as a most serious form of food poisoning, but in reality the microbes are simply laying claim to and protecting a nutrient source and living space.

A few Bacteria and Archaea are **pathogens**; they cause diseases in nearly every other organism. In a sense, there is not that much difference between a decomposer and a pathogen. The latter simply starts decomposition before the victim dies. Some pathogens excrete chemicals that destroy cells and tissues. Others rob hosts of nutrients and may secrete toxic chemicals to ward off competitors, which incidentally sickens or kills the host. To the microbe it makes little difference whether its victim is a host or a carcass. But not all decomposers can be pathogens. Pathogens are highly adapted to the harsh environment they occupy. Living things, unlike carcasses, fight back, as we shall see in a later chapter, by creating, environments hostile to most microbes.

Other heterotrophic Bacteria and Archaea live as mutualistic **symbionts**. They form intimate partnerships with other organisms in which both benefit. Two examples of these relationships are the following:

◆ All living things need a source of nitrogen (why?), which should not be a problem because nitrogen is the most abundant gas in Earth's atmosphere. But most living things cannot make direct use of gaseous nitrogen, and thus in many environments, it is a severely limiting nutrient. An exception is certain bacteria that can transform free nitrogen into compounds, especially ammonia and oxides of nitrogen. The process is called **nitrogen fixation**, and the products can be used by other organisms. Now, some plants, especially legumes (peas, beans, alfalfa), have formed symbiotic relationships with certain nitrogen-fixing bacteria. Legumes supply the bacteria with places to live, in **root nodules**, and with photosynthetic and other products in exchange for a steady supply of usable nitrogen (Figure 8-16).

◆ Populations of the bacteria *Escherichia coli* reside inside the intestines of all healthy humans. There they help us absorb water and manufacture vitamins. They also protect us by defending the space they occupy in our bodies from other, potentially harmful microbes. In return, we give them space, protection, and a habitable environment in which to live and steady supplies of nutrients.

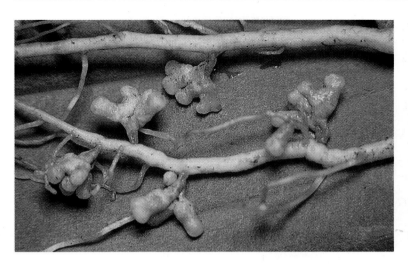

Figure 8-16. Legumes, including clover peas, beans, soybeans, and alfalfa, have symbionts (partners) that supply them with a limited resource. Nodules in their roots house populations of specialized bacteria that convert free nitrogen from the atmosphere, which plants cannot use, into compounds such as ammonia, which plants can use.

An Unlikely Friend If *E. coli* are normally found in everyone's intestines, why do public health officials get excited when they turn up in drinking water? And what about the *E. coli* that taint hamburgers and cause serious diseases? What is being done to control pathogenic *E. coli* ?

Some Bacteria and Archaea are autotrophs; that is, they produce their own food and do not depend on other organisms for basic nutrients. In these groups, too, microbial diversity is extensive. Some, for example, the blue-green algae, use light as their basic source of energy, which they trap with chlorophyll as do plants (except, of course, the chlorophyll is not contained in chloroplasts). Others depend on light as an energy source, but they capture it using pigments other than chlorophyll. Still other autotrophs in these domains use energy sources other than sunlight, especially chemicals in hot springs or those that are spewed forth in underwater vents. These are the **chemoautotrophs**.

Bacteria and Archaea come in four basic shapes: rods (bacilli), spheres (cocci), spirals (spirochetes), and filaments. Traditionally, shape, color, and a few other characteristics have formed the basis of their taxonomy.

8.4.2 A Brief Look at Domain Eukarya, Kingdom Protista

Fast Find 8.4d Protista

Roughly 2 billion years ago, events occurred that are as important to life as we know it as the evolution of cells. Prokaryotic cells evolved into single-celled eukaryotic cells, creating the first protists. Prokaryotic life had already evolved and was (and continues to be) successful. But prokaryotes are somewhat limited. For one thing, they use energy inefficiently. About 2 billion years ago, natural selection apparently favored cells that could get more and faster energy from their nutrients. Eukaryotic cells evolved, although exactly how this might have occurred is unclear.

There are three basic types of protists: **algae**; **protozoans**, or "first animals"; and **slime molds** (Figure 8-17). Algae are autotrophic and, as do green plants, trap light with chlorophyll contained in chloroplasts. They are found in almost all naturally occurring aquatic environments, where they perform an essential ecological function: in the open waters of oceans, lakes, ponds, and slow-moving rivers and streams, floating algae, called **phytoplankton**, convert sunlight into usable energy. Although some of these algae are bacteria and others are simple plants, many are protists. Closer to shore, the importance of green plants as photosynthesizers increases (for example, seaweeds in coastal oceans; submerged, floating, and emergent green plants in freshwater bodies), but even here protistans are sometimes important.

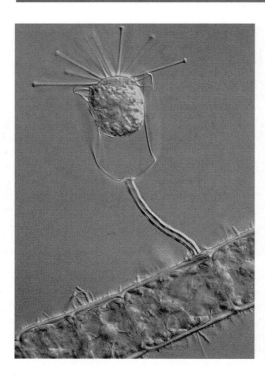

Figure 8-17. Protists are single-celled and simple multicelled organisms whose cells are eukaryotic, meaning that their cells have membrane-bound, internal organelles. Here the protozoan *Paracineta* is attached to a green alga, *Spongomorpha* (magnified 368 times).

Protozoans, another type of protist, are heterotrophs. These single-celled eukaryotes have no cell walls and no chloroplasts, indeed, no chlorophyll. They are commonly found in fresh and marine waters and muds. Most are highly mobile. In the microscopic world, these are tiny grazers, feeding on algae; tiny carnivores, feeding on bacteria and other protists; and tiny scavengers, feeding on dead material and contributing to decomposition. A few protozoans are parasites and cause serious diseases to animals. Human protozoan diseases include malaria, African sleeping sickness, and amebic dysentery.

Still other heterotrophic protists are commonly called slime molds. These have some of the most fascinating life cycles of any organisms. All start life as single-celled, free-living organisms resembling **amoeboid** protozoans (so called because they move by **pseudopods**). Amazingly, at odd times during their life history, these free-living cells organize themselves into multicellular organisms. In one form, the **acellular** or **plasmodial slime molds**, cells aggregate, nuclei divide repeatedly, and the resulting multinucleated masses, called **plasmodia**, stream through soil feeding on bacteria and cellular debris. If moisture or nutrients become lacking, the aggregates break up to form fruiting bodies that resemble tiny mushrooms. These fruiting bodies form spores (by sexual reproduction) that can withstand hostile environmental conditions.

 Another *Plasmodium* What other organisms are called *Plasmodium*? Why are they called that? Why are they important and how are they being controlled?

Cellular slime molds do things a bit differently. Most of their life is spent as amoeboids. When environmental conditions become hostile, they aggregate into sluglike forms that can move rapidly (more or less) in search of better conditions. They, too, occasionally form tiny mushroomlike fruiting bodies when they reproduce sexually.

So, what are these cellular slime molds? First, are they single celled or multicelled? We are so accustomed to seeing these as either–or conditions, we are surprised to encounter organisms that are clearly in between, sometimes one, sometimes the other. Then, in which kingdom do we put them? Convincing cases can be developed for Protista, Fungi, and Animalia. We will leave them in Protista for now, but it is not a clear-cut decision.

Notice that within the protist kingdom are algae that resemble plants, protozoans that resemble animals, and slime molds that resemble fungi. Systematists believe that each of these kingdoms evolved from their respective protistan ancestors. You can visit the CD for further examples of the diversity of protists.

8.4.3 A Brief Look at Domain Eukarya, Kingdom Fungi

Fast Find 8.4e Fungi

Molds, rusts, smuts, blights, yeasts, and mushrooms—there is imaginative diversity even in the common names of Kingdom Fungi. The tone of some of these names suggests we don't like some members of this group very much. Yet they are important in terms of ecosystem function and human society.

Because their eukaryotic cells are surrounded with thick walls, biologists previously considered fungi to be plants that had lost chloroplasts and photosynthetic capabilities. Their cell walls are not made of cellulose, as in plants, but rather chitin, a material more commonly found in exoskeletons. Today, they are placed in their own kingdom, Fungi (Figure 8-18). Some, principally the yeasts, are single celled and microscopic, whereas most are multicellular and macroscopic.

Multicellular fungi that start life as tiny spores may persist for years and be carried vast distances from their points of origin. When conditions become favorable, spore cells divide rapidly. Typically, several threadlike structures, called **hyphae**, grow out in several directions from the spore. As they grow, cells release enzymes that break down organic compounds in the environment into absorbable nutrients. Soon enough, nutrients near the hyphae are exhausted. By this time, the tips of hyphae have grown into new territory, rich in potential nutrients that can be transported back to older cells. The older cells lose walls and membranes between adjacent cells, perhaps to facilitate nutrient transfer. Thus, except for the tip cells, the mass of hyphae, or fungal **mycelia**, become multinucleated protoplasmic masses.

Most reproduction in fungi is asexual. Yeasts reproduce by simple cell division or budding. Hyphae that have broken off of multicellular fungi simply grow, through mitosis, into new individuals. Many fungi form fruiting bodies that produce and release asexual spores. Although some fungi never do so, others resort to sexual reproduction, at least occasionally. Typically, gametes form zygotes that immediately undergo meiosis and grow into new individuals. Note that, unlike animals, in which only the gametes are **haploid**, all cells of these mature fungi are haploid.

Especially in terrestrial ecosystems, fungi are important decomposers, competing actively with bacteria for "king of the composters." As such, they recycle vital nutrients on which other organisms depend. Fungi are extensive and numerous. For example, only about 10 percent of the cells of a living tree are actually alive at any one time. The rest are dead cells in roots, trunks, and branches that support and protect vital, but less exten-

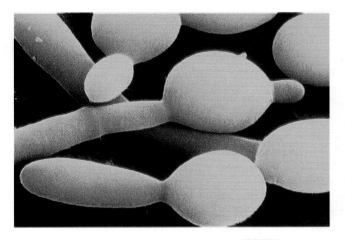

Figure 8-18. Like plants, the cells of fungi are thick walled. Unlike plants, they are heterotrophic; that is, they cannot manufacture their own food. Fungi are important decomposers, recycling nutrients to other organisms. A few, like this *Candida albicans* that causes thrush, are parasitic (magnified 4,730 times).

sive, living cells. Once dead, the tree becomes fair game for fungi. In a rotting log, as much as 30 percent of its mass may be living fungal hyphae. Thus, in a sense, more of the tree is alive when it is dead than when it was alive.

Not only are fungi found in dead trees, they are the primary decomposers of dead leaves. The soil beneath our feet is laced with microscopically thin hyphae that decompose organic matter and are nearly as numerous as in rotting logs. Any individual mycelium may be meters long. Obviously, competition with other fungi and bacteria is often keen. Like bacteria, many fungi produce poisons that they can tolerate and that discourage competitors and predators. These are the basis of much of our antibiotic industry. We collect, purify, and encapsulate fungal poisons and feed them to the sick in controlled doses concentrated enough to kill pathogens but not endanger patients. Penicillin is an example.

Some soil fungi have abandoned the decomposer lifestyle to form symbiotic relationships with other organisms. One of the most important of these is the **mycorrhizal associations** formed with trees. Fungal hyphae surround the tree's root hairs and extend out into neighboring soil. In exchange for photosynthetic products, especially glucose, mycorrhizal fungi provide trees with soil nutrients and water. In some cases, healthy tree growth totally depends on healthy associations with symbiotic fungal partners.

Lichens are the product of another symbiotic relationship involving fungi and blue-green algae. Typically, lichens grow on rocks, tree trunks, or at high latitudes and altitudes—harsh environments in which soil nutrients are lacking or unavailable. Algae provide lichens with photosynthetic products. Fungi provide protection and anchorage, and they release powerful chemicals that etch nutrients out of rock. In some lichens, the symbioses have become so intimate that individual algal and fungal cells are no longer distinguishable.

Some fungi are parasites, especially on plants. Farm crops are particularly susceptible to fungal parasites because the crops occur in large fields and are tightly packed populations of potential victims. Rusts, smuts, and blights are the best known and most loathed by farmers. Humans, too, have fungal diseases. Athlete's foot, ringworm, and vaginal yeast infections are best known.

Fungi also provide important human foods. Certain kinds of fungi convert soybeans into soy sauce and, along with specific bacteria, milk into cheese. (Those little spots in blue cheese are colonies of edible mold.) Yeasts produce alcohol and carbon dioxide in their cellular respiration. In bread making, this carbon dioxide is used to "raise" the dough, while the alcohol is boiled off by baking. In beer and wine making, it's the alcohol that is preserved, while the gas is vented off. See the CD for further discussions on Kingdom Fungi.

8.4.5 A Brief Look at Domain Eukarya, Kingdom Plantae

Fast Find 8.4f Plantae

At last, we come to kingdoms people are most familiar with, first plants, then animals. As mentioned previously, all members of the Kingdom Plantae share certain characteristics (Figure 8-19). Their eukaryotic cells are surrounded with thick cell walls. They are autotrophic. Green is the color most often associated with plants, because at least some of their cells contain chloroplasts laced with chlorophyll, that trap and store the energy of sunlight in molecules of glucose.

Members of this kingdom were among the first organisms to live out of water. The transition from aquatic to terrestrial life was not easy. Several important barriers had to be overcome. First and most obvious, all cells must have water. In aquatic environments, cells are surrounded with water; in terrestrial environments, water is limited. The thick wall surrounding plant cells resists water loss.

Gravity is another barrier to terrestrial life, especially for large organisms. In water, organisms can float, but in terrestrial environments they need support. Thick cell walls are again important in this regard. Their strength combined with **turgor pressure** (the internal pressure of water against cell walls) resists gravity. Another accommodation to gravity is the extensive root systems found in the most successful plants, which provide anchorage. Thick cell walls and extensive roots have allowed some terrestrial plants (for example, redwood trees) to become the largest organisms that have ever lived—in spite of gravity.

Third, all cells must have nutrients and oxygen. In aquatic environments, organisms can simply wait until both come to them in solution. In terrestrial environments, organisms must seek out both. Not only do the root systems of the most successful plants provide anchorage, they facilitate nutrient and water uptake.

The fourth and most difficult barrier to overcome involves reproduction. All organisms are at their most vulnerable in the gamete stage, but these problems are minimized in aquatic environments. Some aquatic organisms simply spew gametes into the water, and eventually, they find each other. In terrestrial environments, such a strategy would expose the gametes to the ever present threat of drying out. One solution might be to surround gametes with thick protective shells, but then, how would they unite?

These are the challenges of terrestrial life: (1) to survive where water is limited, (2) to overcome gravity, (3) to obtain nutrients and oxygen in relatively dry environments,

(a)

Figure 8-19. Plants are multicelled autotrophs; that is, they manufacture their own food through photosynthesis. They are a dominant form of life in the terrestrial world. *(a)* Ferns, *(b)* flowering plants.

(b)

and (4) to protect gametes. As you work with specific plant groups on the CD, try to determine how each group of terrestrial plant overcomes each of these challenges.

In near-shore freshwater environments, submerged, floating, and emergent plants assume the essential ecological roles played by floating algae in the deep-water environments. Through photosynthesis, they make the sun's energy available to other organisms. In freshwater environments, **green algae**, principally as phytoplankton, are the primary energy producers, a role assumed by **brown** and **red algae** near shore in the world's oceans. **Mosses**, **ferns**, **gymnosperms** (so-called "naked seed plants"), and especially **flowering plants** are the energy producers of terrestrial environments, and for humans, too. Nearly everything we eat is either a flowering plant or an animal that eats primarily flowering plants.

The CD contains more information on the diversity of plants.

8.4.5 A Brief Look at Domain Eukarya, Kingdom Animalia

Fast Find 8.4g Animalia

What does it take to be an animal, a member of the Kingdom Animalia (Figure 8-20)? It's not easy to come up with characteristics true of all animals that are not also shared with other organisms. All animals are multicellular (but so are most plants and fungi). All are heterotrophs (so are all fungi and some protists, Bacteria, and Archaea). Animal cells are not surrounded with thick walls (also true of some protists). To be an animal, then, is to have this combination of characteristics: animals are multicellular, eukaryotic heterotrophs, lacking thick cell walls.

There are other characteristics generally true of most animals. Remember that these characteristics are not inviolate—exceptions can and will be found. Generally, animals

Figure 8-20. Animals such as these Japanese cranes are multicelled heterotrophs. They are dominant forms of life everywhere on Earth.

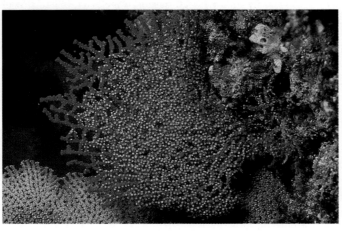

Figure 8-21. At first glance, these red seafans, *Gorgonia adamsii*, may not appear to be animals. But because they are multicellular, with cells that have no walls, and they are heterotrophic, they are considered animals.

are more responsive than plants or fungi; that is, they can respond much more quickly to sudden changes in their immediate environments (Figure 8-21). Push a tree and nothing happens. Push a sleeping tiger and much may happen.

As a result of increased responsiveness, animals move around more than other multicellular organisms. Indeed, much of their evolutionary history has involved developing structures that facilitate movement: legs, wings, surface muscles that change their shape and result in movement (think of a jellyfish pulsing in the ocean). In this latter category are limbless forms such as the three phyla of worms (**flat**, **round**, and **segmented**) that have no limbs. Indeed, limblessness is a popular body plan among animals and is not restricted to worms; think of snakes and eels. Not all animals are motile. **Sponges** and **mollusks** such as mussels and barnacles stay put throughout much of their lives. Can you think of others?

The structure and organization of animals are generally more fixed than other multicellular organisms. All mammals have four limbs. All insects have six. But how many limbs does a tree have? How many hyphae does a fungus have? There is no fixed number. Also, whereas plants and fungi continue to grow a little bit each year, whenever conditions warrant, throughout their lives, animals tend to have spurts of growth when young, and little or no growth when adults. Again, you may be able to think of exceptions.

Most animals have predictable body plans; that is, individuals of any particular group all have roughly the same shape, although sponges are an exception. Furthermore, animal bodies tend to be symmetrical. Most animals are **bilaterally symmetrical**, with a left side that is the mirror image of a right side (Figure 8-22). Some animals are **radially symmetrical**. In these animals, such as **jellyfish** and their relatives and adult **spiny-skinned animals**, any number of lines can be passed through the body dividing right and left mirror images.

All terrestrial organisms face problems associated with limited water. Thick cell walls, an asset to terrestrial plants and fungi that tend to hold water within the cells, are lacking in animal cells, which quickly lose water when exposed to dry air. The solution is animal tissues are bathed in water and their bodies surrounded with an impervious barrier: shell, hard outer skeleton, or waterproof skin. Their extracellular water is high in dissolved sodium chloride. Indeed, the salinity of this water approaches that of seawater, leading some to observe, perhaps fancifully, that as animals evolved into terrestrial environments, they "carried their ocean with them."

(a)

(b)

Figure 8-22. (a) Most animals like this butterfly from Japan, *Papilia xuthus*, are bilaterally symmetrical. One side of their bodies is the mirror image of the other. (b) A few animals, like this starfish and other spiny-skinned animals, are radially symmetrical. Similar-shaped body parts seem to radiate from a central point. In what other animal phylum do members show radial symmetry?

(a) *(b)* *(c)*

Figure 8-23. The transition from water to land occurred several times. *(a)* The land snail *Allogona*; *(b)* the mud worm *Phascolosoma gould*; and *(c)* the amphibian *Ambystoma opacum*.

Animals, unlike fungi and plants, apparently made the transition from water to land several times during their evolutionary history. Presumably, snails and slugs (**mollusks**) became terrestrial independently of insects (**jointed-legged animals**) and amphibians (**back-boned animals**) (Figure 8-23).

Many of the evolutionary trends in animals involved the refinement of their physiological systems, which are much more extensive in animals than other multicellular organisms. For example, the nervous system is nonexistent in the simple sponges, but it becomes netlike in jellyfish, ladderlike in flatworms, and a cord with off-branching lateral nerves in all others. Brains do not appear until the segmented worms, then become increasingly complex in other phyla, and are most complex in backboned animals. Similar progressions describe digestive, endocrine, and other physiological systems. There are exceptions, of course. Excretory and respiratory systems seem to have evolved independently in many animal phyla.

These are the characteristics typical of animals. To be an animal, an organism possesses some, but not all, of them. Within the animal kingdom, diversity is extensive. Each major group can be explored in more detail on the CD.

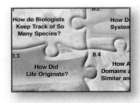

Piecing It Together

As you review the major taxa of Earth's biodiversity, review the principal characteristics associated with each domain and kingdom. Divisions and Phyla are discussed further on the CD.

Domain Archaea (prokaryotic cells)

Domain Bacteria (prokaryotic cells)

Domain Eukarya (eukaryotic cells)

    Kingdom Protista (mostly single-celled and simple mulicelled organisms)

    Kingdom Fungi (mostly decomposers, thick cell walls, no chloroplasts)

    Kingdom Plantae (multicellular, photosynthetic, thick cell walls)

    Kingdom Animalia (multicellular, nonphotosynthetic, no cell walls)

Where Are We Now?

Taxonomy and systematics are currently two of the most active and exciting areas in contemporary biology. There are at least three reasons why:

1. New technologies are being adopted by systematists that allow more precise determinations of relationships between species. Borrowing techniques widely used in biochemistry, systematists can now determine (a) the order of amino acids in proteins, (b) the order of nucleotides in ribosomal RNA and mitochondrial DNA, and (c) sequences of partial and complete genomes of organisms. These are the most basic levels at which existing species become new ones. Species with similar protein and nucleic acid structures are assumed to be more closely related than those with dissimilarities. Furthermore, because the rate at which mutations change these factors is thought to be constant, the length of time two species have been evolutionarily separated can be estimated. Old assumptions are being challenged, some verified, others revised.

2. The growing awareness that biological classification is only just beginning has stimulated increased interest in the field. Estimates now place the total number of Earth's species at 100 million, of which only about 1.5 million have been described. Biologists are finding new species every day, and opportunities to do so may continue throughout most of the next century.

3. At the same time, there is a sense of urgency within the field. Human activities, particularly in the tropics, are negatively affecting biodiversity (Figure 8-24). Indeed, some authorities believe that Earth is entering a new "mass extinction event" that may mean the loss of millions of species. At present, species are becoming extinct every day, some before scientists can describe them. Unless they leave fossils, these species will never be added to the list of Earth's species. Their importance to ecosystems will never be fully appreciated. Perhaps most important of all, as they pass into extinction, their contribution to human welfare and the biosphere is forever lost. Some may contain cures for human diseases, be potential new foods for humans, or produce otherwise useful products. Taxonomists are being urged to go into areas where extinctions are thought to be occurring most rapidly in sometimes desperate efforts to catalog species and save remnant populations before they are lost.

Stemming the Tide What is being done to reverse the tide of extinction that seems to be affecting species everywhere? What more needs to be done? How much will it cost? Are the benefits worth the cost?

Figure 8-24. Human activities are having a negative impact on Earth's biodiversity. As more and more forests are cut to be turned into farms, shopping malls, housing developments, and the like, more and more species are becoming extinct.

REVIEW QUESTIONS

1. Approximately how many species have been described by biologists? How is that number likely to change in the future?

2. How are taxonomy and systematics similar? How are they different?

3. The classification system used by biologists is said to be hierarchical and comparative. What does this mean?

4. Arrange the following taxa in order, starting with the most inclusive (largest):

   class   family   kingdom   phylum
   domain   genus   order   species

5. Why is the scientific name of humans, *Homo sapiens*, written the way it is?

6. What are the purposes of biological classification?

7. Today, biologists recognize three domains. What are they? What are the distinguishing characteristics of each?

8. For a long time there were only two kingdoms, plants and animals. Why was it necessary to create more kingdoms?

9. What are the names of the major taxa in which humans are grouped?

10. Where might life on Earth have originated? (Hint: there are several possible answers to this question.)

11. Describe the major milestones in the evolution of life on Earth.

12. What are the characteristics shared by all bacteria? Briefly discuss the range of variations that exist among bacteria.

13. What characteristics are shared by all protists? What are the major groups of protists? What distinguishes one group from another?

14. What are the characteristics shared by all fungi?

15. What are lichens? Why are they difficult to classify?

16. What are some of the characteristics shared by all plants? What are the major divisions of plants? What distinguishes each division from the others?

17. What characteristics are shared by all animals? Briefly discuss the major animal phyla. How is each phylum different from the others?

18. Think of each of the kingdoms and the major phyla within each. Which of these groups of organisms made the transition from water to land?

Chapter 9

Bioenergetics: How Do Organisms Acquire and Use Energy?

Definition of life: "A living organism is characterized by the ability to effect a temporary and local decrease in entropy by means of enzyme-catalyzed chemical reactions." —Isaac Asimov, 1962

—Overview—

Amid the fantastic array of living forms described in Chapter 8, one feature of life is so obvious that we rarely stop to consider it: living things are distinct entities, separate from their environment. What is not so obvious is that that distinction comes at a price, which is paid in the currency of energy. Every living organism requires a constant supply of energy to stay alive. That energy ultimately comes from the sun. If we fail to replenish the energy that we use, we die and slowly deteriorate until the differences between us and our surroundings disappear.

For many centuries, it was believed that the energy of life was somehow different from other forms of energy in the universe. Life was defined and characterized by the existence of a special *vital force*. It was mistakenly believed that the vital force followed its own set of rules, different from the rules that govern the flow of energy in the inanimate world. We now recognize that energy, in all its various forms, is the same in both the living and nonliving worlds. The rules that govern energy apply universally. In this chapter, we will examine what those rules are and how the chemistry of life has evolved to capture and use energy without breaking the rules.

9.1 HOW DOES ENERGY BEHAVE IN THE UNIVERSE?

Can you distinguish between something that is alive and something that is not? The question may seem trivial at first. Of course animals breathe, move, eat, and reproduce—

Chapter opening photo—Every living thing requires a constant supply of energy to stay alive. Plants use the energy of sunlight to make organic molecules. These hikers obtain energy from their food, which is made up of organic molecules produced by plants and other animals that ate plants.

all things that inanimate objects cannot do. When they cease to be alive, those activities also cease. Plants grow, reproduce, absorb sunlight, consume carbon dioxide, and turn it into complex carbohydrates. They, too, are clearly alive. But what of a virus? A virus is a particle of organic material—mainly protein and nucleic acid—that utterly depends upon a living host to reproduce. Some, including the authors of this book, do not consider viruses living. But others argue quite convincingly otherwise. Even more perplexing, is the tiny nematode *Aphelenchus*, a microscopic worm that lives in seasonal ponds. In the winter, the worm crawls and wriggles, clearly alive. But in summer, when the pond dries up, the worm forms a dry coil that could turn into so much dust with a puff of air. When the pond again fills with water, the worm untangles itself and squiggles away none the worse for wear. Is *Aphelenchus* alive when it is no more than a coiled pile of dust? The task of defining life becomes more complicated the closer we look.

9.1.1 The Energy of Life Is Not Unique

For centuries the idea of a special life energy or vital force was so widespread that virtually every culture had a name for it. The Chinese *ch'i*, the *ruh* of the Arabs, the *prana* of the Indians, the *pneuma* of the Greeks are all roughly translated to mean "breath of life." The idea that breathing and life are connected is an ancient one. Indeed, the exchange of gases between living things and the environment is part of the modern conception of what it means to be alive. But for living organisms, the exchange of gases is an outward sign of a more fundamental process of life—that of metabolism. **Metabolism** includes all of the chemical reactions that occur in cells. As a result of metabolism, organic substances are made or converted to other organic molecules, and with many conversions, energy is transformed. Some of that energy can be used to do the work of living. Metabolism and the chemical reactions that constitute it are the means by which organisms extract and use energy to stay alive. Metabolism is a unique feature of life.

But how does metabolism work? How do chemical reactions that change one molecule into another provide the energy needed to stay alive? To understand, we must first learn the rules that determine how energy behaves in any system, living or not.

9.1.2 Energy Can Take Different Forms, But It Is Always Conserved

Fast Find 9.1 Thermodynamics

The word "energy" was not introduced until the relatively late date of 1807. Prior to that time, physicists had carefully studied different forms of energy, especially heat and the mechanical energy of objects in motion. But the idea that heat and movement, and many other phenomena including magnetism, electricity and light, were different manifestations of the same thing—namely energy—was new. We now use the term **energy** to denote anything that can do work.

Work can have many definitions. To a physicist, work is done when a force is applied to an object and the object moves. To a biologist, however, the very act of staying alive is a form of work. To a biologist, work is done when something is moved, such as a muscle, or when something is made, such as a protein or a strand of DNA. Bioelectrical work is done when a nerve cell carries signals from one part of the body to another. Biological work is also done when some sort of gradient is established. Water, for example, is more concentrated in our bodies than in the atmosphere in which we live. Maintaining this gradient is a form of work that requires energy. Regardless of the type of work to be done, energy is always required.

In the late 19th century, studies of mechanical devices showed that the different forms of energy were interchangeable. Light could be converted to heat, for example, or motion could be converted to electricity. Yet even when one form of energy was converted to a different form, the *total* energy always added up to the same amount. These experiments led to the formulation of a natural law that is one of the most important ideas in

❶ An ice cube in a steaming hot cup of tea represents an ordered distribution of energy. Most of the heat energy is concentrated in the tea and far less energy is in the ice cube.

❷ A spontaneous change – the ice cube melting – more randomly distributes the energy into the teacup. As the ice cube melts, the tea becomes lukewarm and entropy has increased.

Figure 9-1. Ice melting in a cup of hot tea illustrates the second law of thermodynamics.

our understanding of energy. We call this law the first law of thermodynamics, or sometimes the law of conservation of energy. Formally stated, the first law of thermodynamics says that energy may change form, but it may be neither created nor destroyed.[1] In other words, there is as much energy in the universe now as there ever was or ever will be.

It soon became apparent that living organisms can no more create energy from nothing than mechanical devices can. The energy for life comes from the radiant heat and light of the sun. The realization that the first law applied to both inanimate matter and living things was the beginning of the end for vitalism, the ancient idea that living things contained a special kind of energy. No longer could life be exempt from the laws of the universe. That which is true of the flow of energy in the nonliving world is true in the living world as well.

Although no one has ever really explained why energy cannot be created or destroyed, the first law of thermodynamics has never been contradicted. But one thing the first law does not tell us is the direction that energy spontaneously moves and changes. That we learn from the second law of thermodynamics.

9.1.3 In Any Energy Transformation, Entropy Increases

Picture an ice cube in a cup of hot tea. In only a moment the ice cube melts, leaving a lukewarm cup of tea. At the start, there are areas in the teacup system in which the heat energy is concentrated, namely the hot tea, and areas with far less energy, namely the ice cube. A spontaneous change occurs in which the energy in the teacup becomes more randomly distributed, or more disordered (Figure 9-1). This is an illustration of the second law of thermodynamics. The second law states that changes always occur in a direction in which the energy of the universe becomes more disordered.

Physicists have given us a name for the amount of disorder in the universe: **entropy**. Systems with high entropy are highly disordered; those with low entropy are highly ordered. The greater the entropy of a system, the more difficult it is to distinguish one part of that system from any other part. In theory, at least, the second law tells us that entropy will continue to increase until every area of the universe has exactly the same amount of energy and exactly the same composition as every other area. Given

---

[1]Albert Einstein later showed that, under some circumstances, mass and energy are interchangeable. These kinds of mass and energy conversions are not characteristic of living systems, however, and for our purposes can be ignored.

enough time, entropy will reach its maximum, and there will be no ordered regions in the universe—no hot stars like our sun, no blue planets like our Earth, no life, just a vast, even space in which no part can be distinguished from any other. This is a rather bleak prognosis, but one that is not expected to be realized for a long, long time.

Though it is difficult to imagine a universe with maximum entropy, we can envision small parts of the universe, small systems like our teacup, in which entropy has reached a maximum. Such a system is said to be in **equilibrium**, or a condition of maximum entropy. The second law tells us that, without an input of energy from outside the system, all systems are spontaneously moving closer to equilibrium at all times. When a system reaches equilibrium, no further net changes can occur unless energy is supplied from the outside.

How, then, is life possible? Surely when a fetus develops from a fertilized egg, the entropy of the proteins, carbohydrates, lipids, and nucleic acids that compose that being is decreased. Humans, indeed all living things, are highly ordered combinations of organic substances. Is this a violation of the second law? When an entire ecosystem, rich with life, colonizes a newly plowed field, has the second law been violated? The very origin of life, described in Chapter 8, is a case wherein the entropy in the primordial soup decreased. These are systems that have moved away from equilibrium. Why is the second law not violated?

If you look closely at the second law you will find the answers. Each time a life process imposes order on a small part of the universe, there is an even greater increase in entropy somewhere else. In the case of life, we can trace the increase in entropy to the burning of the sun. The sun radiates heat and light energy, some of which strikes Earth. Living plants on Earth capture some of that energy and convert it into complex, ordered molecules, which then supply energy to the other life forms.

From the second law, we begin to see that life is an uphill struggle; living things must do biological work to keep the forces of the universe from dismantling their highly ordered bodies and driving them toward equilibrium. To accomplish this work, a constant supply of energy is needed. But many of Earth's living forms cannot tap directly into the sun, which is the main source. However long we bask in the sun's radiant energy at the beach, we are utterly incapable of extracting that energy directly to fuel our life processes. However, some organisms can carry on a process called **photosynthesis**, in which they capture light energy from the sun and use it to build high-energy organic molecules from low-energy inorganic precursors. Such organisms are called **autotrophs**. Those incapable of photosynthesis, or the **heterotrophs**, must consume the organic molecules built by the autotrophs. The energy used for the work of living is stored in the chemical bonds between atoms in food molecules. The chemical reactions of metabolism have evolved such that autotrophs make food and both autotrophs and heterotrophs have access to that rich source of energy.

Life in the Dark There are a few exceptions to the rule that all life forms ultimately derive their energy from the sun. For example, a community of deep-sea organisms relies upon the energy from oxidation of sulfur-containing compounds. What is known about this strange ecosystem, and when was it first discovered? What kinds of organisms are involved?

Piecing It Together

The energy of living things is not unique. It follows all of the same rules that govern the behavior of energy in the inanimate world. Here those rules are reviewed:

1. Life requires a constant input of energy. This energy is used to do biological work, including building large molecules from small precursors, moving, growing and reproducing, and maintaining an ordered gradient of many substances between the living organism and its environment.

2. Energy cannot be created nor destroyed. It can only change from one form to another. This natural law, derived from experimental observations, is called the first law of thermodynamics, or the law of conservation of energy.

3. When energy changes from one form to another, the entropy, or disorder, of the universe increases. Although changes that increase the order of some small part of the universe can and do occur naturally, elsewhere in the universe entropy has increased. This natural law, also derived from experimental observations, is called the second law of thermodynamics.

4. When some small part of the universe has attained maximum entropy, it is said to be at equilibrium. No further changes occur spontaneously in a system at equilibrium. A system at equilibrium can only undergo change if energy is added from outside the system, moving it away from equilibrium. Also, systems at equilibrium are not capable of doing work.

5. Metabolism is all of the chemical reactions that occur in living things

9.2 HOW IS ENERGY TRANSFORMED IN THE BIOSPHERE?

Fire is a remarkable example of an energy transformation. With a rapid release of light and heat energy, chemical bonds between atoms in the fire's fuel are broken. Living things carry out a similar process, but rather than losing all of the energy as light and heat as foodstuffs burn, organisms capture the energy that is released and use it to do biological work.

9.2.1 Animal Respiration, Like Fire, Is Oxidation

For centuries it was mistakenly believed that flammable substances contained a substance called "phlogiston." It was thought that phlogiston flowed into the air when substances burned, and heat and light were a manifestation of that movement. This idea was championed by Joseph Priestley. In the 1770s, Priestley put a candle under a bell jar and found that in that enclosed space the flame made the air unfit to support fire; the flame went out (Figure 9-2). Likewise, a mouse under a bell jar died much as the flame had. The mouse had made the air unfit for breathing. When a candle and a mouse were put under the same bell jar, they both died. Burning, Priestley correctly reasoned, changed the air in the jar in the same way that breathing did. His explanation, however, that both the flame and the mouse filled the jar with poison phlogiston, was not as convincing.

Priestley's contemporary, Antoine Laurent Lavoisier, disproved the phlogiston theory. Some regard Lavoisier's discoveries as among the most important experiments in animal biology. His reasoning went like this: if a burning object emits phlogiston, then as it burns, its weight should decrease. But when he did an experiment in which he carefully accounted for all the products of the fire, just the opposite happened: the total weight increased. Lavoisier correctly reasoned that burning does not add phlogiston to the air, but instead removes something from the air. He called that something oxygen. Mice and flames both remove oxygen from the air, and in so doing, make the air unfit for animals and fire.

Lavoisier was the first to recognize that both fire and breathing are forms of burn-

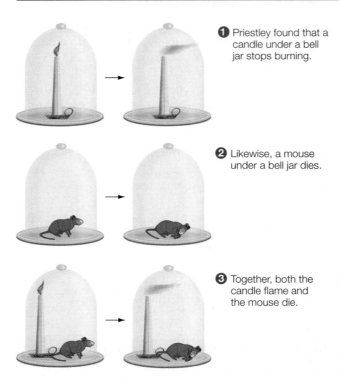

1 Priestley found that a candle under a bell jar stops burning.

2 Likewise, a mouse under a bell jar dies.

3 Together, both the candle flame and the mouse die.

Figure 9-2. Joseph Priestley's experiment proved that both fire and animal metabolism cause the same kinds of changes to the air. His contemporary, Antoine Laurent Lavoisier, proved they both consume oxygen.

ing: they both involve the consumption of oxygen. His further experiments showed that for both the candle and the mouse, the oxygen from the air combines with carbon and hydrogen from the fuel to form carbon dioxide, CO_2, and water, H_2O. This is one example of a process called **oxidation**, or a chemical reaction in which oxygen is combined with some other substance.[2] In both situations, we can write a generalized chemical reaction representing oxidation:

$$C_6H_{12}O_6 + 6\ O_2 \rightarrow 6\ CO_2 + 6\ H_2O$$

The first term in this balanced chemical reaction represents the organic fuel. In this case, we started with glucose, whose chemical formula is $C_6H_{12}O_6$. This is a good starting fuel because cellulose, one of the main constituents of wood, is a polymer of glucose (although because of the way the glucose units are joined, animals cannot use wood for food). Glucose is also the building block of starch and other carbohydrates that we consume as food. The second term of the equation is molecular oxygen gas, represented as O_2. In the course of the chemical reactions in both fire and animal metabolism, the chemical bonds between carbon and hydrogen are broken, and new ones between carbon and oxygen, and also between hydrogen and oxygen, are formed.

By recognizing the similarities between fire and metabolism, we have seen that in both situations old chemical bonds are broken and new ones are formed. Most rearrangements of chemical bonds are accompanied by an energy transformation. In some reactions, energy is absorbed from the environment to form the new chemical bonds. In others, for example, oxidation of glucose, energy is released to do work. Energy in action, that is, energy that is doing work or causing an effect on matter, is called **kinetic energy**. When you move or lift something, you are using kinetic energy to do work. Before fuel molecules are oxidized, however, they contain a different kind of energy, called **potential energy**. Potential energy is energy that is stored or inactive. A pile of wood con-

[2]A more general definition of oxidation is a reaction in which oxygen is added to a compound or atom, or hydrogen or electrons are taken away from a compound or atom.

Figure 9-3. A bonfire is an example of kinetic energy released in the form of light and heat. The pile of wood contains potential energy.

tains potential energy. Before it is ignited, wood exerts no effect on matter. When it burns, the potential energy in wood is converted to kinetic energy, in the form of heat and light (Figure 9-3). Similarly, the molecules in the food we eat contain potential energy. That energy is converted to kinetic energy and used to do biological work when those molecules are oxidized in metabolism.

Thus far we have spent considerable time describing the similarities between the oxidation that takes place in a flame and the oxidation of food in a living organism. But now we must focus on the differences, because it is the differences that define life. There are two main differences. First, fire is a chaotic process that oxidizes everything in its path, and metabolism is an organized series of individual chemical reactions in which only specific fuel molecules are oxidized. Second, fire releases excess energy as heat and light, and living organisms stay at (or close to) the same temperature even though they are undergoing oxidation reactions all the time. In metabolism, the energy of oxidation is captured in other potentially useful chemical bonds. One compound in which living things capture the energy of oxidation is called **adenosine triphosphate,** or **ATP**. ATP will be described a bit later in this chapter. But first, we must examine both of these important differences more closely.

9.2.2 Metabolism Is Both Efficient and Highly Specific

By the time you push your chair away from the dinner table, you have taken in enough fuel to keep entropy at bay, at least for a while. Your stomach is full of metabolic fuel, and you will use it to do the biological work of living. In the process of digestion (discussed in Chapter 11), the food passes from your intestines to your bloodstream. Much of it travels in the bloodstream as glucose, which can enter each of the cells of your body. How do your cells convert the potential energy of glucose into work?

To burn glucose in the same way that wood burns would be impractical, to say the least. Sudden oxidation of so much fuel would mean a massive loss of energy as heat and light, not to mention the destruction that fire would wreak on your delicate tissues. Your cells must stay at or near the same temperature. They must extract energy from your newly acquired food in a controlled manner so that it can be captured to do biological work. The method is elegant. Cells take the high-energy glucose down to low-energy water and carbon dioxide step by miniature step so that there is the best possible chance of capturing the most possible energy (Figure 9-4). The process must be highly *efficient* so as not to lose too much energy as heat, a form in which it cannot be used. Energy in the form of heat rapidly becomes disordered, increasing the entropy of the

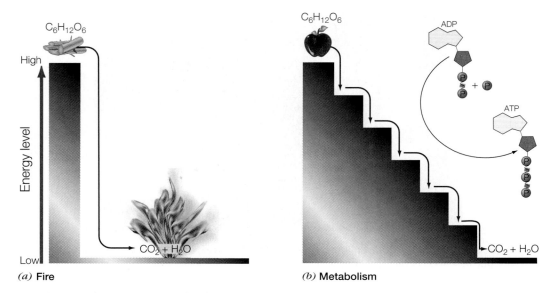

Figure 9-4. (*a*) Fire releases the energy of the wood in an uncontrolled manner. Most of the energy is converted to heat and light. (*b*) The stepwise release of energy that occurs in metabolism allows more energy to be captured in the form of chemical bonds, especially between the phosphates of ATP.

living system. Also, the process must be highly *specific*. In other words, the chemical reactions and their accompanying energy transformations must involve only appropriate fuel molecules and not the other molecules in the cell, so as not to destroy any of the carefully constructed organic molecules of the cell's structure. Only the fuel must burn, not the cellular machinery.

Efficiency and specificity are the hallmarks of cellular metabolism. Both are achieved by **enzymes**, a class of proteins that catalyze, or speed up, the steps of metabolism. It is worth taking a look at how enzymes accomplish both efficiency and specificity in cellular metabolism.

Enzymes

Fast Find 9.2b Enzymes

Enzymes increase the rate at which chemical reactions in cells occur, but even an enzyme cannot force a reaction to go in a direction that is not consistent with the laws of thermodynamics. Yet some reactions that appear to be inconsistent with the second law occur quite readily. Sometimes large, energy-rich compounds are synthesized from smaller, energy-poor compounds. Protein synthesis, discussed in Chapter 6, is an example of this type of chemical process. Enzymes can couple such seemingly unfavorable reactions with other favorable reactions that provide the energy to make them happen.

In other cases, reactions that are highly consistent with the second law apparently do not proceed at all. For example, the paper in this book is a highly ordered complex of cellulose and other materials that contains a great deal of concentrated chemical energy within its covalent bonds. Yet your book is not becoming oxidized to carbon dioxide and water, at least not at a rate that you can detect. (If left for an infinitely long time, this book would indeed begin to show signs of oxidation, but you should be finished with your course in biology before that happens.)

The reason a thermodynamically favorable event may not happen at a discernible rate is that even favorable reactions may need a bit of a push to get them started. Before new chemical bonds can form, the old chemical bonds must first be broken. While the overall reaction represents a net decrease in energy, this first step presents an energy barrier preventing equilibrium from being realized. That initial energy barrier is called the *energy of activation* (Figure 9-5). A touch with a lighted match is all it would take to push the paper in this book over the energy of activation and downhill toward equilibrium.

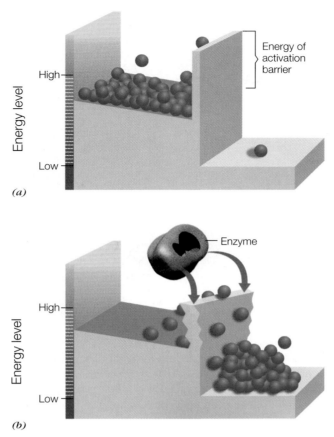

Figure 9-5. (*a*) The energy of activation is a barrier that prevents molecules from undergoing otherwise favorable reactions. (*b*) Enzymes lower the energy of activation barrier, allowing the reaction to proceed.

The glucose molecules in your cells are like the pages in this book. They need a little help to get them past the energy of activation barrier that is keeping them from moving toward equilibrium. But an enzyme is nothing like a match. A match pushes paper toward equilibrium by raising the overall energy of the paper molecules enough to get them over the hump, but an enzyme uses a different tactic. Instead of raising the overall energy of all the glucose molecules in a cell, a single enzyme molecule joins with a single glucose molecule and stretches and bends its chemical bonds until the energy of activation is actually made lower. Enzymes speed up thermodynamically favorable reactions by lowering the energy of activation.

Metabolic Efficiency

The first hallmark of enzyme-catalyzed reactions is metabolic *efficiency*. The most efficient reaction is the one that either captures as much energy as possible from an energy-yielding reaction, or spends as little energy as possible to catalyze a reaction that consumes energy. For example, when glucose is oxidized, enzymes ensure that the energy released is captured in the form of ATP. Other enzymes can use the energy of ATP to drive thermodynamically unfavorable reactions, such as synthesis of large proteins (other enzymes, for example) or new strands of DNA prior to cell division.[3]

---

[3]Even in the most efficient enzyme-catalyzed reactions, a significant portion of the available energy is converted to heat. Animals and even some plants use this heat to keep their body temperatures above that of the environment.

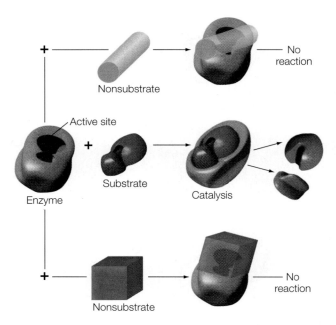

Figure 9-6. Enzymes catalyze only those reactions involving their specific substrates. Other molecules do not react because they do not fit into the active site on the enzyme.

For the energy of glucose to be captured in metabolism, it must be released step-wise, a little at a time. There may be as many as 25 different chemical reactions, each catalyzed by a different enzyme, before glucose is fully oxidized to carbon dioxide and water. We could rewrite the balanced chemical equation for the process like this:

$$C_6H_{12}O_6 \rightarrow A \rightarrow B \rightarrow C \rightarrow (20 \text{ reactions}) \rightarrow X + 6\,O_2 \rightarrow 6\,CO_2 + 6\,H_2O$$

where each of the letters A through X represents a **metabolic intermediate**, or a short-lived compound that occurs at a point in the pathway somewhere between the starting point and the end. From the point of view of thermodynamics, its not how we get from glucose to carbon dioxide and water that matters. There can be, and indeed there are, many intermediate steps. As long as the overall energy change from original reactants to final products is favorable, the process will occur spontaneously and yield energy. But from the point of view of the cell, it is essential that the overall reaction happen in small steps, with many metabolic intermediates, and not in one large explosion. Only in that way can energy be captured efficiently so that it can be used elsewhere in the cell.

The energy-consuming reactions in cells also occur as sequences of small steps, for the same reason energy–yielding reactions do, only in reverse. It is easier to supply the energy to drive an energy-consuming process a little at a time. All of cellular metabolism, therefore, is characterized by **metabolic pathways**, or sequences of enzyme-catalyzed reactions in which the product of one reaction serves as the reactant of the next. Often pathways are linear, but they can be circular; they can interconnect with each other; they can even be branched, in which the product of one reaction can go in either of two or more subsequent directions depending on the needs of the cell at any particular time.

Metabolic Specificity

The second hallmark of enzyme-catalyzed reactions is *specificity*. Remarkably, a given enzyme only binds to a specific kind of molecule, called its **substrate** (Figure 9-6). An enzyme that catalyzes a reaction involving glucose only binds to glucose, not to DNA, a lipid, or even a sugar that is similar to glucose, such as fructose. Likewise, an enzyme that catalyzes a reaction involving DNA does not bind glucose. Enzymes can only lower the energy of activation for those specific molecules to which they bind. A different enzyme catalyzes each of the chemical reactions of metabolism. Thus for each step, only the appropriate substrate reacts.

Cells contain thousands of enzymes shuttling various organic substances from one form to another and changing the concentrations of cellular constituents in response to the demands of staying alive. We will take a look at some of the more common metabolic pathways in Section 9.3. But first let's examine the energy economy of the cell beginning with the energy currency, ATP.

9.2.3 ATP Is the Energy Currency of Life

Fast Find 9.2g ATP

Just as money serves as the currency linking the work we do with the goods and services we purchase, ATP is "earned," or assembled, by the energy-yielding metabolic pathways of the cell and "spent," or broken down, to drive the energy-consuming pathways. But what exactly is ATP, and why has it evolved to be the currency of choice in cells?

ATP stands for **adenosine triphosphate**. It is a ribonucleotide containing three phosphates (Figure 4-12). Adenosine with one phosphate made its first appearance in this story in Chapter 4 as one of the four ribonucleotides that compose RNA. In 1929, a German biochemist first isolated ATP from rabbit muscle. As the years passed, the substance was found to be nearly ubiquitous in living tissue. ATP was soon implicated in almost every metabolic pathway, either as a source of energy for energy-requiring pathways or as a product of the energy-yielding pathways. What is it about ATP that makes it so suitable as an energy currency?

Recall from our discussion of nucleotides in Chapter 4, and the Chemistry Review on the CD, that adenosine is a combination of a nitrogen-containing base called adenine and a five-carbon sugar (Figure 4-12). When adenosine in cells is not part of a chain of RNA, it can connect with one phosphate forming adenosine monophosphate (AMP), with two phosphates making adenosine diphosphate (ADP), or with three phosphates making ATP.

The unique feature of the third and final phosphate on ATP is that it can be broken off relatively easily, leaving behind ADP and a free phosphate. When it does so, a packet of energy is released that the cell is exquisitely designed to capture and use. Thus when we say that ATP is "broken down" to drive other energy-consuming reactions, we do not mean that the entire nucleotide is dismantled to its individual atoms, but that ATP is separated from its terminal phosphate to make ADP and a single free phosphate molecule, a process called **dephosphorylation**. Likewise, when ATP is synthesized in the pathways that release energy, an ADP is joined to a phosphate by means of a reaction called **phosphorylation** to form ATP.

Cells contain a finite amount of adenosine that can be cycled back and forth between energy-rich ATP and its poorer relative, ADP. When you are rested and well fed, most of your adenosine is ATP; as you move or exercise, your adenosine scale tips toward ADP (Figure 9-7). But even when you are on an ATP-spending spree, such as running a marathon or pulling an all-nighter, your cells are constantly replenishing the ATP from spent ADP and phosphate by capturing the energy from metabolic pathways that break down fuels like glucose.

The overall scheme of metabolism and the central role that ATP plays is shown in Figure 9-8. Energy-releasing pathways produce ATP; energy-consuming pathways break down ATP and use the energy released to do biological work; and the amount of adenosine in all its forms adds up to pretty much the same total all the time. That is not to say that living things cannot store extra energy for an emergency, starch, oils, glycogen, and fat are all molecules in which organisms keep a ready supply of energy on

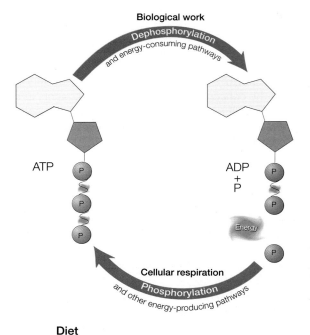

Figure 9-7. The ATP cycle. Biological work uses the energy that is released when ATP is dephosphorylated; the chemical bond between the last two phosphates on ATP is broken, yielding ADP and a free phosphate. In cellular respiration, ADP is again phosphorylated to ATP. The total amount of adenosine stays constant.

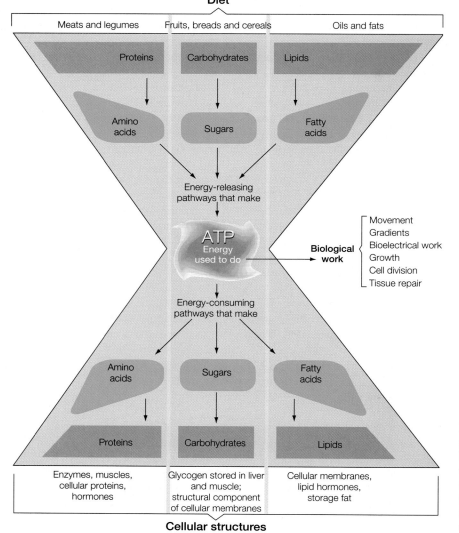

Figure 9-8. An overall view of animal metabolism. Energy-releasing pathways break down foods to produce ATP. ATP is used to do biological work and to fuel the energy-consuming pathways that synthesize the macromolecules that form cellular structures. How would this figure change if we were to show plant metabolism?

hand. But ATP is not accumulated in great abundance. (No one ever goes on a diet to lose ATP.) The energy of all these storage products must first be converted to ATP before it can be tapped.

9.2.4 Other Nucleotide-Based Compounds Shuttle Hydrogen

**Fast Find
9.2m
Energy
Compounds**

Just as the ADP/ATP pair are alternately phosphorylated by energy-producing pathways and dephosphorylated by energy-consuming pathways, three other pairs of nucleotide-based compounds are cycled in metabolism. They are called NAD^+/ $NADH$, $FAD/FADH_2$, and $NADP^+/NADPH$. The important thing to remember about NAD^+, FAD, and $NADP^+$ is that each of them can alternately pick up hydrogen atoms from one metabolic intermediate to become $NADH$, $FADH_2$, or $NADPH$, and later relinquish those hydrogens to a different hydrogen-accepting molecule becoming again NAD^+, FAD, or $NADP^+$ in the process. When organic fuels such as glucose are converted to water and carbon dioxide, most of the energy comes from breaking carbon–hydrogen bonds and forming oxygen–hydrogen bonds in the form of water and carbon–oxygen bonds in carbon dioxide. But this process does not occur directly. First the hydrogens are transferred from organic fuels to the hydrogen carriers, either NAD^+ or FAD, to make $NADH$ or $FADH_2$, respectively (Figure 9-9). Later, the $NADH$ and $FADH_2$ relinquish their hydrogens to oxygen to form water (although that process, too, occurs indirectly, as we will see in the next section). Likewise in biosynthetic

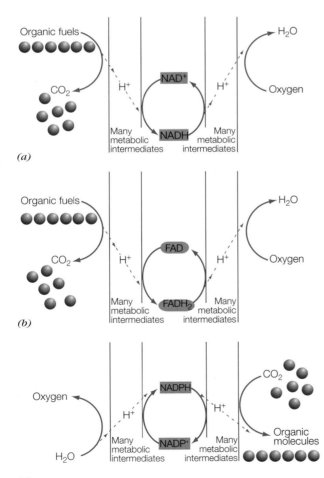

Figure 9-9. Cells use three compounds to shuttle hydrogen atoms between organic molecules and oxygen. (*a*) NAD^+ picks up hydrogen atoms from organic fuels becoming NADH, and through many metabolic intermediates, the hydrogens are transferred to oxygen, creating water (H_2O). (*b*) FAD also shuttles hydrogen atoms from Krebs cycle intermediates. (*c*) $NADP^+$ indirectly acquires hydrogens from water and transfers them through many metabolic intermediates to carbon to make organic molecules.

pathways, where energy is required to forge new carbon–hydrogen bonds, NADPH carries hydrogen to the site of synthesis. There, hydrogens are transferred from NADPH to carbon to make new proteins, lipids, or other cellular constituents. As we take a closer look at some metabolic pathways, these hydrogen-shuttling pairs will appear more than once.

The time has come to look at some of the metabolic pathways in cells that transform energy into work. There are hundreds of them, far too many to examine each one here. Some are highly specific to only certain kinds of organisms or certain kinds of cells; other pathways are found in nearly all living cells—a sure sign that they evolved early in the history of life and were handed down to all life forms that evolved later. But certain features are common to all metabolic pathways: they transform energy from one form to another, they obey the laws of thermodynamics, and they proceed by the stepwise conversion of one molecule to another by means of enzyme catalysts. We can learn much about metabolic pathways in general by looking closely at just a few. We will begin with a few common pathways, the ones that oxidize glucose in cellular respiration.

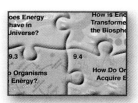

Piecing It Together

The discovery that the energetics of fire and animal respiration have common features was the first breakthrough in explaining how energy is converted in living systems. Here are some generalizations we have learned about energy and life:

1. When fuel is oxidized, either in a fire or by a living organism, chemical bonds between atoms of the fuel are broken and new bonds are formed between carbon and oxygen, and between hydrogen and oxygen. When this occurs, some energy is released. One salient feature of life is the ability to release that energy a little at a time and capture it in a form that can be used to fuel energy-requiring processes of life.

2. Cellular metabolism is all of the chemical reactions in cells. Metabolism consists of many individual metabolic pathways; each one is a series of sequential chemical reactions that results in the production of something needed by the cell. The tiny stepwise conversions that occur in metabolic pathways enable the cell to control the flow of energy and minimize the amount that becomes unusable due to entropy. In this sense, metabolism is highly efficient.

3. Enzymes are proteins that catalyze the individual steps of metabolic pathways. They speed up these chemical reactions by binding to their substrate molecules and lowering the energy of activation. But enzymes cannot force reactions to proceed that do not abide by the laws of thermodynamics. Each enzyme reacts only with its substrate. It is this feature of enzymes that ensures that metabolism is highly specific.

4. ATP acts as the energy currency in cells. When ATP is relieved of its terminal phosphate to become ADP and an unbound phosphate, a small packet of energy is released. Cells are exquisitely designed to capture and use that energy to drive energy-requiring processes. Likewise, ATP is synthesized from ADP and an independent phosphate using energy from other, energy-yielding metabolic pathways.

5. Three other pairs of compounds are also widely used in cells: $NAD^+/NADH$, $FAD/FADH_2$, and $NADP^+/NADPH$. These compounds shuttle hydrogens from

one place to another and from one compound to another in cells. Like ATP, these pairs play a central role in cellular metabolism.

9.3 HOW DO ORGANISMS USE ENERGY?

The morning of the marathon has arrived. The runner begins her day by eating a large helping of pancakes and syrup, which her digestive system quickly converts to glucose. The glucose is sent by means of her circulatory system to her muscles and liver, where some of it is linked together in temporary storage in the form of glycogen. She will need the energy of all that glucose, and more, to finish the 26-mile race.

The starting gun is fired and she takes her first few steps. In short order, her muscle cells spend all of the available ATP, converting the energy into motion. (More on how muscles accomplish this in Chapter 12. For now we will focus on making more ATP so that she can continue the race.) Now the existing glucose and glycogen must be broken down to supply more ATP. The process involves many steps, but they happen quickly and the runner stays near the front of the pack.

9.3.1 Cellular Respiration Occurs in Three Stages

Cellular respiration is the name given to the metabolic pathways in which cells harvest the energy from the metabolism of food molecules. The breakdown of glucose is one of the main pathways of cellular respiration, and one our marathon runner relies on heavily to supply energy for the race. Glucose breakdown occurs within cells as a continuous sequence of more than 25 steps. But when we look at this continuous process carefully, it is apparent that these steps can be grouped into three separate stages. We call these stages **glycolysis**, the **Krebs cycle**, and the **electron transport system**. In the next three sections, we will take a broad view of each of these stages of metabolism separately, but first an overview of all three (Figure 9-10).

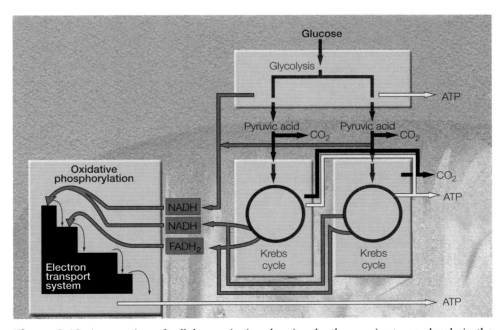

Figure 9-10. An overview of cellular respiration showing the three main stages: glycolysis, the Krebs cycle, and the electron transport system. The synthesis of ATP in the mitochondrion is called oxidative phosphorylation. These are the major pathways organisms use to synthesize ATP.

Overview of Cellular Respiration

The first of these three stages, glycolysis, has been called the universal energy-harvesting process of life. It is the most widespread metabolic pathway in the living world, common to every living thing, and similar—often identical—in every type of cell. Given its widespread occurrence and broad similarities, this pathway is very primitive indeed. It probably evolved early in the history of life. It may have been present in the first living cell and passed on to all the successful progeny of that first cell. In glycolysis (which means "*sugar splitting*"), the six-carbon sugar, glucose, is split in half to make two three-carbon compounds, and hydrogens are stripped from carbons. The energy of the broken bonds is captured, and there is a net yield of two molecules of ATP. Unlike the next two stages of respiration, glycolysis occurs in the cytoplasm of the cell and requires no organelles. Glycolysis also uses no oxygen. Hence it is sometimes called the **anaerobic** (not requiring oxygen) pathway of cellular respiration.

The next two stages of cellular respiration, the Krebs cycle and the electron transport system, are **aerobic**: they are processes that require oxygen. They both take place in mitochondria. The Krebs cycle completes the breakdown of glucose to single carbon molecules by taking derivatives of the two three-carbon products of glycolysis, breaking them apart, and stripping the carbons of their hydrogen atoms, leaving behind six carbon dioxide molecules. While the Krebs cycle itself does not require oxygen, it is a prelude to the next stage of cellular respiration in which oxygen is required: the electron transport system.

Both glycolysis and the Krebs cycle transfer hydrogen atoms from the carbons of glucose to molecules of NAD^+ to make NADH. There is still a lot of energy to be gleaned by transferring those hydrogens from NADH to oxygen (making H_2O). The third and final stage of cellular respiration does just that. The electron transport chain is a series of enzymes that transfers hydrogen from NADH to O_2, step by step, and captures the energy that is released in the process. The energy is captured when ADP and phosphate are joined to make ATP. Because this process requires oxygen, the process is sometimes called **oxidative phosphorylation**. It is in this process that the oxygen we breathe is consumed—converted, actually, to water.

9.3.2 Glycolysis Is the Anaerobic Stage of Cellular Respiration

**Fast Find
9.3b
Glycolysis**

Our runner has just finished the first few miles of the marathon, and the going has been easy. Her conditioned lungs, heart, and circulation have drawn in oxygen from the air and delivered it to the muscles by means of the blood. The prompt delivery of oxygen has permitted her cells to fully oxidize glucose to CO_2 and H_2O using all three stages of cellular respiration. As we will see, the overall aerobic yield of ATP for each glucose that is fully metabolized is 36 ATPs. Glycolysis provides only two of those ATPs, but it prepares the way for the later stages.

Glycolysis occurs in the cytoplasm, in our example, the cytoplasm of the muscle cells (although all cells have the cytoplasmic enzymes to perform glycolysis). The net overall reaction is the following:

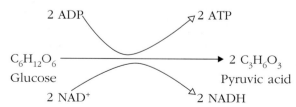

The starting fuel is a six-carbon glucose, $C_6H_{12}O_6$. The end result is two molecules of a three-carbon compound called pyruvic acid, also referred to as pyruvate. In addition, two ATPs and two NADHs are formed. Recall, however, that metabolic pathways occur as series of single steps, each catalyzed by a single, highly specific enzyme. In fact, the splitting of glucose into two molecules of pyruvic acid occurs in nine separate steps, with at least nine different metabolic intermediates and the help of at least nine enzymes. The entire scheme is illustrated in Figure 9-11 and in more detail on the compact disc.

When Oxygen Is Limited—Lactic Acid Fermentation Take Over

As long as the runner can stay at a steady pace and the demands she makes on her muscles do not exceed the capacity of her lungs and circulatory system to deliver oxygen, glycolysis continues to produce ATP to power movement, and NADH and pyruvic acid are used in the next two stages of cellular respiration, by which more ATP is made. But soon she encounters a long uphill stretch, and the work of overcoming gravity exceeds the capacity of her heart and circulation to deliver oxygen to the muscles. Her cells must depend on anaerobic means; that is, they must rely on glycolysis alone to supply the ATP needed to continue the race.

Relying on anaerobic metabolism alone presents two problems. First, the two ATPs that glycolysis can supply for each glucose metabolized are enough for a short sprint, but that level of activity cannot be maintained for long. Some organisms with low

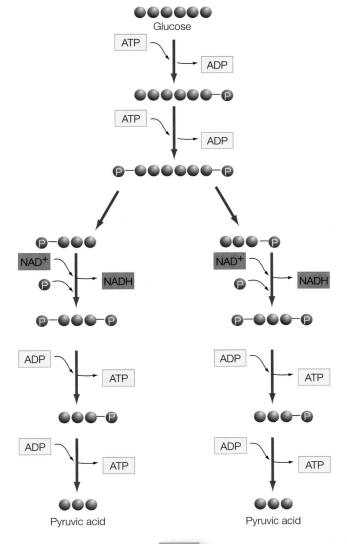

1 In the early steps, two ATPs are "spent" and their terminal phosphates are added to the carbon backbone of glucose. These steps make the sugar more reactive for the later steps in the pathway.

2 The doubly phosphorylated, 6-carbon sugar is split into two 3-carbon sugars, each with one phosphate group.

3 Each 3-carbon sugar joins with a free phosphate, making the sugar even more reactive.

4 In the final steps, the two phosphates on each sugar are transferred to 2 ADPs, making 2 ATPs for each sugar. These steps yield a total of 4 ATPs. The net gain in glycolysis, however, is only 2 ATPs because two were spent getting the pathway started (Step 1).

Figure 9-11. The steps of glycolysis. In this metabolic pathway, the six-carbon sugar, glucose, is broken down into two three-carbon pyruvic acid molecules, with a net production of two molecules of ATP.

energy demands can survive on the meager ATP produced by glycolysis—anaerobic bacteria, for example, and yeast in an oxygen-poor environment—but most organisms have energy demands that exceed the ATP provided by glycolysis alone. You have undoubtedly experienced anaerobic metabolism in activities such as sprinting or lifting weights. You know that you cannot sprint for as long or as far as you can jog; you cannot haul a heavy load for as long as you can carry a light one. Second, in the absence of oxygen, glycolysis converts all of the limited NAD^+ to NADH. When no more NAD^+ is available, glycolysis comes to a screeching halt. The solution to the first problem is to stop and rest, or to slow down until metabolism switches back to aerobic conditions. Cells have evolved an elegant way, however, of solving the second problem.

When oxygen is limited, glycolysis doesn't end with pyruvic acid, but it includes an additional, tenth step designed to regenerate NAD^+ from NADH. In this extra anaerobic step, hydrogen is taken away from NADH and added back to pyruvic acid, regenerating NAD^+. The product made from pyruvic acid is lactic acid (Figure 9-12); hence the process is called **lactic acid fermentation**. Unlike pyruvic acid, lactic acid cannot enter the second and third stages of respiration. Eventually, it accumulates in the muscles, and the corresponding acidity gives our runner the sensation of fatigue.

With the hill behind her, the runner's lungs and heart can again deliver sufficient oxygen to the muscles to maintain aerobic metabolism. At this point, the products of glycolysis are shunted to the mitochondria of the muscle cells for further oxidation and a better yield of ATP.

Brewing Energy In metabolic terms, different organisms deal with limited oxygen in different ways. Yeasts, for example, use a process called **alcoholic fermentation** to regenerate NAD^+ in anaerobic conditions. How does this differ from lactic acid fermentation? What important industries rely on this metabolic activity of yeast?

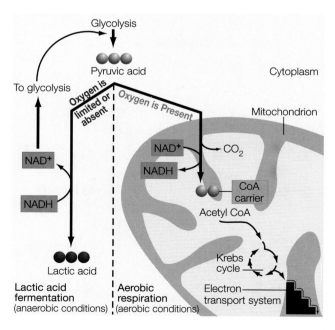

Figure 9-12. The fate of pyruvic acid depends on whether or not oxygen is present in the cell. When oxygen is absent, pyruvic acid is converted to lactic acid, and NAD^+ is regenerated from NADH. When oxygen is present, pyruvic acid enters the mitochondrion. In the mitochondrion, an enzyme removes one carbon from pyruvic acid, and hydrogen is transferred to NAD^+ to make NADH. The resulting two-carbon molecule is attached to a carrier forming acetyl CoA, which takes the carbons into the next stage of respiration, the Krebs cycle.

9.3.3 The Krebs Cycle and Electron Transport Provide Energy Aerobically

**Fast Find
9.3c
Krebs
Cycle**

So far we have considered only carbohydrate fuel, glucose in particular. Fatty acids and amino acids also form a significant part of our diets, and can also be metabolized to make ATP. Indeed, just a few miles into the race, glucose and glycogen stores within muscle cells are depleted, and the marathon runner must switch to burning fatty acids released into the blood from adipose tissue. All three major foodstuffs—carbohydrates, lipids, and proteins—are ultimately oxidized by the enzymes of the mitochondria that constitute the Krebs cycle and the electron transport chain. But other metabolic pathways must first prepare them much as glucose was prepared by glycolysis. In this sense, the Krebs cycle is the central pathway in cellular metabolism. Let's follow the products of glycolysis into the mitochondria of our runner's muscle cells, where they will be fully oxidized and their energy used to make even more ATP. We will see that NADH, a product of glycolysis, is one of the main products of the Krebs cycle, as well.

During the nine enzyme-catalyzed steps of glycolysis, each molecule of glucose is converted into two molecules of pyruvic acid. Each pyruvic acid molecule moves into the mitochondria, where an additional enzyme-catalyzed reaction prepares it for the Krebs cycle. This transitional step accomplishes three things: (1) hydrogen atoms are stripped from pyruvic acid and transferred to NAD$^+$ to make NADH, (2) a carbon atom is removed from the three-carbon pyruvic acid molecule and is lost as CO_2, and (3) the resulting two-carbon compound is attached to a large carrier molecule called coenzyme A or CoA. The two-carbon compound attached to its large carrier is called **acetyl CoA**. All three of these tasks are accomplished by one large enzyme found inside the mitochondria.

In the Krebs cycle, two carbons enter as acetyl CoA, two carbons leave as CO_2, and three NAD$^+$s pick up hydrogens from the Krebs cycle intermediates to become NADHs. In addition, FAD also picks up hydrogens in the course of the cycle. Finally, the equivalent of one ATP is formed for each turn of the cycle, that is, for each acetyl CoA molecule that enters into the Krebs cycle. The details of the Krebs cycle are shown in Figure 9-13.

❶ Before entering the Krebs cycle, pyruvic acid loses one carbon in the form of CO_2. The resulting 2-carbon molecule is attached to a carrier, making acetyl CoA.

❷ As the 2-carbon molecule enters the Krebs cycle, the CoA carrier is lost and a 6-carbon molecule is formed.

❸ During the Krebs cycle, the two carbons are lost as CO_2 and the equivalent of one ATP is formed from ADP and phosphate. Also, hydrogens are transferred to three molecules of NAD$^+$ and one molecule of FAD, making three NADHs and one FADH$_2$.

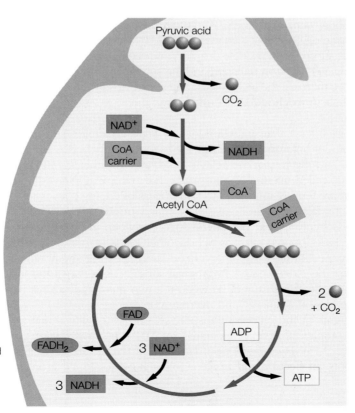

Figure 9-13. The steps of the Krebs cycle. Pyruvic acid enters the mitochondrion, where it loses a molecule of CO_2 and is linked to a carrier molecule called CoA. The resulting compound, acetyl CoA joins with a four-carbon compound and, through several enzyme-catalyzed steps, the result is one molecule of ATP, three NADH molecules, and one FADH$_2$ molecule.

Rather than becoming overwhelmed by the many steps, let's concentrate on what goes in and what comes out. One two-carbon compound, three NAD^+s, one FAD, and one ADP plus phosphate go in, and three NADHs, one $FADH_2$, one ATP (or its energetic equivalent), and two CO_2s come out. When we view every metabolic pathway as a way for the cell to transform energy from one form to another, we can ask ourselves two questions: in what form did energy enter the pathway, and where is the energy at the end of the pathway? For the Krebs cycle, the answers are first, chemical energy entered the pathway in the form of acetyl CoA, and second, chemical energy left the Krebs cycle in the form of ATP, NADH, and $FADH_2$. All that remains is to see how the cell makes ATP using energy extracted from NADH and $FADH_2$. That is the role of the third metabolic pathway of cellular respiration, the electron transport system.

Electron Transport

Fast Find 9.3d Electron Transport

Let's review what has happened so far. Starting with glucose, two molecules of ATP were made in glycolysis, and two more in the Krebs cycle. (Remember each turn of the Krebs cycle makes only one ATP, but glycolysis feeds *two* two-carbon molecules into the cycle for each glucose that it processes. Consequently the cycle turns twice for every molecule of glucose that enters glycolysis.) The Krebs cycle by itself has not delivered the big ATP payoff that a strictly energetic analysis tells us is possible, but it does bring us one step closer to that payoff. The biggest release of energy comes when NADH and $FADH_2$ pass their hydrogens to oxygen to make H_2O. As in all metabolic pathways, this transfer is stepwise. Unlike the other metabolic pathways we've seen so far, however, the electron transport chain and the subsequent synthesis of ATP both depend on the mitochondrial membrane.

At each point in the breakdown of glucose, we have noted where in the cell the pathway is occurring. Glycolysis takes place in the cytoplasm; the Krebs cycle enzymes and intermediates are dissolved in the fluid within the mitochondria. But given the right combination of substrates, enzymes, **cofactors** (the nonprotein substances that help enzymes function), NAD^+, FAD, ADP, and phosphate, either of these pathways can easily happen in a test tube, with no pieces of the cell at all. Indeed, the experiments that elucidated the individual steps of glycolysis and the Krebs cycle were done in just that way, in a test tube containing only dissolved materials. Each step was worked out individually by adding a single substrate, or a single enzyme, and seeing what disappeared and what accumulated. However, electron transport and the synthesis of ATP from ADP and phosphate require more than just the cell's juices. These two closely linked processes require the presence of intact mitochondria. To understand them, we must look again at mitochondrial structure.

Recall from Chapter 4 that mitochondria are large organelles surrounded by two membranes: a smooth outer membrane that forms a boundary between the contents of the mitochondria and the rest of the cellular fluid, and an inner membrane characterized by deep folds called **cristae** (Figure 9-14). The two membranes differ biochemically in ways that are terribly important to the process of energy conversion. The outer membrane contains large protein-lined pores, called **porins**, that permit most small substances to pass freely in and out of the space between the two membranes. The inner membrane lacks porins, but it is rich in other kinds of proteins, many of them part of the energy-transforming machinery. In addition, the inner membrane is an effective barrier against the movement of even small substances between the inside of the mitochondrion and the compartment between the two membranes. This barrier plays an important role in storing energy and making ATP.

The components of the electron transport system are enzymes that are grouped into four large complexes, deeply integrated into the inner membrane of the mitochondrion (Figure 9-15). Hydrogens from NADH and $FADH_2$ are separated into their atomic components: an electron and a proton for each hydrogen. Electrons from NADH

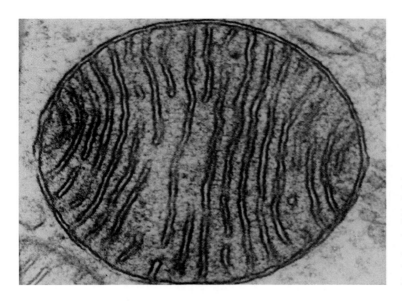

Figure 9-14. An electron micrograph of a mitochondrion. The folds of the inner membrane are called cristae. They increase the area of inner membrane where the important reactions of the electron transport and oxidative phosphorylation occur (magnified 374,000 times).

and $FADH_2$ are shuttled through the complexes and ultimately join with oxygen and other protons on the inside of the mitochondria to make water. The protons from NADH and $FADH_2$, meanwhile, are released into the space between the two mitochondrial membranes.

Each of the four complexes contain several different components, each one capable of picking up one or two electrons and passing them on to the next component. But the substrate that enters the electron transport system is hydrogen taken from NADH and $FADH_2$, not electrons. How do we get from hydrogen to electrons? Recall from the Chemistry Review on the CD that hydrogen is the simplest atom, composed of a single

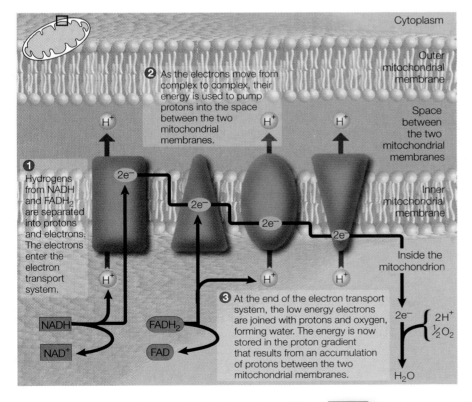

Figure 9-15. The four enzyme complexes of the electron transport system are embedded in the inner membrane of the mitochondrion. NADH gives up hydrogens, which split into their respective electrons and protons. The electrons enter the electron transport system, where they are passed from carrier to carrier within the four complexes, giving up energy at each step. The energy is used to pump protons from inside the mitochondrion to the space between the two mitochondrial membranes. More protons are moved out of the mitochondrion by the other enzyme complexes. Finally, the electrons join with protons and oxygen to make water. The end result is a gradient of protons across the inner mitochondrial membrane.

proton and a single electron. In the electron transport system, electrons are stripped away from their protons and shuttled from carrier to carrier on the four electron transport system complexes within the membrane. As they move, they go from a condition of high energy to one of low energy. Meanwhile, the protons (abbreviated H⁺) are deposited in the space between the two membranes (Figure 9-15). As electrons move through the transport system, more protons from inside the mitochondrion are transported to the space between the membranes. Finally, as the electrons emerge from the end of the electron transport chain, they are handed over to oxygen and, along with two more protons from the inside of the mitochondrion, form water.

The end product of the electron transport system is a gradient of protons across the inner membrane of the mitochondrion and water, but what about ATP? Where is the energy? Where is the big ATP payoff that aerobic metabolism can produce?

9.3.4 ATP Is Made Using Energy from a Proton Gradient

A gradient of any kind is a sure sign that a system is not at equilibrium, and systems that are not at equilibrium can be exploited to do work. Picture a large hydroelectric dam across a fast-moving river. The dam serves to prevent water from flowing downhill toward equilibrium—a direction in which it is being pushed by the relentless force of gravity. Long ago, humans discovered how to capture the energy of a flowing river to do work, such as milling grain or, more recently, powering electrical turbines. The principle is simple: hold the water back until you need to do work, then open a pathway for water to flow (Figure 9-16a). As the water rushes down its gradient toward equilibrium, use a coupling mechanism—a water wheel or a turbine—to put that energy to work for you (Figure 9-16b). The basic components are (1) potential energy in the form of a water gradient; (2) an opening, or conduit, that directs the flow of water in a specific path; and (3) a coupling mechanism to do the work. The synthesis of ATP in mitochondria uses the same basic components.

The energy of electron transport moves protons from the inside of the mitochondrion to the space between the two mitochondrial membranes forming a steep proton gradient. The force of diffusion drives protons inward so as to equalize the concentration gradient across the membrane (recall that diffusion is a passive force that moves substances from regions of high concentration to regions of low concentration). The inner mitochondrial membrane, like a dam, is preventing the protons from moving in the direction that diffusion is driving them. All that is needed to tap this source of potential energy is a passageway through the membrane and a coupling mechanism to do the work. Mitochondria have both.

Among the other proteins embedded in the inner mitochondrial membrane is a large, lollipop-shaped enzyme that is so big it can be seen using the powerful electron microscope (Figure 9-16c). The "stick" of the lollipop is embedded in the core of the inner membrane; the sphere-shaped "candy" part of the protein projects into the inside of the mitochondrion. For many years, the role of this abundant protein was unknown. Then in the mid-1970s, a scientist from Cornell University, Efraim Racker, used biochemical techniques to isolate the sphere-shaped subunit of the protein. When Racker added ATP to his spheres, they catalyzed its breakdown to ADP and phosphate. He interpreted this to mean that the spheres have on their surface binding sites for ATP, and for ADP and phosphate. He correctly reasoned that the stick and the sphere subunits could act together as both the proton passageway through the membrane (the stick), and the coupling mechanism for ATP synthesis (the sphere). This membrane-bound enzyme is the **mitochondrial ATP synthase**.

The stick part of the mitochondrial ATP synthase, called the F_o subunit, is actually a hollow cylinder that forms an opening through the mitochondrial membrane through which only protons can pass. It is the conduit that allows protons to flow in the direction that diffusion pushes them. The sphere part, called the F_1 subunit, sits on top of the hole and readily binds to both ADP and phosphate. As protons move through the

(a)

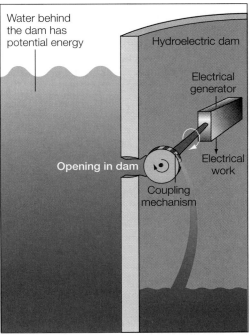

(b)

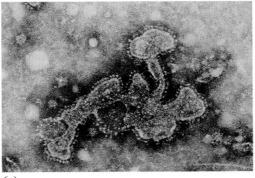

(c)

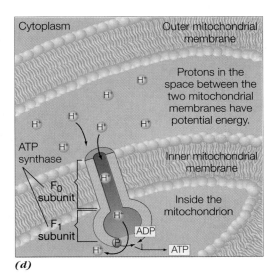

(d)

Figure 9-16. Analogy between a hydroelectric dam and the mechanism of oxidative phosphorylation.

(a) Potential energy is stored in the water held behind the dam.

(b) By opening a hole in the dam, the flow of water can be coupled to a generator that makes electricity. Electricity is used to do work.

(c) The mitochondrial ATP synthase is a large protein embedded in the inner mitochondrial membrane. It is shown here as bumps on the surface of membrane fragments taken from the inner mitochondrial membrane (magnified 262,500 times).

(d) Potential energy is stored in the proton gradient across the inner mitochondrial membrane. The F_0 subunit of the ATP synthase opens a hole in the membrane, allowing protons (yellow balls) to flow through. The F_1 subunit couples the flow of protons to the synthesis of ATP.

hole, the energy of their passing pushes ADP and phosphate in such close proximity that they join together to make ATP (Figure 9-16*d*). This mechanism of ATP synthesis is called **chemiosmosis**. It differs from the other ways that the cell makes ATP. In glycolysis and the Krebs cycle, phosphates are directly connected to the metabolic intermediates of the pathways, then they are transferred to a waiting ADP molecule. In chemiosmosis, the synthesis of ATP is indirect because it is coupled to the movement of protons down their gradient.

It is here, in the inner mitochondrial membrane, that the big ATP payoff from cellular respiration is realized. For each glucose molecule that enters glycolysis and is fully oxidized to carbon dioxide and water, 36 ATPs are formed. As long as our runner has energy reserves in the form of organic fuel molecules in her body, she can continue to oxidize it to make ATP. The ATP is used to power muscular contractions, and she finishes the race.

These same processes provide ATP for all kinds of biological work, in humans and other living things. These are among the most widespread metabolic pathways in the living world. They depend on an outside source of metabolic fuel. In other words, organisms need food. In the next section, we will take a brief look at photosynthesis, that is, the metabolic pathways that create organic molecules from sunlight and carbon dioxide.

Changing Air Life on Earth originated about 3.5 billion years ago, at a time when the atmosphere was very different from today. How did that primitive atmosphere differ from the modern atmosphere? What important event was responsible for altering Earth's atmosphere? When did that event take place? How did the atmospheric change influence the evolution of cellular respiration?

Piecing It Together

Cellular respiration is a series of metabolic pathways that extracts energy from organic molecules and uses it to make ATP. The three stages of cellular respiration occur in nearly all types of cells, making them among the most widespread metabolic pathways in the living world. Here is a summary of the three stages of cellular respiration:

1. Glycolysis is the first stage of cellular respiration. In nine separate steps, each catalyzed by a different enzyme, a six-carbon glucose is split into two three-carbon compounds called pyruvic acid. The pathway nets two molecules of ATP; two molecules of NADH are also formed from NAD^+ plus hydrogens that are stripped from the glucose. NADH can be further oxidized to yield more ATP, but that happens in a later stage of cellular respiration. Glycolysis is an anaerobic process that takes place in the cell's cytoplasm.

2. When oxygen is present, pyruvic acid from glycolysis is trimmed to a two-carbon compound and shunted into the mitochondrion, where it enters the second stage of cellular respiration, the Krebs cycle. In this pathway, more hydrogens are transferred to NAD^+ making NADH, and to FAD making $FADH_2$. The remaining carbons from the original glucose are lost as carbon dioxide. The Krebs cycle nets three NADHs, one $FADH_2$, and one more ATP for each acetyl CoA molecule introduced into the cycle. The cycle turns twice for each glucose entering glycolysis.

3. More ATP is made when both NADH and $FADH_2$ give up their hydrogens to the electron transport chain. The electron transport chain comprises many enzymes and electron carriers clumped into four complexes, all embedded in the inner mitochondrial membrane. As electrons move from one complex to the next, protons are moved from inside the mitochondrion to the space between the two mitochondrial membranes, until a steep gradient of protons forms across the inner mitochondrial membrane.

4. The mitochondrial ATP synthase is a large lollipop-shaped protein that acts as both a passageway allowing protons to move down their gradient, and an enzyme that can capture the energy of proton movement and use it to link ADP with phosphate making ATP. The net overall yield of cellular respiration is 36 ATPs for each glucose that enters glycolysis.

9.4 HOW DO ORGANISMS ACQUIRE ENERGY?

The cells of living organisms extract energy from glucose and other organic compounds, but only photosynthetic organisms make organic molecules from the simple low-energy compounds, CO_2 and H_2O. They supply the food and oxygen other organisms need to survive. Organic compounds contain far more potential energy than either CO_2 or H_2O. Where do plants get the energy to form organic compounds? The answer seems obvious to us, but in the 17th century, the role of sunlight in powering life was not so obvious.

9.4.1 Photosynthesis Uses Light Energy to Make Food

Plants grow in soil, and in 1660 the idea that all of the raw material for plant growth could be found in the soil was still untested, that is, until a Flemish chemist named Jan Baptista van Helmont decided to test it. Van Helmont grew a willow tree in a pot of soil for five years, carefully weighing the pot every year. The willow tree grew large and sturdy, gaining 150 pounds of living tissue. The soil lost only a few ounces. Clearly the soil did not provide the nourishment for the tree; it had to be coming from somewhere else. Van Helmont concluded that the tree was using the daily water he provided to fuel its growth. After all, he argued, plants deprived of water die.

Van Helmont's conclusion was logical, but not entirely true. While it is true that plant tissue, just as animal tissue, is mainly composed of water, water alone cannot fuel the growth of any living being. The energy of the chemical bonds in water cannot fuel life. In the 1770s, Joseph Priestley tried growing a plant under a jar. As long as the plant and the jar were kept in the light, the plant grew, that is, until all of the carbon dioxide from the air under the jar was exhausted and the plant stopped growing. If a mouse were made to live under the jar with the plant, both of them, the mouse and the plant, could live and grow for a longer time than either could alone. The mouse was providing carbon dioxide that the plant used, and the plant was providing oxygen that the mouse used.

The pivotal role that light plays in this relationship was recognized just a few years after Priestley's experiment. Light is absolutely essential for the formation of organic molecules from water and carbon dioxide. The process is called **photosynthesis** (literally "put together by light"). Animals depend not just on the oxygen gas produced by plants in the light, but on the carbon-containing materials that plants produce.

On a global scale, carbon, hydrogen, and oxygen move back and forth between Earth's autotrophs and heterotrophs without being used up or destroyed. The overall process is called the **carbon cycle**. We have spent much of this chapter examining the breakdown of glucose in the presence of oxygen. Now we will focus on the synthesis of organic compounds from carbon dioxide and water, that is, the process called photosynthesis.

9.4.2 Pigments Absorb the Energy of Light

Light is a form of energy called **electromagnetic radiation**. This energy travels through space in the form of rhythmic waves. Electromagnetic radiation occurs in a vast spectrum of sizes and energies (Figure 9-17). Shorter wavelength radiation has more energy than longer wavelength radiation. X rays are an example of short-wave, high-energy electromagnetic radiation; heat, microwaves, and radio waves are long-length, lower-energy radiation. One small region of the spectrum is visible to our eyes. It is called the visible spectrum, or **light**. Even within the visible spectrum, there are differences in the wavelengths of radiation. These differences are interpreted by our brain as colors. Only a few colors of light—a tiny fraction of the spectrum of electromagnetic radiation—provide the energy upon which life on Earth ultimately depends. How do autotrophs capture the energy of just a few wavelengths of visible light and transform it into the energy of chemical bonds?

Greenness is the single trait shared by nearly all organisms that use light energy to make carbohydrates. Many of the details of photosynthesis were not known until the 20th century, but long before the days of sophisticated techniques for studying biochemistry, the color green was thought to hold the key. Photosynthetic tissues appear green because they contain **pigments**, molecules that absorb some wavelengths of light and reflect others. In green plants, the most important pigment is **chlorophyll**, a complex organic molecule that absorbs light in the blue and red parts of the spectrum, but reflects the green wavelength back to our eyes.

It is possible to take green leaves and grind them vigorously enough to break open the cells. If a solvent such as alcohol is added to the mash, chlorophyll dissolves and floats away from the other insoluble parts of the plant. Dissolved in alcohol, the chemical properties of chlorophyll can be studied, including the way in which it absorbs light. When a beam of blue light is aimed at a test tube containing dissolved chlorophyll, the solution strongly *fluoresces*. In other words, the light is briefly absorbed, then emitted back to our eye at a different wavelength (Figure 9-18). Each chlorophyll molecule is absorbing the energy of blue light. That energy raises some of chlorophyll's electrons to a higher energy level. But separated from the other parts of the plant, the energized electrons have nowhere to go, and they fall back to their original energy levels. In so doing, some of their energy is lost as heat, and some is lost as fluorescent light. Because some of the energy is converted to heat, the lost light has less than the total energy absorbed, and light is emitted by the solution at a longer (lower-energy) wavelength. (This is the same principle that operates in "glow-in-the-dark" paints used to make street signs and some toys and novelties.) Such an experiment tells us that chlorophyll absorbs light, which raises electrons to a higher energy level. But if the electrons have nowhere to go, they quickly lose the added energy.

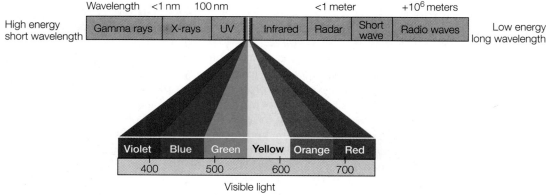

Figure 9-17. The wavelengths of electromagnetic radiation. Our eyes can detect only a small segment of the spectrum, the portion called *visible light*. Chlorophyll absorbs light in the blue and red regions of the visible spectrum.

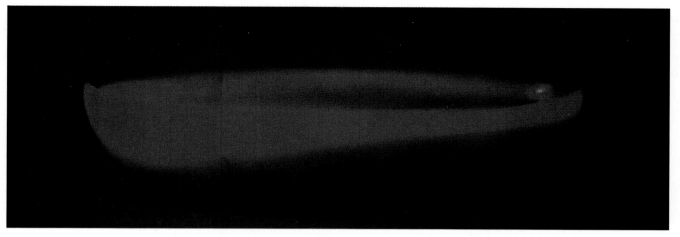

Figure 9-18. A chlorophyll solution. The energy of blue light is briefly absorbed by the chlorophyll and released as red light.

If we repeat this experiment, this time using intact plant cells, there is almost no fluorescence. Electrons are energized, then sent along a metabolic pathway that ultimately captures that energy in the form of organic compounds. The first two forms in which light energy is captured are the familiar energy currency, ATP, and the hydrogen carrier, NADPH. These are the two products of photosynthesis that require light. They are made in the first phase of photosynthesis, called the **light-dependent reactions**. These two products, in turn, are used to make carbohydrates in the second phase of photosynthesis, a series of reactions that are independent of solar radiation, called the Calvin-Benson cycle or **light-independent reactions**.

Hidden Colors What other pigments do autotrophs use to capture light energy? Why do the leaves of deciduous plants turn orange, red, and yellow in the fall?

9.4.3 Light Reactions Make ATP and NADPH

Just as we needed to review the structure of mitochondria to understand the electron transport system, we must revisit the structure of chloroplasts to understand the light-dependent reactions of photosynthesis. Chloroplasts are large, green, membrane-bound organelles that are the cellular sites of photosynthesis (Figure 9-19). Like mitochondria, they are surrounded by two membranes: an inner membrane and an outer membrane. Unlike mitochondria, chloroplasts have a third system of membranes within the internal space. This internal system of membranes is arranged in disk-shaped sacks called **thylakoids**. Thylakoids occur in the internal space of the chloroplast, the space called the **stroma**. The two important features to remember about this complex internal structure are (1) the thylakoid membranes contain the light-harvesting pigments of photosynthesis and the enzymes that capture the energy of light-energized electrons; and (2) the internal membranes define a space, called a **lumen**, that is separate from the rest of the stroma. This separate compartment plays an important role in the light reactions of photosynthesis.

The sun comes up on a summer's day and shines down brightly on the green valley. Chlorophyll molecules, so numerous that the entire valley looks as if it is carpeted in green, are embedded in the membranes of thylakoids in the plants that cover the landscape. They soak up some of the blue and red wavelengths of the bright sun and reflect back the green. The energy of the blue and red light bumps certain electrons from chlorophyll to higher energy levels. In this activated state, the electrons are passed to other molecules in the thylakoid membrane that readily accept electrons, but only when

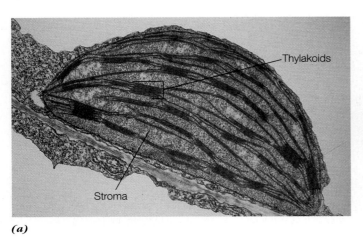

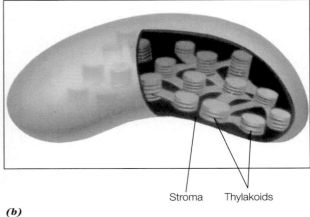

(a) *(b)*

Figure 9-19. *(a)* Electron micrograph of a chloroplast. Chloroplasts have outer and inner membranes and a third system of internal membranes forming thylakoids. The light-dependent reactions of photosynthesis take place in the thylakoid membranes. The dark spots are sites where starch, produced in photosynthesis, is stored (magnified 20,790 times). *(b)* A generalized chloroplast.

**Fast Find
9.4b
Light-Dependent
Reactions**

they are highly energized. Step by step, electrons are passed from one carrier to another, giving up a little bit of their energy with each step. The electron carriers capture that energy to be used later to make ATP.

Meanwhile, the chlorophyll is missing electrons. Without them, chlorophyll is highly unstable; it readily reacts with molecules in the vicinity that can fill the void. One such molecule, always in abundance in cells, is water. The unstable chlorophyll is powerful enough to strip electrons away from water molecules. In so doing, water from the *lumen* of the thylakoid is broken apart into its components: electrons for chlorophyll, electron-deficient hydrogens (which are nothing more than lone protons), and oxygen. The protons remain in the lumen of the thylakoid. The oxygen diffuses out into the surrounding air. Atmospheric oxygen that we (and most living things) use in our own aerobic pathways is generated in this first critical step of photosynthesis. We couldn't survive without it.

The travels of the energized electrons, however, have only just begun. The electrons are picked up and passed on by several enzymes embedded in the thylakoid membrane (Figure 9-20). They require an extra boost of energy, provided by more light-absorbing chlorophylls, before they ultimately join with protons and $NADP^+$ to make NADPH on the *stromal side* of the thylakoid membrane. Along the way, electron carriers move more protons from the stroma into the thylakoid lumen. At the end of the path, the electrons are passed to $NADP^+$ to make higher-energy NADPH. In addition to the production of oxygen, the end result is NADPH and an accumulation of protons inside the lumen. This uneven distribution of protons—high concentrations in the thylakoid lumen and lower concentrations in the stroma—is a form of stored energy. The energy of light has thus far been captured in two forms: the synthesis of NADPH from $NADP^+$ and the proton gradient across the thylakoid membrane. But a proton gradient cannot be directly used to make food. It must first be converted to ATP. That is the role of the chloroplast ATP synthase enzyme.

Chloroplast ATP Synthase

It takes energy to make ATP from its components, ADP and phosphate. All aerobic organisms, plants and animals included, accomplish this by chemiosmosis, using the ATP synthase enzyme embedded in the inner membrane of the mitochondria to couple the energy of a flow of protons with the synthesis of ATP (see Section 9.3.4). The scenario in chloroplasts is similar to that in mitochondria. In the chloroplast, a gradient of protons is formed across the thylakoid membrane, and the energy of that gradient is captured by ATP synthase. Chloroplast ATP synthase is very much like mitochondrial ATP synthase, so similar, in fact, that the two synthases are thought to have evolved from a common ancestor protein. Mitochondrial and chloroplast ATP synthases are molecular "cousins." Both bring ADP and phosphate close together. A chemical bond is formed between them when protons are allowed to flow past these two compounds through a protein-lined hole in the membrane. As the protons move, the gradient is dissipated and ATP is formed.

The reactions of the light-dependent pathways of photosynthesis are complete. Meanwhile, more light energy has been absorbed, and more electrons have been energized. As long as the sun is shining, the light-dependent reactions are making NADPH and ATP. The overall reactions of light-dependent photosynthesis are summarized as follows:

$$2 \ H_2O + 2 \ NADP^+ + ADP + phosphate \xrightarrow{\text{light energy}} O_2 + 2 \ NADPH + ATP$$

Taken together, these reactions are called **noncyclic photophosphorylation** because the electrons follow a linear, noncyclic pathway from H_2O to NADPH.

Cyclic Photophosphorylation

The light reactions of noncyclic photophosphorylation make a bit more NADPH than ATP. However, the Calvin-Benson cycle, in which NADPH and ATP are used to make carbohydrate, requires three ATPs for every two NADPHs used. Plants compensate for the difference between supply and demand for these high energy products by using an alternative pathway that generates a proton gradient to make ATP, but generates neither NADPH nor oxygen. This alternative is called **cyclic photophosphorylation**.

Depending upon the need for ATP, electrons can bypass $NADP^+$ and be passed back to the chlorophyll molecule from which they originally came (Figure 9-21). Many of the steps that are responsible for moving protons across the thylakoid mem-

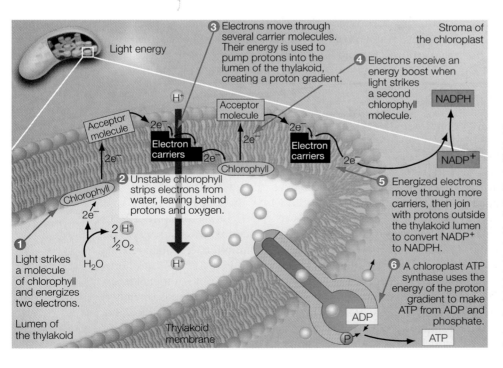

Figure 9-20. The light-dependent reactions of photosynthesis. Light raises the energy of electrons, which then pass from carrier to carrier, releasing energy as they move. The energy is used to pump protons into the lumen of the thylakoid, creating a gradient of protons across the thylakoid membrane. The electrons finally join $NADP^+$ and, along with protons, form NADPH. The chloroplast ATP synthase uses the potential energy in the proton gradient to make ATP. Both NADPH and ATP are used in the light-independent reactions to make sugars.

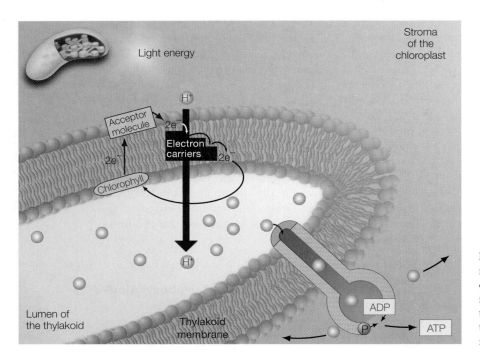

Figure 9-21. In cyclic photophosphorylation, electrons energized by light leave chlorophyll and pass through several carriers, ending up where they started. As electrons move, protons are pumped into the thylakoids. The proton gradient is used to make ATP.

brane still occur, and a proton gradient is still created. Consequently, this pathway can capture enough energy to phosphorylate ADP to ATP, but the electrons end up exactly where they started. For this reason, it is called **cyclic photophosphorylation**. The primitive Archaebacteria use a version of cyclic photophosphorylation to make ATP. For these simple single-celled organisms, this cyclic mechanism can supply enough ATP to sustain life. But for other autotrophs, a combination of cyclic and noncyclic photosynthesis is followed by the light-independent reactions that fix carbon from low-energy carbon dioxide into high-energy organic molecules by means of the Calvin-Benson cycle.

Breath from the Sea Most of the oxygen we breathe comes not from green land plants, but from photosynthetic organisms that inhabit the oceans. What are these organisms? How might pollution in the ocean affect us all?

9.4.4 The Calvin-Benson Cycle Uses CO_2, ATP, and NADPH to Make Food

Fast Find 9.4c Light-Independent Reactions

Melvin Calvin and A. A. Benson were able to learn each of the steps of carbon fixation because of two important research tools developed in the late 1940s and early 1950s. One tool was radioactive carbon, called ^{14}C ("carbon-14"), that could be added to a mash of plant cells in the form of $^{14}CO_2$; the other was paper chromatography, a technique used to separate small organic compounds from a mixture. Calvin and Benson were interested in following the carbon atoms from CO_2 through each single step as they were incorporated into sugar. They used a green alga called *Chlorella*. Cultures of *Chlorella* were exposed to radioactive CO_2, some for several minutes, others for just a few seconds. The cells were killed and the carbon-containing compounds separated using paper chromatography (Figure 9-22). The compounds that were radioactive were the ones that incorporated $^{14}CO_2$; that is, they are the metabolic intermediates of photosynthesis. Calvin and Benson found that the first compound to incorporate ^{14}C was a three-carbon acid called PGA. In fact, later studies showed that carbon dioxide

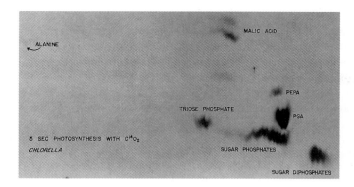

Figure 9-22. Paper chromaography showing the metabolic intermediates of the Calvin-Benson cycle. The black spots represent the positions of compounds containing radioactive carbon.

is actually first combined with a five-carbon compound called RuBP to make a six-carbon sugar, but that sugar immediately splinters into two three-carbon PGAs. Thereafter, radioactive carbon was found in two other three-carbon compounds, one of which was shunted from the stroma of the chloroplast to the cytoplasm of the cell, where it was made into sugar. The individual steps of the Calvin-Benson cycle are illustrated in Figure 9-23.

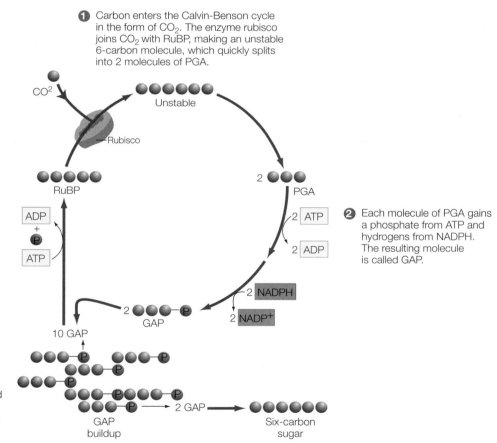

Figure 9-23. The steps of the Calvin-Benson cycle. Carbons enter the cycle as carbon dioxide. The enzyme rubisco combines each CO_2 with a five-carbon compound called RuBP. The resulting six-carbon compound quickly splits into two three-carbon PGA molecules. Each PGA gains a phosphate from ATP and hydrogens from NADPH, making an energized three-carbon compound, called GAP. When the cycle has turned six times and 12 molecules of GAP have accumulated (GAP buildup), two GAPs are joined to make a six-carbon sugar. The other 10 GAP molecules are converted back into RuBP to continue the cycle.

273

The first step of the Calvin-Benson cycle is catalyzed by an enzyme called ribulose bisphosphate carboxylase, or **rubisco**. It is this enzyme that brings carbon into the biosphere. Rubisco is arguably the most important enzyme on Earth; it is certainly the most abundant. It has been estimated that there are 22 pounds of rubisco for every person on Earth. Yet the enzyme is remarkably slow and not very efficient as enzymes go. Rubisco is only capable of fixing about three molecules of CO_2 per second; most enzymes catalyze their particular reaction thousands of times per second. In addition, rubisco is slowed in the presence of oxygen, the very product of the light-dependent reactions occurring in the chloroplast. It is not surprising that agronomists are working hard to improve the speed and efficiency of this critical enzyme upon which all life depends.

Building Better Plants What is photorespiration? How is the energy of photosynthesis wasted in this process? What new technologies are being developed to improve the efficiency of photosynthesis in crop plants?

Piecing It Together

Photosynthesis takes low-energy carbon in the form of carbon dioxide and brings it into the biosphere by fixing it into organic compounds. In turn, heterotrophs, including humans, rely upon fixed, or organic, carbon as fuel. Pigments, the light-dependent reactions, and the light-independent reactions convert the energy of light into chemical-bond energy.

1. Chloroplasts are the site of photosynthesis in plants. A chloroplast is surrounded by two outer membranes and has an extensive network of internal membranes arranged in thylakoids. The thylakoid membranes house the protein machinery for photosynthesis and enclose a space, or lumen.

2. The thylakoid membranes contain pigments, which are large, colorful organic compounds capable of absorbing certain wavelengths of light. Chlorophyll is the main light-absorbing pigment of green plants. It absorbs red and blue light and reflects green light. When light is absorbed by pigments, electrons are pushed to a higher energy level and captured by an electron acceptor that can only pick up energized electrons. In losing electrons, chlorophyll becomes unstable; it fills the electron void by stealing electrons from water, leaving behind oxygen and protons.

3. Energized electrons originating from chlorophyll are passed through a series of electron carriers, ultimately to $NADP^+$ to make NADPH. As the electrons move from carrier to carrier, their energy is used to move protons from the outside of the thylakoid lumen to the inside. The end products of this electron pathway are NADPH and a proton gradient across the thylakoid membrane.

4. Thylakoid membranes also contain an ATP synthase, nearly identical to the ATP synthase in mitochondria. This enzyme creates a pathway for protons to flow down their gradient (from the thylakoid lumen to the stroma) and couples that flow to the synthesis of ATP.

5. The light-independent reactions of photosynthesis, also called the Calvin-Benson cycle, make sugar using energy from the ATP and NADPH made in the light-dependent reactions.

Where Are We Now?

We rely on the chemical reactions of cellular metabolism to provide the energy we need to stay alive, but some unwanted byproducts of these same reactions may be contributing to the physical deterioration that accompanies aging. Recent studies on why our bodies age have led researchers to look closely at one of the byproducts of aerobic metabolism, the **free radical**.

Free radicals are varieties of atoms or molecules that are very unstable and highly reactive. In nonradical molecules, the electrons that orbit an atom's nucleus occur in pairs. This pairing makes these nonradical molecules stable. A free radical, however, is a molecule or atom that has a single, unpaired electron in its outermost orbit. A free radical will "steal" an electron from a neighboring molecule to even out the number of electrons in its outer shell. Such thievery transforms the neighboring molecule into a free radical, which may, in turn, steal an electron from yet another neighbor, setting in motion a kind of biological chain reaction.

Electron exchanges of this sort are at the heart of many of the reactions in the metabolic pathways we learned about in this chapter. For the most part, electron exchanges are highly controlled, and only the appropriate substrate is stripped of its electrons. The chain reaction is stopped at the proper time, and damage is prevented.

In some metabolic reactions, particularly those involving oxygen, free radicals escape. The body has two ways of dealing with dangerous free radicals. Certain compounds in the body can act as "antioxidants"; in other words, they can absorb the extra electron from oxygen radicals and convert these radicals to harmless products. Substances such as beta-carotene and selenium have antioxidant properties. In addition, cells make an enzyme that converts harmful radicals into harmless products. This enzyme, called superoxide dismutase, has been found in all organisms that have been studied.

Some researchers believe that such age-related symptoms as wrinkles, arthritis, loss of flexibility, and even some diseases such as cancer and degenerative diseases of the nervous system are signs of accumulated damage caused by free radicals. Convincing evidence for the role of free radicals in neurodegenerative diseases came to light in 1993, when researchers discovered that a large proportion of afflicted individuals carried a mutation that causes a defective form of the enzyme superoxide dismutase. Although we are still a long way from a cure for neurodegenerative disorders, understanding the mechanism of the disease is a first step toward finding effective treatment.

So how can you stay younger longer? We cannot stop the formation of free radicals because they are a normal part of our metabolism. But we can increase the intake of compounds that can act as antioxidants. A diet rich in fruits and vegetables has lots of zinc; selenium; beta-carotene; and vitamins A, C, and E, all of which have been shown to have antioxidant properties.

REVIEW QUESTIONS

1. Pretend you are a patent agent and an inventor comes to you saying that he has invented a self-powered engine that requires no fuel to run. Will you agree to patent this invention? Why or why not? Use the first law of thermodynamics to justify your answer.

2. You are still a patent agent, but now another inventor comes to you saying she has invented a perpetual motion machine. Will you agree to patent this invention? Why or why not? Use the second law of thermodynamics to justify your answer.

3. Compare and contrast entropy and equilibrium. Begin by giving a definition of each. How are these two concepts related?

4. What evidence did Lavoisier use to reject Priestley's contention that breathing and fire added poisonous "phlogiston" to the air? How do these two activities really change the quality of air? What general term is given to the chemical reactions of both fire and metabolism?

5. You are debating a person who believes in vitalism, or the existence of a special kind of energy peculiar to life. Your opponent claims that the second law of thermodynamics cannot apply to living things because when organisms grow, they become more ordered (entropy decreases), and the second law says that entropy in the universe is always increasing. What argument can you make that the laws of thermodynamics apply to both living and nonliving things?

6. This same opponent argues that there must be a vital force because cells have enzymes that make reactions occur that would not normally occur. How do you answer this argument?

7. Give an example of how metabolic oxidation is more specific than oxidation that occurs in a fire. Give an example of how metabolic oxidation is more efficient than oxidation that occurs in a fire.

8. Why do metabolic pathways have so many small steps instead of just one big step?

9. Many reactions that are very favorable in a thermodynamic sense do not seem to occur at all. Explain.

10. Four pairs of nucleotide-based compounds are cycled in cellular metabolism: ATP/ADP, NADH/NAD$^+$, FAD/FADH$_2$, and NADPH/NADP$^+$. For each of these four pairs, which form of the compound has the highest energy?

11. Some scientists argue that glycolysis was the first metabolic pathway to evolve. What kinds of evidence are used to support this argument?

12. Yeast cells can live anaerobically (without oxygen) indefinitely. What metabolic pathway do yeast depend on for energy in the absence of oxygen? What differences between yeast and humans account for the fact that we cannot survive without oxygen?

13. What goes into glycolysis? What comes out? Where is the energy of glucose at the end of glycolysis?

14. One way of getting an overview of cellular respiration is to take an accounting of carbon atoms as they move through the three stages. Six carbons enter glycolysis at the beginning of cellular respiration. Where are those carbons at the end of glycolysis? Where are they at the beginning of the Krebs cycle? Where are they at the end of the Krebs cycle?

15. In the analogy made between oxidative phosphorylation and a hydroelectric dam, what part of the mitochondrion is analogous to the dam? What is analogous to the water behind the dam? What parts of the dam are analogous to the mitochondrial ATP synthase?

16. When chlorophyll molecules are dissolved in a flask of alcohol and a beam of blue light is directed at the flask, the solution glows red. Explain.

17. The light-independent reactions of photosynthesis that produce sugars can occur in a test tube using only the plant cells' juices and no parts of the cell. Can the light-dependent reactions also occur using only the cells' juices? Why or why not?

18. What is the significance of cyclic photophosphorylation?

19. The enzyme rubisco has been called the most important enzyme in the biosphere. What does this enzyme do?

20. The chloroplast ATP synthase and the mitochondrial ATP synthase have been called "molecular cousins." What does this mean?

Chapter 10

Animal Physiology: How Do Organisms Respond to Change?

The present state of the earth and the organisms inhabiting it is but the last stage of a long and uninterrupted series of changes...
—Alfred Russel Wallace (1823–1913)

—Overview—

Change is the one constant all life faces. Day becomes night. Sunny days become cloudy. Fall moves into winter. Consumed nutrients are used up and must be replenished. Some changes occur with predictable regularity; others are more random, rare, or capricious. Some changes are sudden, demanding immediate attention; others occur in minute increments over long periods of time. The most obvious changes occur in organisms' external environments, but others, equally important, occur internally. These changes—predictable and unpredictable, short term and long, external and internal—can occur simultaneously. If organisms are to survive, all must be tolerated, adjusted to, and compensated for.

How organisms cope with constant change is the topic of the next three chapters. We shall see that there are nearly as many different responses as there are species. Fortunately, there are patterns to these responses, and these we will stress. In this chapter, we will look at physiological processes as expressed in microbes, fungi, and simple animals. We will look at the structure and function of these organisms: how their various parts work, alone and together; how their parts and processes are integrated; how they respond to environmental changes, both internal and external; how organisms change with age; and how they reproduce. This is the study of **physiology**.

Life on Earth has been a journey that first moved from water to land and then along two divergent but parallel paths. One path stressed **autotrophy** and led to flowering plants; their physiology will be discussed in Chapter 12. The other path stressed **heterotropy** and led to birds and mammals. In Chapter 11, we will focus on humans as examples of quintessential, highly adapted, terrestrial animals.

Chapter opening photo—Weasels change color seasonally.

Our task is not quite as daunting as it might appear because closely related species have similar physiological processes. For example, desert kit foxes live in environments very different from Arctic foxes, but both are foxes. The kidneys of both are similarly adapted to conserve and recycle water, which is lacking in the desert and frozen and unavailable in the Arctic. Both use their ears to regulate their body temperature. The ears of kit foxes are huge and radiate unwanted body heat to their slightly cooler environments. Those of the Arctic fox are small, fur covered, too well insulated to lose much-needed body heat to a consistently colder environment. In other ways, too, the two foxes have similar mechanisms to cope with changing but very different environments.

So, let us plunge into a fascinating, complicated, relevant exploration of the physiology of living things.

10.1 WHAT ARE THE UNDERLYING PRINCIPLES OF PHYSIOLOGY?

All multicellular organisms evolved from single-celled organisms. How they might have done so is not entirely clear. Generally speaking, when bacteria reproduce, one cell becomes two, which separate and lead totally independent existences. Often, however, daughter cells do not separate; they stay connected, as do their offspring. Eventually, what started out as a single cell becomes a long string of cells, or a colony—a multicellular entity, to be sure, but not a multicellular organism. Why not? If any cell of the colony becomes separated from the rest, it is fully capable of an independent existence.

This independence is not true of the cells of multicellular organisms. With the possible exception of its gametes, there is no way that an individual shark cell could possibly exist under natural conditions independent of other shark cells. Shark muscle cells can contract, but they are totally dependent on shark pancreas cells for certain digestive enzymes. Shark pancreas cells provide those enzymes and in turn are totally dependent on shark muscle cells to get from place to place. This mutual interdependence among cells is what distinguishes multicellular organisms from multicellular colonies (Figure 10-1).

Multicellular organisms are also distinguished by certain features and responses, which are discussed in the following sections.

10.1.1 The Cells of Multicellular Organisms Are Hierarchically Organized

Multicellular organisms are more than simple masses of interdependent cells; their cells instead are highly organized. First, cells are organized into **tissues**, which are groups of similar cells performing a similar function. The number of tissues an organism has depends on the complexity of the organism. Among animals, jellyfish have only a few tissues, whereas mammals such as humans have more than 200. Fortunately for those studying animal tissues, they can be classified into a smaller number of types. Frequently encountered examples include the following:

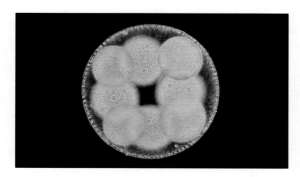

Figure 10-1. *Volvox* is perhaps the simplest multicellular organism; that is, its cells show a definite division of labor. Cells on the inside of the organism are responsible for reproduction. Cells on the outside are responsible for all other functions (magnified 161 times).

◆ **Epithelial tissues** cover surfaces throughout the body, inside and out, and provide protection to what they cover. Typically, epithelial cells reproduce rapidly to replace those damaged by injury or disease. There are three general types of epithelial tissue based on cell shape: **columnar** (column-shaped), **cuboidial** (cube-shaped), and **squamous** (flattened).

◆ **Connective tissues** are also found throughout the body and generally reproduce much more slowly than epithelial tissue. Connective tissues are usually composed of living cells surrounded by nonliving matrices. The functions of connective tissues vary. **Cartilage** and **bone** are connective tissues that form skeletons. **Fibrous connective tissues** form **tendons**, which bind muscles to bones, or **ligaments**, which bind bones to other bones. **Loose connective tissue** forms delicate sheets throughout the body that surround and protect other parts and fill spaces. **Adipose tissue** stores fat. **Blood** is connective tissue that transports materials around the body.

◆ **Muscle tissues** are highly contractile. Most reproduce more slowly than epithelial tissue. Their major function is to move internal parts or, along with bones, to move the body as a whole. There are three types: **striated**, associated with body movements; **smooth**, associated with movements of internal body parts; and **cardiac**, associated with the heart.

◆ **Nerve tissues** are also found throughout the body and are especially numerous in brains and nerve cords. Normally, in birds and mammals, the cells of nerve tissue lose their ability to reproduce prior to an animal's birth. The primary function of nerve tissue is to relay information from the external and internal environments to the brain, process that and other information, and relay messages from the brain to all internal body parts.

The cells of plants, too, are organized into tissues, albeit fewer in number; we will forgo their discussion until Chapter 12.

The **hierarchical** organization of cells in multicellular organisms does not stop with tissues (Figure 10-2). The tissues, in turn, are organized into **organs**, complex systems of tissues that work together to perform common functions. Organs are the obvious parts of a body. For instance, the stomach is an organ surrounded by connective tissue and containing smooth muscles, nerves, other connective tissues, and blood vessels, which are themselves made of epithelial, connective, muscle, and nerve tissues. The forearm is an organ. So is the skin.

Organs are further organized into **organ systems**. Although physiologists are interested in tissues and organs, they focus most of their attention on the functioning of organ systems. Plant organ systems include root, shoots (stems), and leaves and will be discussed in Chapter 12. Table 10-1 lists the animal organ systems that will be discussed further in this chapter.

Table 10.1 Principal Organ Systems of Complex Animals

Organ system	Function	Discussed in section
Digestive	Absorption of nutrients	10.2.1 and 10.2.2
Respiratory	Exchange of oxygen and carbon dioxide	10.2.3
Circulatory	Transport of materials throughout body	10.2.4
Excretory	Elimination of metabolic wastes	10.2.5 and 10.2.6
Skeletal	Body form and movement	103.1 and 10.3.2
Muscular	Movement	10.3.2
Endocrine	Communication	10.3.4 and 10.3.5
Nervous	Communication	10.3.6
Reproductive	Reproduction	10.4

Finally, organ systems are organized into individuals. Notice that the organization in multicellular organisms is hierarchical; one layer builds into the next: cells into tissues into organs into organ systems into whole organisms.

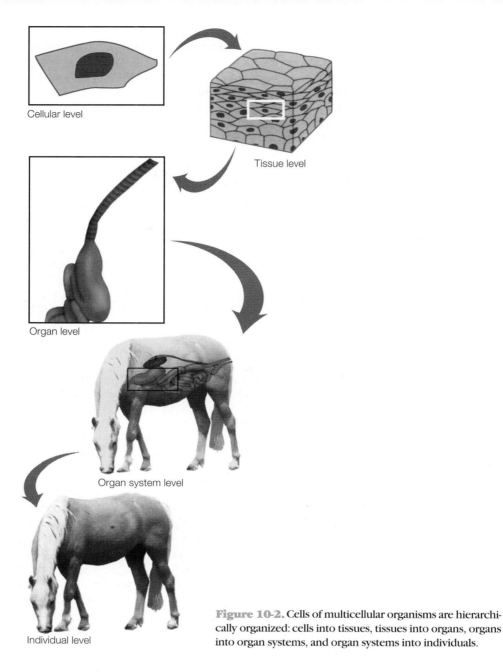

Cellular level

Tissue level

Organ level

Organ system level

Individual level

Figure 10-2. Cells of multicellular organisms are hierarchically organized: cells into tissues, tissues into organs, organs into organ systems, and organ systems into individuals.

10.1.2 Multicellular Organisms Are More Than the Sum of Their Parts

Being a multicellular organism means that parts work together, seamlessly. Even simple actions illustrate the point. Move your arm. What organ systems were involved? The muscle system certainly. And the skeletal system, because to move an arm, muscles move bones. And the nervous system, because, under normal conditions, muscles of the body do not contract unless ordered to do so by nerves (the heart is an exception).

More complex actions involve more systems. You're hungry, so you eat an apple. What systems are involved? Nerves of the stomach inform the brain that the stomach is empty; this is part of the information processed by the brain in deciding that you are hungry. Muscles, bones, and nerves are used to pick up the apple, bite into it, and chew and swallow it. Other nerves tell the stomach food is coming, stimulating stomach muscle cells to increase contractions and other stomach cells to release gastric juices. Other nerves dilate blood vessels to bring more blood and oxygen to the stomach and intestines in anticipation of increased activity and eventual nutrients. Meanwhile, the pancreas is monitoring blood sugar levels. If it detects that levels are high, it releases hormones stimulating the liver to store glucose coming from the apple. If not, other pancreatic hormones may be released, ensuring that the glucose goes directly into the blood and hence to body cells needing glucose for energy. Generating the energy for all this activity requires oxygen, brought to the cells of the organ systems by the respiratory and circulatory systems. Eating an apple, then, requires nervous, muscular, skeletal, circulatory, respiratory, endocrine, and, oh yes, digestive systems, all acting and reacting to each other and the situation in a seamless and completely integrated fashion. By themselves, none of these systems is capable of processing an apple into useful nutrients and energy. Together they can.

Physiologically, multicellular organisms are **synergisms**, or systems that are more than the sum of their parts. Why, then, don't we study the physiology of whole individuals? The short answer is, we can't. Individual multicellular organisms, especially when they get to the level of complexity of humans, are simply too complex to contemplate as a single unit. Whenever scientists are confronted with overwhelming complexity, they tend to become reductionists; by breaking complexity into small, manageable units, it can be more conveniently studied. Later, perhaps, after we have come to understand how components work, we'll put the components together and see how the system functions as a unit. We study physiology system by system as a matter of convenience, but we shouldn't forget that all these systems interact synergistically. We, and all multicellular organisms, are, indeed, more than the sum of our parts.

10.1.3 Multicellular Organisms Function Best with Stable Internal Environments

Often, an organism's coping with external and internal environments is geared to provide either (1) survival or (2) nearly constant internal conditions. The need to survive is obvious. All organisms must cope with predators, elusive nutrients, limited space, competitors, and so on. Those that can successfully cope may garner enough excess energy, nutrients, and other resources to be able to reproduce. They are the ones that pass individual characteristics on to offspring.

Maintaining nearly constant internal conditions—known as **homeostasis**, meaning "steady state"—is perhaps a less obvious but equally important need. All organisms do best within a rather narrow range of internal conditions, including body temperature, blood glucose levels, and water. Maintaining homeostasis in an ever-changing environment often involves considerable effort on the part of the organism.

Let's look at some examples of homeostasis. Snakes are cold-blooded (**ectothermic**) animals; that is, heat and thus their body temperature is absorbed from their environment. Early on a cool morning, snakes are likely to be found basking on sunny rocks, soaking up solar rays, warming their internal temperatures to within optimal ranges. Later, as the day warms, snakes seek shade to avoid overheating. By their behavior, snakes regulate their internal temperatures to keep within a rather narrow range (Figure 10-3).

Snakes that fed the previous night received a shot of nutrients, including glucose, as the meal was digested. All their cells need glucose to fuel activities, but too much in the blood could have serious consequences. One of the functions of the snake's liver is to store glucose. Over the next several days, it gradually releases glucose, first to the blood, and then to the cells, so that there is always enough but never too much. Body temperature and blood glucose levels are two of the many factors involved in animal homeostasis.

For humans and other warm-blooded (**endothermic**) animals, body temperature is generated internally, and maintaining near-constant internal temperatures in ever-changing thermal environments is much more complex, involving both behavioral and physiological processes, as discussed below.

Figure 10-3. A California king snake suns on a cool morning and thus regulates its internal body temperatures behaviorally.

Sugar in Blood Humans regulate blood sugar levels homeostatically. How do they do it?

10.1.4 Feedback Systems Control Many Physiological Processes

Fast Find 10.1b Feedback

Homeostasis is an energetically costly challenge faced by animals. First, they must constantly monitor their environments and then respond appropriately to changes, always with the goal of maintaining near-constant internal conditions. Frequently, such systems are controlled by **feedback**.

Basically, there are two kinds of feedback systems: **positive** and **negative**. A good mechanical example of positive feedback is a microphone on a stage connected to speakers. If the microphone is too sensitive, it may begin to pick up and broadcast background sounds. The system begins to whine. The more sound it picks up, the louder it broadcasts. The louder it broadcasts, the more it picks up, until someone grabs the microphone or pulls the plug. Positive feedback leads to increasing instability (as in the ever-louder noise) and some kind of climactic event. Hysteria, childbirth, and sexual arousal are examples of human activities controlled by positive feedback.

More frequently seen and generally more useful to an organism are negative feedback systems. To see how they work, let's again look at a mechanical example: the home thermostat. This device, usually located at some point in the house's interior that experiences minimal temperature change, senses temperature and controls the furnace. If the temperature falls below a predetermined critical temperature, called the **set point**, the furnace is turned on. When the temperature gets above the set point, the furnace is turned off. Notice that the thermostat measures only a portion of the house's total heat energy. The temperature in an isolated bedroom can be freezing cold, but if the thermostat is warm, the furnace is not turned on. What makes this a negative feedback system is that when the house gets *cold*, the thermostat causes it to *warm*. Notice that negative feedback systems generally lead to stability.

Notice, too, how closely this mechanical system parallels the human body's system of maintaining nearly constant internal temperatures. The body's thermostat is a small portion of the **hypothalamus**, located at the base of the brain about midway between the ears and richly surrounded with blood vessels. It's the temperature of the blood that the hypothalamus monitors. Like the thermostat, the hypothalamus measures only a small portion of the body's total heat. If the temperature of the blood flowing through the hypothalamus falls only a few tenths of a degree below the body's set point at that location (about a degree warmer than the temperature found under the tongue), the hypothalamus sends out signals that restrict blood flow to the extremities, cause the body to shiver, draw in the limbs, become restless, or any of the other things we normally do when cold (Figure 10-4).

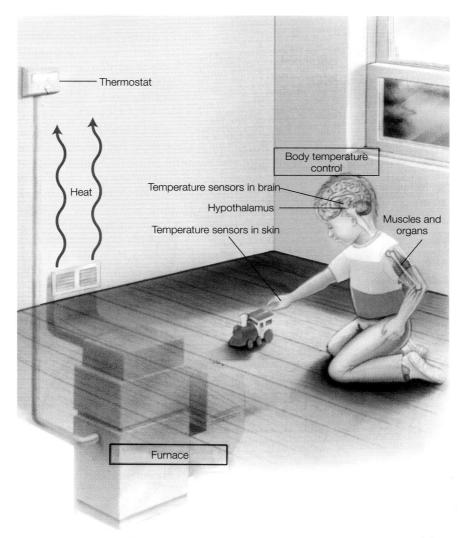

Figure 10-4. Two examples of negative feedback systems. Thermostats control heat output from a furnace by monitoring room temperature. When room temperature exceeds some critical level, termed the set point, the thermostat turns the furnace off. When the temperature falls below the set point, the thermostat turns the furnace on. In a similar fashion, the hypothalamus, located in the brain, monitors and controls body temperature. The hypothalamus receives multiple inputs from temperature sensors in the brain and skin. It monitors the temperature of the blood and receives information from peripheral nerves measuring skin temperatures. In a system similar to the furnace, when body temperature falls below the body's set point (98.6°F or 37°C in humans), the hypothalamus causes muscles to shiver, thus creating heat and raising the body's core temperature. When the body gets too warm, the hypothalamus causes the body to sweat, which cools the body.

In many houses, when it is too warm in summer, the thermostat controls an air conditioner. Similarly, when the body gets too warm, the hypothalamus causes the body to sweat. These functions, too, are controlled by feedback systems, more complex than the simple one discussed above. Indeed, negative feedback systems can get quite complex. Go to the CD to see how some of them work.

10.1.5   Some Additional Points to Keep In Mind

As we explore further the physiology of multicellular organisms, look for additional examples of hierarchical organization, homeostasis, and feedback. Also keep in mind five additional points.

1. Ultimately, *all physiological processes are cellular phenomena*. Nutrients are first absorbed by cells in the small intestine. The force that moves nutrients from intestines to other cells of the body are contractions of heart muscle cells that pump blood. The plasma membranes of cells are of particular importance to physiologic processes. Hormones fasten onto target cells by binding to specific plasma membrane proteins. Nutrients enter and wastes leave cells through their plasma membranes. You may want to review cell membrane structures and functions in Chapter 4, particularly Sections 4.2.2 and 4.2.3.

2. Generally speaking, *there is a relationship between form and function*. With some notable exceptions, the shape and structure (form) of an organ, tissue, or cell is related to what it does (its function). Nerves connect various parts of the body with the brain and with other nerves. Their long, threadlike form facilitates this function. This relationship between form and function is seen at all levels of the physiologic hierarchy. In the more complex animals, the digestive system is long and hollow, which facilitates its functions as a processor of food and collector of nutrients. The relationship between form and function is a result of evolution. As natural selection favors those individuals with adaptive traits, it favors cells, organs, and organ systems best suited for specific functions. In a sense, natural selection molds form to fit function.

3. *Tolerable ranges of internal and external conditions often vary from one organism to another.* For example, certain species of algae do best in nearly boiling hot springs (Figure 10-5). Other algae thrive in snow where environmental temperatures are consistently at or below freezing. Neither can tolerate the temperatures that are optimal for the other. Within a species there may be similar individual differences. Among humans, for example, certain individuals are better able to cope with cold than others.

4. *Evolution often results in increasing complexity, but not always.* Consider the digestive systems of animals. Those of simple animals such as jellyfish are considerably less complex than those of more advanced animals such as chordates. Free-living flatworms—more advanced than jellyfish but less advanced than

Fast Find 10.1g Surface-to-Volume Ratio

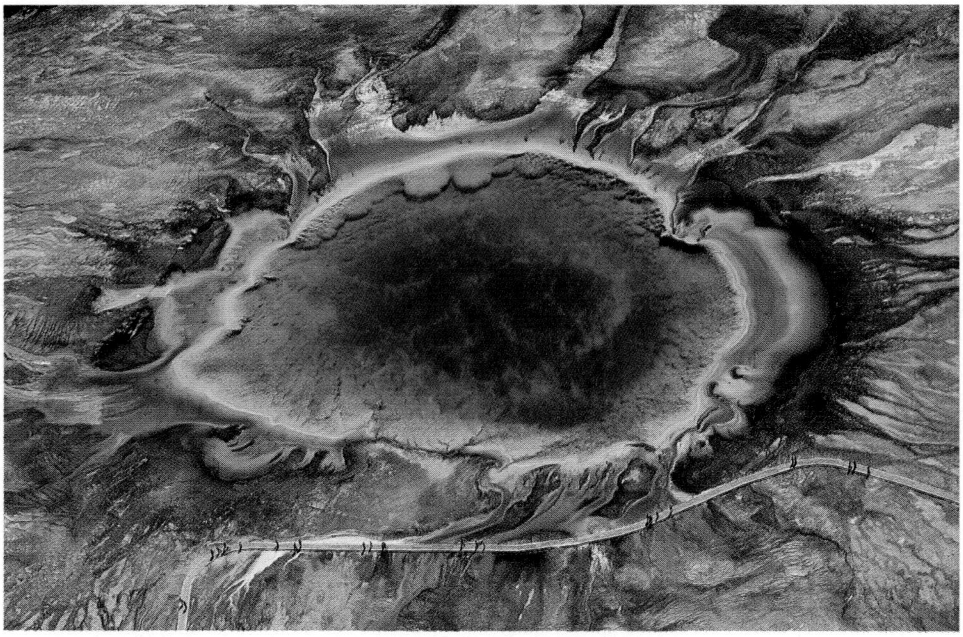

Figure 10-5. Some algae do best in hot water. In Prismatic Spring, Yellowstone National Park, different colored algae tolerate different temperatures. Some tolerate temperatures that approach the boiling point of water.

chordates—have digestive systems of intermediate complexity. But there are exceptions. Although not common, there are instances of evolution leading to simplicity rather than complexity. For example, tapeworms are closely related to and obviously evolved from free-living flatworms. But they lack digestive systems altogether.

5. *Plants are structurally and physiologically less complex than animals.* The most advanced plants, those that produce flowers, basically consist of only three organ systems (roots, shoots, and leaves) comprising roughly a dozen tissues. By contrast, chordates have 9 principal organ systems and over 200 tissue types. It is important to remember that, in spite of these great differences in complexity, both plants and animals have been successful in coping with changing environments. Because of their differences, plants will be considered separately in Chapter 12.

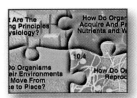

Piecing It Together

The major underlying principles of physiology can be summarized as follows:

1. Physiologically, the cells of multicellular organisms are organized hierarchically: cells into tissues, tissues into organs, organs into organ systems, and organ systems into organisms.
2. Common tissue types found in animals include epithelial, connective, muscular, and nervous.
3. Multicellular organisms are synergisms; that is, they are more than the sum of their individual parts.
4. Multicellular organisms, especially the more complex ones, function best under homeostatic (nearly steady-state) internal conditions.
5. Many physiologic processes are controlled by negative feedback systems.
6. Ultimately, all physiological processes are cellular phenomena.
7. Generally speaking, there is a close relationship between form and function.
8. The range of conditions that can be tolerated often vary from one organism to another.
9. Evolution often results in increasing complexity, but not always.
10. Plants are structurally and physiologically less complex than animals.

10.2 HOW DO ORGANISMS ACQUIRE AND PROCESS NUTRIENTS AND WASTES?

In 1911, John Muir pointed out that when we try to pick out anything by itself, we find it hitched to everything else in the universe. The statement is certainly true of individual organisms. Each of us is intimately and continuously connected to our external environments. Indeed, we are not just connected; we are totally dependent on our surroundings. We and all organisms need energy that is obtained, directly or indirectly, from the sun (or, in rare exceptions, the Earth). Energy is stored in nutrients, which also provide raw materials for growth, maintenance, and other functions. These nutrients come from the atmosphere, the Earth, or both. For a variety of reasons, all organisms—indeed, all cells—continuously require water. As we consume energy and process nutrients, we produce potentially toxic wastes, which we return back into our surroundings where they become another organism's nutrients. Our continuous needs for energy, nutrients, water, and waste disposal keep us continuously connected with and

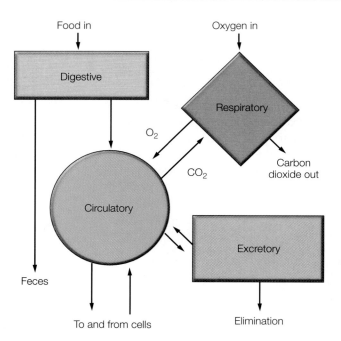

Figure 10-6. Four major organ systems are centrally involved in acquiring and processing nutrients and metabolic wastes in complex multicellular organisms. In the digestive system, useful nutrients pass from the food into the circulatory system and indigestible wastes pass from the body as feces. In the respiratory system, oxygen passes from inhaled air to the circulatory system. The circulatory system carries nutrients and oxygen to cells throughout the body and exchanges them for metabolic wastes and carbon dioxide. The metabolic wastes are transported to the excretory system, where they are eliminated from the body. Carbon dioxide is transported by the circulatory system to the respiratory system, where it is exhaled from the body.

dependent on sun, Earth, atmosphere, and other organisms. In Chapters 14 and 15 we will stress interactions between organisms and their environments. Here, let us focus on how organisms acquire and process their nutrients and wastes. Four organ systems are involved, namely, the digestive, respiratory, circulatory, and excretory systems (Figure 10-6).

10.2.1 Bacteria, Protists, and Fungi Excrete Enzymes to Break Down Complex Compounds and Obtain Nutrients

Frequently, nutrients required by bacteria, protozoan protists, and fungi are not readily available. For the most part, these organisms are decomposers or pathogens, living on carcasses or bodies of larger plants or animals. Initially, the victim's compounds are much too large to move through the decomposers' cell walls and membranes. First, they must be broken down into simpler compounds that can transported into the cells.

As needed, bacteria and fungi produce enzymes that are secreted into the environment to accomplish these tasks. In effect, complex compounds are digested outside the cells. As a result, the carbohydrates are broken down into simple sugars, proteins into amino acids, and fats into fatty acids—compounds much smaller in size than their predecessors. What is the next step? By what processes do digested nutrients pass from the environment to the inside of cells? If you can't answer these questions, you may want to review Section 4.2.3.

Feeding Fungi As a result of their extracellular digestion, microbes and fungi soon find themselves in environments in which nutrients are depleted. Microbes must move and fungi must grow into new territories, rich in undigested potential nutrients. Fungi obtain nutrients primarily from the tips of their hyphae. By what process are older fungal cells (that is, those not at the tips of the hyphae) fed? What special characteristics of fungi facilitate this process? Note, once again, that form fits function.

10.2.2 Multicellular Animals Have Digestive Systems with Which to Obtain Nutrients

Multicellular animals obtain nutrients by ingesting portions of the external environment, internally breaking down complex compounds into simpler ones, absorbing useful nutrients, and returning unused portions to the environment. Perhaps more than any other physiological system, digestive systems show an evolutionary sequence from the simplest animals (sponges) through the most complex (birds and mammals). Notice as you read the next few paragraphs, how the digestive systems of each group are related to and are more complex than those of their predecessors.

Sponges, the simplest of animals, don't really have digestive systems. Flagellated cells lining their internal spaces create currents that suck water into and through their bodies. Food items are picked up by endocytosis and digested intracellularly. Smaller compounds diffuse into cells or are picked up by active transport.

More Filter Feeders The way sponges feed is an example of **filter feeding**. Among aquatic animals, filter feeding is quite common. Describe at least one example of filter feeding in each of the common animal phyla (see Chapter 8 if you need to). Hint: There may be some phyla with no examples, but most phyla have some examples.

Fast Find 10.2b Hydra Food Capture

Jellyfish and flatworms have digestive systems with only one opening. Food is swallowed into a large cavity, which is highly branched in flatworms. Digestive enzymes secreted into the cavity break down complex compounds into simpler ones that can be absorbed (by what processes?). After a while, undigested material and metabolic wastes are expelled through the mouth into the environment.

Roundworms and all other animal phyla have **complete digestive systems**, that is, tubes equipped with two openings, mouth and anus. In roundworms the tube is simple and uniform throughout its length. As swallowed food passes along the tube, it is mixed with enzymes and its usable nutrients are absorbed. Eventually, undigested material passes out the anus as feces. A complete digestive system allows the processing of more than one meal at a time. As nutrients from one meal are being absorbed farther along the tube, a new meal can be swallowed at the head end.

In segmented worms, the digestive tube runs the length of the body and is divided into a series of specialized regions: pharynx, esophagus, crop, gizzard, and intestine. As food passes along the digestive tract, it is chewed, stored, digested, absorbed, and finally excreted as though moving along an assembly line.

The digestive systems of mollusks, arthropods, and vertebrates are refined further. They have long, differentiated tubes that also process food in assembly-line fashion. But unlike segmented worms, these animal's digestive tubes are coiled and twisted, their lengths much greater than the length of the animal as a whole. This allows for more complete and efficient digestion. Accessory organs, such as livers, gallbladders, and pancreases, also participate; each enhances digestion further (Figure 10-7).

Fast Find 10.2c Termite Gut Symbiosis

Some animals establish **symbiotic relationships** with other organisms to improve their chances of obtaining nutrients. Some corals house green algae whose photosynthetic products feed both the coral and the algae. Animals that feed on cellulose are in particular need of assistance. Cellulose, a major component of algal and plant cell walls, is abundant in many ecosystems. But few organisms produce enzymes capable of breaking down this complex carbohydrate into simple sugars that can diffuse through cell membranes. A major exception are the assorted bacteria and protozoa that can digest cellulose. How do cows, rabbits, and termites seemingly thrive on nothing more than dead grass or wood? Each supports large populations of symbiotic microbes in their digestive tracts. Humans, too, have their intestinal symbionts. Benign strains of

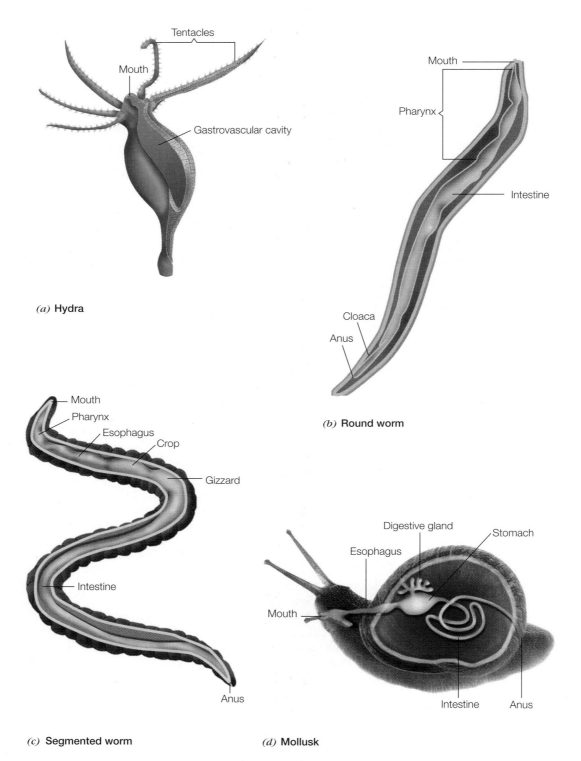

(a) **Hydra**

(b) **Round worm**

(c) **Segmented worm**

(d) **Mollusk**

Figure 10-7. Evolution of the digestive system in animals is a story of increasing complexity. *(a)* In jellyfish, such as *Hydra*, the digestive system consists of a large, internal chamber with but one opening to the outside, through which food enters and wastes leave. *(b)* In roundworms the digestive system consists of a simple, undifferentiated tube with two openings (mouth and anus) that runs the length of the worm's body. *(c)* In seg-mented worms, the digestive tube is differentiated into regions (esophagus, stomach, and so on) that process food rather like an assembly line. *(d)* In mollusks such as the snail, the highly differentiated digestive tube is twisted and looped so that its length is greater than that of the animal as a whole. In addition, there are accessory organs that assist in digestion.

Escheriscia coli crowd our large intestines, help absorb water, and produce certain vitamins, among other functions. The CD illustrates details of the relationships and provides other examples.

10.2.3 Animals Use Respiration to Obtain the Gases They Need

For most organisms, nutrients from soil or water are not enough. They require reliable sources of oxygen. As we learned in Chapter 9, oxygen is needed for chemical reactions that release energy stored in complex compounds. In the process, animals produce carbon dioxide, a waste product that they must get rid of.

There are exceptions, of course (in biology, there nearly always are). Many Archaebacteria and some primitive fungi (yeasts) find oxygen downright toxic. In Earth's initial atmosphere, with oxygen lacking, life forms evolved that were totally intolerant of this most reactive of elements. Their direct descendants still are intolerant, living in bottom muds, carcasses, intestinal tracts, and other environments where oxygen can be avoided. More modern organisms need oxygen.

How do organisms obtain oxygen and get rid of carbon dioxide? The short answer is by diffusion. Indeed, at some point, all organisms that require oxygen depend on diffusion. For single-celled and smaller multicelled organisms, nothing else is needed. Jellyfish and flatworms attain considerable size without need for specialized respiratory systems. With their large internal spaces and, for flatworms, body shape, none of their internal cells are very far from the environment. Oxygen diffuses in and carbon dioxide out to meet cell needs. But as organisms became larger and bulkier, as their activity levels and thus energy requirements increased, and as their bodies became covered with impermeable coverings (bark, shells, cuticle, scales, bony plates, and skin), the need grew for specialized, efficient respiratory systems to assist diffusion.

Among animals, there are basically three types of respiratory systems: **skin**, the outer body covering; **gills**, outpockets of tissue that work best in water; and **lungs**, inpockets of tissue that work best in air. All share two characteristics—extensive surface areas and thin, moist cell layers—that facilitate diffusion. The evolutionary relationships between respiratory systems are considerably less straightforward than was the case with digestive systems. Indeed, respiratory systems may have evolved independently at least three times among terrestrial animals.

Many invertebrates and some vertebrates, namely amphibians, breathe through their skin. Small blood vessels or capillaries bring blood rich in carbon dioxide and deficient in oxygen close to the skin's surface, where gaseous exchange takes place by diffusion. In most frogs and salamanders, respiration through the skin is only a backup system; it meets their needs during periods of inactivity, say, when they are hibernating. With increased activity levels, needs increase and lungs or gills, when present, become essential. Interestingly, one rather large group of salamanders have neither gills nor lungs; they respire only through skin. In all these organisms, skin must be kept moist to work as a respiratory organ.

Primitive gills, complex outpocketings of tissue that are in contact with water, are especially common in aquatic insect larvae and some salamanders. In salamanders, these gills consist of thin coverings and thin walled blood vessels. Diffusion of dissolved gases in and out is straightforward. Problems arise because these delicate tissues are unprotected and difficult to move. Movement of gills becomes necessary as immediately surrounding water becomes deficient in oxygen and as carbon dioxide builds up.

In aquatic mollusks and fish, gills are internalized and covered with protective body parts. This creates the immediate problem of how to move replenishing water, at least periodically, in past the gills. Clams use ciliated tissues to move water in through **incurrent siphons**, past gills (and incidentally, past mouths, by which they obtain food; and excretory pores, reproductive pores, and anuses, into which they deposit wastes and gametes) and out through **excurrent siphons**. Squids and octopuses suck water into

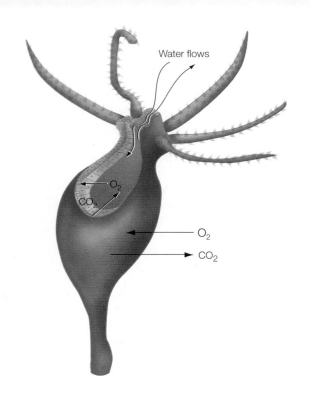

(a) **Hydra**

Gastrovascular cavity

O_2 CO_2

Flatworm

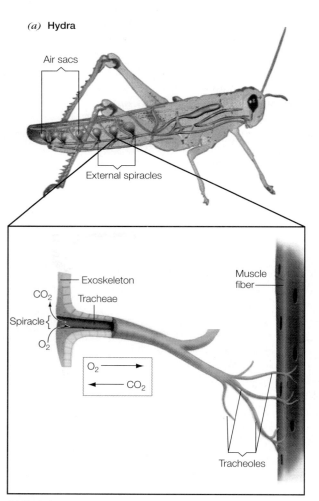

Air sacs

External spiracles

Exoskeleton

CO_2 Tracheae

Spiracle

O_2

$O_2 \rightarrow$

$\leftarrow CO_2$

Muscle fiber

Tracheoles

(b) **Grasshopper**

Figure 10-8. Animals show a variety of different respiratory systems.

(a) Simple animals, such as jellyfish and flatworms, have no respiratory systems. With simple diffusion through their skin, they exchange gases with their environments.

(b) Insects such as this grasshopper have a system of internal tubes (tracheae) that open to the outside through spiracles and connect to internal air sacs. In some insects, movement of gases into and out of the tracheae is passive, depending on diffusion. In others, muscles pump gases in and out.

(c) Right page. In bony fish, water containing oxygen is brought into the mouth and directed back, over and past the gills. The structure of the gills vastly increases the surface area through which gaseous exchange takes place. Water, relatively high in carbon dioxide, is expelled from the body through the gill slit.

their mantle cavities, where gills are located. In highly active sharks and bony fish (tunas, for example), water flows in through continuously open mouths, past gills, and out through gill slits (several in sharks; only one large one in bony fish). Most bony fish continuously gulp water, forcing it past gills and out through gill slits. In this way, they supply gills with fresh water during periods of relative inactivity. In all these animals, as refreshed water passes over gills, oxygen diffuses in and carbon dioxide diffuses out (Figure 10-8).

Lungs are the respiratory tissue most often seen in terrestrial animals. One of the major problems faced by terrestrial animals is the need to protect respiratory tissues from drying out. Lungs do this by being internal structures into which air enters through one (or relatively few) openings, but this only minimizes water loss. Indeed, replenishing water lost through respiratory surfaces is a continuous price paid by all terrestrial organisms.

The most primitive lungs are simple internal spaces connected to the surface by openings. No muscles move gases in and out. Rather, oxygen diffuses in and carbon dioxide diffuses out. These are the lungs found in terrestrial snails and primitive arthropods.

Insects have **tracheae**, complex systems of tubes that lead from the external world to internal regions of the body. The external openings of the tracheae, called **spiracles**, are found mainly on the insect's abdomen. Gaseous exchange takes place in the smallest of these tubes generally located furthest from the spiracles. In most insects, the flow is passive as gases diffuse in and out. In the more active insects, grasshoppers, for example, muscles pump air in and out of tracheae.

In the **pulmonary lungs** of vertebrates, including lungfish, amphibians, reptiles, birds, and mammals, special muscles move air in and out. These are muscles of the mouth region in fish and amphibians, rib muscles in reptiles, both rib muscles and air sacs in birds, and rib muscles and diaphragms in mammals. The lungs of lungfish and amphibians are structurally rather simple, consisting of little more than simple sacs and

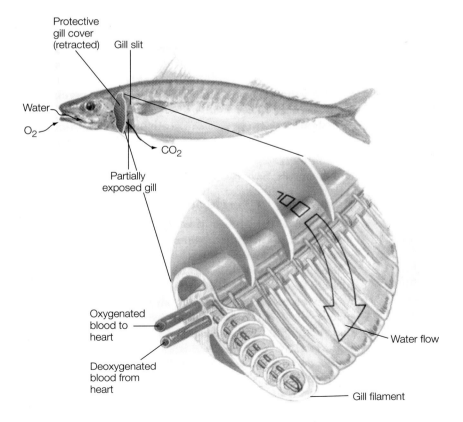

(c) Bony Fish

connecting tubes (larger **tracheae** and smaller **bronchi**). Internal structure is much more complex in birds and mammals. The lungs of birds are connected to internal spaces that increase buoyancy (useful when flying) and work like bellows, pumping air into lungs. In mammals, lungs are composed of myriad small pockets, **alveoli**, that vastly increase the surface area through which gaseous exchanges take place.

10.2.4 Circulatory Systems Move Nutrients, Gases, and Other Materials through the Bodies of Complex Animals

Once materials from the outside environment are internalized, the next problem is to distribute them to needy cells and tissues. Small animals have little need for special circulatory systems. In sponges, for example, flagellated cells circulate water throughout their internal spaces, allowing them to exchange nutrients for metabolic wastes and oxygen for carbon dioxide, using nothing more than diffusion and osmosis. Jellyfish and flatworms are another case. Their digestive systems are large or complexly branched internal spaces, and as a result, no cell is far removed from either its external environment or the internal digestive system, so materials and gases are exchanged by diffusion and osmosis.

For larger, more complex animals, moving materials through their bodies is a challenge. Their internal cells are so far removed from the external environment that simple diffusion does not suffice. From the segmented worms on, special transport systems carry nutrients and oxygen from the environment to their cells and metabolic wastes, including carbon dioxide, to the environment from their cells. These are the major functions of the circulatory system, and there are basically two types. In an **open circulatory system**, found in arthropods, the blood periodically leaves the blood vessels, bathes the tissues, and is recollected into the heart. In a **closed circulatory system**, found in many other animals, including humans, blood never leaves the blood vessels. In the more complex animals, the circulatory system also carries products made by the animals' cells and tissues, such as metabolic wastes, hormones, and disease-fighting proteins and cells (Figure 10-9).

Fast Find 10.2d Heart Evolution

Circulatory systems have three major components: vessels, blood, and hearts. Each evolved from rather simple components in primitive worms to highly complex ones in chordates. The CD explores their development in more graphic detail.

10.2.5 Animals Must Maintain Proper Water Balance

For all cells, water is a precious necessity. Life on Earth is not possible without it. The reason is rather simple. Imagine that an organism needs iron and finds a nail in its environment. The iron, locked tightly as solid metal, is totally inaccessible to the organism. To become accessible to organisms, iron usually reacts first with oxygen (it rusts)—a reaction facilitated by water. More importantly, atoms of iron, as an element or in a compound, cannot move into the organism's cells until they are first taken into **solution**, that is, dissolved in water. The same can be said for all other minerals and chemicals required by the organism.

Too much of a good thing is inevitably bad; water is no exception. In too much water, the needed nutrients become too dilute, and acquiring them then becomes difficult, if not impossible. There is not much the organism can do to control the amount of water in its external environment, but it can control water in its internal environment. How it does so is the next part of our story.

---

[1]Remember that as a result of osmosis there is an inverse relationship between solute and water concentration. That is, where solute concentrations are high, water concentration is low.

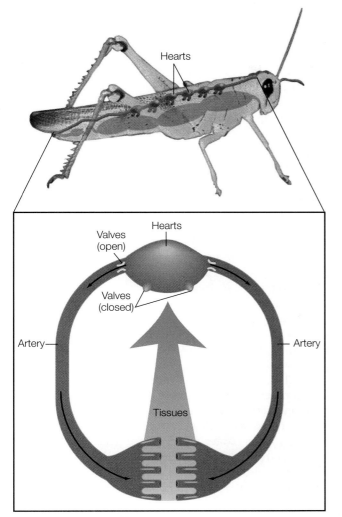

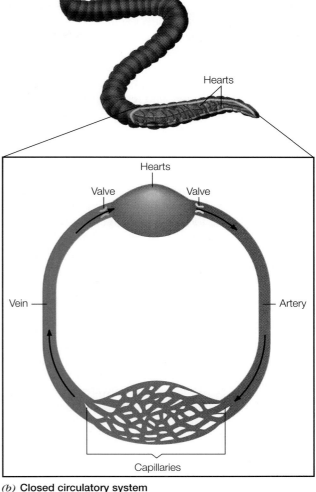

(a) Open circulatory system

(b) Closed circulatory system

Figure 10-9. There are two basic types of circulatory systems in the more complex animals. (a) Insects have an open circulatory system in which blood leaves the blood vessels, washes over tissues and cells, and then regathers in the hearts. (b) Segmented worms demonstrate closed circulatory systems. In these systems, blood never leaves the blood vessels.

The proper concentration of solutes and water in an animal's body fluids is quite often different from those concentrations in the environment.[1] That diffusion and osmosis constantly tend to minimize these imbalances creates a major problem for most animals. The mechanisms they employ to overcome this problem depend partly on evolutionary history (generally, mechanisms available to vertebrates are more advanced than those found in invertebrates) and environment (problems experienced in salt water differ from those in fresh water and from those experienced on land).

Cells and tissues of many marine invertebrates and in sharks and rays are **isotonic** with respect to their environments, that is, the concentrations of solutes and water inside matches concentrations outside. This means they are osmotically balanced and experience no net increase or decrease in water. In sharks and rays, this balance is achieved by controlling the amino acid concentration within cells. Thus, by reducing amino acid concentrations slightly, perhaps through protein synthesis, the cell voids water. Increasing amino acid concentrations has the opposite affect. Some marine invertebrates, notably the spiny-skinned animals (echinoderms), are iosotonic because they lack the ability to regulate many specific ions. In these animals, concentrations of water and solutes balance those of their environment.

Other marine invertebrates live in **hypertonic** environments, in which concentrations of solutes are higher (and concentration of water lower) in the environment than in their tissues. They show marked concentration gradients between their tissues and the environment of at least a few solutes. Often these differences are related to their environments and individual lifestyles. For instance, the jellyfish *Aurelia* regulates its sulfate ions at levels considerably below those of the environment. Sulfate is a relatively heavy ion, and reducing its concentration in tissues reduces overall density, making the animal more buoyant. Many marine arthropods that have the ability to move quickly show lower than environmental concentrations of magnesium, an ion that decreases muscular activity.

Most marine vertebrates, especially bony fish, also live in hypertonic environments. They tend to lose water by osmosis through their gills (the rest of their body is protected from water loss by scales, skin, or layers of slime) and in their urine. This they replace by drinking copious quantities of salty water, creating the problem of eliminating the accompanying solutes. Gills rid fish of some solutes by active transport; other solutes are concentrated in and voided with the urine.

Generally, invertebrates in brackish and fresh water live in **hypotonic** environments, where solute concentrations are lower in the environment than in their tissues, and show lower overall concentrations of solutes than their marine counterparts. This lessens diffusion and osmotic gradients and simplifies the challenge of maintaining balance with their environments. Still, because their environments are hypotonic, they tend to gain water and lose solutes. Excess water is generally voided by special excretory organs, such as kidneys in the vertebrates and similar structures in the invertebrates. Keeping the proper amount of solutes is more of a problem. In general, these are recaptured from the environment by active transport through gills or intestines (Figure 10-10).

In general, freshwater fish face the same problems as invertebrates; that is, they, too, live in hypotonic environments and tend to gain water. Water enters by the gills and is eliminated by the kidneys. Typically, the urine of freshwater fish is highly diluted; still, some solutes are lost in this manner. Some lost solutes are replaced in the food these fish eat, by their normal digestion, and by active transport. Gills recover solutes by active transport.

Terrestrial animals face a double-edged problem with respect to water. Generally, water is limiting in terrestrial habitats and because of chronic dryness, the tendency to lose water is intensified. Animals lose water through evaporation from skin and respiratory surfaces, and through elimination in feces, urine, or special secretions. To obtain water, they drink it, soak it up through skin, extract it from food (even the driest grain is over half water; Figure 10-11), or produce it by metabolism (remember the products of respiration from Chapter 9).

In general, terrestrial animals have five ways they can reduce water loss:

1. Many seek out moist environments. Thus, earthworms do best in damp soil. Frogs, salamanders, and many invertebrates live close to water, easing the necessity of maintaining moist skin.

2. Other terrestrial animals seek out habitats where humidity is high, which lessens losses by evaporation. Such areas include burrows in soil or wood, underneath logs or rock, and within leaf litter. Such areas often teem with invertebrates and more than a few vertebrates.

3. Other terrestrial animals are active mainly at night when humidity is highest.

4. Still others have body coverings that protect them from water loss. Snails have shells. Arthropods have exoskeletons. Reptiles have scaly skin. Birds have feathers and impermeable skin. Mammals have fur and impermeable skin. Many insects and a few amphibians have special glands that secrete waxy substances, which cover bodies.

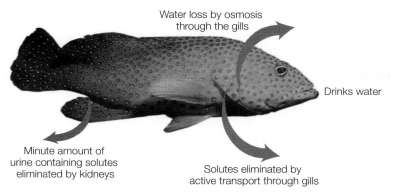

Water loss by osmosis
through the gills

Drinks water

Minute amount of
urine containing solutes
eliminated by kidneys

Solutes eliminated by
active transport through gills

(a) **Marine fish in hypertonic environment**

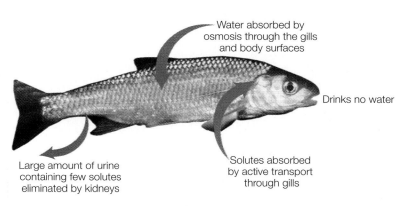

Water absorbed by
osmosis through the gills
and body surfaces

Drinks no water

Large amount of urine
containing few solutes
eliminated by kidneys

Solutes absorbed
by active transport
through gills

(b) **Freshwater fish in hypotonic environment**

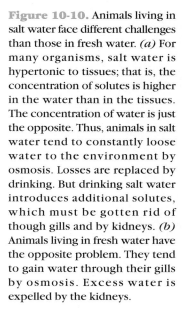

Figure 10-10. Animals living in salt water face different challenges than those in fresh water. *(a)* For many organisms, salt water is hypertonic to tissues; that is, the concentration of solutes is higher in the water than in the tissues. The concentration of water is just the opposite. Thus, animals in salt water tend to constantly loose water to the environment by osmosis. Losses are replaced by drinking. But drinking salt water introduces additional solutes, which must be gotten rid of though gills and by kidneys. *(b)* Animals living in fresh water have the opposite problem. They tend to gain water through their gills by osmosis. Excess water is expelled by the kidneys.

Figure 10-11. Kangaroo rats tolerate dry conditions, yet never drink free water. They obtain all of the water they need from the food they eat. Their kidneys are specifically adapted to recycle and conserve water.

5. Finally, animals that live in the driest environments, from meal worms in dry flour to desert mammals such as kangaroo rats, often have special mechanisms for reducing water loss in feces and concentrating wastes in urine.

Can you think of other examples in each of the above categories?

10.2.6 Organisms Must Get Rid of Metabolic Wastes

Cellular metabolism produces waste compounds that must be disposed of. One of the most serious of these is ammonia, a nitrogen-containing compound formed when pro-

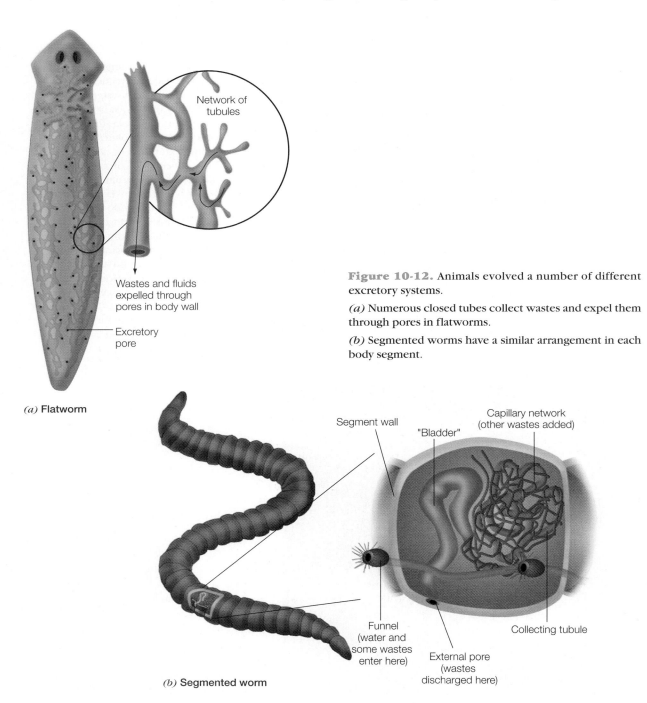

Figure 10-12. Animals evolved a number of different excretory systems.

(a) Numerous closed tubes collect wastes and expel them through pores in flatworms.

(b) Segmented worms have a similar arrangement in each body segment.

teins and amino acids are metabolized. Ammonia is highly poisonous, and most cells must dispose of it quickly. Other nitrogen-containing waste compounds are not particularly toxic, but their buildup could interfere with normal cellular processes. If nothing else, accumulation of these compounds could cause osmotic imbalances.

Individual cells and small organisms rely solely on diffusion to rid themselves of unwanted wastes. Similarly, marine sponges, jellyfish, tapeworms, and spiny-skinned animals have no special excretory organs and depend entirely on diffusion of wastes through body surfaces. For larger organisms, getting rid of metabolic wastes is a more complex challenge and requires special excretory systems. Apparently, excretory organs evolved at least three separate times, in worms, mollusks, and vertebrates.

Flat-, round-, and segmented worms have numerous open or closed tubes that may be simple or branched. These collect wastes and fluids and eventually dispose of them through numerous pores (Figure 10-12). Mollusks have paired **excretory organs**, tube-like affairs that collect wastes and dump them into the mantle cavity and ultimately into the environment.

The excretory systems of insects consist of **Malpighian tubules**, two to several hundred tubules that lie in the body cavity and open into the intestine. Much more is known about how these systems work than other invertebrate excretory systems. Waste products, especially uric acid, enter the tubules from the body cavity by active transport, simple diffusion, and osmosis. Once in the intestine, uric acid precipitates into a semisolid material, much water and some solutes are reabsorbed, and what remains is excreted along with feces.

Insects that eat mainly fresh vegetation that is high in water content produce copious amounts of watery wastes. Insects that eat primarily dry food produce little waste water. Extreme among insects are meal worms, such as *Tenebrio*, that eat nothing but dry flour, yet receive all the water they need. They are extremely efficient at reabsorbing water. (But where does their water come from initially?)

Vertebrates have **kidneys** to get rid of their metabolic wastes. To function properly, kidneys depend primarily on four processes: diffusion, osmosis, active transport, and **filtration**. How they work will be discussed in greater detail in Chapter 11.

Piecing It Together

1. Organisms obtain energy, nutrients, and water from and return wastes back into their environments.

2. Ultimately, digestive, circulatory, respiratory, and excretory systems depend on diffusion and osmosis to move nutrients and water between cells, tissues, and organ systems and into and out of the environment.

3. Primitive organisms obtain nutrients by excreting enzymes onto their food; the enzymes break down complex compounds into simpler ones.

4. Among animals, digestive systems (absent in sponges) evolved from those with single openings (jellyfish and flatworms), into simple, undifferentiated tubes with two openings (roundworms), to complex, convoluted tubes and accessory organs (all other animals).

5. Among animals, there are basically three types of respiratory systems: skin, gills, and lungs. All share two characteristics that facilitate diffusion: extensive surface areas and thin, moist cell layers.

6. Circulatory systems have three major components: vessels, blood, and hearts. Each evolved from rather simple components in primitive worms to highly complex ones in chordates.

7. A variety of excretory systems maintain water balance and rid animals of potentially harmful metabolic wastes.

10.3 HOW DO ORGANISMS MONITOR THEIR ENVIRONMENTS AND MOVE FROM PLACE TO PLACE?

External environments are frequently challenging. Predators and competitors come and go. Surroundings become too hot, too dry, or contaminated. Often overlooked, gravity, too, is an environmental challenge to all life, imposing limits on how large organisms can become and how efficiently they can move. It should not be surprising, then, that organisms have evolved a variety of mechanisms with which to minimize, avoid, work with, or overcome external challenges.

Fast Find 10.3a How Organisms Move

Basically, organisms have four options in the face of such challenges: (1) *isolate* themselves within thick shells or similar structures; (2) *seek shelter*, (3) *adjust* to changing conditions, or (4) *move* to more favorable environments. Each option has limits. Isolation cannot be so extreme that organisms lose complete contact with the environment. Similarly, organisms can stay sheltered only so long; eventually, they must at least seek nutrients. Adjustments to environmental changes often take time and require that environments change slowly (Figure 10-13). Often, these options work against each other; for example, thick body coverings protect but also impede

(a)   *(b)*

Figure 10-13. Rock ptarmigan show all four options available to counter environmental challenges. *Isolate:* Their feathers form a protective shield from their sometimes harsh environment. *Seek shelter:* During the worst of winter weather, ptarmigan seek shelter by forming coveys amid dense vegetation. *Adjust:* In summer *(a)*, ptarmigan are basically brown colored; in winter *(b)* they are basically white. *Move:* In summer, ptarmigan live at high altitudes, whereas in winter, they migrate into more protected valleys.

movement. Occasionally, as we shall see in the case of body coverings and movement, these options work together.

10.3.1 Microbes Evolved Thick Cell Walls and Special Organelles for Movement

Early on, bacteria evolved thick cell walls, barriers that hold environments at bay. Thick cell walls are also barriers that assists the cell membrane to hold water, nutrients, and other chemicals within cells. Bacteria are also surprisingly mobile. Their flagella and the ability to seemingly glide (by mechanisms that are not well understood) provide movement. Their ability to move rapidly even with thick walls appears to be helped by their small size, which in watery surroundings minimizes gravity. Compared to those of other microbes, the flagella of bacteria are relatively short, thin, and stiff. Bacterial flagella rotate at the base, close to where they join the cell wall. As flagella rotate, the tips spin like propellers, pulling the bacteria from place to place.

Many protistan algae share thick cell walls and methods of movement with their bacterial ancestors; that is, they glide or use flagella, but protistan flagella are structurally quite different from those of bacteria. To see their structure and how they may work, visit the CD. Typically, protistan algae, too, are found in aquatic environments, where the effects of gravity are minimized.

Unlike bacteria and algae, protozoan protists have no cell walls and are even more highly mobile. There are basically four groups of protozoa (discussed in more detail in Chapter 8), differentiated by their means of locomotion. *Zoomastigina* move by flagella that are structurally and functionally similar to those of algae. *Ciliophora* move, as their name implies, by cilia, which can be thought of as numerous, short flagella. Cilia beat in unison to produce complex motion—backward motion, forward motion, twists, and spirals are typical. Amoeboid protozoans, the *Sarcodina*, move by means of **pseudopodia** (false feet). When amoebas move, their cytoplasm streams in a particular direction, forming first a bulge that becomes a footlike extension into which the rest of the cytoplasm streams. Pseudopodia don't last long; new ones form whenever the amoeba adjusts direction of movement (Figure 10-14; see section 10.3 of the CD). Compared with other microbes, amoebas are often larger in size and their movements slower. Most live in bottom muds and damp soils. Interestingly, some live suspended in water where, because of their relatively large size, gravity should be a problem; however, these protists have evolved shells of silicon dioxide (glass) or calcium carbonate, some gas filled, some with elaborate spines, that keep them suspended. Finally, one group of protozoans, the *Sporozoa*, have lost all means of locomotion. All are parasites, depending on other organisms to carry them from place to place.

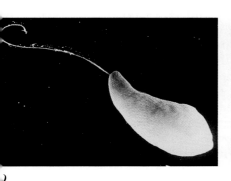
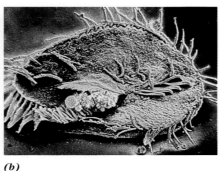
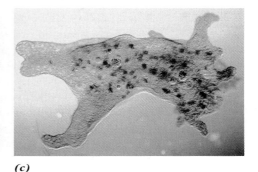

(b)   *(c)*

Figure 10-14. Protozoa move in a variety of ways, including *(a)* flagella (magnified 1,608 times), *(b)* cilia (magnified 3,600 times), and *(c)* pseudopodia (magnified 60 times).

10.3.2   Animal Movements Involve Interactions between Muscles and Skeletons

Animals are generally quite sensitive and highly responsive to environmental change. Typically, they are highly mobile. Indeed, the ability to move is one of the characteristics used to distinguish animals from plants. How do animals do it?

First, animals retained and built on methods of movements inherited from protozoan-like ancestors. Ciliated and flagellated tissues are found in all animal phyla. For example, sperm are flagellated cells. Ciliated tissues move eggs through reproductive systems, move foreign objects out of respiratory systems, and are associated with many animal sensory systems. Cilia along the ventral (belly) surface propel free-living flatworms in smooth, gliding movements. Amoeboid cells are ambulatory tissues in many animals, including sponges, in which they apparently transport materials from digestive systems to other cells. In vertebrates, amoeboid cells include white blood cells that fight disease.

For most animals, movements require additional specialized tissues, namely, muscles. In these tissues, cells rich in the proteins actin and myosin are arranged as fibers. When actin and myosin interact, they slide over each other, which, in effect, contracts the cell. To produce motion, muscles must be connected to something. Typically, long, fibrous muscles are connected to other body parts by connective tissue. One end of the muscle is anchored while the other is connected to the part of the body that moves. Go to the CD to see more of the physiology of muscle contraction.

One of the simplest types of gross muscle movement is seen in roundworms. Longitudinal muscles run lengthwise and are attached at various locations along their cylindrical bodies. Contracting the muscles along one side shortens the cylinder on that side only, twisting the worm into a **C** shape. If, as those muscles relax, those of the opposite side of the body contract, the worm is twisted into a backward **C** shape. If the muscles of the anterior (head) end are contracted opposite those of the posterior (tail) end, the body undulates in an **S** shape, first in one direction, then in the other. A water-filled cavity within the worm's body keeps it from collapsing. Water's incompressibility provides, then, a kind of skeleton, termed a **hydrostatic skeleton**, useful to all soft-bodied animals and some soft tissues. (Can you think of examples of human tissues that depend on hydrostatic pressure for support?) Obviously, even this simple motion requires careful coordination and timing of muscle action, which is provided by the nervous system.

Notice two important points in the roundworm example. Muscles only contract; they do not push. Also notice that different sets of muscles work as *antagonists*; that is, one set of muscles moves a portion of the worm's body in one direction, and another set moves it in the opposite direction. Both of these points are typical of muscle action in all multicellular animals. (Think of examples of antagonism involving your muscles.)

Earthworms have longitudinal muscles whose antagonists are circular muscles that run around the body and are associated with each body segment. When longitudinal muscles contract, the worm's body shortens. When the circular muscles contract, the worm lengthens. Bristles on the ventral (belly) surface point toward the tail, dig into the substratum, and keep the worm moving forward. (See this movement in Section 10.3 of the CD).

About 600 million years ago, some multicellular animals evolved hard parts: shells and skeletons. Hard parts offer great protection from a variety of environmental challenges. Hard parts also contribute to animal movements by providing convenient places for muscles to attach. On the debit side, hard parts are heavy and difficult to move.

Some of the first animals to evolve shells were mollusks (clams, mussels, and snails). Whenever their environments become threatening, mollusks close their shells with special muscles. Not all mollusks have obvious shells. Those of slugs, squid,

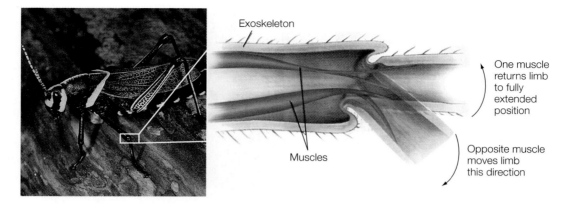

Figure 10-15. The exoskeletons of insects and other jointed-legged animals provide numerous attachment sites for muscles. As in vertebrates, their muscles often work in pairs.

and octopuses are greatly reduced, minimizing weight, sacrificing tissue protection, but still providing convenient sites for attachment of key muscle groups. Not surprisingly, these mollusks have much greater mobility than their more fully shelled counterparts.

Much of the success of arthropods, perhaps the most successful of all animal phyla, is due in large part to their **exoskeleton** (external skeleton). Not only does an exoskeleton provide great protection, it provides a nearly infinite number of potential sites on which to anchor muscles (Figure 10-15). As a result, various arthropods swim, walk, fly, jump, or burrow on legs, wings, or fins, while other appendages are used as pincers, claspers, antennae, and mouth parts. The movements of all these appendages are controlled with delicate, complex muscle groups fastened at strategic locations to exoskeletons. Of course, exoskeletons are also heavy and impose rather severe limits as to how large arthropods can become.

In most chordates, particularly vertebrates, skeletons are internal. What are the costs of these **endoskeletons**? More soft tissue is directly exposed to the environment. The benefits? Endoskeletons weigh less and are easier to move, permitting individuals to become larger in size. Endoskeletons facilitate growth but still provide numerous attachment sites for muscles, which facilitates complex body movements.

 Evolving Flight The ability to fly evolved independently in three groups of terrestrial animals—insects, birds, and mammals. Describe the basic mechanisms by which each of these methods of flight work. How are they similar? What features of flight are unique to each group of animals?

10.3.3 Organisms Need Information about Their Environments

In order to live successfully in ever-changing, often unpredictable, potentially dangerous environments, in which resources are limiting and competition keen, organisms need information. Three questions are constantly important: What is the current state of the environment? What changes are taking place? What adjustments need to be made, internally and externally, to adapt to these changes? For multicellular organisms, these questions are equally valid for both environments in which they live, their external surroundings and their internal conditions. There are basically two physiologic systems devoted to answering these questions—one is chemical and the only system available to microbes, fungi, and plants; the other is cellular, that is, the nervous system found in animals.

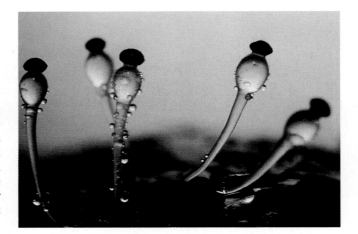

Figure 10-16. The fungal mold *Philobolus* is sensitive to light. Its fruiting bodies grow toward the sun and away from the ground. When they burst, spores are spewed away from their immediate environment (magnified 11 times).

10.3.4 Microbes and Fungi May Monitor and Respond to Their Environments Chemically

The most primitive environmental monitoring system available to organisms is chemical. All cells probably respond to their environments chemically. For example, some microbes respond to the sudden introduction of lactose into their environments by producing appropriate enzymes with which to digest it. When lactose is no longer present, the production of these enzymes ceases. Microbes assess the presence or absence of lactose chemically.

Other means by which microbes and fungi assess and respond to their environments are largely unstudied. Two additional examples suggest the possible importance of chemicals:

◆ Certain microbes practice a form of primitive sexual reproduction. Periodically, two cells come together, form a cytoplasmic bridge (a **pilus**, see below) between them, and exchange nuclear material. How does a microbe know that it is contacting an appropriate partner? Surface proteins in the cell membrane are apparently key. If their surface proteins are compatible, partners are appropriate; if not, microbes move on.

◆ The fungal mold *Philobolus* is particularly common in cow pastures where it feasts on feces (Figure 10-16). After a time, it forms fruiting bodies, stalks that spew spores up to two meters. The spewing event is far from random. Apparently, photosensitive cells control the growth of stalks so that they point toward the sun, that is, up and away from the ground, ensuring that the spores are spewed well away from the cow patty. The control agent is probably a chemical sensitive to light that controls stalk growth. Now, those spewed spores lucky enough to land on grass, be eaten by another herbivore, and survive transit through the herbivore's digestive system are deposited with a new universe of delectable resources.

10.3.5 Hormones Are Important Chemically Based Monitor and Control Systems in Animals

As humans, we are so conscious of the importance of nervous systems in gathering and assessing information about our environments that we are scarcely aware of the importance of our hormones. It turns out that hormones produced by endocrine systems are extremely important in the lives of most animals. In general, hormones monitor and control functions that, compared with nerves, happen slowly, over a period of several days or weeks, such as growth, maturity, reproduction, and many metabolic functions.

Many hormones produced by humans and other vertebrates are also found in invertebrates and protists. For example, adrenaline (involved in human flight or fight responses) and endorphins (natural painkillers) are found in protozoans. Insulin (controls blood sugar levels in humans) and cholecystokinin (controls human gallbladder activity) are found in many invertebrates. How these chemicals are useful to protists and invertebrates has not been studied, but their presence suggests two important points. (1) Vertebrate hormones, even those of humans, are of ancient origin. Vertebrates inherited their hormones from their ancestors, although often they have redefined the physiological contexts in which they are useful. (2) Life is conservative. That is, a chemical found useful by one organism is likely to be useful to another.

Hormones often work together to control complex phenomena in animals. For example, the hormone ecdysone controls molting in arthropods. As these animals mature, they outgrow their exoskeletons, which periodically must be shed and replaced with new ones. In insects, each life stage between molts is an **instar**, and insects may go through several before becoming adults. Later instars not only grow and molt, but take on adult characteristics such as wings and the ability to reproduce. While molting is controlled by ecdysone, maturation is controlled by juvenile hormone. Its presence in early instars inhibits maturation. In later instars, juvenile hormone is not secreted and maturation occurs.

Occasionally, hormonal control becomes downright complicated. In humans, five interacting hormones control blood sugar levels, and female reproductive activities involve no less than seven interacting hormones. Both of these topics will be studied in greater detail in Chapter 11.

10.3.6 In Addition to Endocrine Systems, Animals Have Nervous Systems

For animals, quick adjustment to environmental conditions is often essential to survival. Nervous systems evolved to handle such situations. In complex animals such as vertebrates, billions of **neurons**, the basic cellular unit of the nervous system, connect all parts of the body with the animal's brain. The brain assesses sensory inputs from the animal's internal and external environment and initiates appropriate responses. How animals respond to perceived environmental changes is often controlled by motor nerves that carry information from the brain to the body. Most animals have the same modes of sensing as humans: touch, sight, hearing, taste, and smell. Some animals have additional sensory systems, such as the lateral line system in fish that senses pressure, the facial pits of some snakes that sense infrared radiation, and special skin receptors in certain fish (electric eels, for example) and amphibians that detect electric fields.

As with the digestive system, the evolution of nervous systems within the animal kingdom is a study of increasing complexity. Sponges, which are thought to be the simplest, most ancient of animals, lack nervous systems altogether, yet their cells and tissues can still communicate. On the inside of their bodies specialized, flagellated cells, called collar cells, create currents that suck water in through myriad incurrent pores; the water circulates through the sponge's body and exits through a few larger excurrent pores. As water travels through the body, nutrients and oxygen are extracted by the cells, and wastes, carbon dioxide, and gametes are added for removal from the body. When weak acids or other potentially dangerous substances are placed just outside the sponge they will start to enter. Then the flagella stop, reverse direction, and they force the noxious substance out, protecting the sponge. How the sponge detects noxious substances and controls flagella is not well understood. What is clear is that the cells of the sponge are coordinated without benefit of a nervous system.

Jellyfish and their relatives have a primitive nervous system consisting of a netlike pattern of neurons that control muscles. Unlike more complex animals, impulses in jellyfish nerves flow in both directions. They control pulsing movements of medusal (bell-shaped) jellyfish and complex crawling movements of polyp (barrel-shaped) jellyfish. There is nothing like a brain in these animals, but the number of nerves increases near the mouth.

Free-living flatworms, such as *Planaria*. have ladder-shaped nervous systems running the length of their otherwise unsegmented bodies. Where the long, lateral nerves of these ladders meet crosspieces, knobs of neurons, called **ganglia**, form. Near the head end of the worm, ganglia connected to eyespots are unusually large, the evolutionary precursor of brains. Flatworms can learn to find food in a simple T-shaped maze; they are the simplest animals in which learning can be demonstrated.

Segmented worms, such as earthworms, have a single nerve cord running the length of their bodies close to the ventral (belly) surface. In every body segment, lateral nerves branch off, controlling muscle and other organ activities in that segment. At the anterior (head) end, enlarged ganglia in several segments form a proper brain.

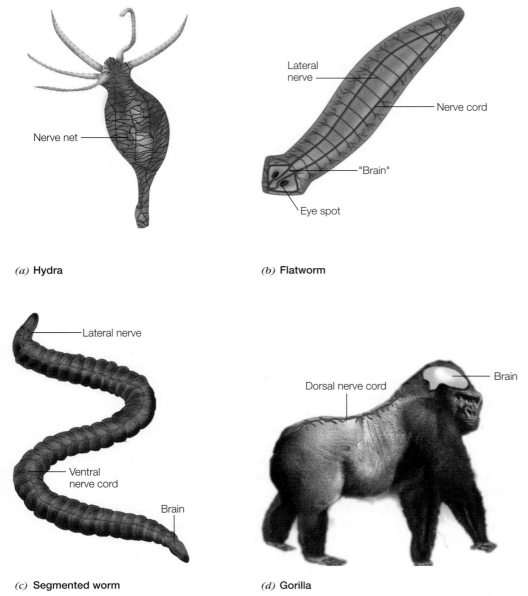

(a) **Hydra**

(b) **Flatworm**

(c) **Segmented worm**

(d) **Gorilla**

Figure 10-17.

(a) The nervous system of the hydra consists of a nerve net centered around the mouth.

(b) In flatworms, the nervous system is ladder shaped. Near the head end, two enlarged areas form a primitive brain.

(c) The nervous system of segmented worms consists of a nerve cord running along the ventral (belly) surface with lateral nerves branching off in each segment. Compared with flatworms, the brains of segmented worms are more highly developed.

(d) In the vertebrates, the nervous system is even more fully refined. A nerve cord, with numerous branching lateral nerves, runs along the dorsal (back) surface. The brains of vertebrates are the most highly developed of all animals.

Invertebrates more complex than segmented worms have basically the same kind of nervous system. Lateral nerves regularly branch off of a ventral nerve cord. As complexity of the animal group increases, so does that of the lateral nerves and, in particular, the brain.

Vertebrates have the most complex nervous systems of all, especially their brains. Note that the pattern is basically the same as in advanced invertebrates: lateral nerves regularly branch off of a central nerve cord, and the brain is located near the head end. In vertebrates, the nerve cord runs near the dorsal (back) surface (Figure 10-17). Details of this group's nervous system will be stressed in Chapter 11 and on the CD.

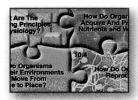

Piecing It Together

1. Thick cell walls and shells protect many microbes from potentially hostile environments. Within such environments, microbes move by means of flagella, cilia, and pseudopodia.

2. Gross body movements among animals involve interactions between muscles and shells or skeletons.

3. Among animals, there are three types of skeletons: hydrostatic skeletons, exoskeletons, and endoskeletons.

4. Endocrine and nervous systems gather and process information about an animal's internal and external environment, and integrate mechanisms that maintain homeostasis.

5. In general, hormones, which may be important in all multicellular organisms, respond to long-term changes in an organism's environment. They are key in such processes as growth, maturity, reproduction, and metabolism.

6. Nervous systems, found only in animals, allow quick adjustments to environmental change.

7. Within the animal kingdom, nervous systems evolved from rather simple networks into complex systems comprised of numerous lateral nerves that feed information into and out of centralized nerve cords and brains.

10.4 HOW DO ORGANISMS REPRODUCE?

What is the purpose of life? To biologists, this perplexing, age-old question is easy to answer. The purpose of life is to reproduce. This is certainly an oversimplification, but the answer makes an interesting point. Life can persist only through reproduction. Only species that can consistently produce enough offspring to replace losses due to disease, predation, accidents, aging, and so forth can persist. Those that do not become extinct.

Individuals alive today came into existence through some form of reproduction. Their parents were successful in the sense that not only did they survive for a time, they garnered enough resources and energy to be able to produce offspring. And the offspring survived. Evolutionary theory tells us that these offspring tend to possess characteristics appropriate to their particular environments. Soon, it will be the offspring's turn to pass genes and characteristics on to the next generation. And so life goes on in successful species.

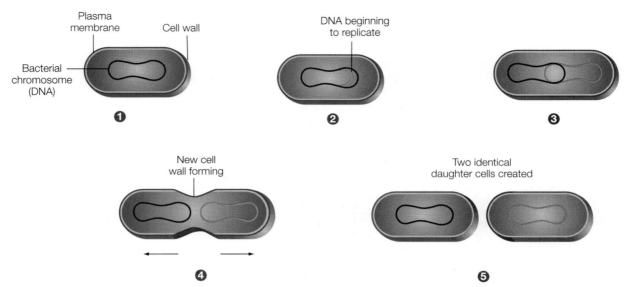

Figure 10-18. Most bacterial reproduction is by binary fission. Contrast the steps in this process with those of mitosis and meiosis (see Figure 5.7 and Section 5.1 of the CD).

10.4.1 The Most Common Method of Reproduction Is Asexual

Fast Find 10.4b Asexual Reproduction

For most single-celled organisms, reproduction is asexual. Under favorable environmental conditions, other activities are briefly curtailed while one cell becomes two—a process called **binary fission** in prokaryotic cells (Figure 10-18) and **mitosis** in all others. In some bacteria, most single-celled protists, and single-celled fungi, the resultant cells immediately separate and become totally independent organisms. Those cells that fail to separate become colonies of individuals joined by shared cell walls. Note that even in colonies each individual cell is capable of independent existence.

All organisms, no matter how sophisticated, at least partially depend on asexual reproduction. Their cells reproduce asexually and by this process multicellular organisms pass from a single-celled stage (fertilized **gametes**, or **zygotes**) to multicellular adults. Through a series of mitotic events, one cell becomes two, two become four, and four eventually become trillions. Growth at least partially involves the asexual reproduction of cells.

Through mitosis multicellular animals replace lost tissues. When lizards lose their tails, they grow new ones by mitosis. Similarly, salamanders replace lost legs. Pity the poor oystermen who, finding oysters being eaten by starfish, chop the starfish into tiny pieces and throw them into the sea. Each piece grows into a new starfish, largely through mitosis of surviving cells.

As humans, we have lost much of our ability to replace lost parts. Indeed, our nerve cells totally lose their ability to reproduce, that is, undergo mitosis, long before birth. But other human tissues can reproduce by mitosis. Thus, each day we replace our stomach lining and give rise to a staggering 180 trillion new red blood cells!

10.4.2 Sexual Reproduction Evolved among Bacteria, Protists, and Fungi

Many microbes are not totally dependent on asexual reproduction. In **conjugation**, the simplest form of sexual reproduction, a cytoplasmic bridge, called a **pilus**, develops between two individuals through which a one-way exchange of genetic material occurs (Figure 10-19). In an evolutionary sense, this is the birth of sexual reproduction. Although well studied in the laboratory, not much is known about conjugation in

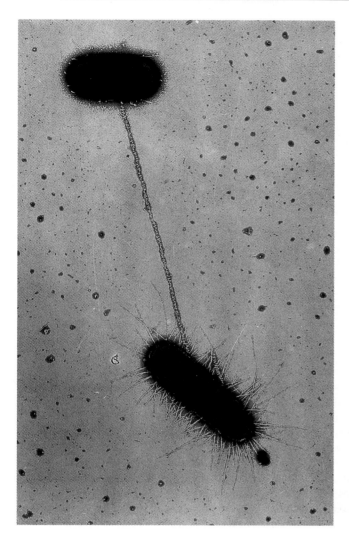

Figure 10-19. Conjugation is a form of primitive sexual reproduction found in some microbes. A cytoplasmic bridge (pilus) forms between adjacent cells through which a one-way exchange of chromosomal material takes place (magnified 23,450 times).

nature. How widely does it occur? How important is it in shaping and changing characteristics in microbes? These are key questions that remain to be answered.

Among the eukaryotes, reproduction becomes very complex. It seems there are almost as many different patterns as there are species. Fortunately, there are three major reproductive life cycle themes around which the variations spin:

1. **Diploid cycle.** This is the most familiar life cycle pattern, the one seen in nearly all animals and some protists and algae. All cells of the organism are *diploid* except for gametes. Gametes remain single-celled, haploid entities until fertilization. The resultant zygote develops into the new individual, which is diploid except for its gametes.

2. **Haploid cycle.** This is the life cycle frequently seen in fungi and some algae. Throughout most of their lives, individual organisms comprise *haploid* cells. Often, **budding** and other types of asexual reproduction result in new haploid individuals. When these individuals reproduce sexually, a limited number of cells from two individuals merge, fuse nuclei, and become diploid cells. Often, they immediately undergo **meiosis**, and the resulting haploid spores are dispersed, usually by wind or water, to grow into new, haploid individuals.

3. **Alternating cycle.** All plants and some algae live this life cycle. Typically, there are two distinct generations, a haploid **gametophyte** (gamete-producing) gen-

eration and a diploid **sporophyte** (spore-producing) generation. In the sporophyte generation, cells undergo meiosis and become spores, which grow into female and male gametophytes, which in turn, produce gametes that fuse and grow into sporophytes. In algae and primitive plants (mosses and ferns), these two generations occur in completely separate individuals. In the more advanced plants the gametophyte is greatly reduced but is still a multicellular entity.

Further details and examples of specific reproductive cycles will be discussed in Chapters 11 and 12 and on the CD.

10.4.3 Sexual Reproduction in Water Is Relatively Easy

**Fast Find
10.4d
Gametes**

As gametes, organisms are virtually defenseless and at their most vulnerable, especially to drying out. Reproducing in water minimizes this problem. Thus, many aquatic organisms simply spew gametes into the environment, where they must find each other so that fertilization can occur (Figure 10-20). Timing is everything, and releasing gametes is usually far from random. In the vast majority of cases, all individuals of a particular species become reproductively active at the same time. This means they release gametes "on cue." But what exactly is the cue? In tropical sponges, jellyfish, squid, clams, and others, the cue appears to be the lunar cycle. Every 28 days, all individuals of a given species release gametes. Each species has its own phase of the cycle to respond to—some release when the moon is full; others during new moon; still others somewhere in between.

Simultaneous release of gametes has at least two advantages. First, it facilitates fertilization, because sperm and eggs are available at the same time and are able to unite. Secondly, it foils predators by "flooding the market." By the time predators detect and respond to the sudden appearance of so much food, the theory goes, the event is over and enough successful fertilizations have occurred to perpetuate the species. Of course, predators can use their own sense of time to anticipate and be ready for such events.

Away from the tropics, the timing and release of gametes are more complex. Gametes are released only during certain times of the year, typically the warm months. Within this time span, precise release of gametes is again often correlated to phases of the moon.

Reproduction among aquatic vertebrates is even more complex. These animals are less responsive to lunar cycles and more responsive to location, time of year, and presence of suitable mates. Typically, at a specific place and time of year, male and female fish come together. One or both choose a location. One or both construct a nest, rang-

Figure 10-20. Some aquatic invertebrates, including these sponges, periodically spew large numbers of gametes into the water. In order for fertilization to take place, these gametes must locate partners and join. In most cases, release of gametes is timed to some environmental phenomenon, such as a full moon.

Figure 10-21. During the breeding season, large groups of salamanders, such as these California newts, gather in a suitable habitat to exchange gametes.

ing from a simple scrape in sand or gravel to an elaborate structure made of seaweed or other materials. After appropriate courtship behaviors (details will be discussed in Chapter 13), eggs are released into the nest by the female, and sperm are released by the male. In some cases, the resulting zygotes are immediately ignored. In others, zygotes are buried or nests otherwise closed. In a few cases, offspring are protected by one or both parents.

Reproduction among amphibians often resembles reproduction among fish. Most still depend on water as a medium in which to reproduce. At appropriate places and times, females and males approach. Male frogs often attach themselves to females for periods ranging from a few minutes to several days. As eggs are laid in the water, sperm are released. Whole populations of salamanders may gather in roiling masses of individuals releasing gametes into the water (Figure 10-21). Reproductively, amphibians are a complex group. Variations on what is typical are common and fascinating.

Among aquatic organisms there is considerable variation in the amounts of yolk, the source of energy and nutrients for developing **embryos**, their eggs contain. The eggs of many spiny-skinned animals contain virtually no yolk, whereas those of other invertebrates have considerable yolk. Between these extremes are many variations.

10.4.4 Many Animals Successfully Reproduce on Land

Independently, several groups of animals, including segmented worms, mollusks, arthropods, and chordates, acquired the ability to survive on land. Each followed similar evolutionary paths. That is, invertebrates originated in the ocean and evolved successively in brackish water (estuaries and saltwater marshes, for example), fresh water (rivers, lakes, and especially wetlands), and finally truly terrestrial habitats. Chordates started their journey in fresh water. As they did so, each group evolved mechanisms whereby they could protect gametes and the developing young from the rigors of increasing dryness.

Many segmented worms, of which earthworms are typical, are **hermaphroditic**, or contain both male and female reproductive systems. Reproduction typically involves two individuals coming together and exchanging sperm, which each individual then stores internally. After a time, each individual earthworm develops a collarlike structure around the outside of the body near the head end into which eggs and sperm are eventually released. Thus, after a true copulation, fertilization occurs externally. Later, the collar is shed as a cocoon in which subsequent development takes place. There are no larval stages; what hatches is essentially a miniature adult.

Mollusks are represented on land by the terrestrial snails and slugs. As in the segmented worms, they are largely hermaphroditic and exchange sperm during mating. External fertilization occurs immediately. Their eggs contain considerable yolk to nourish developing embryos. Development is direct; that is, there is no larval stage. Frequently, embryos have numerous cilia that circulate the internal contents of the yolk, bringing nutrients and oxygen to and wastes away from embryos. Most terrestrial mollusks (snails and slugs) keep eggs in their mantle cavity. Others lay eggs immersed in a bag of slime in damp soil or rotting vegetation.

**Fast Find
10.4f Insect
Metamorphosis**

Arthropods, as was noted earlier, are perhaps the most successful of terrestrial animals, at least in terms of number of species. It should not be surprising, then, that variations and exceptions are the rule in describing any aspect of their biology, including reproduction. Here we will focus on insects, the most successful terrestrial arthropods. Insect eggs have little yolk. In some primitive insects, female gametes and eggs are nourished by neighboring cells that surround them. In more advanced insects, the mature egg contains several germ cells, most of which become support cells that nourish the primary egg cell. Typically, insect eggs are surrounded and protected by a shell that contains a terminal pore through which sperm gain access. Fertilization occurs inside the female's body. Sperm are transferred either in liquid form or as a sperm packet picked up by or injected into the female.

Vertebrates are the best-studied animal to make the transition from water to land. A significant key to their success was evolution of the **amniote eggs**, in which the embryo is encased in a fluid-filled sac and protected from drying out. Remember that the primary problem faced by any developing embryo is to secure nutrients and get rid of accumulating and potentially toxic metabolic wastes without drying out. In marine environments these problems are minimized for vertebrates as with other organisms, because the developing embryo is surrounded by nutrient-rich sea water. Nutrients and water diffuse in and wastes diffuse out with little or no effort expended by the embryo.

Some marine fish eggs are relatively simple structures consisting of little more than the female gamete surrounded by a gelatinous coat through which sperm (or at least its nucleus), nutrients, water, and wastes easily penetrate. Other fish eggs, especially those of freshwater fish, contain considerable yolk—nutrient-and energy-rich material upon which the developing embryo feeds during early development. Still, fish eggs are vulnerable to predators and other environmental hazards. Development is typically rapid; after only a few days, larval fish hatch and can swim, secure food, seek shelter, and avoid at least some predators.

Amphibians usually lay eggs in or near freshwater habitats, which have limited nutrients. In addition to protective layers, their eggs consist of large quantities of yolk. Compared with fish, amphibian development takes longer. Larvae, also known as tadpoles, hatch with huge distended bellies crammed with remnants of yolk that they continue to feed on after hatching. As yolk is consumed, larvae become increasingly more able to secure their own food.

Reptiles were the first vertebrate group to become truly terrestrial, and a key factor in their success was the evolution of the reptilian egg. These are truly remarkable structures. Leathery shells, resistant to water loss yet permeable to oxygen and carbon dioxide, surround and protect contents. Inside, large masses of yolk contain sufficient energy and raw materials to fuel their development until they hatch as small versions of adults, fully responsive to their environments (Figure 10-22). Compared to fish, development within the reptilian egg is slow. Soon after fertilization, a series of membranes form and surround the developing embryo. The liquids they contain provide additional layers of protection. One membrane system forms a sac that accumulates and isolates wastes. Additional layers of protein material (called egg white) surround and protect embryos.

With all of the embryos' developmental needs provided for, these eggs are ready for a truly terrestrial existence. They are still susceptible, of course, to the worst the environment has to offer: predators, direct sunlight, and flooding. Usually, reptiles bury eggs in loose soil or lay them beneath rocks or logs. Alligators and some lizards protect sites where eggs are laid. Some snakes keep eggs inside their bodies until they hatch. These snakes then give birth to living young.

(a)

Figure 10-22. Evolution of the amniote egg was a major milestone in the transition from aquatic to terrestrial life. *(a)* The first to evolve was the reptilian egg, with its leathery shell. *(b)* Bird eggs are similar to reptilian eggs, except their shells are brittle.

(b)

Which came first, the chicken or the egg? The egg did. Birds, including chickens, simply borrowed and made minor modifications to the reptilian egg (leathery shells became brittle, for example).

Mammals, too, initially borrowed the reptilian egg. Today, only two species of egg-laying mammal continue to exist, the platypus and spiny anteater, both found in Australia. Their eggs are essentially identical to those of reptiles. After hatching, additional nourishment is provided by the female's **mammary glands**.

One group of mammals, the marsupials, did away with the reptilian egg entirely. Their embryos are retained inside the female's body for a time. They are born at a very early embryonic stage, usually when they are only a few millimeters or centimeters long. Their first task is to crawl from the birth canal, along the outside of the female's body, and into a special abdominal pouch, **marsupium**, where mammary glands are found (Figure 10-23). The mammary glands provide **fetuses** with the nutrients and energy they need to complete development.

Placental mammals further protect embryos and fetuses. The egg structure, with its membranes and fluids but without shells, is retained inside the female's body. Some surrounding membranes become modified into a **placenta**, a structure that firmly attaches the embryo and later fetus to the female's uterus. Placental blood vessels flow beside the female's blood vessels, and from the female nutrients, oxygen, and a variety of other substances pass to the fetus. Wastes, including carbon dioxide, pass the other way to be disposed of by female's kidneys and lungs. Thus protected and provided for, fetuses remain in their mother's body until they are at least minimally able to withstand external environments.

Figure 10-23. Marsupial mammals, such as the Virginia opossum, house developing young in a specialized abdominal pouch, the marsupium, where the mammary glands are found.

Variations on the Reproductive Theme The reproductive patterns described above are generalizations. Pick some specific vertebrates and describe the details of their reproduction. How are they typical of their group? What features of their reproduction are exceptional?

10.4.5 After Fertilization, Organisms Develop and Grow

Compared with plants, animal growth and development are considerably more complex and have been more widely studied. Much is known about various patterns of embryonic development in animals. Generally, these patterns are variations on a theme. In the initial stage, there are significant increases in number of cells (through a series of mitotic events called **cleavages**) with little or no overall growth in size (Figure 10-24). Subsequently, a second stage involves the early appearance of organ systems, followed by a third stage that entails further refinement of organ systems and significant growth. Invariably, this process starts at the head end and proceeds toward the tail end. Often, there is a considerable lag between anterior (head end) and posterior (tail end) development. Marsupial mammals are an extreme example. At birth, they have fully functional heads and forelimbs, while their hindlegs are still in the limb bud stage. There are almost as many specific patterns as there are species. On the CD we examine two—insect and frog.

Fast Find 10.4e Animal Development

Early cleavage stages are complicated by the amount of yolk eggs contain. Sea urchin eggs contain little yolk; their early cleavages are straightforward and simple. Frog eggs contain considerable yolk. Their first two cleavages result in four eggs, each roughly the same size, containing equal amounts of cytoplasm and yolk. Their third cleavage occurs off center, producing four smaller cells containing most of the active cytoplasm and four larger cells containing most of the yolk. This divides the embryo into two regions, **animal pole** and **vegetal pole**. Thereafter, cleavages occur more frequently toward the animal pole, which becomes the developing embryo. Reptile and bird eggs contain even more yolk. Indeed, what will become the new individual starts out as a small insignificant blob floating on a sea of yolk. It is in this mass that cleavages occur, without dividing the yolk at all.

Yolk is important to most developing embryos. Those organisms that develop from eggs containing little yolk get their nourishment from their watery environments. Typically, their development is rapid, and they become independent as soon as possible. Vertebrates require more development time, and their eggs contain larger amounts of yolk. The most advanced mammals nourish developing young with mammary glands, placentas, or both.

What happens after hatching or birth depends on how well developed the young are. Growth and development in arthropods are complicated by their exoskeletons. In

Figure 10-24. Early cleavages of the embryo result in an increase in the number of cells, but little increase in overall size. Growth comes later. Here a frog's egg has undergone several cleavages to the sixteen cell stage (magnified 34 times).

insects, two patterns of growth occur. Some follow the typical pattern of arthropod development: they hatch as miniature adults, but without wings or reproductive organs, and then undergo a series of molts before becoming adults. Other insects experience complete metamorphosis. What hatches is a wormlike larva, known as a caterpillar, grub, or maggot. After undergoing a series of molts, they go into a period of inactivity (**pupate**) and then emerge completely transformed into the adult form. Caterpillars become butterflies or moths, grubs become beetles, maggots become flies, and so on. Often their dietary preferences also change.

Freshly hatched fish and reptiles are essentially miniature adults. They are fully capable of independent existence, which initially involves finding food and avoiding predators. Those successful at both grow, eventually reach adult size, and become reproductively capable.

Growth and development of amphibians are complicated because they have only partially transformed from aquatic to terrestrial lifestyles. Typically, what hatches is an aquatic creature equipped with gills; a streamlined body shape; a flattened, paddlelike tail; and no other limbs. This creature in no way resembles the adult and is called a tadpole. After a time, which varies from a few days to over a year, larvae undergo metamorphosis, in which they lose gills and gain lungs, lose tails and gain limbs, and otherwise become equipped for terrestrial life. Many change dietary preferences; typically, they stop being herbivores and become carnivores. Another period of growth occurs before sexual maturity is reached. The amphibian group is complex, and many exceptions to this typical pattern are known. For example, salamanders keep their tails. Some larvae become reproductively capable and fail to metamorphose to a terrestrial form. Others grow limbs, keep gills, and continue to be aquatic as adults.

Two patterns of post-hatching growth and development are seen in birds. A few, notably waterfowl, shorebirds, some waders, and chickenlike birds, follow an essentially reptilian pattern. Hatchlings are miniature adults to the extent that they can immediately leave the nest, recognize food, and secure it. Most other birds are much less well developed at hatching—they are blind, featherless, helpless. For several weeks they are totally dependent on adults for food and protection from predators, rain, and excessive heat or cold. In essence, they undergo much of their later stages of embryonic development after hatching (Figure 10-25).

Whether they are hatched from an egg, born into a marsupium, or separated from placentas, mammals are helpless at birth, totally dependent on adults for nourishment and protection. For all baby mammals, initial nourishment is provided by the female's

(a) *(b)*

Figure 10-25. Two patterns of growth and development are seen in birds. *(a)* In some birds, such as these magnolia warblers, hatchlings are virtually helpless, totally dependent on parents for food and protection from predators and the environment.

(b) Other birds, such as these mallard ducklings, are much more highly developed when hatched. These young leave the nest within minutes of hatching, can find their own food, and are partially protected from the environment by downy feathers.

mammary glands. Even after weaning, which occurs after a few weeks to months, adults continue to either bring food to the young or take the young to food. Some baby mammals rapidly get beyond initial helplessness. Within only a few hours, the young of hoofed animals can run almost as fast as adults. Newborn whales are brought to the surface for their first breath but are soon able to swim with adults, to find the surface, and to breathe independently. For other mammals, development is slower. Mammals are unique in that much of what they need to know about their environments—what foods to eat and where to seek food, where to seek shelter, how and where to avoid predators, and how to interact with others of their species—is learned from close associations with parents and adults. As time passes, young become more self-sufficient, until finally they disperse. This process can take considerable time—typically, in humans, 18 to 20 years.

Piecing It Together

1. The most common form of reproduction is asexual, either by binary fission (seen in all prokaryotic organisms) or mitosis (seen in all eukaryotic cells except gametes).

2. Conjugation, a simple form of sexual reproduction, evolved in microbes.

3. Sexual reproduction among the eukaryotes generally follows one of three types: diploid cycle, seen in all animals and some protists, in which all cells except gametes are diploid; haploid cycle, seen in most fungi and some algae, in which the organism's cells are haploid throughout most of their lives; and alternating cycle, seen in all plants and some algae, in which there are two distinct generations, one in which all cells are haploid and another in which all cells are diploid.

4. For organisms that reproduce sexually in water, there is generally less danger of gametes drying out. Their gametes are often less protected (and occasionally released directly into the environment), and their eggs often have less yolk than organisms that reproduce on land.

5. The ability to reproduce on land apparently evolved independently among segmented worms, mollusks, arthropods, and chordates (vertebrates).

6. Among vertebrates, the transition from water to land is seen most dramatically by comparing their eggs. Those of marine fish generally have relatively little yolk and no shells. Eggs of freshwater fish and amphibians, who reproduce in nutrient-poor environments, have considerably more yolk and are often protected by surrounding gelatinous layers. Eggs of reptiles, birds, and primitive mammals are the largest cells known, contain relatively huge amounts of yolk, and are surrounded by thick protective shells. The eggs of advanced mammals have no shells and no yolk; they are retained within the female's body where they are protected and nourished.

7. Growth and development don't stop with hatching or birth, but continue throughout life. Our understanding of how they occur and are controlled is only now being worked out.

Where Are We Now?

Traditionally, the study of plant and animal physiology has lagged behind the study of human physiology, perhaps for obvious reasons. Three trends, particularly in recent decades, have tended to narrow these gaps in our understanding.

First, the physiology of plants and animals is an innately interesting biological topic. Since the early days of biology, a relatively small, often undersupported group of dedicated biologists has been drawn to this study. It was through their efforts that much of what we know about the fascinating physiology of these organisms was documented. Fortunately, their work continues and is, in fact, expanding.

Second, certain procedures and experiments in physiology can be done more easily on animals than on humans. Many intrusive procedures are simply not permitted on humans. As students, surgeons hone skills first on test animals. Only later are results of such efforts extrapolated and extended to humans. As a result, much of what we know about human physiology was first discovered by studying laboratory rats, pigs, monkeys, and rabbits. Similarly, these animals have been used to test drugs for side effects and the toxicity of many new products and pollutants. A relevant question raised by all these studies is, to what extent are these animal responses typical of those occurring in humans? Answering this question has expanded our knowledge of animal physiology.

Finally, in recent decades, the kinds of products animals and plants provide to humans has moved well beyond food, clothing, and companionship. We are fast approaching the day when the genes of any organism can be transplanted into the genome of any other organism. Microbes are being used to produce human proteins, including insulin. Adult sheep and cattle have been cloned from somatic cells, and it is only a matter of time until they will be infused with new genes and used to produce drugs, hormones, and other proteins for human consumption. There are plans to use pigs to grow human organs for transplant. It turns out that the physiology of pigs and humans is remarkably similar. Why not replace an ailing human heart with a pig's heart? In today's world, it would be rejected within days by the human immune system. But, what if a selected number of human genes, perhaps taken from the ailing human, were inserted in a fetal pig? If the ailing human can be kept alive for a few months while the pig grows up, when its heart is transplanted, the human's immune system might recognize the heart not as foreign but as self. Other organs could be produced as well, including livers, kidneys, skin, bones, ears, and so forth. Another method would be to implant fetal pig organ tissue directly into the ailing human. Early results indicate that the resulting organ is not rejected (Figure 10-26).

The numerous procedural questions that exist today with each of these methods

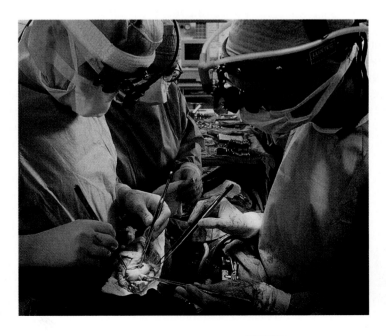

Figure 10-26. In the future, can the organs of animals such as pigs be used to replace ailing organs in humans?

seem answerable, given a little more time and effort. But what about ethical questions? A pig's heart certainly seems to be equivalent to a human heart, but is it really? Does the pig have any rights or interests here? What if pigs harbor a benign virus that in the human becomes highly virulent? Could **transgenic** (across species) transplants trigger another AIDS virus? Answering these and related questions demands that we know as much about animal physiology as we do about human physiology. Indeed, in the early decades of the 21st century, the distinction between the two fields of study may blur entirely.

Spare Parts from Pigs? What is the current status of efforts to use organs from pigs or other animals as human implants?

REVIEW QUESTIONS

1. What is the study of physiology?

2. Internally, multicellular animals are hierarchically organized. What does this statement mean? Name the layers of the internal hierarchy.

3. What are the differences between epithelial, connective, muscle, and nerve tissue? Give examples of each.

4. Give an example of organs working synergistically.

5. Contrast how snakes and humans maintain homeostatic body temperatures.

6. Besides body temperature, what other internal body conditions are maintained homeostatically.

7. What are the differences between negative and positive feedback? How are they similar? Give examples of each.

8. Explain the statement, "there is a relationship between form and function."

9. Contrast the processes used by protists and multicellular organisms in obtaining nutrients.

10. Briefly discuss the evolution of the digestive system in animals.

11. How are the three basic types of respiratory systems similar and different.

12. Contrast the respiratory systems of insects with that of mammals.

13. What are the major functions of circulatory systems?

14. How do animals living in salt water maintain proper water balance in their cells? (Hint: There are two solutions for animals living in salt water)

15. How do animals living in fresh water maintain proper water balance in their cells?

16. What can terrestrial animals do to reduce water loss?

17. Contrast the excretory systems seen in segmented worms with those seen in insects and in vertebrates.

18. By what processes do kidneys transport metabolic wastes from blood to urine?

19. Explain the mechanisms by which muscle cells contract.

20. Contrast the three kinds of skeletons.

21. Two organ systems in animals are involved in monitoring the environment. How are they similar? How are they different?

22. What are the three major reproductive cycles seen in organisms? How are they similar? How are they different?

23. One of the biggest challenges faced by terrestrial animals is protection of gametes. Why is this a problem? How have terrestrial segmented worms solved this problem? How have terrestrial mammals solved this problem?

24. Which came first, the chicken or the egg? Why?

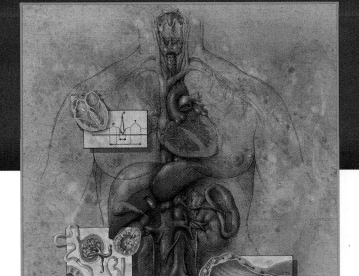

Chapter 11

Human Physiology: How Does the Human Body Work?

The body is better than any machine of human invention.
—Gustav Eckstein, 1969

—Overview—

Now we apply the principles of physiology from the previous chapter to one particular organism: a large, mobile, terrestrial animal called *Homo sapiens*, or a human being. The human body is adapted to survive in many different environments, which partly explains why we can live in nearly every part of the globe. But while the body as a whole can withstand vastly different external environments, the individual cells of the body function best when such factors as temperature; acidity; and salt, water, oxygen, and carbon dioxide concentrations in their immediate vicinity are kept within narrow limits. If local conditions vary too much from these limits, cells begin to die. Human physiology is the study of how the human body as a whole copes with change.

To begin, we must revisit a concept that was introduced in the last chapter, that of **homeostasis**. In Section 10.1.3 we learned that homeostasis means maintaining a stable internal environment. But what exactly is that "internal environment," and where would we look for it in the human body? For practical purposes, we can think of the internal environment as the fluid that surrounds the cells and in which they are constantly bathed. This is the **extracellular fluid**. It is everywhere within the body. Changes in factors outside the body are reflected in the components of extracellular fluid. These compositional changes, in turn, trigger counteracting responses that bring the fluid constituents back to within the narrow limits that are best for cells. It is this tendency to maintain the extracellular fluid within narrow limits that constitutes homeostasis.

Imagine that you are hiking in the high desert, a dry external environment. Relatively few cells of your body actually come in contact with the dry air, namely those

Chapter opening art—The various organ systems of the body help to maintain homeostasis.

of your skin and the layer of cells that lines your lungs and breathing passages. The rest of the body's cells are buffered from the desert air, sheltered by the extracellular fluid in which they are bathed. As you lose water to the dry air by evaporation from the skin and lungs, water from the extracellular fluid diffuses into those cells to replace what is lost. As a result, the volume of the extracellular water decreases. Without your conscious awareness, the kidneys and circulatory system are mobilized to conserve what water remains and replace that which has been lost. At a conscious level, your brain registers a sense of thirst, and your behavior is affected, causing you to drink water. How is such a complex response coordinated?

The mechanisms that control and coordinate the internal environment are called **homeostatic control systems** (Figure 11-1), and they all have the same kinds of elements:

◆ **Sensors** first recognize that the balance of some factor in the internal environment has changed. In our example, a sensor recognizes that the amount of water in the extracellular fluid has decreased.

◆ An **integrating center** receives the signal that something is amiss and coordinates a response that will bring that variable back within the narrow limits that are optimal for the cells. Often parts of the brain and spinal cord act as integrating centers.

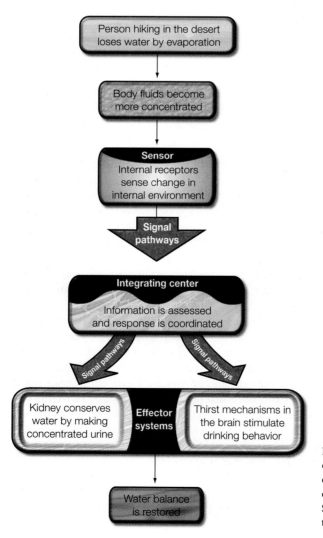

Figure 11-1. An example of a homeostatic control system with the four elements: sensor, integrating center, effector system, and signal pathways. Signal pathways can include both the nervous and endocrine systems.

◆ **Effector systems** are the cells, tissues, or organs that actually carry out the homeostatic response. In our example, the kidneys, circulatory system, and brain are the effector systems.

◆ **Signal pathways** are the body's messenger systems. Information from sensors must be sent to an integrating center. From there, information about the response must be sent to the effector systems. There are two kinds of signal pathways in humans, the nervous system and the endocrine system. The nervous system carries information very rapidly, whereas the endocrine system is generally best suited for slower delivery of more long-lasting messages.

In the sections that follow, we will examine the various homeostatic control systems of the human body, including the one from our example that regulates and balances the body's water content. But first we will take a look at the signal pathways. We begin our discussion of human physiology with the two major avenues of communication for coordinated and integrated responses: the nervous system and the endocrine system.

11.1 HOW DO THE PARTS OF THE HUMAN BODY COMMUNICATE?

How far is your brain from the tips of your toes? If you are very tall, it may be as far as two meters, or over six feet. Even if you are not tall, the distance is much too far for information from the brain to travel to the toes in the form of chemical messengers moving by diffusion. Diffusion is effective for moving substances within a cell and between a cell and the surrounding extracellular fluid. However, diffusion over large distances is slow. Fast, effective communication over distances of more than a few micrometers requires anatomical structures that are specialized for moving information. The nervous system is specialized for this purpose.

11.1.1 The Nervous System Provides Rapid Communication

Marvel at a Mozart concerto; solve differential equations; argue about the meaning of life. Pull your hand away from a hot stove; swallow a mouthful of food; breathe more deeply when the air is thin. These two different kinds of activities—one occurring with a high degree of conscious awareness, the other practically without thought—are both coordinated and controlled by the nervous system.

The nervous system is a system of interconnected nerve cells, called **neurons**, that extend throughout the entire body. Neurons are oriented relative to one another in precise ways that form complex, information-conducting pathways. We distinguish between two different regions of the nervous system, known as the **central nervous system** and the **peripheral nervous system** (Figure 11-2). The central nervous system includes the brain and the spinal cord. The peripheral nervous system includes all of the **nerves**, each composed of bundles of neurons, that carry messages toward and away from the brain and spinal cord. Information about the outside world or about conditions within the body enters the peripheral nervous system from sensory receptors, then travels by means of **sensory neurons** of the peripheral nervous system into the central nervous system, where it is processed and integrated. The central nervous system then sends messages through **motor neurons** of the peripheral nervous system to the tissues and organs that effect a response (effectors).

Higher functions of the nervous system—thought, language, learning, and comprehension—occur in the brain, where billions of neurons form complex, interconnected networks. This form of integration is so complex that even today our understanding of it is minimal. Less complicated integration, such as the reflex that occurs when you pull

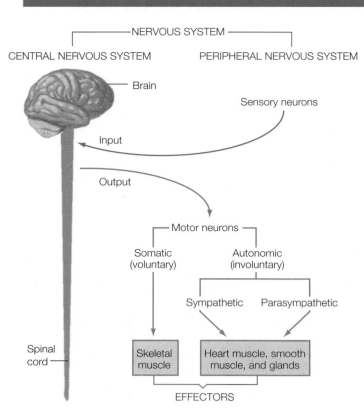

NERVOUS SYSTEM

CENTRAL NERVOUS SYSTEM   PERIPHERAL NERVOUS SYSTEM

Brain

Sensory neurons

Input

Output

Motor neurons

Somatic (voluntary)   Autonomic (involuntary)

Sympathetic   Parasympathetic

Spinal cord

Skeletal muscle

Heart muscle, smooth muscle, and glands

EFFECTORS

Figure 11-2. Organization of the human nervous system. The central nervous system includes the brain and spinal cord; the peripheral nervous system includes sensory neurons that carry information to the central nervous system and the motor neurons that carry information to the effectors. Motor neurons are further characterized as somatic (carrying information to voluntary muscles) and autonomic (carrying information to involuntary effectors such as the heart and other internal organs). The autonomic nervous system is further divided into sympathetic and parasympathetic subdivisions, discussed later in this section.

your hand away from a hot stove, occurs in the spinal cord where signals pass from incoming sensory neurons to outgoing motor neurons that, in turn, signal your muscles to move your arm. (The sensation of pain upon burning your fingers occurs after you have already moved your hand away from the stove. This is a function of the brain and is not a reflex activity.) Regardless of whether we are speaking of the central or peripheral nervous systems, sensory or motor neurons, or responses coordinated in the brain or the spinal cord, the way that neurons encode and carry information is the same. Messages are transmitted along individual neurons as electrochemical signals.

The Neuron

**Fast Find
11.1.c
The
Neuron**

A **neuron**, or nerve cell, is the basic functional unit of the nervous system. Most neurons are cells with elongated processes, as shown in Figure 11-3. The **cell body** contains the nucleus and other organelles. Many long, thin processes called **dendrites** extend outward from the cell body. Dendrites receive incoming signals and conduct them toward the cell body. A single, larger extension leaving the cell body is called the **axon**. Axons conduct signals away from the cell body, toward an effector or another neuron. In some neurons, axons can be up to a meter in length. Because of the way signals are received, signals move in *only one direction*, in any given neuron, arriving at the dendrites, then traveling through the cell body and down the axon, and finally passing to adjacent neurons or other effector cells that reside at the ends of the axon. This unidirectional flow of information is efficient; outgoing and incoming messages travel at the same time along different neurons.

The signals or nerve impulses carried by neurons are electrical in nature. Neurons, as well as muscle cells and some cells of the endocrine system, are in a class called **excitable cells**. Excitable cells use electricity to do their work.

Another important feature of many human neurons, illustrated in Figure 11-3*b*, is insulation. Just as the electrical wires in your house are insulated with plastic coatings, the nerve cells in your body are insulated with a fatty insulating substance called **myelin**,

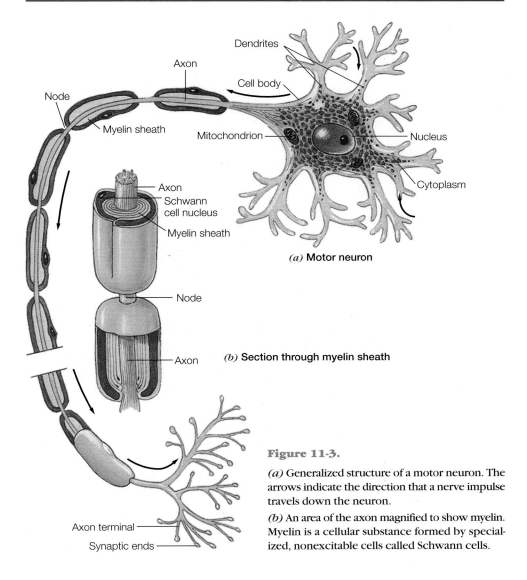

Figure 11-3.

(a) Generalized structure of a motor neuron. The arrows indicate the direction that a nerve impulse travels down the neuron.

(b) An area of the axon magnified to show myelin. Myelin is a cellular substance formed by specialized, nonexcitable cells called Schwann cells.

which is formed by Schwann cells. Schwann cells are not excitable cells; they are specialized accessory cells that grow in a spiral around most axons and some dendrites, forming a sheath of concentric layers of cell membrane. The insulation provided by myelin is important because it increases the speed with which signals travel in neurons and keeps the signal traveling on its own neuronal pathway. Devastating neural disorders, such as multiple sclerosis, occur when nerve cells fail to become fully myelinated or when myelin begins to deteriorate.

Action Potentials

Fast Find 11.1.d Action Potential

Imagine that you have a tiny electrode that is small enough to be inserted into a single cell without causing any damage to the cell. If you connected your electrode to a voltmeter and measured the electrical potential, or **voltage**, between the inside of the cell and the extracellular fluid, you would find that the inside of the cell is negatively charged with respect to the outside. Regardless of what type of cell you measure, you would find a small but measurable voltage across the cell membrane. This voltage is on the order of 10 to 100 millivolts, or nearly 100 times smaller than the voltage between the ends of a typical flashlight battery. The voltage across cell membranes results from an uneven distribution of electrically charged **ions**, that is, charge-carrying atoms or molecules, between a cell and its surroundings. In excitable cells, this voltage is called the **resting potential**. A distinctive feature of excitable cells is that the membrane voltage can rapidly reverse in response to a stimulus, so that the inside of the cell becomes positively charged, and just as rapidly change back again to its original, negatively charged, resting state (Figure 11-4). This rapid, transient reversal in membrane voltage is called an **action potential**. Only excitable cells, mostly neurons and muscle cells, can undergo action potentials.

What feature of excitable cells is responsible for the action potential? The answer is found in the cell membrane. Recall from Chapter 4 that the cell membrane is a fluid mosaic; islands of proteins float in a flexible sheet of lipids. Among the proteins in the membranes of excitable cells are **gated ion channels**, which are proteins capable of opening and closing holes in the membrane that allow specific ions to diffuse into and out of the cell. A neuron at rest has a relatively high concentration of sodium ions on the outside. These positively charged ions (Na^+) are partially responsible for the resting potential. When a neuron is stimulated, ion channels specific to Na^+ open, and positively charged sodium ions rapidly move "downhill" from the outside of the cell to the inside of the cell by diffusion. They are driven inward by both the difference in their concentration and, because opposite charges attract, by the more negative environment inside the cell. When they move into the cell, sodium ions carry their positive charge with them, temporarily reversing the voltage across the cell membrane so that the inside becomes positive with respect to the outside.

The action potential can be seen as a steep upward sweep of the electrical potential, followed by an equally steep downward sweep (Figure 11-4). Sodium ions entering the cell are responsible for the upward sweep, but what causes the downward sweep? To answer this, we must look to a different ion, the potassium ion, and a different kind of protein within the cell membrane, the gated potassium channel. Gated potassium channels coexist with gated sodium channels in the membranes of neurons. These channels,

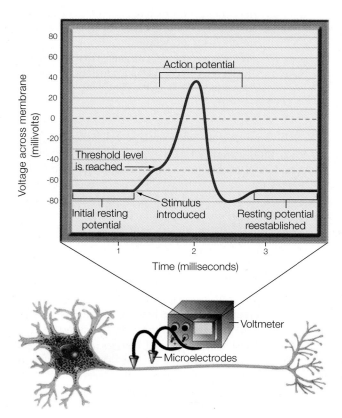

Figure 11-4. Electrical properties of neurons. The resting potential of a neuron is about –70 millivolts. When the neuron is stimulated, the voltage rises until it reaches a threshold. At the threshold voltage, an action potential occurs, consisting of a rapid and transient reversal of the membrane voltage. Within a few milliseconds, the resting potential is reestablished. (This neuron is shown without its myelin sheath.)

too, can open or close in response to certain stimuli, and when they are open, only potassium can diffuse through them. When the inside of a neuron becomes positive, as it does when sodium ions rush inward, potassium channels are stimulated to open. Potassium is also a positively charged ion (K^+). Unlike sodium, it is more concentrated on the inside of cells than it is on the outside. When the inside of the cell is made positive by the entry of Na^+, both the concentration gradient and the electrical attraction for the (temporarily) more negative outside drives potassium out of cells through their open channels. This rush of potassium ions out of the cell brings the voltage back to its resting level and ends the action potential. The entire action potential, summarized in Figure 11-5, is very brief, lasting about 3 milliseconds, or 3 one-thousandths of a second.

What kind of stimulus initiates an action potential? In other words, what causes the sodium channels to open in the first place? Action potentials begin when a signal from outside the cell disturbs the balance of ions between the inside and outside and thus alters the resting membrane voltage. The gates that normally block the sodium channels open in response to changes in voltage across the membrane. Often it is a chemical signal from a connecting neuron, but for some neurons, the signal comes from a

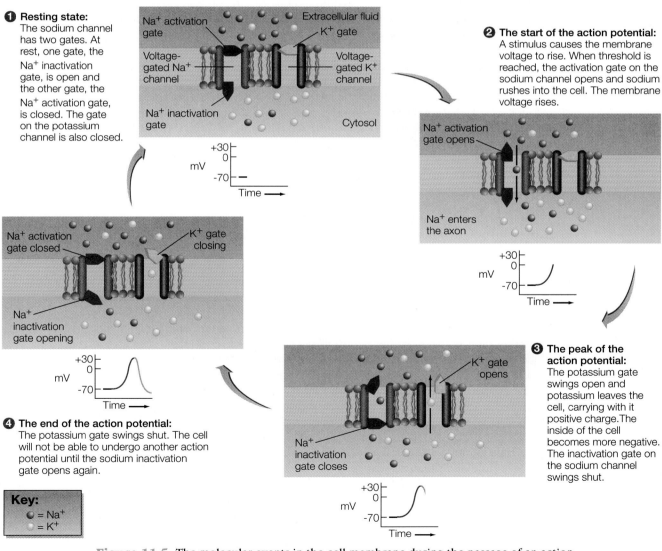

❶ Resting state:
The sodium channel has two gates. At rest, one gate, the Na^+ inactivation gate, is open and the other gate, the Na^+ activation gate, is closed. The gate on the potassium channel is also closed.

❷ The start of the action potential:
A stimulus causes the membrane voltage to rise. When threshold is reached, the activation gate on the sodium channel opens and sodium rushes into the cell. The membrane voltage rises.

❸ The peak of the action potential:
The potassium gate swings open and potassium leaves the cell, carrying with it positive charge. The inside of the cell becomes more negative. The inactivation gate on the sodium channel swings shut.

❹ The end of the action potential:
The potassium gate swings shut. The cell will not be able to undergo another action potential until the sodium inactivation gate opens again.

Key:
● = Na^+
○ = K^+

Figure 11-5. The molecular events in the cell membrane during the passage of an action potential.

sensory receptor or even from a mechanical disturbance of the neuron itself. Touch receptors on the skin, for example, fire action potentials when they are stretched by a touch or change in pressure. These changes cause positively charged ions to leak into the neuron and raise the resting voltage. If the stimulus is weak, only a few ions may leak in, raising the voltage only a little above its normal resting level. Soon the disturbance subsides. As the magnitude of the stimulus gets larger, the disturbance of ion balance, too, increases. If the stimulus is strong enough, the resting voltage is raised to a level sufficient to cause sodium channels to open and thereby set off an action potential. That level, which differs for different neurons, is called the **threshold**.

Below the threshold level, a weak stimulus causes a smaller disturbance in the membrane potential than a strong stimulus, but once a stimulus is strong enough to trigger an action potential, an increase in stimulus strength does not cause a larger action potential. This characteristic of action potentials is called the **all-or-nothing rule**. The size, or magnitude, of the action potentials in your sensory neurons is no larger in response to a sharp pinch than it is in response to a gentle touch, provided threshold has been reached. Instead, we register the strength of a stimulus by both the number of neurons that are triggered and the number of times any neuron fires in succession.

Once the action potential has passed, the voltage across the neuron's membrane is restored to the resting potential. For a brief time, however, the neuron cannot be stimulated to undergo a second action potential. This short interval is called the **refractory period**, and it lasts only a few microseconds. After the refractory period, the cell is ready for another action potential.

Although the movement of sodium and potassium during an action potential is sufficient to reverse the voltage of the membrane for a brief time, the total number of ions that cross the membrane in one action potential is really quite small. The concentrations of each ion on either side of the membrane do not change appreciably with a single action potential. Another way of saying this is that a very small number of ions crossing the membrane can have a large effect on the membrane voltage. A neuron may undergo many action potentials before the concentrations of sodium and potassium on either side are appreciably altered. When that happens, however, yet another protein in the cell's membrane works to move sodium back to the outside and potassium back inside. That protein is called the sodium–potassium pump.

Sodium–potassium pumps are not only found in excitable cells; they are embedded in the membranes of virtually all animal cells. They constantly work to move sodium ions out of the cell and at the same time move potassium ions into the cell (see Figure 4-17). Recall from Chapter 4 that when substances diffuse down their concentration gradients (from an area of high concentration to an area of low concentration), it is similar to rolling a ball downhill; it can happen without an input of energy. When substances are moved against their concentration gradients, however, it is like pushing a ball uphill. The cell must expend energy. For the sodium–potassium pump, that energy is supplied in the form of ATP. The pump breaks the chemical bond that connects the third phosphate to ATP, forming ADP (see Figure 4-12), and uses that energy to move Na^+ and K^+ against their concentration gradients. The sodium–potassium pumps in the membranes of neurons help to establish and maintain the resting potential.

Neurons carry signals from place to place, sometimes a long way from where action potentials are first triggered. The action potential must therefore travel from where it is initiated to where it is passed on to the next cell in the information pathway. Action potentials must be triggered in each successive patch of membrane to the very end of the cell, without diminishing in magnitude.

Remember that neurons are encased in myelin (see Figure 11-3). Along the length of the axon, the insulating myelin sheath is interrupted by unsheathed spaces between adjoining Schwann cells, called nodes. The areas of neuron membrane inside the myelin have no sodium or potassium channels and cannot, therefore, fire action potentials. The nodal areas, however, are rich in these gated ion channels. When an action potential occurs at one node, the sodium that enters the cell diffuses within the neuron, carrying its positive charge. This raises the resting potential at the next adjacent node to a level above threshold, causing the sodium channels in that node to open and triggering an action potential there. The incoming sodium at this node, in turn, diffuses to the next adjacent node and causes an action potential there. In this manner, action potentials move rapidly along myelinated neurons by jumping from node to node until they reach the end of the axon. This jumping action greatly facilitates the speed at which action potentials can travel.

When an action potential has reached the end of a neuron, it passes its signal to the next adjacent cell in the communication pathway. The next cell may be another neuron, or it may be a muscle or gland cell. Regardless of the kind of cell involved, the junction is called a **synapse**.

Synapses

Fast Find 11.1.e Synapse

A synapse is the junction between a neuron and another cell. The signal, carried by the neuron as an action potential, is transmitted to an adjacent cell across the synapse. Most synapses in the human body do not involve physical contact between the two cells; instead they are places where the neuron and the adjacent cell come in very close proximity without actually touching. The space, or cleft, between the neuron and the adjacent cell is tiny—about one one-millionth of an inch (or about 10 nanometers)—but it is still wide enough that an action potential coming from the first cell cannot jump across. Instead its arrival must be signaled to the target cell by means of a chemical messenger called a **neurotransmitter**. This kind of synapse is called a *chemical synapse.*

Packets of neurotransmitter molecules are stored in the axon endings at a chemical synapse. When an action potential reaches the end of the axon, these packets release neurotransmitters into the cleft between the two cells. Because the gap is narrow, molecules of neurotransmitter diffuse quickly across the gap. **Receptor proteins** on the surface of the target cell bind to the molecules of neurotransmitter. You can watch this process on the CD. The response of the target cell depends on several things, including the chemical nature of the neurotransmitter and the kind of receptor proteins present. The neurotransmitter might make the cell more likely to undergo an action potential; in other words, it brings the cell closer to threshold. That kind of response is called an excitatory response. A cell receiving many excitatory inputs would fire an action potential, passing the signal to the next cell in the communication pathway. Alternatively, another effect might be to inhibit the cell, making it less likely that it will undergo an action potential. A cell receiving both excitatory and inhibitory inputs from different neurons puts together, or *integrates*, all of the incoming signals and only fires if the balance of the input is excitatory.

Chemical synapses are the slowest points in any nervous system pathway. The time it takes for the neurotransmitter to be released, diffuse across the cleft, bind to the receptor protein, and bring about a response in the adjacent cell just about doubles the time it would take for a signal to travel if there were no synapse to cross. But the delay caused by the synapse is more than compensated for by the extra flexibility. The ability to send either excitatory or inhibitory signals and to integrate incoming information from different sources forms the basis for the sophistication of the human nervous system, including the higher functions of the brain. Although we are still a long way from understanding how the brain accomplishes such complex tasks as language, emotions, memory, and abstract thought, we can say for certain that without the flexibility provided by chemical synapses, none of these things would be possible.

The Human Nervous System

Fast Find 11.1.f The Brain

Many of our activities are voluntary; that is, they are matters of choice. We choose, for example, to reach for a book, play a game of tennis, or hug a friend. These are our conscious activities in which ideas originate in our brains and signals from the brain are sent by means of peripheral nerves to muscles. Our voluntary actions reflect the coordinated activity of nerves and muscles. The portion of the peripheral nervous system that brings about voluntary actions is called the **somatic nervous system** ("soma" is Greek for *body*). Many internal functions, however, go beyond our voluntary control, such as circulation of the blood or movement of food within our digestive tract. These activities are under the control of a different portion of the peripheral nervous system

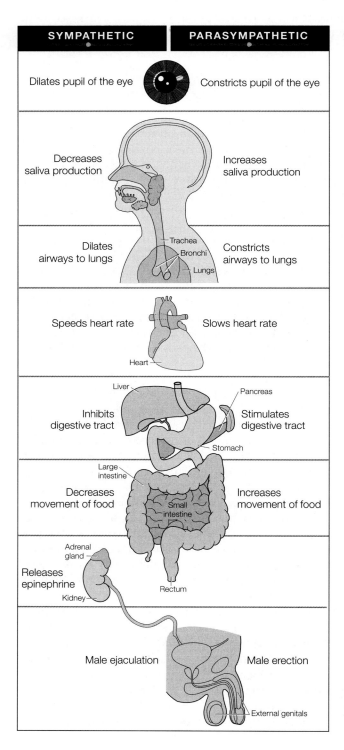

SYMPATHETIC	PARASYMPATHETIC
Dilates pupil of the eye	Constricts pupil of the eye
Decreases saliva production	Increases saliva production
Dilates airways to lungs	Constricts airways to lungs
Speeds heart rate	Slows heart rate
Inhibits digestive tract	Stimulates digestive tract
Decreases movement of food	Increases movement of food
Releases epinephrine	
Male ejaculation	Male erection

Figure 11-6. Sympathetic and parasympathetic nervous systems often have opposing effects on organs that receive input from both.

called the **autonomic nervous system** (from "autos," the Greek for *self*). The autonomic nervous system works to maintain the body's homeostasis *autonomously*, or independent of our conscious will.

The autonomic nervous system is further subdivided into two parts, called the **sympathetic** and the **parasympathetic** subdivisions. Nearly all the involuntary organ systems of the body are connected to both sympathetic and parasympathetic nerve fibers. The input from these two systems is usually antagonistic (Figure 11-6). In general, input from the parasympathetic nervous system tends to keep the body's "housekeeping" systems functioning, including the steady movement of material through the intestine, and the dependable regularity of the heartbeat and breathing. The sympathetic nervous system, on the other hand, leaps into action when certain urgent demands are made on the body. Fear, excitement, fighting for one's life, and fleeing from danger are all mediated by activity of the sympathetic nervous system. When the sympathetic nervous system is activated, some of the humdrum housekeeping functions are temporarily shut down and the body's resources are directed toward dealing with the urgent matter at hand. As shown in Figure 11-6, for some organs, sympathetic input is excitatory and parasympathetic input is inhibitory; for other organs, it is the other way around. At any given moment, the actual response of an organ to these regulatory influences depends on which nerve fibers, sympathetic or parasympathetic, are most active. The autonomic nervous system is one of the ways in which the various parts of the body constantly signal one another to maintain stability within cells, tissues, and organs, and therefore within the body as a whole.

 Mind and Body We are now beginning to appreciate the influence that our conscious lives, particularly our emotional lives, can have on the health and well-being of our bodies. Can you find evidence that the autonomic nervous system responds to our thoughts and emotions? What part of the brain could provide the link between conscious and unconscious activities? Are there ways in which we can learn to be healthier by either directly or indirectly influencing our autonomic systems?

11.1.2 Hormones Coordinate Slower, Longer-Lasting Processes

The human nervous system, with its magnificent brain interpreting information from in and around the body and integrating and coordinating responses to that information, is the main part of what makes us different from other animals. But even our large brains cannot provide all of the integration that is required to keep our bodies in homeostasis. In addition to the nervous system, the cells of animals' bodies communicate by means of chemical messengers called **hormones**, which are substances released into the blood by one group of cells and transported to a distant group of **target cells**, where they exert their effect. While the nervous system coordinates rapid responses to our immediate environment, many of the long-term, ongoing processes of the body, such as growth and maturation, regulation of water balance, and reproduction, are under the control of hormones.

Hormones are a chemically diverse group of signaling molecules. Some hormones are proteins; others are steroids, similar in structure to cholesterol, shown in Figure 4-7*c*; still others are single amino acids that are chemically modified by the cells that make them. Like neurotransmitters, hormones serve as chemical messengers. Unlike neurotransmitters, however, the cells that are influenced by hormones are often a long distance from the cells that produce and secrete the hormone. Consequently, hormones cannot diffuse to their target cells, but instead they must be carried to them in the blood. As we will soon see, all the cells of the body, not just the target cells, are reached by the blood as it circulates throughout the body. But hormones are often quite specific to the cells that they target. How does a hormone "know" which cells to target?

The answer can be found within the target cell itself. Target cells for a specific hormone have receptor proteins that recognize and bind only to that hormone; non–target cells lack the appropriate receptors. In fact, a hormone alone cannot affect any cell. Thus, while all the body's cells come in contact with all the body's hormones, only those cells with the appropriate receptors can respond to a given hormone. The action of a hormone, its unique effect on only its target cells, depends as much on its receptor as it does on the hormone itself.

Once a hormone binds to its receptor, the target cell can respond in one of three basic ways:

1. Some aspect of the cell's metabolism may be affected; for example, some enzyme-catalyzed reactions may be sped up or slowed down.

2. The movement of substances across the cell membrane may be influenced.

3. Gene expression may be turned on or off, resulting in more or less of some specific protein.

The first two kinds of responses are rapid and immediate, and they usually characterize a shorter-term hormonal effect. Epinephrine, also known as adrenaline, is the hormone released in response to fear or exhilaration. It is a good example of a hormone that, bound to its receptor, activates a cellular enzyme. Everyone is familiar with the physical sensations associated with a "rush of adrenaline": the heart rate increases, the mouth feels dry, muscles are mobilized for action. These feelings reflect changes in enzyme activities in epinephrine's target cells, including cells of the heart, salivary glands, and muscles, among others. As soon as the stimulus is gone, the hormone and its cellular effects subside. **Insulin** is an example of a hormone that exerts the second kind of effect, that of influencing membrane transport. When glucose levels in the blood rise after a meal, insulin promotes the transport of glucose across cell membranes and into the cytoplasm of certain cells. The third kind of cellular response, control of gene expression, is slower and usually only occurs in response to hormones whose effects are long-term. The sex hormones—estrogen, progesterone, and testosterone—are hormones with long-term effects. When bound to their receptors in target cells, these hormones influence maturation of the ovaries and testes; production of eggs and sperm; and other sexual characteristics such as growth of the breasts, facial hair, and changes in voice quality.

The many tissues in the body that produce and secrete hormones constitute the **endocrine system**. Figure 11-7 shows some of the more important organs and tissues that produce hormones. It is an impressive list, indeed, but one of them, the **pituitary gland**, found attached to the brain, stands out among the rest. The hormones of the pituitary gland and its partner, the **hypothalamus**, control so many vital functions that the pituitary has been called the master gland of the body, and the pituitary and hypothalamus together are referred to as the central axis.

Fast Find 11.1.h Endocrine System

The Hypothalamus and Pituitary Gland

At the base of the brain, separate from the parts in which higher thought originates, lies a tiny bit of gray matter called the hypothalamus (Figure 11-8). The hypothalamus is only a minute portion of the total brain, about one three-hundredths of its total mass, but this small region has a disproportionate share of the responsibility for keeping the body in homeostasis. The hypothalamus receives neural information from both the conscious centers of the brain (which receive information from the senses) and the organs and tissues within the body. With these various inputs, the hypothalamus gathers and integrates incoming information about the external world and the body itself. The neurons of the hypothalamus help to coordinate the body's responses to changes in water balance, temperature, and events that affect us at an emotional level, such as anger and

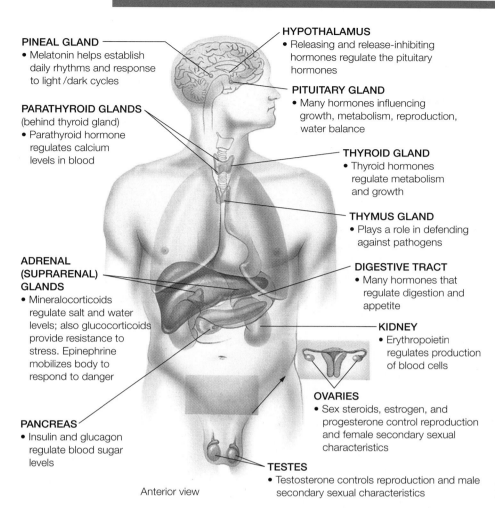

PINEAL GLAND
• Melatonin helps establish daily rhythms and response to light /dark cycles

PARATHYROID GLANDS
(behind thyroid gland)
• Parathyroid hormone regulates calcium levels in blood

ADRENAL (SUPRARENAL) GLANDS
• Mineralocorticoids regulate salt and water levels; also glucocorticoids provide resistance to stress. Epinephrine mobilizes body to respond to danger

PANCREAS
• Insulin and glucagon regulate blood sugar levels

Anterior view

HYPOTHALAMUS
• Releasing and release-inhibiting hormones regulate the pituitary hormones

PITUITARY GLAND
• Many hormones influencing growth, metabolism, reproduction, water balance

THYROID GLAND
• Thyroid hormones regulate metabolism and growth

THYMUS GLAND
• Plays a role in defending against pathogens

DIGESTIVE TRACT
• Many hormones that regulate digestion and appetite

KIDNEY
• Erythropoietin regulates production of blood cells

OVARIES
• Sex steroids, estrogen, and progesterone control reproduction and female secondary sexual characteristics

TESTES
• Testosterone controls reproduction and male secondary sexual characteristics

Figure 11-7. Some of the more important hormone-producing glands and organs and their primary hormones. Many glands produce more than one hormone.

pleasure. The hypothalamus is also involved in appetite, thirst, growth, sleeping and waking, and reproduction. "Is there any pie," asked one physiologist, "into which the hypothalamus does not dip its finger?" The answer would seem to be no.

The hypothalamus is a collection of nerve cells. It is a part of the central nervous system. But not all of the messages that originate within the hypothalamus are coded in the electrical language of the neurons. In fact, the cells of the hypothalamus make and release a number of different peptide hormones. Hormones that originate in nervous tissue, such as those of the hypothalamus, are called **neurohormones**. The neurohormones of the hypothalamus act in conjunction with the pituitary gland.

Nestled in a protective pocket of bone, the pea-sized pituitary gland hangs from the hypothalamus and is connected to it by a thin stalk (Figure 11-8). A cross section of the pituitary reveals that it has two parts: a posterior, or rear, part made up of axon endings of neurons whose cell bodies are in the hypothalamus, and an anterior, or forward, part made of more spherical cells, which are true endocrine cells. The axon endings of the posterior pituitary release two very similar neurohormones with very different effects. One of these neurohormones is called **antidiuretic hormone**, or **ADH**. Its target cells are in the kidney, where, as its name implies, ADH invokes cellular changes that act to conserve water. We will learn more about ADH in Section 11.2.4. The other neurohormone released from the posterior pituitary is called **oxytocin**, and it is involved in reproduction. Oxytocin targets cells in the uterus and breasts of pregnant or nursing females. Oxytocin is the hormone that causes contractions in the uterus that occur during labor and childbirth. When doctors induce labor artificially, they use a drug that is a synthetic form of oxytocin. Another effect of oxytocin is the release of milk into the

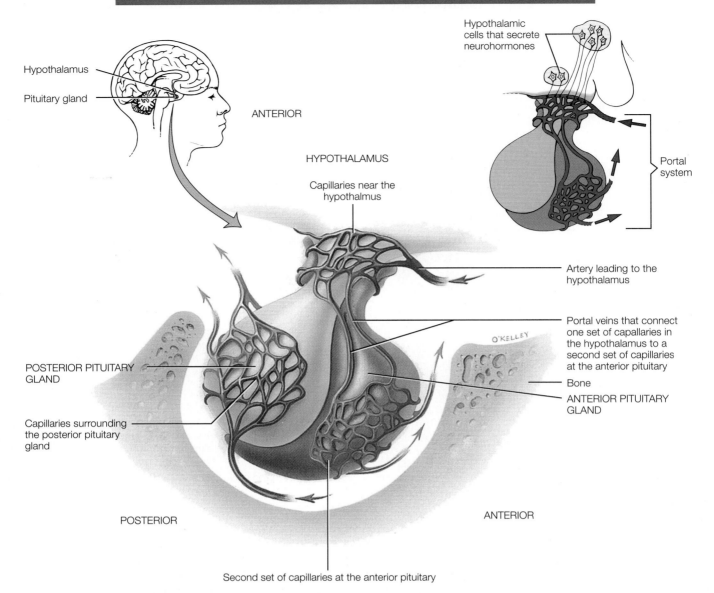

Hypothalamus

Pituitary gland

ANTERIOR

Hypothalamic cells that secrete neurohormones

Portal system

HYPOTHALAMUS

Capillaries near the hypothalmus

Artery leading to the hypothalamus

Portal veins that connect one set of capallaries in the hypothalamus to a second set of capillaries at the anterior pituitary

O'KELLEY

POSTERIOR PITUITARY GLAND

Bone

ANTERIOR PITUITARY GLAND

Capillaries surrounding the posterior pituitary gland

POSTERIOR

ANTERIOR

Second set of capillaries at the anterior pituitary

Figure 11-8. The hypothalamus and the pituitary gland. Note the two sets of capillaries connected in series, the first set at the hypothalamus and the second at the anterior pituitary (right side of figure). The first set of capillaries picks up releasing and release-inhibiting neurohormones from the hypothalamus. These capillaries converge to form a portal vein, which gives rise to the second set of capillaries at the anterior pituitary gland. This arrangement, called a portal system, allows hypothalamic hormones to reach their target cells before they become diluted by entering the general circulation. The smaller diagram at the upper right shows the neurons of the hypothalamus that synthesize neurohormones.

ducts of the breasts during nursing of an infant. Both ADH and oxytocin are synthesized in neural cells of the hypothalamus. They migrate down the axons of these cells to the release sites, the axon endings in the posterior pituitary.

A different group of hypothalamic cells controls the activity of the anterior pituitary. These cells release a different group of neurohormones, called **releasing hormones** and **release-inhibiting hormones**, from axons whose endings are located in the stalk of the pituitary. The neurohormones from the stalk enter the bloodstream but do not have to travel far to reach their target cells in the adjacent anterior pituitary. Once in the anterior pituitary, they stimulate or inhibit the release of yet other hormones from that gland.

As you can see from Figure 11-9, hormones of the anterior pituitary target many tissues and coordinate a wide variety of body functions. Many of these hormones stim-

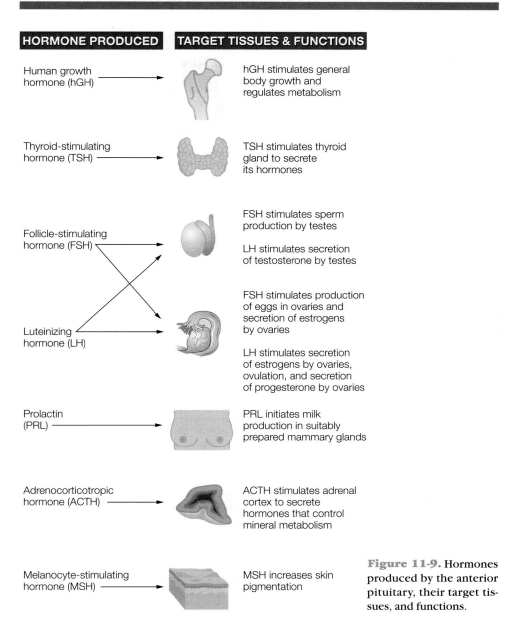

HORMONE PRODUCED **TARGET TISSUES & FUNCTIONS**

Human growth hormone (hGH) — hGH stimulates general body growth and regulates metabolism

Thyroid-stimulating hormone (TSH) — TSH stimulates thyroid gland to secrete its hormones

Follicle-stimulating hormone (FSH) — FSH stimulates sperm production by testes

LH stimulates secretion of testosterone by testes

Luteinizing hormone (LH) — FSH stimulates production of eggs in ovaries and secretion of estrogens by ovaries

LH stimulates secretion of estrogens by ovaries, ovulation, and secretion of progesterone by ovaries

Prolactin (PRL) — PRL initiates milk production in suitably prepared mammary glands

Adrenocorticotropic hormone (ACTH) — ACTH stimulates adrenal cortex to secrete hormones that control mineral metabolism

Melanocyte-stimulating hormone (MSH) — MSH increases skin pigmentation

Figure 11-9. Hormones produced by the anterior pituitary, their target tissues, and functions.

ulate a target endocrine tissue to release yet another hormone (or hormones) to carry out the final response.

Why are there so many layers of hormones controlling other hormones? A stimulus that reaches the brain causes the hypothalamus to secrete releasing neurohormones. These, in turn, stimulate the anterior pituitary to secrete stimulating hormones. Hormones from the anterior pituitary stimulate the release of yet other hormones from endocrine glands elsewhere in the body. The reason for these many layers of control has to do with overall regulation of the body's activities. It requires a great deal of fine-tuning to keep a system as complex as the human body in homeostasis. More layers of control provide more opportunities for fine-tuning. A system that depends on three hormones to bring about a single change can be regulated at three different places. In fact, most of the processes influenced by the hormones of the hypothalamus and pituitary are controlled by **negative feedback**. As the final hormone in the pathway accumulates in the bloodstream, it can "feed back" on the hypothalamus, the pituitary, or both, inhibiting the further release of the hormones that first stimulated it. Without negative feedback, a response designed to bring some aspect of the body back to homeostasis could easily overcompensate, pushing the system too far in the opposite direction. This illustrates one of the characteristics of the endocrine system: it is highly regulated and finely tuned, usually by negative feedback loops.

Blood Sugar Not all of the endocrine functions of the body are under the control of the hypothalamus and pituitary. For example, levels of glucose in the blood are tightly regulated by two hormones produced in the pancreas: insulin and glucagon. What stimulates the release of these hormones? What are their target cells? Diabetes mellitus is actually a group of diseases that result in imbalances in blood glucose. What kinds of malfunctions characterize the different forms of diabetes?

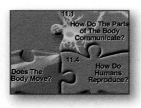

Piecing It Together

The various systems of the human body can be thought of as homeostatic control systems, working together to maintain an internal environment that changes very little and provides an optimal medium for the body's cells. In this first part of the chapter, we examined the body's major avenues of communication: the nervous system and the endocrine system.

1. Neurons are the functional cells of the nervous system. They rapidly transmit information in the form of electrical impulses called action potentials. Action potentials are transient, all-or-nothing reversals in the electrical potential across neuron cell membranes that move very rapidly from dendrites, to cell body, to axon.

2. The junction between a neuron and an adjacent nerve, muscle, or gland cell is called a synapse. At synapses, axon endings release neurotransmitters, or chemicals that diffuse across a small gap and make it more or less likely that the adjacent cell will undergo an action potential. Because a cell averages the sum of many inputs, synapses are responsible for integrating incoming information. Neural integration accounts for the remarkable sophistication of the human nervous system.

3. The human nervous system is divided into the central nervous system (brain and spinal cord) and the peripheral nervous system. The peripheral nervous system is further divided into the somatic, or voluntary, nervous system and the autonomic, or involuntary, nervous system. The autonomic nervous system is subdivided into the sympathetic and parasympathetic nervous systems. The sympathetic nervous system dominates in urgent "fight or flight" situations. The parasympathetic nervous system controls the body's normal everyday "housekeeping" chores.

4. The endocrine system is a second system of communication that works by means of hormones, or chemicals that travel in the bloodstream. The endocrine system coordinates slower, more long-term responses, such as growth, maturation, reproduction, and water balance.

5. Though hormones reach all cells of the body, they affect only target cells, or cells with receptors that can recognize and bind to them. Target cells can respond in one of three ways: by changes in cellular metabolism, in membrane transport properties, or in the expression of certain genes.

6. Many hormones of the body are influenced by hormones of the hypothalamus and pituitary gland, which are found in the brain. Through their hormones, these structures connect the integrating centers of the brain with many of the other endocrine

tissues of the body. This multilayered process allows the levels of hormones to be tightly regulated, mostly by negative feedback.

11.2 HOW DO HUMANS ACQUIRE AND PROCESS NUTRIENTS AND GET RID OF WASTES?

If we were single-celled organisms, we could acquire the things we need—food, water, oxygen—by simple diffusion across our cell membranes. But we are large, multicellular animals, and while these substances are generally abundant in our environment, we require methods better than simple diffusion for getting them into our bodies. Once we have taken in these substances, we also need specialized anatomical structures for distributing them to all of our trillions of cells, and for ridding ourselves of the wastes. These are the topics we now examine—the acquisition of nutrients, their distribution, and the disposal of wastes. We begin with the digestive system.

11.2.1 Food Enters the Body through the Digestive System

On the morning of June 6, 1822, a young French Canadian named Alexis St. Martin became the victim of a terrible accident. Alexis was working as a boatman at a fur-trading post on Lake Michigan's Mackinac Island when someone accidentally discharged a musket into his stomach. The powder and buckshot, fired from a distance of less than 3 feet, tore into Alexis's stomach, opening a gash of several inches. Within minutes, an army surgeon named William Beaumont arrived at the scene. Beaumont could do little more than clean the wound, cover it with a dressing, and wait for Alexis to die.

But two years later, Alexis St. Martin was still alive. The accident had left him with a 2-inch hole, or fistula, in the wall of his stomach. As long as the fistula was kept properly bandaged, he could eat and drink normally. Without the bandage, food and drink poured from the fistula. William Beaumont immediately saw remarkable possibilities in his patient with a window into his stomach. In May 1825, Beaumont began a series of experiments on the unwilling Alexis that would teach us much about the workings of the stomach and digestive tract. Beaumont made careful observations of the workings of Alexis's stomach, examining its movements and tasting the mucous and acid that the stomach produced. He collected stomach fluids through the fistula and, in controlled experiments, proved that these fluids could digest protein-rich foods like meat. He observed the movements that Alexis's stomach made when he was hungry and when he was well-fed. His results, published in 1833 in a book titled *Experiments and Observations on Gastric Juice and the Physiology of Digestion*, form the foundation of much of our current understanding of digestion. Alexis St. Martin survived to the age of 83, fathered four children, and long outlived his doctor.

The Basic Processes of Digestion

**Fast Find
11.2.b
Human
Digestion**

The stomach that Dr. Beaumont came to understand so well is just one part of the **digestive tract**, a long hollow tube, beginning at the mouth and ending at the anus (Figure 11-10). The function of the digestive tract is to take in food, temporarily store it, and prepare it to enter the inside compartment of the body. The hollow tube itself, like the hole in a donut, is not actually inside the body. Usable materials are absorbed while the unused portions exit the tube through the anus. In our study of the digestive tract, we will follow the path of food as it makes its way through the tube. What, in biological terms, is food? To a physiologist, food is a complex of organic macromolecules—mostly carbohydrates, proteins, and lipids—that the digestive system breaks down and the cells use for both raw materials and energy. The foods we eat also contain water, salts, vitamins, and other trace elements, but only the macromolecules must be altered before they can be absorbed into the body. Breaking down and taking in food is accomplished by four processes occurring in the digestive system.

- ◆ **Digestion,** or dissolving food and breaking it down into smaller molecules by means of digestive enzymes.
- ◆ **Secretion,** or the release of digestive enzymes, acids, and other substances that aid in digestion.
- ◆ **Motility,** or the movement of food and digestive secretions through the digestive tract.
- ◆ **Absorption,** or the entry of the simple molecules produced by digestion across the wall of the digestive tract, into the blood, and finally to the cells of the body.

As we wind our way through the human digestive tract, we will see what happens with respect to these four processes. We begin in the mouth.

Mouth and Esophagus

By the time we bite into a morsel of food, some of the processes of digestion have already begun. Our mouth waters and our stomach may growl as we anticipate the entrance of food into the system. Saliva is a secretion from a set of **salivary glands** located near the mouth (Figure 11-10). Saliva contains water and mucous, which dissolves and

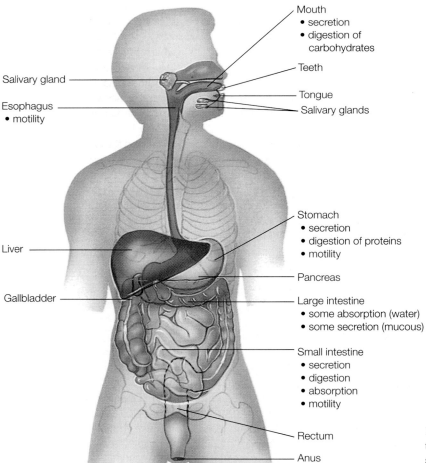

Mouth
- secretion
- digestion of carbohydrates

Teeth

Tongue

Salivary glands

Salivary gland

Esophagus
- motility

Stomach
- secretion
- digestion of proteins
- motility

Pancreas

Liver

Large intestine
- some absorption (water)
- some secretion (mucous)

Gallbladder

Small intestine
- secretion
- digestion
- absorption
- motility

Rectum

Anus

Figure 11-10. Structures of the digestive tract and its accessory glands, followed by their functions.

lubricates the food, and enzymes that begin the process of chemically dismantling carbohydrates, one of the four types of macromolecules. Another function of the mouth is chewing. Chewing is a mechanical process that breaks food into smaller pieces that are not only easier to swallow, but are also more readily dissolved and therefore more accessible to digestive enzymes. Digestive enzymes can only digest food molecules that are dissolved.

From the mouth, food enters the **esophagus**, a tube leading from the mouth to the stomach. Although the esophagus does not contribute to digestion directly, the muscles in and around it control swallowing. Gravity helps to move the food somewhat, but in fact, it is possible to move food to the stomach without the help of gravity. We could, if we wanted, swallow a mouthful of food while standing on our heads.

Stomach

The stomach sits like a big boxing glove (Figure 11-11) in the space beneath the ribs on the left-hand side of the body. The intestine leaving at the bottom forms the arm or wrist of the glove; attached to the curved upper part of the glove is the esophagus. The stomach has several functions, few of which, it turns out, are absolutely essential, but all of which make life easier. First, the stomach stores food for a brief time and moves it to the intestine in small amounts after it is liquefied. In this way, the intestine—the portion of the tube in which most of the food is broken down and absorbed—gets enough food to provide a constant flow of nutrients to the body even though it may have been hours since the last meal. While storing food between meals is not essential for survival, it does free us to do other things besides eat.

A second role of the stomach is digestion of proteins. Secretions of the stomach, called gastric secretions ("gastric" is from the Greek word for *stomach*), include enzymes that dismantle proteins, as well as a powerful acid, called hydrochloric acid, that enables these enzymes to work. Salivary enzymes that have begun dismantling carbohydrates in the mouth are inactivated in the stomach because the salivary enzymes cannot work in such highly acidic conditions. Acidic, enzyme-containing gastric secretions are powerful enough to digest even the stomach wall. They are prevented from doing so, however, by a different kind of gastric secretion, a thick layer of alkaline mucous that coats the stomach wall. The mucous neutralizes stomach enzymes and acid before it comes into contact with the stomach wall.

The acidity of the stomach has another role as well. It acts as a first line of defense against foreign invaders. Bacteria inevitably enter the digestive tract along with the food

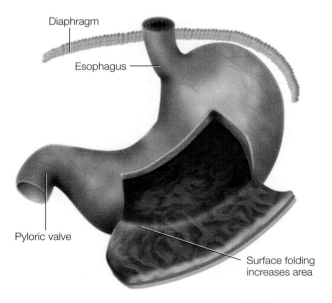

Diaphragm

Esophagus

Pyloric valve

Surface folding increases area

Figure 11-11. The stomach with a portion removed to show features of the inner wall.

we eat, but most of them cannot survive the hostile environment of the stomach. Some, however, do escape to the intestine to take up residence and multiply there, particularly in the terminal part, or the large intestine. Some bacteria are even helpful to us there, but that is only because the digestive tract is technically not inside the body. These same bacteria can be deadly if they enter the body's internal environment.

By the time food is ready to leave the stomach, it is in a liquid form that has little resemblance to what was eaten. For this reason, we give it a different name; we call it **chyme**. Chyme contains fragments of proteins and carbohydrates; droplets of fat; saliva; and the gastric secretions including enzymes, acid, and mucous. None of these things can cross the wall of the stomach, so little absorption takes place here. A few substances, including water and ethanol, can enter the body through the stomach wall, which explains why people feel the effects of drinking rapidly, especially on an empty stomach.

Chyme is pushed from the stomach into the next region of the digestive tract, the **small intestine**, by waves of muscular contraction in the stomach wall called **peristalsis**. These waves move down the length of the stomach like a ring around a flaccid balloon, pushing against the chyme inside and moving it downward (Figure 11-12a). At the juncture between the stomach and the small intestine, a small ring of muscle guards the entrance to the small intestine. This ring, called the **pyloric valve**, opens to allow limited amounts of chyme to enter the intestine, where most of the digestion and absorption of nutrients takes place. Chyme that does not enter the intestine is pushed back up into the stomach, where it is more thoroughly mixed. Stomach growls that you can hear and feel when you are hungry are peristaltic waves in the absence of food.

The Intestines

Named for the Latin word meaning *internal*, the **intestines** are the last part of the digestive tube and carry chyme from the stomach to the anus, changing it profoundly along the way. The intestines are about 14 feet long, but the many folds, twists, and convolutions create a surface area that is far greater than a smooth tube of equal length would have. The surface area of the inner wall of the intestinal tube is over 100 square feet, five times greater than the surface area of the skin—a clue that something very important is occurring on and across the surface of the intestine. That something is **absorption**, the process by which nutrients are taken into the body.

In the small intestine, the digestive process enters its final stages where all three of the major kinds of food molecules—proteins, carbohydrates, and lipids—are dismantled and made ready for transport into the body. Partially digested carbohydrates and proteins are degraded to sugars and amino acids, and fats are broken down to fatty acids and an alcohol called glycerol. These processes are accomplished by enzymes on the inner surface of the intestinal wall and by secretions from two important glands, the **pancreas** and the **liver**.

Secretions from the pancreas empty into the small intestine very close to the pyloric valve. These pancreatic juices contain several enzymes that break down proteins, carbohydrates, lipids, and even the tiny amount of nucleic acid that is found in food. In addition, a chemical called bicarbonate is dissolved in the pancreatic juice. You may be familiar with the solid form of bicarbonate, called baking soda or bicarbonate of soda. A teaspoon or so of baking soda dissolved in water is often used to relieve stomach upset. It works by neutralizing the acid in the stomach. Bicarbonate from the pancreas plays a similar role. It neutralizes the acidity of chyme as it enters the small intestine. This prevents stomach acids from damaging the intestine and creates a more hospitable environment for the pancreatic digestive enzymes to work.

The liver, too, secretes a fluid that aids in digestion. This fluid, called **bile**, is a mixture of bicarbonate and **bile salts** that aid in lipid digestion. One of the problems associated with lipid digestion is that fats do not readily mix with the watery chyme. Instead they tend to form globules that are not accessible to the pancreatic digestive enzymes that break them down. Bile salts act as **emulsifiers**, breaking up large fat globules into smaller droplets so that lipid-digesting enzymes have a larger surface area on which to work. Between meals, bile is still secreted by the liver, but it is not dumped into the small intestine. Instead, it is stored in a small sac under the liver called the **gallbladder**. When chyme enters the intestine after a meal, the gallbladder contracts, injecting bile into the small intestine. Both bile and pancreatic secretions enter through a thin tube, called the **bile duct**.

The slurry of chyme that enters the small intestine is jostled about and moved along the tube at a rate of about one inch each minute. Muscles in the wall of the digestive tract are responsible for moving chyme from end to end. Recall that motility in the stomach occurs as peristaltic waves, or a ring of contraction that moves from one end of the sac to the other. Short spurts of peristalsis may occur in the intestine, but most of the movement along the small intestine occurs by a different mechanism, called **segmentation**. In segmentation, small regions or segments contract and relax in a rhythmic pattern (Figure 11-12*b*). The contents are repeatedly divided and subdivided, and thoroughly mixed with the various secretions from the pancreas and liver. Segmentation also ensures that the chyme repeatedly comes in contact with the cells that line the wall of the small intestine, where various transport proteins move digested materials through cell membranes into the body. The entry of chyme from the stomach and its exit into the large intestine push the contents of the small intestine forward. Also, the occasional, brief waves of peristalsis and the greater frequency of segmental contractions at the stomach end help to move the chyme along.

The walls of the small intestine secrete enzymes and other substances. Mucous, acting as a lubricant, and large amounts of watery fluid containing ions such as sodium, chloride, and bicarbonate mix with the chyme. However, far more absorption takes place than secretion, and the net movement of substances across the wall of the small

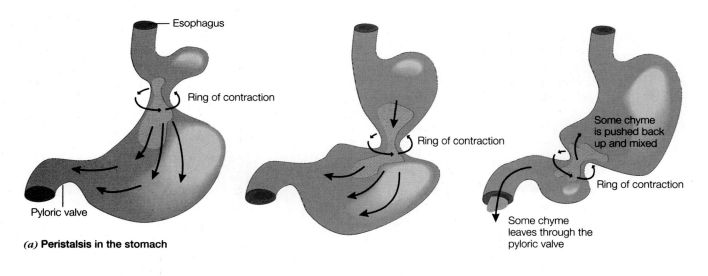

(a) **Peristalsis in the stomach**

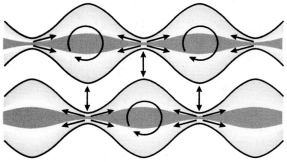

(b) **Segmentation in the small intestine**

Figure 11-12. A comparison of peristalsis and segmentation.

(a) In peristalsis, waves of contraction move in one direction, pushing the chyme toward the pyloric valve.

(b) In segmentation, alternate segments of the small intestine contract and relax, mixing the contents and propelling the chyme in both directions. Chyme moves forward because contractions at the stomach end occur more frequently than those at the end near the large intestine, and because chyme entering from the stomach pushes it forward.

intestine is inward, or into the body. Nearly all of the absorption of food takes place across the wall of the small intestine. We have already seen how its various folds and convolutions create a tube with a disproportionately large internal surface area. Now let's examine some of the finer features of the intestinal wall that are responsible for intestinal absorption.

The series of diagrams in Figure 11-13 illustrates the wall of the intestine in increasingly fine detail. The wall of the tube is characterized by ridges and valleys. On the surfaces of these ridges and valleys are multicellular, fingerlike projections called **villi** (singular, *villus*) that extend into the intestinal space. Each villus, in turn, is covered with a layer of epithelial cells that form the first barrier between the contents of the tube and the inside of the body. The ends of the cells that face the inside of the tube are further convoluted into tiny, hairlike projections called **microvilli**. Because of their brushlike appearance, this layer of microvilli is called the **brush border**. At every level, the anatomy of the intestinal wall is adapted for increasing the surface area across which nutrients are absorbed. To see how nutrients cross into the body, however, we must look at the chemistry of the brush border.

The different products of digestion—amino acids from proteins, simple sugars from carbohydrates, and fatty acids from larger lipids—are moved into the cells of the villi, each in their own way. For sugars and amino acids, the two water-soluble products of digestion, the key to crossing the brush border lies in specific membrane transport proteins (introduced in Chapter 4), or large proteins embedded in brush border membranes that pump these nutrients into the cells. Once inside the cells, different membrane transport proteins on the membrane opposite the microvilli move sugars and amino acids into the extracellular fluid. As shown in Figure 11-13c, each villus contains **capillaries**,

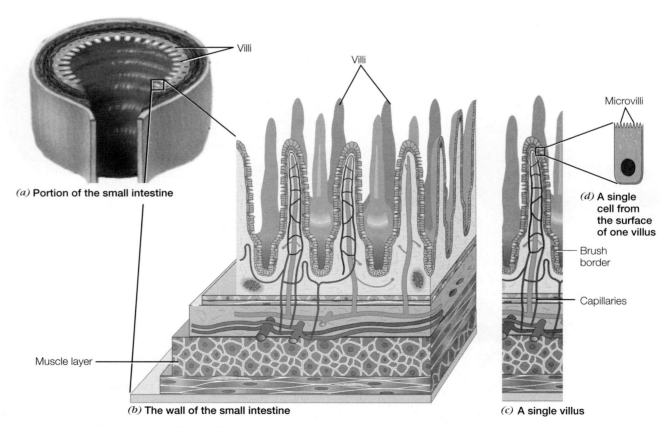

(a) Portion of the small intestine

Villi

Muscle layer

(b) The wall of the small intestine

Villi

(c) A single villus

Microvilli

(d) A single cell from the surface of one villus

Brush border

Capillaries

Figure 11-13. The small intestine, showing *(a)* a cut region of the intestinal tube, *(b)* a portion of the wall of the small intestine with villi, *(c)* a single villus, and *(d)* one cell of the brush border with microvilli.

or tiny blood vessels that absorb nutrients from the extracellular fluid into the bloodstream. Sugars and amino acids diffuse from the extracellular fluid into the capillaries and are carried to the various cells of the body by the circulation of the blood.

Unlike amino acids and sugars, fatty acids are lipid soluble and cross the brush border cell membranes without the help of protein transporters. Inside the cells, fatty acids are reassembled into a larger lipid, called triacylglycerol, and combined with cholesterol and proteins into droplets called **chylomicrons**. Because of their relatively large size, chylomicrons must leave the cells and enter the extracellular fluid by exocytosis. When they ultimately enter the circulatory system, chylomicrons relinquish their triacylglycerols to cells that need fat for energy or to build new membranes. The triacylglycerols that remain in the bloodstream are taken up by fat cells, which store them as fat.

As you can see, the composition of the chyme is constantly changing as it moves along the small intestine. By the time the chyme reaches the end of the small intestine, all that remains is indigestible material, water, bacteria, and the few other assorted things that could not be absorbed. This is the form in which chyme enters the **large intestine**, the terminal 5 feet of the intestine. The primary role of the large intestine is to store fecal matter, absorb water from it, and propel it into the rectum. When the rectum becomes distended by an accumulation of feces, the fecal matter is eliminated by a bowel movement, a process called **defecation**.

Thus we have followed the path of food from mouth to anus and noted the processes that enable us to deliver fuel to the cells of the body. Food, however, is not the only nutrient we need to stay alive. We also need oxygen. In Chapter 9 we learned that oxygen is essential for the energy-producing pathways of cellular metabolism. Here we will see how oxygen enters our bodies by way of the lungs.

 Allergic to Food Recent studies have shown that some small proteins, called peptides, can enter the body through the small intestine. Once in the blood, these peptides are recognized by the body's immune system as foreign, that is, not part of the body's normal complement of proteins, and the body may react with an allergic response. This may account for some of the food allergies suffered by millions of Americans. What are some common kinds of proteins that can cause severe food allergies? Why are infants more susceptible to developing food allergies than adults? What recommendations have health officials made about feeding infants to prevent food allergies in later life?

11.2.2 Oxygen and Carbon Dioxide Are Exchanged in the Lungs

One of the problems associated with the relatively large size of the human body has to do with the amount of surface area that is available for uptake of nutrients. As a body gets larger, the bulk of that body, its volume, increases more than its surface area. A very small organism can provide oxygen to its entire body by absorbing it directly across its surface, but for a larger animal, the surface of the body is not large enough to provide the amount of oxygen needed. In other words, the surface-to-volume ratio of the human is too small.

Another problem associated with getting oxygen is water conservation. As land animals, our natural habitat is dry compared with the wet internal environment that our cells need to survive. Oxygen, however, must enter the body dissolved in water. Hence, the surfaces across which oxygen diffuses must be wet. The human **respiratory system**, that is, the lungs and passages leading to them, is an exquisite adaptation that provides a large, wet surface for exchanging gases while at the same time minimizing water and heat loss to the environment.

The human respiratory system has several functions. The most obvious is the exchange of the gases oxygen and carbon dioxide. Oxygen enters the body through the lungs and diffuses into the blood; it is distributed to all the body's cells, where it is used in cellular respiration. Carbon dioxide, a waste product of cellular metabolism, is carried away from the tissues by the bloodstream to the lungs where it is eliminated from the body. Less obvious is the role that the lungs play in regulating the acidity, or pH, of the body. Carbon dioxide dissolved in the blood helps to balance blood pH, and by regulating the rate at which CO_2 is eliminated, the human respiratory system closely regulates the pH of the blood. The respiratory system is also used for vocalization. Air moving across the vocal cords creates the vibrations used for speech. Finally, the respiratory system protects the body from inhaled irritants and pathogens. Several adaptations ensure that potentially harmful substances are removed or destroyed before they enter the body. For our discussion, we will focus on the first role of the lung, that of gas exchange between the atmosphere and the blood.

Structure of the Respiratory System

The respiratory system brings oxygen-rich air into contact with a deeply internalized, moist surface. If you knew nothing about how humans breathe, you could guess that any such system would need three parts: a pump for creating airflow; a large wet surface for gas exchange; and a series of airways, or pipes, leading from the moist surface to the outside. These are the parts of the respiratory system: the **diaphragm** and other muscles of the **thorax** (chest) act as a pump; the **lungs** constitute the exchange surface; and the **conducting system**, including the nose, mouth, **trachea**, and **bronchi**, are the airways. The anatomical relationship among these structures is shown in Figure 11-14*a*. We shall examine each of these individually.

The lungs are a pair of cone-shaped, spongy organs that occupy most of the space in the thorax. They consist of delicate, pink tissue containing about 300 million tiny air-filled sacs, called **alveoli** (singular, *alveolus*). The inside surfaces of the alveoli are coated with a mixture of water and a detergentlike substance called **surfactant**. The water acts as a solvent in which oxygen and carbon dioxide dissolve as they move into and out of the body. The surfactant keeps the water in a thin sheet on the inner alveolar surface, preventing it from forming beads or droplets. Surfactant eases the expansion of the lung when air enters during breathing.

The outer surface of the alveolar sacs are surrounded by capillaries. **Capillaries**, whose walls are only one cell layer thick, are the only vessels of the circulatory system through which materials can be exchanged between the blood and the tissues. We will discuss capillaries in more detail when we study the circulatory system in the next section. The air in the alveoli and the blood in the capillaries are very close together, separated by one layer of alveolar cells, one layer of capillary cells, and two extremely thin basement membranes through which oxygen and carbon dioxide can readily diffuse. Taking into account all of the alveoli in the human lungs, the surface area for gas exchange is about 40 times greater than the surface area of the skin. The total surface is nearly the size of a tennis court, almost 1,000 square feet. Gas exchange occurs only in the alveoli and not in the tubes of the conducting system or any other parts of the respiratory system.

The conducting tubes of the respiratory system are arranged somewhat like a hollow, inverted tree (Figure 11-14*a*). The main passage, the trunk of the tree, is the **trachea**. The trachea carries air from the mouth and nose into two large branches called **bronchi**, each leading to one of the lungs. In the lungs, the bronchi branch many more times into ever shorter and thinner tubes. They end in clusters of alveoli extending from terminal bronchioles of the conducting system, like clusters of grapes on a hollow vine (Figure 11-14*b*). The internal surfaces of the airways are lined with cilia and coated with mucous; the mucous traps foreign particles, which are moved to the mouth, where they can be eliminated.

The third component of the respiratory system is the pump for moving air between the alveoli and the environment. The lungs lack muscle and are incapable of acting as a pump. Instead, the main muscle for moving air is the **diaphragm**, a sheetlike muscle that forms the floor of the thorax. Muscles of the chest and back also contribute to the

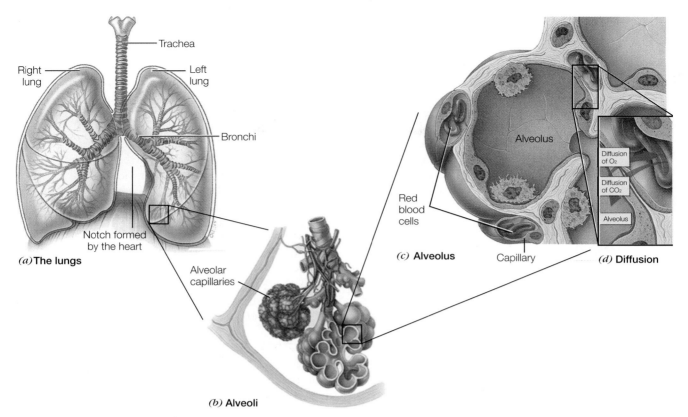

Figure 11-14. The respiratory system, showing *(a)* both lungs with air passageways and flattened diaphragm, *(b)* several clumps of alveoli with some of their capillaries, *(c)* a single alveolus and its relationship to capillaries, and *(d)* diffusion of O_2 and CO_2 in an alveolus.

pumping action of the lungs, especially when breathing is deep, as occurs during exercise or at high altitude where the air is thin. The muscles that act as the pump are best described in the context of how they work.

Ventilation

Ventilation is the movement of air between the alveoli and the environment. The rhythmic act of breathing creates the currents of ventilation. When it is relaxed, the diaphragm is shaped like a dome pushing up into the thoracic cavity. Signals from neurons originating in the brain cause the diaphragm to contract. As it contracts, the diaphragm flattens, pulling down about an inch to form a flat surface. This causes the lungs to expand, creating suction that draws air into the mouth and nose, through the conducting system, and ultimately into the alveoli. The signals from the brain causing the diaphragm to contract come in rhythmic spurts; when the diaphragm is fully contracted, the signals temporarily cease. Without the signals from the brain, the diaphragm relaxes back into its dome shape, pushing upward on the lungs as it does so and forcing air back out through the mouth and nose. As the diaphragm relaxes, the brain signals commence again. In a normal, quiet breath, about 500 milliliters (about 2 cups) of air are exchanged, but this amount can be greatly increased when the need to take in oxygen or to rid the body of carbon dioxide is urgent. The most air that can be moved in a single breath is about 4.8 liters, or over 5 quarts.

Gas Exchange in the Lungs

All parts of the respiratory system—the lungs, airways, and muscles—act to bring air in close proximity to the blood. When air and blood are close enough, gases dissolve in the water that coats the alveoli and move easily between the alveoli and the capillaries by simple diffusion (Figure 11-14*d*). Recall from our discussion of diffusion in Chapter 4 that substances diffuse from regions of high concentration to regions of low concentration, that is, down their concentration gradients. The concentration of oxygen in air is about two and a half times greater than it is in the blood entering the lung capillaries. Although oxygen must move through two layers of cells (the alveolar cell layer and the capillary cell layer), two thin membranes, and also through the extra-cellular fluid, it enters the bloodstream rapidly because the concentration gradient is large, the distance it moves is small, and oxygen easily crosses cell membranes. For the same reasons, carbon dioxide also moves rapidly from the blood into the alveolar spaces. It takes less than one second for the oxygen and carbon dioxide levels in blood and alveoli to reach equilibrium. The blood leaving the lungs is relatively low in carbon dioxide and enriched in oxygen.

Breathe Easy, Breathe Hard Because gas exchange at the lungs occurs by diffusion, any factor that interferes with the rate of diffusion diminishes the ability to exchanges gases. The diffusion rate is influenced by the (1) surface area, (2) thickness of the alveolar membrane, and (3) distance across which gases must move. Name some common respiratory diseases. Describe how these illnesses negatively affect the rate of diffusion of gases across the respiratory surfaces. How are these illnesses treated?

From the lungs, oxygen-rich blood is carried to all the cells of the body. At the cells, the concentration gradients are opposite those in the lungs, so that O_2 diffuses out of the blood and into the cells, and CO_2 waste diffuses from the cells into the blood, which carries it back to the lungs. The same metabolic pathways that consume oxygen also produce the carbon dioxide waste that must be carried away. The task of transporting gases between the tissues and the lungs, and many other tasks, are accomplished by the body's cardiovascular system.

11.2.3 The Cardiovascular System Moves Materials throughout the Body

The **cardiovascular system**, or circulatory system, is made up of the heart and all of the blood vessels through which blood flows. The heart is the pump that moves blood throughout the vessels. The blood vessels include the **arteries**, which carry blood from the heart to the tissues; the **veins**, which carry blood back to the heart from the tissues; and the **capillaries**, the tiny vessels that connect the arteries and veins.

The primary role of the blood is to transport substances to and from all parts of the body. For example, oxygen is picked up at the lungs, and nutrients are picked up at the intestine; these substances are delivered to cells throughout the body. Simultaneously, cellular wastes such as carbon dioxide and other wastes are picked up in the tissues and delivered to the lungs or kidneys to be removed from the body. Some materials are moved from place to place within the body. Glucose, for example, is delivered from the liver to cells that are metabolically active. Fatty acids are transported from the adipose (fat) tissue to cells that need energy-rich fuel. The blood also carries hormones and other cellular messengers to their various target cells. In addition, certain elements of the blood play an important role in defending the body against disease-causing bacteria and viruses, as we will see in Section 11.5.

While the heart's role in pumping blood was known even in ancient times, until the 17th century it was generally believed that the liver provided a constant supply of new blood that was pumped to the tissues by the heart through veins. The tissues were thought to be a dead end, consuming all of the blood that they received. In 1628, how-ever, an English physician named William Harvey published a small, 72-page book in which he described the results of experiments that overturned these ancient beliefs. Using a sheep's heart, Harvey measured how much blood the heart can hold. He then calculated the total amount of blood that was pumped each minute. A sheep's heart, he found, pumped enough blood in a half-hour to equal the entire weight of the animal! The liver could not possibly generate so great a volume each half-hour; instead, he concluded, the blood that leaves the heart must eventually return to it. When Harvey examined the structure of the heart, he realized that all of its intricate valves and chambers were situated so that blood enters the heart at one place and leaves at another. "The blood in the animal body moves around in a circle continuously" wrote Harvey, "and the action or function of the heart is to accomplish this by pumping." *Why* blood moves in a circle was not entirely clear to him, however; "whether for the sake of nourishment or for the communication of heat, is not certain." In fact, circulating blood does both of these things, and more.

As with so many revolutionary ideas in science, Harvey's conclusion was met with vehement disagreement. It was not until later that century that the microscopic vessels, the capillaries, that link the arteries to the veins were observed. But Harvey's many experiments and careful calculations were successfully repeated by others, and eventually the truth of his assertion became evident. Ultimately Harvey's work became the foundation of modern cardiovascular physiology. We will begin with the heart and follow the flow of blood as it moves through the various parts of the human cardiovascular system.

The Heart

For thousands of years, the heart was believed to be the center of our intelligence and our emotions. This ancient belief is reflected in our language: we love with all of our heart, believe in our heart, and speak from the bottom of our heart. In fact, our feelings and emotions are much more closely tied to the hypothalamus than they are to the heart. The real function of the heart, however, is no less remarkable. This small pump, about the size of a fist and weighing less than a pound, moves the entire blood content of the body—nearly 5 liters—every minute of every day, every year that you live.

The heart is a muscular organ that lies at an angle slightly to the left of center in the thorax. It is sandwiched between the two lungs, with its tip resting on the diaphragm. The heart contracts and relaxes about 70 times each minute. These highly coordinated contractions pump the blood out of the heart and into the arteries.

**Fast Find
11.2.d
Human
Heart**

Though we refer to the heart as a pump, it is, in fact, two separate pumps. It is divided along its long axis into a right pump and a left pump. Each of the two pumps is made up of two chambers, a small upper chamber called the **atrium** and a larger, lower chamber called the **ventricle**. Thus a human heart has four chambers in all, two atria and two ventricles (Figure 11-15). Let's follow the path that blood takes through the heart.

Oxygen-poor blood enters the right atrium through the **vena cava**, a large vein that carries blood to the heart from the peripheral tissues of the body. From the right atrium, the blood moves into the right ventricle through a one-way valve that separates the right atrium from the right ventricle. This valve is called the **tricuspid valve**. The right ventricle pumps the blood into a set of **pulmonary arteries** that carry it to the lungs. After oxygen and carbon dioxide are exchanged in the lungs, the blood returns to the heart in **pulmonary veins**, making one complete circuit. This circuit is called the **pulmonary circulation** ("pulmonary" means *lung*). It is just one of the two complete circuits that the blood makes in the body.

The second circuit involves the left side of the heart. Blood returning to the heart by means of the pulmonary veins enters the left atrium. From the left atrium, blood moves through another one-way valve into the left ventricle. This valve is called the **bicuspid valve**. The left ventricle pumps blood into the **aorta**, a large artery that feeds other, smaller arteries. Arteries carry blood throughout the body, where it exchanges nutrients and wastes with the cells. Veins carrying blood from the tissues converge into the vena cava, and the blood is returned to the right atrium of the heart. This circuit is called the **systemic circulation**. Blood in the pulmonary circulation does not mix with blood in the systemic circulation. The two are kept separate so that oxygen-rich blood from the lungs does not mix with oxygen-poor blood from the other tissues.

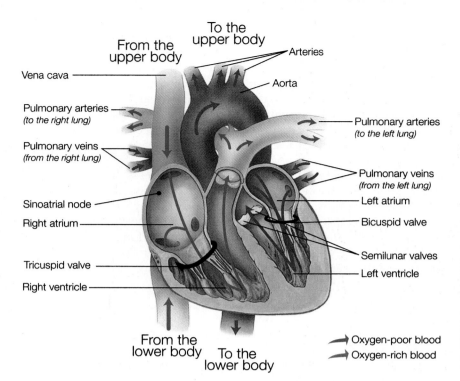

From the upper body

To the upper body

Arteries

Aorta

Vena cava

Pulmonary arteries
(to the right lung)

Pulmonary arteries
(to the left lung)

Pulmonary veins
(from the right lung)

Pulmonary veins
(from the left lung)

Sinoatrial node

Left atrium

Right atrium

Bicuspid valve

Semilunar valves

Tricuspid valve

Left ventricle

Right ventricle

From the lower body

To the lower body

Oxygen-poor blood
Oxygen-rich blood

Figure 11-15. The flow of blood through the heart. Oxygen-poor blood enters the right atrium from the vena cava, then moves into the right ventricle, which pumps it out to the lungs through the pulmonary arteries. Oxygen-rich blood returning from the lungs in pulmonary veins enters the left atrium and moves into the left ventricle. The left ventricle pumps the blood into the aorta, which carries it into the systemic circulation.

The Heartbeat

The pumping action of the heart is accomplished by the **heartbeat**, the rhythmic contraction, or shortening, of muscle cells that make up the walls of the blood-filled chambers. There are two components to the heartbeat: an electrical component and a mechanical component. The electrical component is a widespread action potential that moves through the heart. The mechanical component of the heartbeat is the contraction of heart muscle cells. First we'll take a look at the electrical component, the action potential in the heart.

Muscle cells, like neurons, are excitable cells. Recall from our discussion of neurons that excitable cells can conduct action potentials, or rapid and transient changes in the distribution of electrical charges across their membranes. Unlike neurons, however, certain cells of the heart, called **pacemaker cells**, undergo repeated, rhythmic action potentials in the *absence* of an outside stimulus. The muscle cells of the heart are connected in such a way that an action potential in one cell quickly spreads to the others. The main pacemaker cells of the heart are located in a region called the **sinoatrial node**, a clump of cells which is in the upper portion of the right atrium (Figure 11-15). When these cells fire (that is, undergo an action potential), all the cells of both atria follow suit, firing simultaneously. The action potential in the atrial cells causes them to contract, squeezing the blood in the atria and forcing it into the ventricles.

Meanwhile, a bundle of specialized conducting cells carries the action potential from the atria into the ventricles after a brief delay. The cells of both ventricles simultaneously fire action potentials about one-tenth of a second after those of the atria. As a result, the ventricles contract, pushing on the blood and forcing it out through a second set of one-way **semilunar valves** into the pulmonary arteries (from the right ventricle) and into the aorta (from the left ventricle). When the ventricles contract, the blood pushing on the tricuspid and bicuspid valves force these valves to close. This prevents the blood from flowing back into the atria. Likewise, when the ventricles relax, the semilunar valves between the ventricles and the arteries close, preventing back flow into the ventricles.

Because the heartbeat originates in the pacemaker cells of the heart itself, a heart that has been removed from the body may, under the right conditions, continue to beat for several hours. The heart does, however, receive incoming signals both from hormones and the nervous system. This input can modify the heart rate in response to certain outside stimuli. Fear and excitement, for example, cause the heart to beat faster. Calming influences can slow the heart rate.

The Human Battery These electrical properties of the heart can be followed using an instrument called an electrocardiograph. An electrocardiograph produces a record called an **electrocardiogram**, or ECG. How does the electrocardiograph work? What kinds of information can be learned from an ECG? How has the ECG been used as a diagnostic tool in cardiovascular medicine?

The Vessels of the Circulatory System

**Fast Find
11.2.e
Human
Circulatory
System**

In both the pulmonary and systemic circuits, blood that leaves the heart travels through several kinds of vessels before it makes its way back to the heart. The first of these vessels are the arteries.

Arteries are thick-walled, elastic vessels that carry blood from the heart to the tissues. With each beat of the heart, the arteries bulge under the surge of pressure caused by blood entering from the contracting ventricle. Their elasticity causes the arteries to recoil, forcing blood onward. The result is that, in the arteries, the pulsations of the heart are dampened, but the pressure does not drop. This pressure is necessary to force blood through the rest of the circulatory system.

Arteries branch into smaller-diameter vessels called **arterioles**. Arterioles regulate the amount of blood that flows to any one tissue or organ, increasing the flow to active tissues with high nutrient demands and decreasing the flow to less active tissues. The walls of the arterioles contain a layer of muscle cells. When these muscle cells are relaxed, the diameter of the arteriole increases and more blood flows in that direction. Likewise, when the circular muscle cells of an arteriole are contracted, blood is diverted from that pathway to other tissues whose nutrient requirements are greater. Whether an arteriole is contracted or relaxed depends in part on signals from the autonomic nervous system, and also on changes in local chemical conditions surrounding the arteriole.

Arterioles may branch many times, eventually leading to **capillaries**, or tiny, thin-walled vessels that permeate almost every tissue of the body. The diameter of the capillaries is so small that blood cells must move through them one at a time. The average length of a capillary is only about 1 millimeter, but there are so many of them—an estimated 25,000 miles if they were lined up end to end—that at any given time, nearly 5 percent of the blood is flowing through capillaries. This 5 percent of the blood is carrying out the primary function of the cardiovascular system, that is, the exchange of nutrients and wastes. Exchange between the blood and the tissues occurs only in capillaries and not in any of the other vessels of the cardiovascular system.

Most substances move between the capillaries and the fluid surrounding the cells of tissues by diffusion. Oxygen and carbon dioxide readily diffuse through the cell membranes of the single layer of cells that form the capillary wall. Diffusion is a very effective way of moving substances short distances, and the capillary networks of the body are so extensive that no cell is more than about 10 micrometers from a capillary. Some substances that cannot easily cross cell membranes, such as ions and other water-soluble substances, enter and leave the blood through spaces, or clefts, between the cells of the capillary wall. These spaces are too small for proteins to pass through, however, and the proteins that circulate in the blood generally remain there.

Fluid can also leave the capillaries by a second mechanism, called bulk flow. In *bulk flow*, water and dissolved solutes move in bulk from an area of higher pressure to an area of lower pressure. The pressure of the blood as it enters the capillaries is greater than the pressure of the fluid in the tissues pushing back on the capillaries. Consequently, fluid from the blood, called **plasma**, leaves the capillaries through the spaces between cells of the capillary walls. However, the proteins that remain in the blood because of their large size exert osmotic pressure. Recall from Chapter 4 that osmosis is the movement of water from a region with a higher water concentration to one with a lower water concentration. Fluid in the extracellular region is drawn back into the capillaries by the osmotic pressure exerted by the proteins inside. The two opposing forces, blood pressure pushing fluids out of the capillaries and osmotic pressure pulling fluids back in, help to maintain a fluid balance between the blood and the extracellular spaces.

Capillaries converge to form the smallest veins, or **venules**, which, in turn, merge into larger and larger veins that carry blood back to the heart. Most of the time, about half of the blood can be found in veins. The walls of the veins are thinner than those of arteries, but like arteries, they are surrounded by a layer of muscle. If the body needs to increase the amount of blood flowing to tissues, for example, during profuse bleeding, these muscles contract. This decreases the diameter of the veins and pushes blood toward the heart.

Blood Cells The blood is a complex tissue comprising fluid, solutes, and several different kinds of cells. We have learned about some of the solutes in the blood, including oxygen, carbon dioxide, proteins, and ions. What are the different types of cells found in blood? What role do these different cells play in the body?

We have seen how nutrients are taken in through the digestive system and oxygen through the respiratory system, and how the cardiovascular system transports substances throughout the body. Our discussion of nutrient processing must also include how the body rids itself of undesirable byproducts. Carbon dioxide, we have seen, leaves through the lungs, but what of the other metabolic wastes? Contrary to what you might expect, the digestive tract is not the major avenue for cleansing the body of metabolic wastes; wastes that leave in feces are mostly undigested materials that have never entered the body. The organs for cleansing the blood of metabolic wastes are the kidneys. As we will see in the next section, the kidneys also maintain the water and salt balance of the body.

11.2.4 Metabolic Wastes, Water, and Salts are Processed by the Kidneys

You might have a glass of orange juice, or possibly a cup of coffee or tea, with breakfast. Perhaps you have water, soda, or juice with lunch. During your afternoon exercise you sweat, and throughout the day and night, water evaporates from the surface of your lungs as you breathe. Undoubtedly you will make several trips to the bathroom before your day is done. Your body is constantly gaining and losing fluids. Remarkably, however, the overall fluid volume of the body changes very little. Homeostasis of the body's fluids is one of the primary functions of the **urinary system**. The urinary system includes the **kidneys**, the **urinary bladder**, and the tubes that connect them to each other and to the outside.

Structure of the Kidney

The kidneys are a pair of bean-shaped organs, each about the size of a fist, that lie against the back wall of the body cavity at the level of the lowest ribs. Each kidney is connected to the urinary bladder by a thin tube called a **ureter**. The bladder opens to the outside of the body through a tube called the **urethra**. Fluids from the blood enter the kidneys. The kidneys process the fluid, converting it to urine. From the kidneys, urine moves through the ureters into the bladder and finally to the outside. Next we will see how fluid enters the kidneys and how it is converted into urine. We begin by looking at the structure of the kidney.

Encased in a thin, tough coat of tissue, each kidney is a collection of over a million slender, convoluted tubes called **nephrons**. The nephron is the functional unit of the kidney, that is, the smallest structure that can carry out all of the functions of the kidney. Nephrons cleanse the blood by removing wastes and foreign substances from it and adding a few substances back to it. This is accomplished by three **renal processes** ("renal" is the Greek word for *kidney*). They are (1) *filtration*, (2) *reabsorption*, and (3) *secretion*.

The Renal Processes

Each nephron consists of two parts: blood vessels, called the vascular element, and a hollow, twisted tube called the tubule (Figure 11-16). Blood enters the kidney through an arteriole, which branches into a ball of capillaries called the **glomerulus**. Unlike other capillaries in the body, the glomerulus is enclosed in a capsule formed from one end of the nephron tubule, similar to the fingers of a loosely clenched fist pushed into a balloon. The entire structure—the glomerulus and the capsule—is called the **renal corpuscle**. Filtration occurs here.

The pressure of the blood entering the glomerulus is high enough to force water, salts, and low molecular-weight substances, such as glucose, out of the glomerular capillaries by bulk flow. Larger structures and molecules, such as blood cells and proteins, remain in the capillaries. In other words, the blood is *filtered* at the renal corpuscle. The filtrate that enters the tubule at the renal corpuscle eventually becomes urine, but not until it is further processed.

About 180 liters of fluid pass from the blood into the nephron tubules of the kidneys each day. Considering that the average volume of fluid in the blood (excluding the blood cells) is only about 3 liters, most of this volume must be returned to the blood

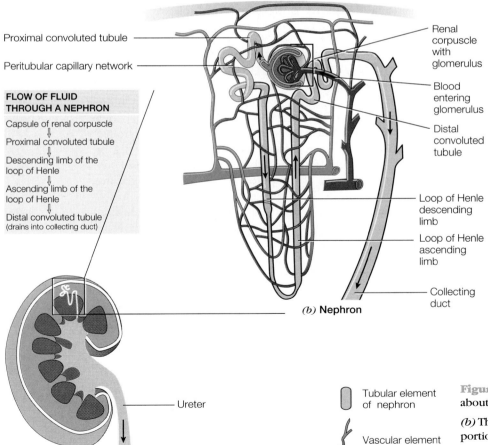

Proximal convoluted tubule

Peritubular capillary network

FLOW OF FLUID THROUGH A NEPHRON

Capsule of renal corpuscle
⇩
Proximal convoluted tubule
⇩
Descending limb of the loop of Henle
⇩
Ascending limb of the loop of Henle
⇩
Distal convoluted tubule
(drains into collecting duct)

Renal corpuscle with glomerulus

Blood entering glomerulus

Distal convoluted tubule

Loop of Henle descending limb

Loop of Henle ascending limb

Collecting duct

(b) **Nephron**

Ureter

To bladder

(a) **Kidney**

Tubular element of nephron

Vascular element of nephron

Figure 11-16. *(a)* The kidney, which has about a million nephrons.

(b) The nephron, which includes a vascular portion (the glomerulus and the peritubular capillaries, in red and blue) and a tubular portion (in gold).

**Fast Find
11.2.f
The
Nephron**

before it makes its way into the bladder as urine. If all the fluid filtered by the renal corpuscle was lost as urine, we would urinate ourselves to death in just 24 minutes. The process of filtration is relatively nonspecific; that is, any substance smaller than a small protein can enter the tubule of the nephron. Many substances, such as sodium ions and glucose, are not wastes but are needed by the body. Therefore, it is essential that many substances be reclaimed, or *reabsorbed*, before they become part of the final fluid that is excreted as urine.

Reabsorption occurs at several places along the length of the tubule, including the first twisted section, called the *proximal convoluted tubule*. In the proximal convoluted tubule, protein pumps found in the cells that form the wall of the proximal convoluted tubule actively transport sodium and chloride. These ions are moved across the wall of the tubule and into the extracellular fluid surrounding the nephron. Glucose is moved out of the tubule and into the blood. The walls of the proximal convoluted tubule are highly permeable to water, so water readily follows the sodium and chloride ions into the extracellular fluid by osmosis. A second set of capillaries, called the peritubular capillaries, envelope the entire nephron tubule and pick up sodium, chloride, and water from the extracellular fluids. By the time the filtrate reaches the end of the proximal convoluted tubule, nearly three-quarters of the sodium, chloride, and water filtered in the renal corpuscle has been reabsorbed into the blood.

From the end of the proximal convoluted tubule, the filtrate enters the loop of Henle, a looped section of the tubule that is shaped like a hairpin. The two arms of the hairpin have different characteristics that contribute to the production of urine. The downward arm, like the proximal convoluted tubule, is water permeable. Water continues to leave the tubule as the filtrate moves down toward the hairpin turn. But the upward arm of the loop of Henle is impermeable to water. Sodium and chloride are pumped out of the tubule into the extracellular fluid, but water is prevented from following. Consequently, the fluid in the tubule becomes dilute as it moves upward toward the last sections of the nephron.

From the loop of Henle, the filtrate enters a second twisted region of the tubule, called the *distal convoluted tubule*. It is in this section that most secretion takes place. Secretion provides a mechanism for removing substances from the body rapidly. Secreted substances are pumped out of the extracellular fluid by transport proteins found in the membranes of the distal convoluted tubule cells. Potassium ions (K^+) and hydrogen ions (H^+) are two examples of substances that are actively secreted into the urine. Controlled secretion of H^+ is an important mechanism for regulating the overall acidity, or pH, of the body's fluids. Secretion also rids the body of foreign substances, such as drugs. Penicillin is an example of a drug that is secreted. The nephron is so efficient at secreting penicillin that, within 3 hours of taking it, nearly 80 percent of the drug is lost in the urine. In the 1930s and 1940s, before it was mass-produced, penicillin was in such short supply that it was a common practice to collect the urine of patients being treated with penicillin so that the drug could be isolated and reused.

In addition to transport proteins that function in secretion, the cells of the distal convoluted tubule contain receptor proteins that recognize certain hormones. **Aldosterone** is a hormone that is produced when the body is low on sodium. Receptor proteins in the cells of the final tubular region, the collecting duct, bind to aldosterone. The aldosterone–receptor complex accelerates sodium pumps, causing these cells to reclaim even more sodium from the fluid in the tubule and return it to the blood.

From the distal convoluted tubule, the fluid enters the *collecting duct*, the final tubular region. The pathway of fluid through the collecting duct is parallel to that through the hairpin loop of Henle, an anatomical feature that has important consequences. Active reabsorption of sodium and other ions in the loop of Henle causes the extracellular fluid in this region to be salty; in other words, the osmolarity of this part of the kidney is quite high. As fluid moves through the collecting duct, there is an osmotic force drawing water out of the tubule and into the extracellular space. But for water to move, the walls of the collecting duct must be permeable to water. Unlike other parts of the nephron tubule, the water permeability of the distal convoluted tubule and collecting duct walls can change. When we are dehydrated, as might happen while hiking in the desert, the walls of these tubular regions are highly permeable to water; when we are fully hydrated, the walls become impermeable. What regulates the permeability of the distal convoluted tubule and collecting duct walls?

Recall from our discussion of the endocrine system that the cells of the posterior pituitary secrete two hormones, one of which is called **antidiuretic hormone**, or **ADH**. A diuretic is an agent that causes water to be lost in the urine. (Caffeine is a diuretic.) Antidiuretic hormone causes water to be retained by the body, preventing it from passing into the urine. ADH is secreted by the posterior pituitary into the bloodstream when the body's water balance leans toward dehydration. The cells that form the wall of the distal convoluted tubule and the collecting duct have receptors for ADH. When ADH binds to its receptors, a series of metabolic responses causes an increase in the number of water channels that line the distal tubule and collecting duct, providing a pathway for water to move by osmosis into the extracellular fluid. This water diffuses into the capillaries that engulf the nephron and is returned to the body by means of circulation. When water is abundant, the number of water channels declines, and the water permeability of the distal convoluted tubule and collecting duct walls is low. Water moves with the rest of the filtrate into the bladder and is lost in the urine.

From the renal corpuscle where filtration occurs, the different parts of the nephron modify the filtrate, creating urine with a composition that is carefully regulated according to the needs of the body. In this manner, the kidney is responsible for the fluid, salt, and acid homeostasis of the body.

Piecing It Together

Because humans are large animals, we cannot rely on diffusion alone to exchange nutrients and wastes with our environment. Humans require specialized organ systems to accomplish these tasks, and a system for moving substances throughout our bodies. Several of the body's organ systems have evolved to acquire and process the essential nutrients for life.

1. The digestive system is a hollow tube that takes in and processes food. Different parts of the digestive tube are specialized for different purposes. Food enters at the mouth, where chewing helps to break it down and dissolve it, and enzymes begin digesting carbohydrates. The stomach stores food and begins the process of protein digestion. From the stomach, food moves into the small intestine, where pancreatic enzymes continue digesting carbohydrates and proteins, and bile from the liver aids in the digestion of fats. The products of digestion are absorbed across the wall of the intestine. Undigested material leaves the body through the anus.

2. Oxygen and carbon dioxide are exchanged across the surfaces of the lungs. Muscles of the thorax, especially the diaphragm, create pressure changes that bring oxygen-rich air from the outside into close proximity with the lung capillaries. Oxygen moves by diffusion into the oxygen-poor blood, and carbon dioxide diffuses from the blood into the air spaces of the lung and is expelled when we exhale.

3. The cardiovascular system is adapted to transport substances to all parts of the body. The system consists of the heart, which acts as a pump; the blood, which acts as a medium for transporting substances; and the various blood vessels that carry the blood to and from the body cells. The cardiovascular system moves nutrients and wastes, and it also serves as the transport system for many types of cellular messengers, including hormones.

4. Metabolic wastes are removed from the blood by the kidneys. The functional unit of the kidney is the nephron, of which there are over 2 million. A protein-free, cell-free filtrate of the blood enters the nephron tubule at one end. As this filtrate moves through the nephron tubule on its way to becoming urine, substances such as sodium, chloride, water, and glucose are removed from it, and other substances such as potassium, hydrogen ions, and drugs are added to it from the blood. Hormones control the amounts of water and sodium that are reclaimed from the filtrate before it becomes urine, and in this way, the kidney maintains the water and salt balance of the body.

11.3 HOW DOES THE HUMAN BODY MOVE?

The atmosphere was tense. Two dozen students were trying out for just five starting positions on the varsity basketball team. The students sat ready, with their hearts pounding and their stomachs tight. One by one, the students jumped up as their names were called and began shooting baskets from the foul line. The coach looked on critically. Success or failure would depend on the coordinated action of each student's skeletal and muscular systems. Even the pounding hearts and knotted stomachs were the result of muscle activity.

Muscle is the tissue that generates force or movement by shortening. There are three types of muscle in the human body. (1) **Skeletal muscle** is attached to bones. Its contraction supports and moves the bones of the skeleton. Skeletal muscles contract in response to signals from motor neurons that originate with conscious thought. For this reason, skeletal muscles are often called voluntary muscles. (2) **Cardiac muscle** is the muscle of the heart. The rhythmic contractions of cardiac muscle constitute the heartbeat, which propels the blood through the circulatory system. (3) Sheets of **smooth muscle** surround the hollow internal organs, such as the digestive tract, blood vessels, and the larger air passageways leading to the lung. Cardiac muscle and smooth muscle contract without conscious direction and are sometimes referred to as involuntary muscles. Some involuntary muscles contract spontaneously without nervous input, but the frequency and strength of contraction of these muscles is often modified by the activity of the autonomic nervous system and by certain hormones

In this section we will focus primarily on skeletal muscle. Although the three types of muscle have differences, the force-generating mechanism in all of them is very similar.

11.3.1 The Skeleton Provides the Framework

The human skeleton is a combination of rigidity and flexibility. It must provide a strong anchor for skeletal muscles, yet it must move when those muscles contract. The competing needs for stiffness and freedom of movement are met by a structure containing different kinds of materials. **Cartilage** is a firm, flexible material that is hard, but not brittle (Figure 11-17a). In babies, cartilage is the primary component of the skeleton, but with age, much of the cartilage is replaced with the more brittle, inorganic matrix called bone (Figure 11-17b). Cartilage persists in adults in places where firmness and flexibility are needed, such as the connections between the individual bones of the spine, at the ends of the long bones and ribs, in the walls of the trachea, at the tip of the nose, and in the outer parts of the ear.

Bone is mostly made of a calcium-rich matrix that exists outside of cells, but certain cells that reside within the bone matrix secrete the matrix (Figure 11-17b). In a long bone, such as those of the appendages, the bone shaft has a central cavity containing a fatty, cellular substance called yellow bone marrow. The matrix at the swollen ends of the bone is porous and contains red bone marrow. Red bone marrow consists of cells that divide by mitosis, giving rise to red blood cells and white blood cells.

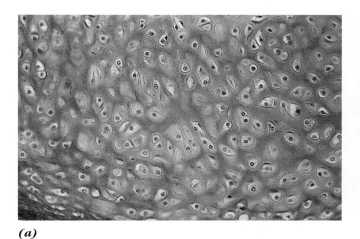

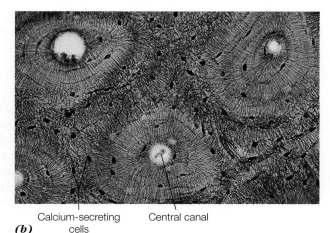

Calcium-secreting cells   Central canal

(a)   *(b)*

Figure 11-17. *(a)* Cartilage from the human trachea. Cartilage consists of a shiny, bluish-white substance containing protein fibers. Numerous cells that secrete the protein fibers are scattered throughout (magnified 102 times). *(b)* Cross section of a portion of a human thigh bone. This type of bone consists of cylinders of hard, calcium-rich matrix surrounding a central canal. The dark spots are spaces containing cells that secrete the bone matrix (magnified 158 times).

Ligaments are strong, flexible bands that connect cartilages or bones together at points of attachment called **joints**. Some joints permit little movement of bones, such as those of the skull and those that hold the teeth in place. Others are filled with cartilage and permit some slight movement, such as those joints between the individual vertebrae of the spine. Still others are highly flexible, such as the elbow and knee joints and the joints of the hand. The human skeleton is an elegant compromise between flexibility and strength.

**Fast Find
11.3.c
Skeletal/Muscle
Movement**

Skeletal muscles attach to bones by means of **tendons**, which are tough bands of white connective tissue. By exerting a force through the tendons, muscle contraction causes bones to move and the skeleton to bend at the joints. The force of muscle on bone is always a pulling force and never a pushing force, but the arrangement of different pairs of muscles around certain bones allows a joint to be bent or straightened, depending on which muscle is contracting and the direction of the pulling force. On the CD, you can bend a virtual limb and see what muscles are involved. Movable bones act like a lever system, with the fulcrum of the lever at the joint. Both the nature of the joint and the positions and numbers of different muscles on the bones surrounding the joint determine how the skeleton bends and the possible range of motion.

The skeleton provides the framework for posture and motion, but it is the skeletal muscle that exerts the force that moves the bones and cartilage. In the next section, we examine the cells of a muscle to see how that force is generated.

The Weakest Link Movement would be impossible without joints, but joints are also the weakest link in the human skeleton. What are some common joint injuries? What are some diseases that affect joints? What features of joints make them susceptible to these problems? How are some of these joint malfunctions currently being treated?

11.3.2 Muscles Cause Movement

Muscles cause movement by becoming shorter, or contracting. For many years it was believed that the proteins that make up muscle must, themselves, shrink in order to shorten a whole muscle, but careful microscopic observations comparing contracted and relaxed muscle tissue revealed that the proteins in muscles do not shrink. Instead, long, fibrous proteins slide past one another, much like the parts of a telescope, shortening the whole muscle while the length of the individual proteins remains unchanged. To appreciate how this works, we must first examine the molecular structure of the muscle.

Structure of Skeletal Muscle

Look closely at Figure 11-18. One thing that is immediately apparent is that skeletal muscle tissue is highly organized at many different levels. Muscles are composed of bundles of fibers. Each fiber is a single muscle cell, although muscle fibers are the largest cells of the body and may have several hundred nuclei. Nerves and blood vessels are embedded between the bundles of fibers so that each fiber has a ready supply of nutrients and a nerve that stimulates it to contract. Skeletal muscle fibers have very little cytoplasm, but instead they are filled with hundreds or even thousands of bundles of parallel cylinders called myofibrils. Each myofibril is surrounded by a network of membrane called the **sarcoplasmic reticulum**, abbreviated SR. The SR encloses a space that is separate from the space containing the myofibrils, called the lumen of the SR. The lumen of the SR has a high concentration of calcium ions that, we will soon see, play an important role in contraction.

Looking even more closely, we see that the myofibrils are composed of bundles of proteins. There are two main types of long, fibrous proteins that do the work of contraction and several other kinds of associated proteins. Of the two main proteins, there is one thicker protein called **myosin** and one thinner protein called **actin**. With a microscope,

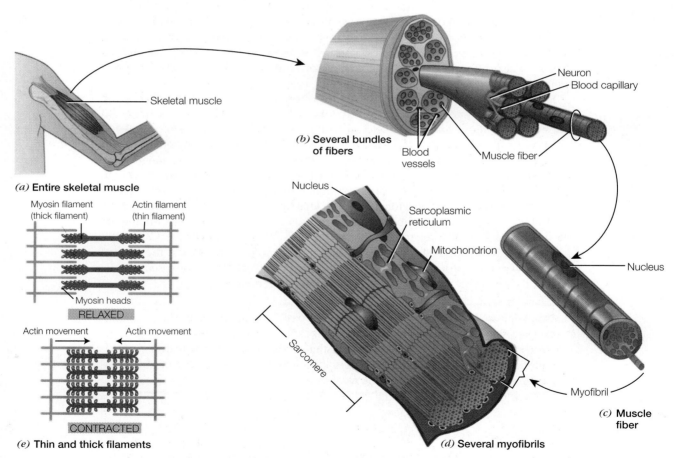

Figure 11-18. The organization of skeletal muscle. (*a*) Entire skeletal muscle. (*b*) Bundles of skeletal muscle fibers. (*c*) A single muscle fiber. (*d*) Several myofibrils, showing their relationship with the sarcoplasmic reticulum. (*e*) Thin and thick filaments of the myofibrils, showing how they interact when a muscle is relaxed and when it contracts.

Fast Find
11.3.b
Skeletal
Muscle

the myofibrils appear as long cylinders composed of shorter segments connected in series, or end to end. Each short segment is called a sarcomere. Figure 11-18, parts *d* and *e*, and the CD show that each sarcomere has a central region containing many longitudinally arranged thick filaments, or bundles of myosin molecules. The edges of the sarcomere contain many longitudinally arranged thin filaments, or twisted chains of actin molecules. Whereas the center of the sarcomere has only thick filaments and the outer edges of the sarcomere has only thin filaments, the two kinds of filaments overlap near their ends.

The first clues about the mechanism of muscle contraction came when sarcomeres from contracted muscle and relaxed muscle were compared under the microscope. It was found that neither type of filament gets shorter in contraction, but the area in which they overlap gets larger when the muscle is contracted. In other words, when a muscle contracts, the thin filaments and the thick filaments slide past each other, bringing the borders of the sarcomere closer together. This remarkable discovery, made in the 1950s, is called the **sliding filament theory of muscle contraction**. You can watch how the filiments slide past each other on the CD.

What causes the filaments to slide past each other during muscle contraction? Anyone who has exercised knows that one essential component of contraction is *energy*. The energy for muscle contraction comes from ATP. The machinery for capturing that energy and converting it into sliding filaments is found on the myosin proteins of the thick filaments.

Each thick filament is made of about 250 molecules of myosin. A single myosin molecule is a long, club-shaped protein with a stiff, rodlike tail and two flexible, hinged heads. The heads of myosin have spaces that can bind to two different things. First, myosin heads have sites that fit onto the surface of actin molecules (the thin filaments). Likewise, actin has sites that fit myosin. Second, myosin heads have pockets for binding to ATP. When a myosin head binds to an ATP, part of the myosin head acts as an enzyme, cleaving the ATP into ADP and a single phosphate. (We learned in Chapter 4 that this is a common cellular strategy for attaining energy.) Energy from the broken chemical bond in ATP causes the myosin head to swing to a new position. In this new position, the myosin head attaches to a nearby actin molecule on one of the thin filaments. When the ADP and phosphate leave, the myosin head swings back to its original position, pulling on the actin and moving the thin filament as it does so. Whereas each individual myosin head moves the actin only a minute distance, the many myosin molecules in each sarcomere and the many sarcomeres found in each muscle are responsible for the full range of motion of the human body.

The ingredients for contraction—actin, myosin, and ATP—are normally quite abundant in muscle fibers, yet muscles are not always contracted. In fact, a voluntary muscle contracts only when the appropriate signal, that is, an action potential, arrives from the motor nerve that connects with it. How do action potentials in motor nerves regulate muscle contraction?

Calcium Regulates Contraction

Skeletal muscle cells, like neurons and heart muscle cells, are excitable cells capable of undergoing action potentials (Section 11.1). When an action potential originating in the central nervous system arrives at a skeletal muscle fiber by means of a motor neuron, an action potential is triggered in the muscle fiber. This electrical event, in turn, initiates the mechanical events described above that cause the muscle to contract. It does so by means of a chemical intermediary: calcium ions (Figure 11-19).

In a relaxed muscle cell, calcium ions are restricted to a separate compartment of the cell, the lumen of the SR. Recall that this compartment is separated from the myofibrils by the SR membrane. In a relaxed state, the concentration of Ca^{2+} in and around the sarcomeres is very low, but within the SR, the Ca^{2+} concentration is relatively high. When an action potential reaches a muscle cell membrane, the electrical signal travels deep into the cell and opens calcium channels on the SR membrane. Calcium ions rush down their concentration gradient by diffusion, invading the space containing the actin and myosin,

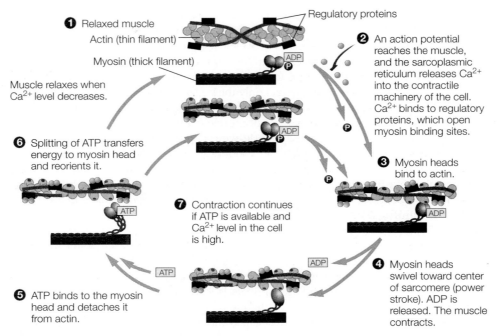

① Relaxed muscle

Actin (thin filament)

Regulatory proteins

Myosin (thick filament)

Muscle relaxes when Ca²⁺ level decreases.

② An action potential reaches the muscle, and the sarcoplasmic reticulum releases Ca^{2+} into the contractile machinery of the cell. Ca^{2+} binds to regulatory proteins, which open myosin binding sites.

⑥ Splitting of ATP transfers energy to myosin head and reorients it.

③ Myosin heads bind to actin.

⑦ Contraction continues if ATP is available and Ca^{2+} level in the cell is high.

⑤ ATP binds to the myosin head and detaches it from actin.

④ Myosin heads swivel toward center of sarcomere (power stroke). ADP is released. The muscle contracts.

Figure 11-19. Muscle contraction is regulated by the interaction between calcium and regulatory proteins on actin filaments.

raising the calcium concentration 100-fold. This is the first step in coupling an action potential with a muscle contraction.

The next step occurs on the thin filaments. In addition to actin proteins, the thin filaments have regulatory proteins that cover the binding sites to which myosin heads attach. In the presence of Ca^{2+}, the position of the regulatory proteins shifts. The myosin binding sites on actin are uncovered and myosin heads bind, triggering a muscle contraction.

Immediately after Ca^{2+} leaves the SR, protein pumps on the SR membrane begin moving it back in. Unless there is another action potential from the motor neuron, the Ca^{2+} levels of the contraction machinery drop rapidly and the muscle relaxes. Prolonged muscle contractions result from repeated action potentials entering the muscle fiber from motor neurons.

Fibers and Fitness Not all skeletal muscle fibers are alike. Some contract rapidly but also tire quickly. Other types of fibers contract slowly but can continue working for longer periods before they fatigue. What characteristic of muscle fibers accounts for these differences in performance? What kind of fibers would be abundant in the muscles of a trained weight lifter? A distance runner? Why?

Piecing It Together

Wherever there is movement of or in the body, there is muscle tissue exerting a force that causes that movement. The three kinds of muscle found in the body—skeletal muscle, cardiac muscle, and smooth muscle—differ in appearance but employ the same basic mechanisms to generate the force to move.

1. The skeleton is a compromise between rigidity and flexibility. The calcium-based structure of bone is a strong, rigid matrix that provides support and a framework for muscle action. The joints are the areas of the skeleton that can bend and flex.

2. Muscles generate force by contracting, or shortening. At the molecular level, muscle contraction occurs when protein filaments in the sarcomere slide past one another. Thick proteins called myosin slide past thinner proteins called actin. The process is fueled by ATP.

3. All of the ingredients for skeletal muscle contraction are usually present in a muscle cell, yet contraction occurs only when there is an incoming action potential from a motor neuron. An incoming action potential causes the muscle cell to release calcium ions from the lumen of the sarcoplasmic reticulum. Calcium interacts with the contraction machinery to uncover myosin binding sites on actin and permit myosin to bind, resulting in contraction.

11.4 HOW DO HUMANS REPRODUCE?

Unlike any other organ system of the body, the human reproductive system contributes almost nothing to the body's homeostasis and survival. Without it, however, our species would soon disappear from Earth, for it is the human reproductive system that enables us to form new individuals by sexual means.

Unlike many other kinds of organisms, humans reproduce *only* by sexual means. Sexual reproduction, as we learned in Chapter 3, has the advantage over asexual reproduction of producing offspring with new combinations of genes. The variety of genetic combinations in the offspring of sexual species increases the likelihood that at least some offspring will survive in our ever-changing environment.

There are, however, costs associated with sexual reproduction. Sexually reproducing species must have two kinds of individuals, male and female, each with a different kind of reproductive machinery. In species such as humans, in which fertilization takes place inside the mother's body, two individuals must come together in a coordinated way, a process that has led to the evolution of complex and energetically expensive courtship behaviors. Delicate gametes—egg and sperm—must come together in a hospitable environment to ensure their survival and facilitate their union. The entire process of producing gametes, fertilization, and nurturing the newly formed embryo must be highly regulated to ensure that gametes find one another and that the embryo has the greatest chance of surviving. These are some of the issues addressed in the study of human reproductive physiology.

11.4.1 Gametes Are Produced in Gonads

Recall from our discussion of meiosis in Chapter 5 that all human cells *except* the gametes contain 46 chromosomes: 22 pairs of homologous chromosomes and one pair of sex chromosomes. When cells divide by mitosis, all of the 46 chromosomes are duplicated, and each daughter cell receives one full set of 46. Gametes arise by meiosis, a different type of cell division in which daughter cells receive only one member of each homologous pair of chromosomes, or 23 chromosomes altogether. When an egg and a sperm join in fertilization, the full diploid complement of 46 chromosomes is restored in the zygote.

Cells that are destined to become gametes are the only cells in the human body that divide by meiosis. These cells are found in the gonads: the **testes** in males and the **ovaries** in females. In both sexes, the process of making gametes, which includes meiosis and maturation of the daughter cells of meiosis, is called **gametogenesis**.

The timing of gametogenesis is different in males and females. In females, meiosis begins even before birth, and a newborn girl has in her ovaries all of the eggs she will ever have. These eggs mature and are released from the ovaries one at a time about every 28 days during her reproductive years. The reproductive years for a female begin with puberty, at about age 12 or 13, and end with menopause, which occurs at an average age of 50. In males, meiosis and sperm maturation occur continuously in certain cells of the testes from the onset of puberty at about the age of 12 or 13 until death.

Gametogenesis in both sexes is influenced by a group of hormones called sex hormones. Most of these hormones are common to both males and females, although it is different levels of sex hormones that are responsible for many of the differences between the sexes, as well.

Hormonal Control

Hormonal control of reproduction begins in the brain. The diagram in Figure 11-20 illustrates that the hypothalamus and pituitary glands regulate gametogenesis in both males and females. In both sexes, a releasing hormone from the hypothalamus causes the anterior pituitary gland to secrete the hormones LH and FSH (luteinizing hormone and follicle-stimulating hormone). These hormones travel in the bloodstream to the gonads. There, FSH stimulates gametogenesis and maturation of the gametes. LH stimulates accessory (nongamete) cells of the gonads to produce another set of hormones called the sex steroids.

The sex steroids have names that may be familiar to you: testosterone, estrogen, and progesterone are the main ones. While both sexes produce all of the sex steroids, testosterone is more abundant in males and estrogen and progesterone are more abundant in females. Estrogen and testosterone are responsible for secondary sexual characteristics. For example, in females, estrogen causes proliferation of breast tissue and the pattern of fat deposition that characterizes the female body form. In males, testosterone causes changes in voice quality, growth of facial and pubic hair, growth of muscle and

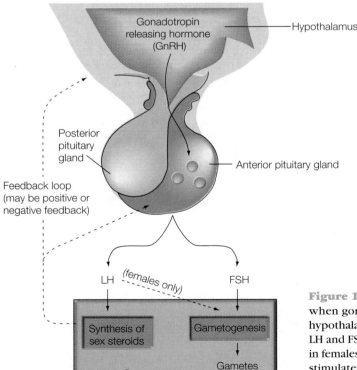

Figure 11-20. Hormonal control of reproduction begins in the brain, when gonadotropin-releasing hormone (GnRH) is released from the hypothalamus. GnRH stimulates the anterior pituitary gland to release LH and FSH. These hormones target the testes in males and the ovaries in females, where LH stimulates the production of sex steroids and FSH stimulates gametogenesis. The sex steroids provide feedback to both the hypothalamus and the anterior pituitary gland.

bone tissue, and the sex drive. Sex steroids also play a role in maturation of gametes of both sexes, and as we will see next, sex steroids circulating in the blood act on the pituitary gland and the hypothalamus to regulate the release of both LH and FSH from the pituitary. In many cases, this occurs by negative feedback.

Although the kinds of sex hormones and their effects are similar in both males and females, the timing of their release is very different in the two sexes.

11.4.2 The Reproductive Male

The testes are paired organs that lie outside of the male abdominal cavity in a sac called the **scrotum** (Figure 11-21). This arrangement keeps the temperature of the testes lower than the average temperature of the body, a condition that is essential for the production of viable sperm. In the male embryo, the testes begin development inside the abdominal cavity and normally descend into the scrotum during the seventh month of pregnancy. Individuals whose testes do not descend must be treated with hormones or undergo surgery to move them. Otherwise, the high temperature of the abdominal cavity causes infertility in adult life.

Each testis is packed with tightly coiled **seminiferous tubules** in which the sperm are produced. Seminiferous tubules constitute nearly 80 percent of the testicular mass. Cells embedded in the lining of the seminiferous tubules undergo meiosis and mature into sperm under the influence of hormones (mainly testosterone) from neighboring accessory cells.

Sperm that leave the lining of the seminiferous tubules and enter the tubular space (called the lumen) are not yet fully mature. From the seminiferous tubules, sperm enter another coiled tubule, called the **epididymis**, that lies just outside of the testis. In the 10- to 12-day period during which they are in the seminiferous tubules and the epididymis, sperm mature and develop their characteristic swimming ability.

From the epididymis, sperm travel into the **vas deferens** (plural, *vasa deferentia*), a duct that loops up behind the bladder in the abdominal cavity and connects the epididymis to the urethra. The urethra is the tube that connects both the vasa deferentia and the urinary bladder to the outside. In males, the urethra passes through the penis. Sperm are stored in the vasa deferentia.

During intense sexual arousal, sperm enter the urethra from each of the two vasa deferentia and are mixed with secretions of glands that are part of the male reproductive system. The mixture of sperm and glandular secretions is called **semen**. As sexual

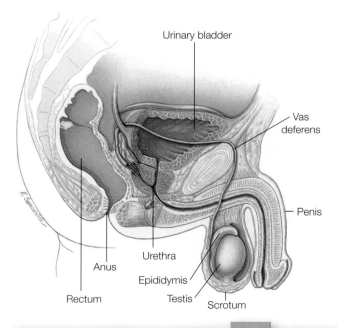

Figure 11-21. Structures of the male reproductive system. Sperm produced in the testes move into the epididymis where they mature. Sperm are stored in the vasa deferentia. During sexual arousal, sperm move into the urethra where they are combined with secretions from accessory glands to make semen. Semen is ejected from the penis during ejaculation.

stimulation intensifies, the semen is ejected from the urethra of the erect penis by rhythmic contractions, a process called **ejaculation**. Ejaculation is just part of a total sexual response called **orgasm**. In addition to ejaculation, orgasm involves changes in other physiological systems: blood pressure and heart rate increase, skeletal muscles tense, and the process is accompanied by intense physical pleasure. Orgasm is followed by muscular and mental relaxation and a refractory period during which a second orgasm is not possible.

11.4.3 The Reproductive Female

The female reproductive organs include the ovaries, where eggs develop; the **uterus**, or womb, where the embryo implants and develops during pregnancy; the oviducts, or **fallopian tubes**, through which the egg travels from the ovary to the uterus; and the **vagina**. During sexual union, or **copulation**, the orgasm in males is accompanied by sperm delivery, or ejaculation. The female orgasm is similar to that of males but is not accompanied by gamete delivery: rhythmic contractions of uterine muscle may assist sperm in their journey to the oviducts, where fertilization takes place. The relationships among the anatomical structures of the female reproductive system are shown in Figure 11-22.

The ovaries contain many **follicles**, small saclike structures containing one egg, called an oocyte, surrounded by accessory cells. A newborn girl may have as many as 2 million follicles, but a woman produces only one mature egg each month during her reproductive years, so only a small number of eggs actually mature. Because the eggs are present at birth, they age as the woman ages. The age of the eggs may be one reason why the chances of a woman giving birth to an infant with genetic abnormalities increases as she ages.

Unlike males, whose gametes are produced continuously, female gametes mature in a monthly cycle called the **menstrual cycle**. The most salient feature of the female menstrual cycle is **menstruation**, which is a 3- to 7-day period each month during which there is a bloody discharge from the uterus. The menstrual cycle involves cyclic changes in the ovaries and the uterus, and it is regulated by the rise and fall of hormones. These hormones exert secondary effects on other tissues as well, including the breasts and the brain. Figure 11-23 shows the events in a single menstrual cycle.

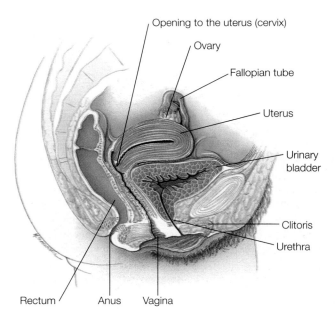

Opening to the uterus (cervix)

Ovary

Fallopian tube

Uterus

Urinary bladder

Clitoris

Urethra

Rectum　Anus　Vagina

Figure 11-22. Structures of the female reproductive system. Eggs are produced in the ovaries and released, usually one every 28 days, into the fallopian tubes. If an egg encounters sperm in the oviduct, it may be fertilized resulting in a zygote. The zygote begins to divide in the oviduct, becoming an embryo that may implant in the thickened endometrium of the uterus. If it is not fertilized, the egg passes through the uterus and is expelled from the vagina.

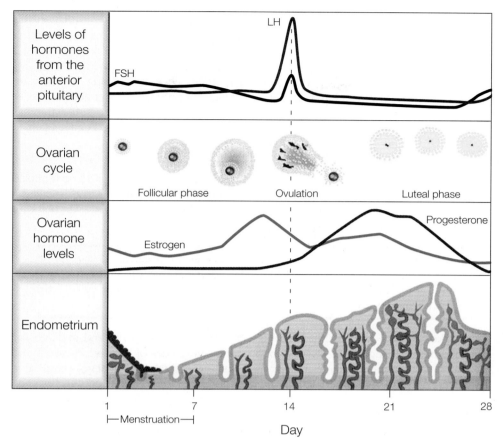

Figure 11-23. The female menstrual cycle in the absence of fertilization. During the first half of the cycle, called the follicular phase, FSH stimulates several ovarian follicles to mature, and LH stimulates estrogen secretion from the ovaries. High estrogen levels cause midcycle surges in LH and FSH, which, in turn, cause ovulation. During the second half of the cycle, called the luteal phase, the follicle develops into a corpus luteum and begins to secrete progesterone. Progesterone maintains the endometrium, but the corpus luteum degenerates toward the end of the cycle, and progesterone levels fall. As the cycle begins again at day 1, the endometrium is sloughed off during menstruation.

The menstrual cycle illustrated here begins with the first day of menstruation. During this time, the uterus sheds its thickened lining. This corresponds with release of LH and FSH from the pituitary gland. As the cycle progresses, FSH causes about a dozen follicles in the ovaries to develop and mature. The accessory cells of the follicle begin to secrete estrogen, which gradually increases during the first 10 days of the cycle. By the time estrogen levels reach a peak, most of the follicles have ceased to develop, and usually only one dominant follicle matures fully. For most of this first phase of the cycle, estrogen exerts a negative feedback on FSH and LH, but at its peak, estrogen's impact on FSH and LH reverses. In other words, the type of feedback that estrogen exerts on FSH and LH differs according to its concentration: low levels of estrogen inhibit the release of sex hormones from the pituitary, whereas high levels stimulate the release of sex hormones. Because of this, there is a marked surge in both FSH and LH from the pituitary at the midpoint of the cycle. The surge in LH in particular causes the final maturation of the egg. About 20 hours after the surge, the egg breaks free of its follicle and of the ovary, and enters the oviduct. This process is called **ovulation**. Ovulation marks the beginning of the second phase of the female menstrual cycle.

In addition to promoting the LH and FSH surge, estrogen helps prepare the lining of the uterus to accept an embryo in the event that the egg is fertilized. As estrogen levels increase, so does the thickness of the inner layer of the uterine wall, called the **endometrium**.

Following ovulation, the ovarian follicle that was left behind after ovulation fills with cells and is now called the **corpus luteum**. These cells begin to secrete the other female sex steroid, progesterone. This hormone is the dominant steroid during the last 2 weeks of the female cycle. Progesterone, along with estrogen, inhibits the release of FSH and LH by negative feedback, thus preventing another egg from ovulating. It also maintains the thickness of the endometrium in preparation for pregnancy. After about 10 days, the corpus luteum disintegrates and ceases to produce progesterone. If no fertilization has occurred, the endometrium is sloughed from the uterus, and the cycle begins again.

**Fast Find
11.4.a
Reproductive
Technologies**

Pregnancy

On its journey through the oviduct, the egg may encounter sperm and be fertilized by one of them. The fertilized egg is the first cell of the new individual and is called a **zygote**. The zygote immediately begins a series of mitotic cell divisions, even before it reaches the uterus, becoming an embryo.

In the Womb The human gestation period (length of pregnancy) lasts for 38 to 40 weeks. What are some of the major changes that the embryo undergoes during gestation? What are the events that occur during childbirth? What hormones are involved?

It takes about 7 days for the embryo to reach the uterine cavity and implant into the endometrial lining of the uterus. Upon implantation, cells of the endometrium and the embryo form a connective disk of tissue between mother and developing baby, known as the **placenta**. The developing embryo is supplied with nutrients and oxygen from the mother's blood through the placenta, and wastes from the embryo pass through the placenta to the mother. In addition, the placenta acts as an endocrine tissue, secreting a hormone called human chorionic gonadotropin, or hCG. This hormone acts on the ovaries to stimulate the continued secretion of progesterone. In this way, the endometrium is prevented from breaking down. Instead, it persists as a rich medium for the growing embryo. Thus menstruation does not occur again until after the pregnancy is over. Human chorionic gonadotropin is present in the blood within 24 hours of implantation, and it is freely filtered by the nephrons of the kidney. The presence of hCG in the urine is used as an indicator of pregnancy even before a pregnant woman has missed a menstrual period.

Birth Control Several birth control methods have been developed that function by interfering with normal hormone levels in females. Which hormones could act to prevent pregnancy if administered to a sexually active woman? Is it possible to develop a birth control pill that acts by interfering with normal hormone levels in males? Aside from hormonal methods, what other strategies can be used to prevent pregnancy? What is their relative effectiveness?

Piecing It Together

Humans reproduce sexually, by a process that requires the coordinated activities of the male and female reproductive systems.

1. Gametes, or egg and sperm, are produced in ovaries in females and testes in males, respectively. Gametogenesis is regulated by hormones from the brain and from cells of the gonads.

2. In males, gametogenesis occurs continually from puberty until death. LH and FSH, sex hormones from the pituitary, stimulate the gonads to produce both gametes and testosterone. Testosterone is required for gametes to mature. It is also responsible for secondary sexual characteristics in males.

3. In females, gametogenesis begins before birth. Usually only one egg reaches maturity during each monthly cycle of a woman's reproductive years. The female's menstrual cycle begins with menstruation, or sloughing of the endometrium, and is marked by a midcycle event, called ovulation, in which a single egg is released from the ovary. All of these events are regulated by changes in the level of different sex hormones.

4. An egg may be fertilized by a sperm cell on its way to the uterus, forming a zygote. The zygote begins to divide, becoming an embryo. The embryo may implant in the endometrium, resulting in pregnancy.

11.5 HOW DOES THE BODY FIGHT DISEASE?

The human body is constantly challenged by a wide assortment of would-be invaders. In the United States, bacteria and viruses are the most common types of **pathogens**, or disease-causing microbes. In other parts of the world, certain parasites are a global threat to human health. For example, it is estimated that as many as 100 million people in the world are infected with the protozoan parasite that causes malaria. Other invaders, including multicellular parasites like hookworms and tapeworms and certain kinds of fungi, also present a threat to human health. But the body is not without defenses. The ability of the body to protect itself from disease-causing entities is called **immunity**, from the Latin word *immunis*, meaning "exempt." The system responsible for immunity is the called the **immune system**.

The immune system does more than protect the body from outside invaders, however. Cells of the immune system patrol the body for other cells that are damaged or diseased, or that have simply grown old and deteriorated. Cells of the immune system also recognize and remove abnormal cells that can result from genetic mutations. Many cancers arise from genetically abnormal cells that have escaped the watchful guard of the immune system.

The best way to prevent infection from an invader is to prevent the invader from entering the body in the first place. Although not technically part of the immune system, the human body has several barriers against invaders. The skin is a tough barrier that most pathogens cannot cross unless it is damaged. Oil glands of the skin secrete chemicals that weaken or kill bacteria. Some of the sites that are most vulnerable to invasion are the digestive tract and the respiratory system. Here the barrier between the internal compartment of the body and the environment is necessarily thin to expedite the exchange of nutrients. The lining of much of the respiratory tract is covered with ciliated cells that sweep mucous and trapped particles, including pathogens, into the throat, where they are subsequently swallowed. Many pathogens are killed by the acidic secretions of the stomach. Pathogens that enter the intestine can be held in check by the mixture of nonpathogenic bacteria that normally reside there. Pathogens that succeed in entering the body are dealt with by the immune defenses.

11.5.1 The Immune System Is Integrated throughout the Body

Unlike most other physiological systems, the immune system is not restricted to a few defined organs in one area of the body. Instead, the tissues that play a role in immunity, called **lymphoid tissues**, are diffuse (Figure 11-24). They are integrated into the tissues of other systems, such as the skin, intestine, and bones. There are two primary lymphoid tissues and several other secondary ones. The primary lymphoid tissues are red bone marrow and the **thymus gland** (Figure 11-24). Red bone marrow, found in the ends of the long bones, is the site of production of all the different kinds of cells that circulate in the bloodstream. Among these cells are the family of **leukocytes**, or white blood cells, which play an important role in recognizing and fighting invaders. The thymus is a lobed gland found in the thorax just above the heart. During infancy, certain kinds of leukocytes travel from the bone marrow where they are produced to the thymus where they mature.

In addition to the thymus gland and red bone marrow, **lymph nodes** are lymphoid tissues found scattered throughout the body. Lymph nodes trap invaders and prevent them

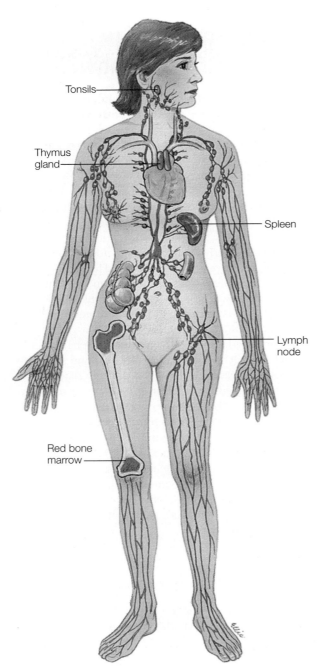

Tonsils

Thymus gland

Spleen

Lymph node

Red bone marrow

Figure 11-24. Lymphoid tissues of the body include the thymus gland; the red bone marrow; the spleen; and many small, widely scattered lymph nodes. A network of lymph vessels running throughout the body connects the thymus gland, spleen, and lymph nodes. There are also many loose aggregates of lymph cells strategically positioned near the intestine, respiratory tract, tonsils, and skin, where they can intercept invaders before they enter the body (not shown).

from spreading through the body. Often when your throat is sore, you are able to feel swollen "glands" in your neck. These are lymph nodes filled with leukocytes engaged in fighting trapped bacteria or viruses. The **spleen** is a lymphoid tissue that removes aging or damaged blood cells and destroys them. Finally, there are several sites in the body in which loose aggregations of lymphoid tissue are strategically assembled to fight foreign invaders. The tonsils are one such place, and collections of lymphoid tissue are associated with the intestines, respiratory tract, and skin—all places that are particularly vulnerable to entry by foreign entities.

Cells of the Immune System

Fast Find 11.5.a How Humans Fight Disease

Much of the body's fight against invaders is carried out by individual cells acting as free agents, not as part of a gland or organ. These cells, the leukocytes, patrol the body from their vantage point in the bloodstream and the tissues. Certain kinds of leukocytes can leave the capillaries at sites of infection or injury by squeezing through the narrow spaces between the cells that form the capillary walls. Here we will inventory the different types of leukocytes and some of their most important characteristics. In the next section, we will see how the individual types of leukocytes fight disease.

There are six categories of leukocytes that each play a different role in the immune response. Figure 11-25 gives some of their general characteristics and illustrates differences in their appearance. Several of these cell types have different names in the immature form; others have different names when they occur in the blood as opposed to the

Neutrophils 	60 - 70% of all leukocytes	Phagocytic cells. Destroy bacteria.
Eosinophils 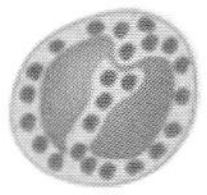	2 - 4% of all leukocytes	Phagocytize antigen-antibody complexes. Also destroy some types of parasitic worms.
Basophils 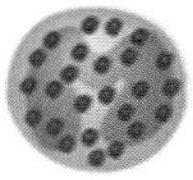	$1/2$ - 1% of all leukocytes	Release histamine and other chemical defenses. Play a roll in allergic reactions. When basophils leave the bloodstream and enter the tissues, they are called **mast** cells.
Lymphocytes 	20 - 25% of all leukocytes	Several subtypes exist. Two subtypes, B cells and T cells, make antibodies as part of the specific immune response. Other subtypes kill a wide variety of microbes. Others are helper cells, aiding in antibody production.
Monocytes 	3 - 8% of all leukocytes	Leave the bloodstream and enter the tissues, where they are called **macrophages**. Primarily act as phagocytic cells.

Figure 11-25. Types of leukocytes in the blood and their roles in the immune response.

tissues. Still others, particularly the lymphocytes, occur as several different subtypes; each subtype is identical in appearance but different in its chemical characteristics and its exact role in fighting invaders.

11.5.2 Nonspecific Responses Are the Second Line of Defense

At the start of this section, we learned that the body's first line of defense is the physical barriers (skin, mucous, and the acidity of the stomach) that prevent foreign invaders from entering the body. When these barriers fail, the second line of defense involves the **phagocytes**, meaning "eating cells," of the immune system. The main phagocytes are **neutrophils** in the blood and **macrophages** in the tissues. (Macrophages, before they leave the bloodstream, are called monocytes.) These cells recognize invaders and ingest them in a process called **phagocytosis**. A phagocyte engulfs the invading particle, bringing it into the phagocyte's cytoplasm within a membrane-enclosed vesicle. Once inside the cell, the vesicle fuses with **lysosomes** (Chapter 4) containing powerful enzymes, and the particle is killed and digested.

Another role of the phagocytes is to attract additional immune cells to an area of infection. This is accomplished by chemical attractants called **cytokines**. When a large number of immune cells are congregated in an area of infection, they may form a whitish-green substance called **pus**. Pus is a collection of living and dead macrophages and neutrophils, as well as bacteria, fluids, and other cell debris.

The second line of defense against invaders involves chemicals as well as phagocytic cells. Several kinds of proteins that are normally found dissolved in the blood in an inactive form are converted into deadly poisons at the site of an infection, where they kill invaders on the spot. Other chemicals are released by activated cells that have migrated to the battle site. **Basophils** in the blood (renamed **mast cells** when they enter the tissues) contain large granules filled with histamine (Figure 11-25). Cytokines released by phagocytic cells trigger the release of histamine into the infected area. Histamine opens pores in the walls of local capillaries, allowing other immune cells to enter the area, but also allowing fluid and proteins to leak out of the bloodstream. The result is a hot, swollen region surrounding the infection. Mast cells are abundant under the mucous membranes that line the respiratory airways and the digestive tract. When certain particles—pollens, for example—are inhaled or ingested, histamine may be released in certain people. This is the basis of the congestion that many people with seasonal allergies suffer. Antihistamine drugs act by blocking the release of histamine from mast cells or preventing histamines from acting on cells of the capillary wall.

All of these second-line defenses cause an infected region of the body to become red, hot, tender, and swollen—a response known as **inflammation**. If the infection is severe, the inflammatory response may raise the core temperature of the body, causing fever. These second-line defenses are nonspecific, meaning they do not require that the body recognize the invader from a previous attack. If the invader has been encountered previously, however, these nonspecific responses are accompanied by a series of specific defenses, targeted to that particular invader. These specific responses, called **acquired immunity**, are the third and most powerful weapon in the body's defense arsenal.

11.5.3 Acquired Immune Responses Recognize Familiar Invaders

Some foreign substances or microbes are prevalent enough in our environment that we may encounter them many times in the course of our lives. Acquired immune responses are the third line of defense, directed specifically against repeat invaders. The cells that carry out the acquired immune responses are the **lymphocytes** (Figure 11-25).

When viewed under the microscope, lymphocytes all look very much the same. But in fact, lymphocytes can be distinguished from one another by the specific kinds of proteins that are embedded in their cell membranes. These membrane proteins, called **antibodies**, recognize and bind to certain bits of foreign material, usually proteins on the surface of an invader, but sometimes carbohydrates or other materials that identify an invader as foreign. Any substance that is recognized by the immune system as foreign is called an **antigen**. The first step of the acquired immune response occurs when antibodies from lymphocytes bind to antigens on invaders. But how does the body make lymphocytes with enough different kinds of antibodies to handle all the possible invaders it may encounter during its life? In other words, what accounts for the vast diversity of human antibodies?

For many years, it was believed that the acquired immune response occurred because an invader somehow instructed lymphocytes to produce antibodies that could bind specifically to that invader. This theory stated that before a lymphocyte encountered an invader for the first time, its antibody proteins were generalized and very nonspecific. In the 1950s, Niels Jerne, a Danish immunologist, proposed a different theory for the production of highly specific antibodies. Jerne, and later F. MacFarlane Burnet of Australia, proposed that antigens do not *instruct* lymphocytes to make specific antibodies, but instead, existing lymphocytes with fully formed antibodies on their surfaces are *selected* to grow and proliferate when they encounter a complementary antigen. This theory, which is universally accepted by immunologists today, is called the **clonal selection theory**.

Let's take a step-by-step look at the way in which clonal selection of lymphocytes provides a specific immune response to an invader:

1. *Each lymphocyte becomes committed to producing only one kind of antibody.* This occurs in the early stages of lymphocyte development, so that the entire array of antibody-producing cells a person will ever have is present in the lymphoid tissue, before any antigen is encountered. Within a lymphocyte, the DNA that codes for the antibody protein is irreversibly rearranged in a way that causes that cell to make just one kind of antibody protein.

2. *An invader enters the body, and some portion of that invader acts as an antigen.* Invaders come in contact with the entire repertoire of lymphocytes, usually in the lymphoid tissues where the lymphocytes are concentrated. If it is the first time the body has encountered the invader, only one or a few of these lymphocytes display an antibody that can interact with the invader. The antibody on the lymphocyte interacts by recognizing some part of the invader that serves as an antigen.

3. *The lymphocyte (or lymphocytes) that interact with antigens are stimulated to grow and divide or to make more antibody proteins to be exported from the cell.* The first time an antigen is encountered, this process takes several days. In most cases, the activation of a lymphocyte requires the assistance of other kinds of lymphocytes nearby. Some of these are called helper cells.

4. *Antibodies and lymphocytes facilitate the destruction of the invader.* Invaders whose surfaces are coated with antibody molecules are more readily recognized by the phagocytic cells of the immune system, the macrophages and neutrophils, which ingest and destroy them. Also, antibodies activate the chemical defenses that are part of the nonspecific immune response described above, such as histamines and cytokines.

5. *Some of the progeny of an activated lymphocyte become memory cells.* **Memory cells** are specialized lymphocytes that remain in the lymphoid tissues, ready to respond quickly should that same antigen be encountered later. After the first encounter with an antigen, subsequent encounters produce specific immune responses that are stronger (more specific antibodies are made) and quicker, occurring in a matter of hours.

In the late 18th century, before much was known about immunology, an English physician named Edward Jenner carried out a daring and dangerous experiment. Smallpox was a widespread and often deadly disease at that time, but Jenner observed that the incidence of smallpox was very low among girls who tended cows. These girls often suffered from a much milder disease, cowpox, at an early age. Jenner concluded that cowpox rendered milkmaids "immune" to the ravages of smallpox. Jenner purposely infected an eight-year-old boy with cowpox by injecting him with pus taken from a cowpox blister of a milkmaid. When the boy had recovered, Jenner injected him with pus taken from a blister on a smallpox sufferer. Ordinarily, the second injection would have caused the dreaded disease in the child, but this boy showed no signs of the disease. Jenner had given the first vaccine.

The principles of vaccination are based on the specific immune response: either an attenuated (noninfectious or only mildly infectious) form of a pathogen or some fragment of the disease-causing entity that serves as an antigen is injected into a person, usually at an early age. This first exposure to the antigen stimulates the immune system to produce a specific immune response. Memory cells take up long-term residence in the lymphoid tissues, ready to spring into action if that antigen is encountered at a later date. Vaccinations against diseases such as tetanus, measles, mumps, rubella, smallpox, hepatitis B, and recently chickenpox are administered to children worldwide. In the last decade, scientists have searched for a vaccine that is effective against the AIDS virus, but without success. An effective vaccine has been elusive because the proteins on the virus that might have antigenic properties change extremely rapidly by mutation. Even if a person has been vaccinated with one form of the AIDS virus, it is likely that another form of the virus will arise that is different from the form that was used to vaccinate and will not be recognized by memory cells.

The Antibody Family What does an antibody protein look like? What features are shared by all antibodies? How do antibodies differ from one another?

Distinguishing Between Self and Nonself

The effectiveness of acquired immunity depends on the ability of immune cells to recognize antigens, that is, molecules that are not part of the body's normal chemistry. In other words, specific immunity depends on distinguishing between self, or that which belongs in the body, and nonself, or that which is foreign. During maturation of the lymphocytes, some of them inevitably acquire antibodies that are capable of binding to self proteins. If these cells were allowed to survive, the body would mount a specific immune response to itself, causing widespread destruction of otherwise healthy tissues.

Clearly some mechanism exists for eliminating these self-reactive immune cells. This occurs mainly in the thymus. As lymphocytes mature, the thymus gland screens them for their reactivity with peptides derived from the body's own proteins. Those that react are induced to commit suicide. But sometimes these mechanisms fail.

Autoimmune diseases are a group of diseases that result when the body mounts an immune response against its own tissues. Each disease is brought about by the existence of antibodies to a different kind of self protein. Some examples are: rheumatoid arthritis, in which the body makes antibodies to its own collagen; insulin-dependent diabetes mellitus, in which antibodies are produced that attack the insulin-producing cells of the pancreas; multiple sclerosis, a crippling disease in which antibodies attack the myelin sheath that surrounds neurons; and lupus (systemic lupus erythematosus), in which antibodies attack proteins in the nuclei of the body's own cells.

Why some self-reactive lymphocytes manage to escape the screening process in the thymus is unknown, but autoimmune diseases appear to begin with an infection. One theory states that autoimmune diseases are triggered when a foreign antigen is very similar to a self protein. Antibodies that are produced in response to an infection may cross-react with self proteins, causing an autoimmune disease.

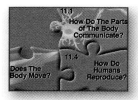

Piecing It Together

The immune system protects the body from invaders that manage to cross the barriers that enclose and protect the internal environment. Immunity is a combination of generalized, nonspecific defenses and highly specific defenses directed against particular invaders.

1. The tissues that effect the immune response, unlike other organ systems, are found scattered in strategic places throughout the body. The main cells of the immune response, the leukocytes, are produced in the bone marrow. Many of these cells migrate to the thymus gland where they mature.

2. The first line of defense against invaders is the shield created by barriers such as the skin, stomach acid, and mucous layers that line the respiratory and digestive tract. These defenses make it difficult for invaders to gain access to the internal compartment of the body.

3. The second line of defense is a nonspecific inflammatory response that occurs at the site of an infection. Phagocytic leukocytes engulf invaders and attract more immune cells to the site. Chemical defenses include histamines and other chemicals that kill cells directly.

4. Specific immune responses are the third and most powerful line of defense directed at pathogens that have been encountered more than once. On the first encounter, lymphocytes that produce antibodies that recognize an invader antigen are stimulated to proliferate and produce more antibodies. Some of the progeny become memory cells that remain in the lymphoid tissues, ready to make antibodies in the event that the same antigen is again encountered.

Where Are We Now?

Until recently, there have been two general strategies for developing vaccines. In the first, a solution containing a killed pathogen or part of a pathogen is injected into a patient. In the second, a solution containing a living but weakened (attenuated) pathogen is used. In both strategies, the vaccine tricks the patient's immune system into behaving as if it were being invaded. The pathogen or its parts are recognized as foreign, and antibodies are synthesized to combat it. Even more importantly, memory cells are created that can repel the same pathogen in the future.

Vaccines of the first type, based on antigens derived from killed pathogens or their parts, include the hepatitis A and B vaccines and the injected (Salk) polio vaccine. These vaccines introduce antigenic proteins into the blood, where they are recognized by a class of immune cells called B lymphocytes. The B lymphocytes lead the immune attack by making specific antibodies that bind to the antigen and tag it for destruction. This approach has the advantage of employing a harmless antigen, but there are disadvantages as well. Because the killed pathogens or their parts do not actually enter cells and infect them, vaccines based on this strategy are ineffective against many kinds of microbes that do infiltrate cells when they invade. Also, the protection afforded by these vaccines often wears off after a time, making periodic booster shots necessary.

Vaccines of the second type use attenuated strains of living pathogens, usually viruses. The viruses enter some of the patient's cells, but because they are attenuated, they cause no damage. Once inside, viruses induce the cell to make viral proteins using the cell's own metabolic machinery. Some of these proteins enter the blood, where they act as antigens and stimulate B lymphocytes to produce antibodies. Other viral proteins are displayed on the surfaces of infected cells, where they stimulate a second group of lymphocytes called killer T lymphocytes. Killer T lymphocytes are the part of the immune arsenal that attacks intracellular pathogens. When killer T lymphocytes recognize an antigen on the surface of an infected cell, they often destroy the cell, and the invader dies with it. This two-pronged attack is essential for fighting many viruses and other pathogens that wreak damage by entering cells. Additionally, live attenuated vaccines confer lifetime immunity. For these reasons, live vaccines are often preferred over killed-pathogen vaccines. Some examples of live vaccines are the MMR vaccine against measles, mumps, and rubella; the oral (Sabin) polio vaccine; and smallpox vaccines.

Live vaccines, however, have their own shortcomings. People whose immune systems have been compromised—cancer patients undergoing chemotherapy, AIDS patients, or the elderly, for example—cannot be given live vaccines because they may develop full-blown illness even though the pathogen has been weakened. Moreover, attenuated pathogens have been reported to mutate spontaneously into virulent forms. An ideal vaccine is one that has the effectiveness of a live vaccine without the associated dangers.

Recently, scientists have been testing a new group of vaccines, called DNA vaccines. DNA vaccines are made from small rings of DNA called **plasmids**. These plasmids contain genes for one or more proteins normally found in a pathogen, but they lack the genes necessary for the pathogen to cause an infection. DNA vaccines are delivered to patients either by injection or by a device called a *gene gun*, which propels them into cells near the surface of the body. Once inside the patient's cells, some of the plasmids make their way into the nucleus, where they instruct the host cell to synthesize the proteins encoded in the plasmid DNA. These proteins act as antigens, eliciting both B-lymphocyte immunity when they escape from the cells and the potent killer-T-lymphocyte immunity when they are displayed on the cell surface. They also confer long-term protection against would-be invaders bearing the same proteins. Thus DNA vaccines overcome the disadvantages of both killed-pathogen and live-attenuated vaccines.

Studies on the effectiveness of DNA vaccines have been promising. Several reports indicate that DNA vaccines stimulate both B-lymphocyte and killer-T-lymphocyte responses of rodents and primates against many different pathogens and even some cancers. The first human trials began in 1995, in which AIDS patients were vaccinated with plasmids containing genes from the AIDS-causing HIV virus. These were primarily safety trials designed to determine if DNA vaccines are toxic, or if they elicit any inappropriate autoimmune responses against the patient's own DNA. So far, these studies have not identified any serious side effects of DNA vaccines.

Currently, human trials are being conducted to assess the effects of DNA vaccines on the immune system. Early results hint that useful immune responses can be achieved against many disease-causing agents, including HIV, hepatitis B virus, and *Plasmodium* (the microbe that causes malaria), but it is still too early to tell if these responses are strong enough to provide immunity to these devastating diseases. Many researchers are optimistic, however, that DNA vaccines will be a potent weapon in the fight against some of the world's most persistent diseases.

REVIEW QUESTIONS

1. What are the elements of a homeostatic control system? Suppose you have accidentally touched a hot stove. Describe your response in terms of a homeostatic control system. Name the structures of your body that correspond to each of the system's elements.

2. If you inserted a tiny electrode into a neuron and placed another electrode outside the neuron, then measured the voltage between the two electrodes with a voltmeter, what would you see on your voltmeter? What is this called?

3. Describe the movements of ions that occur across the membrane of a neuron during an action potential. Which ions are involved? In which direction, into the cell or out of the cell, do they move? What force moves these ions?

4. What is myelin? What is it made of? What role does myelin play in neurons?

5. Describe what happens when an action potential reaches a chemical synapse.

6. Give an example of a voluntary activity. Now give an example of an involuntary activity. Which parts of the human nervous system mediate these two kinds of activities?

7. Suppose you are walking to class when a large, growling dog blocks your path. Describe some physiological responses you might have to this emergency. Which subdivision of your autonomic nervous system would dominate in this scenario?

8. How is the role of hormones similar to that of the nervous system? How is it different?

9. Hormones circulate in the blood and reach every cell of the body, yet only certain cells respond to certain hormones. Explain.

10. Compare the functions of the anterior and posterior pituitary gland. How are the hormones of the posterior pituitary regulated by the hypothalamus? How are the hormones of the anterior pituitary regulated by the hypothalamus? What are the advantages to having one gland regulated by hormones from another gland?

11. The four digestive processes are digestion, secretion, motility, and absorption. Which parts of the digestive system are involved in digestion? What is secreted in the stomach? Where in the digestive tract does most absorption occur? What are the two main kinds of motility?

12. How is the absorption of fat in the small intestine different from the absorption of proteins and carbohydrates?

13. What role does the liver play in digestion? What does the pancreas do?

14. Compare the structural features of the small intestine and the lungs that increase their surface areas. Why must the surface areas of each of these two organs be large? Neither the air spaces in the lungs nor the lumen of the intestine are considered to be inside the body. Explain.

15. How does the diaphragm function in respiration?

16. Describe the flow of blood through the body, starting in the right atrium and ending at the vena cava, and name the structures of the cardiovascular system through which the blood moves.

17. What are the electrical and mechanical components of the heartbeat? Where does the heartbeat originate?

18. What features of capillaries enable them to serve as the sites of exchange between the blood and the extracellular fluid? What kinds of substances move out of the capillaries? What stays in the capillaries? What moves into the capillaries?

19. Describe the relationship between the blood vessels of the nephron and the nephron tubule. Follow a drop of fluid from the renal corpuscle through the tubule, and to the bladder. Name the parts of the tubule through which the fluid moves.

20. What are the three basic renal processes? Which one occurs in the renal corpuscle? Which occurs in the proximal convoluted tubule?

21. Which renal processes are responsible for removing wastes from the blood? Which one is responsible for retrieving substances before they are excreted in the urine? Which process removes foreign substances, such as penicillin, from the blood?

22. How do changes in the permeability of the distal convoluted tubule and collecting duct act to conserve water? What hormone regulates this response?

23. The force of muscle on bone is always a pulling force and never a pushing force, yet we are capable of pushing on an outside load. Explain this.

24. Muscles cause movement by becoming shorter. If you examined a contracted muscle cell and a relaxed muscle cell under the microscope, would the proteins of the contracted cell be shorter compared with those of the relaxed cell? Why or why not?

25. Muscle contraction has both an electrical component and a mechanical component. What constitutes the electrical component? What constitutes the mechanical component? How are they related?

26. What is gametogenesis? How does it differ in males and females? How is it the same?

27. Vasectomy is a means of sterilization for males in which a portion of each vas deferens is removed. How does this prevent pregnancy? Does this procedure affect secondary male sexual characteristics, such as beard growth or voice quality? Why or why not?

28. If the first day of menstruation is arbitrarily assigned as day 1 of the menstrual cycle, on what days are the levels of FSH and LH the highest? What ovarian event do these peaks correspond to? When is the level of progesterone the highest?

29. In one kind of female birth control pill, progesterone levels are kept artificially high for most days of the female menstrual cycle. How does this prevent pregnancy? Can you propose altering the levels of other sex hormones as a method of birth control?

How does your proposed hormone treatment prevent pregnancy?

30. What role does the endometrium play in pregnancy?

31. The body has several avenues for warding off would-be invaders, including physical barriers, nonspecific immune defenses, and specific immune responses. Give an example of each. Which of these avenues of defense are mobilized by vaccines?

32. What is the significance of memory cells? If your classmate sneezes on you, exposing you to a novel virus, will the virus stimulate existing memory cells? Why or why not?

33. What happens when the immune system fails to distinguish between proteins that are a normal part of the body's repertoire and foreign proteins? Give an example of a disease in which this happens.

Plant Form and Function: How Do Plants Live in the World?

All flesh is grass, and the
goodliness thereof is as the flower of the field.
—Book of Isaiah 40:6

—Overview—

Wherever you live, if you look out your window on a summer's day, you will probably spot something green. If you live in the country, most of the landscape may be covered in green. Even the places you cannot see from your window, such as lakes, streams, and oceans, are teeming with green life. Plants and other photosynthetic organisms are among the most successful living things. Their diversity and abundance are unsurpassed. The secret of their success is their ability to adapt to nearly every kind of environment and to exploit nearly every kind of habitat.

The challenges faced by plants are not all that different from those faced by animals. Regardless of where they live, plants need food and water; they need to exchange gases with the environment; they need to grow, develop, and reproduce, just as animals do. This chapter tells how plants meet the challenges of living. We will see that, although the needs of plants are not that different from those of animals, the strategies plants use to meet those needs are unique. Even among the different kinds of plants, there are many styles for coping, so many, in fact, that it is necessary to organize our study of plant physiology within a meaningful framework. Though the plant kingdom is vast and varied, the connections in their life strategies are more obvious if we consider the important events in the history of plants that have molded and influenced their form and function. Thus we begin our study of plant life by looking at some key events in plant evolution.

Chapter opening photo—For the past 400 million years, plants have dominated the terrestrial landscape, as they do in this Brazilian tropical rainforest.

12.1 WHAT IMPORTANT EVENTS DEFINE THE HISTORY OF PLANT LIFE?

Over 3.5 billion years ago, the first cells became established on this hot, young planet. Prior to that, nearly a billion years of volcanic eruptions, intense heat, and relentless ultraviolet radiation from the sun had provided sufficient energy to create simple organic compounds from the various elements found on Earth. Earth's early atmosphere was devoid of oxygen, and these compounds were able to accumulate without being oxidized. This **abiotic** process, that is, occurring in the absence of life, enriched the primordial seas with organic compounds including amino acids, lipids, nucleotides, and sugars. The earliest cells were probably **heterotrophic**, meaning they could not make their own food. They depended upon ready-made organic materials for food, absorbing these compounds directly from their environment. As these early cells divided and their numbers increased, competition for this finite source of chemical energy became keen. A cell with the ability to provide its own organic food, an **autotroph**, would surely have had an advantage. Autotrophic cells undoubtedly developed early in the history of life.

12.1.1 Photosynthesis Changed the World

Most evidence indicates that the first autotrophs were **chemoautotrophs**, or cells that could extract energy and build organic compounds from inorganic acids that were plentiful in primordial seas. A few present-day prokaryotes, for example, those living around the deep-sea hydrothermal vents, are chemoautotrophs. They probably have much in common with those first autotrophic cells. But chemoautotrophs did not become widely successful. Instead it was the **photoautotrophs**, or cells that could tap the energy of sunlight to make food, that gave rise to the vast and successful plant kingdom.

Today's photoautotrophs come in all shapes and sizes. Some, the cyanobacteria, for example, are prokaryotic; others, the plants and algae, are eukaryotic. Some are tiny, single cells; others are large and multicellular. Some live in water, others on land, but almost without exception, they share the following property: photoautotrophs capture the energy of sunlight by means of the pigment **chlorophyll**, or a chlorophyll-like molecule (Figure 12-1). Because chlorophyll in one form or another is common to nearly all present-day photosynthesizers, scientists believe that the ability to make it must have evolved early in the history of life. The *ancestral* photoautotrophs from some 3.5 billion years ago contained the genes required to make chlorophyll. Those cells passed their genes on to their offspring and, over the eons, eventually gave rise to the photosynthetic bacteria, algae, and plants that exist today. As we will see, this evolutionary step changed the world.

When one thinks of photosynthesis, it is usually trees, bushes, grasses, and other land-dwelling plants that first to come to mind. However, the earliest photosynthesizers were simple, water-dwelling, prokaryotic cells. Their direct descendants are modern photosynthetic bacteria, but most of these bacteria do not generate oxygen and, in fact, do not even consume oxygen. They are the anaerobic photoautotrophs. For them, the oxygen in the atmosphere is a deadly poison. Consequently, these prokaryotes are restricted to habitats that have no oxygen, such as the mud at the bottom of lakes and oceans. We rarely see them. However, there is one group of photosynthetic bacteria, called the **cyanobacteria**, that generates oxygen during photosynthesis. These prokaryotes (you might know them from the scummy layer they form on the surface of stagnant ponds) gave rise to the entire array of modern algae and plants.

The highly successful cyanobacteria prospered and diversified in the primitive oceans. Their kind of photosynthesis, in which oxygen is produced as a waste product, had profound effects on the planet and on the life it supported. Between the appearance of the first life forms some 3.8 billion years ago and the beginning of the Precambrian era nearly 1.3 billion years later (Figure 12-2), cyanobacteria released sufficient oxygen to raise the oxygen content of the atmosphere to nearly 2 percent of its current

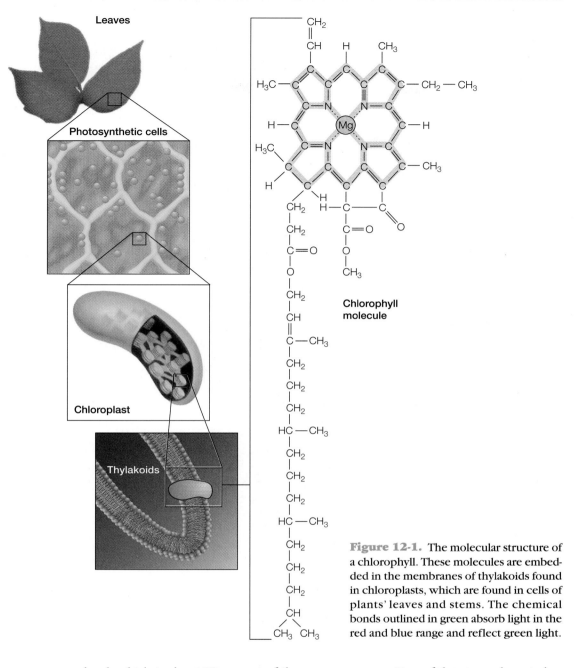

Figure 12-1. The molecular structure of a chlorophyll. These molecules are embedded in the membranes of thylakoids found in chloroplasts, which are found in cells of plants' leaves and stems. The chemical bonds outlined in green absorb light in the red and blue range and reflect green light.

level, which is about 20 percent of the gaseous composition of the atmosphere today. This "pollution" had major consequences for the life forms that existed at the time. Very few of them could tolerate this reactive gas. Earth became a place where only those few who could either tolerate atmospheric oxygen or escape it survived. With few exceptions, modern organisms are well adapted to the oxygen-rich atmosphere that is the legacy of the early cyanobacteria.

12.1.2 Eukaryotic Cells and Multicellularity Enabled Plants to Diversify

The evolutionary pathway from the first prokaryotic photosynthetic cells to today's eukaryotic plants was neither simple nor straightforward. One important step was the evolution of eukaryotic cells from prokaryotic ancestors. The differences between prokaryotes and

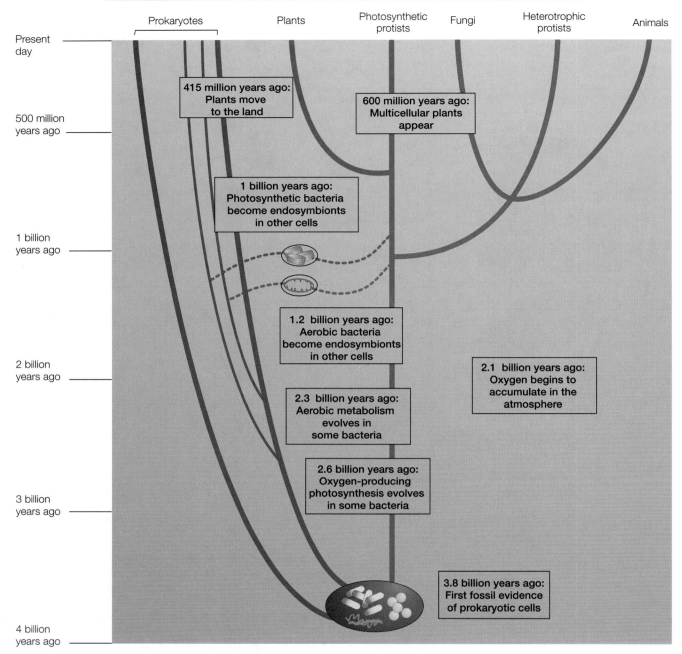

Prokaryotes   Plants   Photosynthetic protists   Fungi   Heterotrophic protists   Animals

Present day

500 million years ago

1 billion years ago

2 billion years ago

3 billion years ago

4 billion years ago

415 million years ago: Plants move to the land

600 million years ago: Multicellular plants appear

1 billion years ago: Photosynthetic bacteria become endosymbionts in other cells

1.2 billion years ago: Aerobic bacteria become endosymbionts in other cells

2.1 billion years ago: Oxygen begins to accumulate in the atmosphere

2.3 billion years ago: Aerobic metabolism evolves in some bacteria

2.6 billion years ago: Oxygen-producing photosynthesis evolves in some bacteria

3.8 billion years ago: First fossil evidence of prokaryotic cells

Figure 12-2. Some important events in the evolution of plants. These events are explained in Section 12.1.

eukaryotes are profound. The distinction between these two types of cells is the most fundamental dichotomy in the living world. Recall that prokaryotic organisms are typically single cells that are surrounded by a cell membrane and contain DNA, but they lack complex internal structures (Chapter 4). Eukaryotes may be either single celled or multicellular. All truly multicellular organisms, including today's plants, fungi, and animals, are eukaryotic. Eukaryotic cells are generally about 10 times larger than prokaryotic cells. Eukaryotic DNA is contained in chromosomes, which are enclosed in a membrane-bound nucleus. Other membrane-bound organelles are specialized for aerobic respiration (mitochondria), photosynthesis (chloroplasts), and intracellular transport (endoplasmic reticulum and the Golgi apparatus). Eukaryotic cells, unlike prokaryotes, have an internal protein cytoskeleton, which determines the shape of the cell and plays a role in cell motion and cell division. The events by which prokaryotes gave rise to eukaryotes were for many years unclear. How did that happen? How were the first eukaryotic cells formed?

There is a curious similarity between certain prokaryotic cells and the larger organelles of eukaryotic cells, particularly mitochondria and chloroplasts. For example, mitochondria and chloroplasts are about the same size as bacteria. Certain features of their structure and chemistry, too, are similar to bacteria. In the 1880s, it was proposed that today's eukaryotic cells may have arisen several billion years ago when one prokaryote engulfed another (or several other) prokaryotes, and the second cell became a friendly trespasser inside the cytoplasm of the first. This idea, now known as the **endosymbiotic theory** of the origin of eukaryotic cells, was largely ignored for nearly a century. In 1967, a young scientist named Lynn Margulis decided to investigate the idea in a systematic way. Her groundbreaking article, which helped to bring the endosymbiotic theory into the mainstream of scientific thinking, was submitted to several journals and was rejected by nearly all of them. Finally, the avant-garde *Journal of Theoretical Biology* agreed to publish Margulis's article, entitled "On the Origin of Mitosing Cells." Today, largely due to the overwhelming evidence brought forward by Lynn Margulis, the endosymbiotic theory of the origin of eukaryotic cells is widely accepted by biologists everywhere. That evidence, set out by Margulis in her first article and later in her book entitled *Symbiosis in Cell Evolution*, includes the following observations:

1. Mitochondria and chloroplasts have their own DNA, which is separate from the DNA of the nucleus in the cells in which they reside. Mitochondrial and chloroplast DNA occurs as a single loop that is not wrapped around histone proteins. This is unlike nuclear DNA, which occurs as chromosomes that are tightly affiliated with histone proteins. The DNA of prokaryotic cells is more like that of mitochondria and chloroplasts: a single DNA loop without histone proteins.

2. Mitochondria and chloroplasts are self-replicating organelles; that is, they divide within the cytoplasm of the cell by a process that resembles the binary fission that prokaryotic cells use to divide.

3. Mitochondria and chloroplasts have their own protein-synthesizing machinery, including ribosomes, transfer RNAs, and enzymes (see Chapter 6 and the CD to review these structures and the process of translation, or protein synthesis). The translation machinery in mitochondria and chloroplasts resembles that in prokaryotes more closely than it does the translation machinery in the cytoplasm of eukaryotic cells.

4. There are examples of living prokaryotes that share many features with both mitochondria and chloroplasts of eukaryotic cells. For example, the cyanobacteria are prokaryotes that have a system of photosynthesis that closely resembles that of chloroplasts in higher plants. There are aerobic bacteria whose metabolic processes resemble those of mitochondria in eukaryotic cells, as well.

Although it is impossible to observe events that happened in the ancient past, it is entirely possible to design and do experiments that test hypotheses about what might have happened during the course of evolution. What kinds of experiments can you devise to test the hypothesis that eukaryotic cells arose by means of endosymbiosis among ancient prokaryotes?

A second important step in the evolutionary pathway leading to the higher organisms, both plants and animals, was the development of large, complex, multicellular forms from single-celled ancestors. Whereas some prokaryotic cells remain in contact after they divide, forming filaments or loosely affiliated colonies of cells, an organism is truly multicellular only when several neighboring cells adhere, interact, and communicate with one another and are specialized to perform separate roles in the life of the organism. It was only after the evolution of eukaryotic cells that truly multicellular organisms developed. As we will see, multicellularity was a necessary condition for the successful invasion of the land.

What are the advantages of multicellularity? As we learned in Chapter 10, multicellular organisms have the opportunity for cellular specialization. In other words, within the

**Fast Find
12.1b
Multi-
cellularity**

same individual, some cells can provide food—by photosynthesis, for example. Others can become specialized for reproduction, and still other cells can become specialized for structural support or conducting substances throughout the organism. In unicellular species, there can be no specialization; each cell must carry out all the life processes. An additional benefit of multicellularity is a decreased vulnerability to changes in temperature, humidity, and nutrient availability that comes with increased size. There is even a decreased vulnerability to death from predation or disease. When one part of a multicellular organism is attacked, it can be cut off from the rest of the organism, allowing the individual to survive. Such adaptations were essential for survival on the land because, unlike the relatively stable environment of the sea, conditions on land are constantly changing.

12.1.3 Plants Moved from Water to Land

Long before animals moved from the early seas to dry land, plants had already successfully adapted to life out of water. Indeed, animals could not have survived on land without the food and shelter provided by green plants. The greening of the terrestrial landscape is one more of the critically important ways in which plants have changed the very character of the planet. What adaptations were necessary for life on land? What distinguishes a land plant from those that came before?

To be a land plant, in the strictest sense, means not only surviving out of the water but successfully reproducing on dry land as well. Life on land means life in dry air; thriving on the land means coping with desiccation. As we will see in the remainder of this chapter, much of plant form and physiology is about acquiring, transporting, and holding on to water. All the varied forms of land plants share two important characteristics to cope with dessication. First, land plants have adaptations to minimize water loss across their surfaces. All land plants have an **epidermis**, or a transparent layer of cells on the outer surface that helps to retard water loss. Most also have a waterproof waxy coating on their surfaces, called the **cuticle**. Second, the delicate egg and developing embryo of land plants are enclosed in a structure that prevents them from drying out.

Land plants share another characteristic, as well. Their life cycle is divided into two stages, a system called **alternation of generations** (Figure 12-3). How are the two stages different? Land plants can reproduce sexually, as all eukaryotic species can. Therefore, the life cycle includes both meiosis (cell division that reduces the chromosome number to one-half of the diploid number; see Chapter 5), and fertilization (fusion of gametes producing a zygote with chromosomes from both egg and sperm). In the human life cycle, the haploid phase (*n* chromosomes) that results from meiosis is only one cell, either an egg in the female or a sperm in the male. The human organism grows by mitosis only *after* the egg and sperm fuse, creating a diploid cell (2*n* chromosomes). But in land plants, both the haploid cells that result from meiosis *and* the diploid cells that result from fertilization grow into larger structures by means of mitosis. In other words, during part of the life cycle there is a multicellular structure called the **gametophyte** that is haploid, and during another part of the life cycle there is a multicellular structure called the **sporophyte** that is diploid.

Throughout the course of evolution, two different lineages of land plants have followed two different strategies for life on land. One strategy has resulted in a group of plants with a conspicuous gametophyte generation, that is, a haploid structure that is nutritionally independent of the sporophyte. The gametophyte can be as large as or even larger than the diploid sporophyte. The plants with dominant gametophytes are the **bryophytes**, which include the mosses, liverworts, and hornworts. Though most of these plants are small and grow close to the ground, they have been around a long time, and there are many different kinds. The second evolutionary strategy has resulted in a group of plants with a dominant sporophyte. In this group, the smaller gametophyte generation remains attached to the sporophyte and depends on it for nutrition. Plants in this group also have specialized conducting tissues, called vascular tissues, for moving food and water. This group, referred to as the **vascular plants**, includes the evergreens and flowering plants,

**Fast Find
12.1c
Alternation
of Generations**

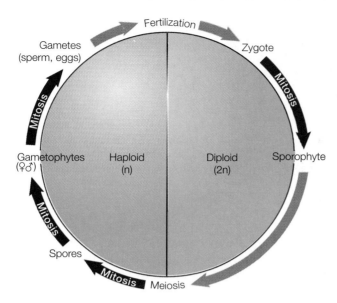

Figure 12-3. Alternation of generations. Plant life cycles are characterized by the presence of both a diploid sporophyte generation and a haploid gametophyte generation. Both sporophyte and gametophyte are multicellular.

among others. In some vascular plants, the gametophyte generation is reduced to just a few cells that stay enclosed in a structure on a large sporophyte (Figure 12-4). The vascular plants are the most successful and diverse group of plants alive today.

The most vulnerable stages in the life cycle of any organism, especially land-dwelling species, are the gametes and the embryos. These tiny structures are more subject to desiccation, damage, and attack by predators or pathogens than the larger, more mature stages. Another important evolutionary development that occurred in most of the vascular plants, which is certainly partly responsible for their phenomenal success, is the

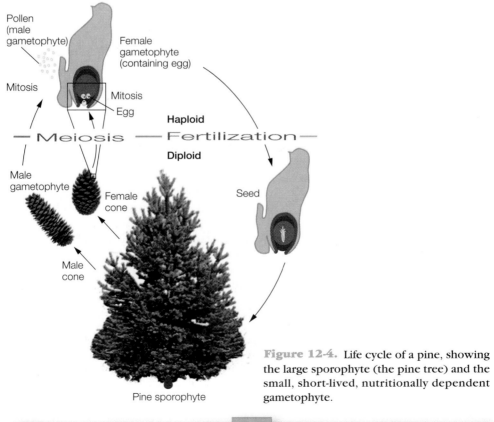

Figure 12-4. Life cycle of a pine, showing the large sporophyte (the pine tree) and the small, short-lived, nutritionally dependent gametophyte.

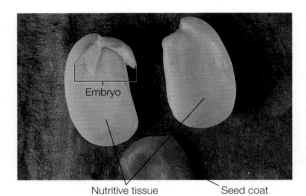

Figure 12-5. Bean seed, separated into halves. The left half contains the delicate embryo, and both halves have nutritive tissue. The tough, resistant seed coat is shown nearby.

seed. A **seed** is a delicate plant embryo encased in a tough, drought-resistant coat (Figure 12-5). Some nutritive tissue is often present as well to nourish the embryo until it can provide its own nourishment. The seed not only protects and nourishes the developing embryo, it is exquisitely structured to disseminate that embryo to new places where it can grow and flourish. This is especially important because plants do not move from place to place; often the seed is the only phase of the life cycle that can move about (or more accurately *be moved* about by wind, water, and animals) and colonize new territory. Whereas not all vascular plants make seeds (ferns, for example, are seedless vascular plants), the most successful land plants are those that do.

12.1.4 **Vascular Plants Dominate the Terrain**

The first plants to colonize the land about 450 million years ago were small and simple by comparison with modern vascular plants. Lacking leaves and strong supportive tissues, they could not grow very large and were probably limited to the shores of lakes and primitive oceans. About 425 million years ago, these early land colonizers gave rise to the two distinct evolutionary lines that became the bryophytes and the vascular plants. Like their early ancestors, today's bryophytes lack vascular tissues and are therefore small and restricted to moist habitats. By contrast, the modern vascular plants, which account for about 95 percent of all living plants, have exploited a variety of different habitats.

Figure 12-6 illustrates the three major evolutionary lines of vascular plants: the seedless vascular plants, the gymnosperms, and the angiosperms. The seedless vascular plants have a long history. About 300 million years ago, at about the time the reptiles were first emerging, much of the land that is now Asia and North America was covered with swamps. Vast forests of seedless vascular plants grew in these swampy areas. When these plants died, the stagnant swamp water preserved them and prevented them from decaying. The result was an accumulation of black sediment derived from fossilized plant material, or coal. Coal, oil, and natural gas are all types of **fossil fuels**, so named because they were formed from the fossils of extinct plants. Today we burn fossil fuels in our cars and electricity-generating power plants. Fossil fuels are a finite natural resource. When they are depleted, we will need to find new ways of providing energy to meet our ever-growing needs.

The **gymnosperms**, or cone-bearing plants, made their first appearance about 300 million years ago. These plants, the first to produce seeds, shared the terrestrial landscape with the dinosaurs for 240 million years. The dominant gymnosperms of today are the conifers, or evergreens. Pines, firs, spruces, and the giant redwood trees of California are all conifers. Conifers form the vast forests of the northern temperate regions of the globe, where the wood they provide is of great commercial value.

Needles Are Leaves One reason the conifers are successful in the northern temperate regions has to do with their leaves. What leaf features make conifers particularly well adapted to cold, dry climates?

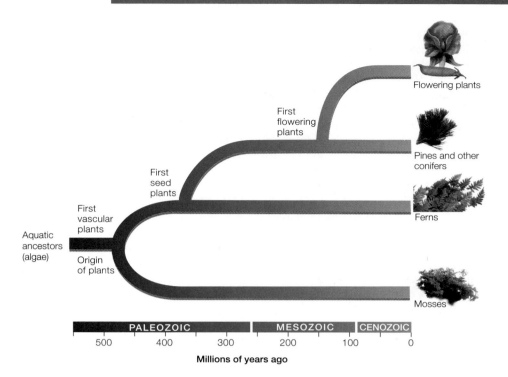

Figure 12-6. Evolution of land plants. The earliest land plants arose from aquatic algal ancestors about 415 million years ago. The bryophytes, such as mosses, are the oldest surviving land plants, and the angiosperms, or flowering plants, are the youngest.

Sometime during the Jurassic or Cretaceous period, just about the time the dinosaurs were ruling the animal kingdom, a new group of terrestrial plants arose: the **angiosperms**, or flowering plants. Unlike the dinosaurs, the angiosperms have survived to the present day and have become the most diverse and successful kind of land plant on Earth.

The defining characteristic of angiosperms is the **flower**. Horticulturists and gardeners have used artificial selection to produce such a vast array of beautiful flower types and colors that it is sometimes difficult to remember just what the flower is for: it is a structure that produces seeds. Flowers are, in fact, nothing more than stems bearing leaves, although the leaves are often highly colored and modified into petals, sepals, and reproductive structures called **stamens** and **carpels** (Figure 12-7). Stamens and

Figure 12-7. Flower of the Michigan lily. The petals, sepals (not shown in this photo), and reproductive structures of flowers are modified leaves.

carpels are the places where the angiosperm gametophytes are enclosed and where the gametes are produced, analogous to the gonads of animals (although animals have nothing analogous to the plant gametophyte). The vast array of flower types reflects the many different strategies that plants have evolved that ensure that sperm are delivered from one plant to another. In short, flowers are the sex organs of the angiosperms.

Have you ever admired a friend's houseplant and tried to cultivate one for yourself using a small cutting? Growth of a new plant from part of another is a kind of reproduction that does not entail seeds. This type of reproduction, which takes place without seeds, is known as *vegetative propagation* and is a form of asexual reproduction. The ability to propagate asexually was another feature of the angiosperms that undoubtedly contributed to their evolutionary success. Vegetative propagation is a way for the size of the population to increase rapidly. When environmental conditions are stable, plants that reproduce vegetatively can outcompete plants that must rely on slower sexual reproduction to increase their numbers. The disadvantage of vegetative propagation is that all of the new individuals are genetically identical. Each is derived from a parent plant by mitosis without the advantage of genetic recombination that occurs with sexual reproduction. Genetic recombination provides for variety, and variety increases the likelihood that some members of the population will survive if conditions change (to review genetic recombination and the advantages of variety, see Chapters 3 and 7). When the environment is fairly stable, vegetative propagation is advantageous; when conditions fluctuate, sexual reproduction is advantageous. The flowering plants have both options, the best of both worlds.

For the remainder of this chapter, we will examine the manner in which individual plants meet the requirements for life. Our emphasis will be on the flowering plants both because they are the most successful plant group and because they are of the greatest economic importance to humans.

Piecing It Together

A look at some of the key events in the history of plant life provides a framework for understanding the diverse plant kingdom.

1. Plants use chlorophyll to capture the energy of sunlight for use in photosynthesis; oxygen is released as a byproduct. The atmospheric oxygen that has accumulated from photosynthesis over the past 2.5 billion years has changed the character of the planet and all life forms that occupy it.

2. The evolution of eukaryotic cells occurred before plants became multicellular. Eukaryotic cells are thought to have evolved when one or more prokaryotes became endosymbionts, living within the cytoplasm of another cell. A great deal of scientific evidence supports the endosymbiotic theory of the origins of eukaryotic cells.

3. The invasion of land was an important step in plant evolution. Several evolutionary adaptations enabled plants to tolerate life on dry land, including multicellularity, a waxy cuticle, an epidermal layer, and structures that protect and enclose the delicate gametes and embryo, including the seed.

4. Alternation of generations is a characteristic feature of the plant kingdom. Both the diploid stage resulting from fertilization and the haploid stage resulting from meiosis divide by mitosis. The multicellular diploid form is the sporophyte and the multicellular haploid form is the gametophyte.

5. Flowering plants, called angiosperms, have been wildly successful and dominate the modern terrestrial landscape.

12.2 WHAT DO PLANTS NEED AND HOW DO THEY GET IT?

Plants, like animals, need air, water, and nutrients, but as we will see, their strategies for meeting these needs are far different from our own. While we *collect* our food (directly or indirectly) from plants, plants *produce* their food from carbon dioxide and water using energy from sunlight. We shall begin by considering some of the characteristics of solar radiation and the manner in which plants collect it.

12.2.1 Plants Need Light

Why is the sky blue and the Earth green? Bright colors indicate that some of the wavelengths of light from the sun are absorbed; those that are not are reflected back to our eyes. The sky is blue because the atmosphere absorbs some wavelengths and bends and scatters others. Plants are green because pigments found in chloroplasts absorb strongly in the red and blue parts of the spectrum and reflect the green wavelengths. (Colors other than green are reflected as well, but the predominant pigment, green chlorophyll, masks them. In the fall, chlorophyll is the first pigment to disappear. Only then do we see orange, yellow, or red light reflected by other pigments.) The energy of light in the red and blue wavelengths is used to drive photosynthesis, the result of which is high-energy sugars (see Chapter 9). The chemistry of photosynthesis, however, does not tell the whole story of the elegant way in which plants have evolved to optimize the amount of light they can absorb.

The leaf is the place where most photosynthesis occurs, and it is exquisitely adapted for this important task. Most leaves are broad, flat, and thin. Sunlight is absorbed across the surface of leaves, and a broad flat surface provides for maximum light absorption with a minimum of leaf volume. Next time you are resting under a tree, look up at the arrangement of branches emerging from the trunk. Branches are arranged in an asymmetric fashion (Figure 12-8), an architectural feature that ensures that the leaves on

Figure 12-8. View looking upward at the branches of a tree. Branches emerge from the trunk in an asymmetric pattern, an arrangement that exposes the greatest number of leaves to the greatest amount of sunlight.

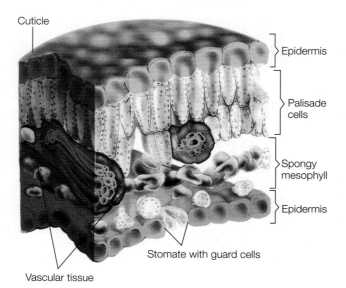

Cuticle

Epidermis

Palisade cells

Spongy mesophyll

Epidermis

Stomate with guard cells

Vascular tissue

Figure 12-9. Cross section of a typical leaf. Palisade cells and spongy mesophyll are sandwiched between layers of epidermal cells. The surfaces are coated with a waxy cuticle.

lower branches are not shaded by those directly above them, and that the greatest number of leaves are exposed to the most sunlight. But the sun is not always in the same position in the sky, and some leaves adjust their position to follow the sun throughout the day, a process called **solar tracking**.

The leaf interior, too, functions beautifully to maximize light absorption (Figure 12-9). The upper and lower surfaces of the leaf are covered with a layer of clear epidermis cells coated with a waxy cuticle that prevents water loss. Immediately below the upper epidermis is a layer of photosynthetic cells called the **palisade layer**. These cells are tall and thin and rich with chloroplasts, the chlorophyll-containing organelles. The palisade layer does much of the leaf's photosynthesizing. Palisade cells are tightly packed with very narrow spaces in between. The narrow spaces act much like optical fibers; they direct any light that is not absorbed by the palisade layer down to the next layer of cells, called the **spongy mesophyll**. The cells of the spongy mesophyll also photosynthesize, and they, too, are rich in chloroplasts. These cells are not as tightly packed as the palisade cells. Because the light they use must first pass through the palisade layer, these cells need an extra boost of light. The presence of air spaces creates lots of surfaces from which light entering through the "optical fiber" pathways can bounce and reflect. This gives these deeper cells more opportunities to absorb light. The whole leaf is an extremely efficient light trap with a structure that maximizes its potential for photosynthesis. The structure of leaves, however, functions for more than just light absorption; it enables them to absorb another essential ingredient for photosynthesis: carbon dioxide gas.

12.2.2 Plants Need Gases

Light provides the energy that drives photosynthesis, but carbon dioxide and water are the raw materials. Carbon atoms from gaseous CO_2 are assembled into the energy-rich sugars used by both plants and animals for food. For photosynthesis to work, the photosynthesizing cells under the epidermis need a constant supply of CO_2. Carbon dioxide constitutes only a small fraction of the different gases in the air. The largest component of air, about 80 percent, is nitrogen (as N_2). Oxygen, as O_2, is the next most abundant gas at just a bit less than 20 percent. Carbon dioxide constitutes about 0.03 percent of air, making it relatively rare. Air must diffuse into the cells of the palisade and spongy layers of the leaf, and once there, the CO_2 must diffuse into the chloroplasts where it is fixed into organic compounds.

This presents a potential dilemma for plants because the same adaptations that prevent desiccation—a waxy cuticle and a cellular epidermis—also act as barriers to the dif-

Figure 12-10. Scanning electron micrograph of a stomate (magnified 2800 times). Stomates are pores on the underside of leaves formed by two guard cells.

Guard cells

Stomate

Fast Find 12.2b Stomata

fusion of gases, including CO_2. The solution is the presence of pores, called **stomates** (or **stomata**), on the underside of the leaf (Figure 12-10). Stomates act as passageways for the diffusion of air, much as our mouths allow air to enter our lungs. The word *stoma* is Greek for "mouth." Stomates are composed of two cells, called **guard cells**, which surround and define an opening, or pore. The pore opens when the plant's greatest need is to acquire CO_2 and closes when the greatest need is to conserve water. When water is plentiful, guard cells swell with moisture. When they swell, they bend in a manner that creates a gap between them. When water is scarce, guard cells lose some of their rigidity and come together, closing the gap and preventing water from escaping. If a plant is well watered, all of the thousands of stomates on the underside of an average leaf remain open most of the day.

Most of the land plants—about 95 percent of species—acquire CO_2 this way and use it directly in photosynthesis. A small percentage of plants, however, have evolved an elegant, very efficient way of capturing CO_2, even on hot, dry days when stomates are closed most of the time. These plants, called **C_4 plants**, use a kind of chemical trap to capture and stockpile lots of CO_2 during the brief time when the stomates are open. The carbon dioxide is chemically combined with a three-carbon compound already present in the plant's cells to make a four-carbon compound, which is where these plants got the name C_4. The four-carbon compound is shuttled from the palisade and spongy cells to a specialized layer of cells that surround the leaf's veins, called the **bundle sheath** cells. Here, far away from the open stomates, the CO_2 is released from its four-carbon carrier and used to make sugars; the three-carbon compound is shuttled back to the palisade and spongy cells. Because of the bundle sheath layer, the anatomy of leaves from C_4 plants (Figure 12-11) is conspicuously different from the more common C_3 type of leaf (Figure 12-9).

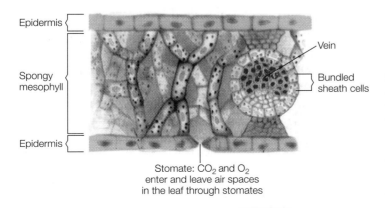

Epidermis

Spongy mesophyll

Epidermis

Vein

Bundled sheath cells

Stomate: CO_2 and O_2 enter and leave air spaces in the leaf through stomates

Figure 12-11. Cross section of a leaf from a C_4 plant. The vascular tissues of C_4 plants are surrounded by a layer of bundle sheath cells, which store carbon dioxide in the form of a four-carbon compound.

Only about 3 percent of plant species are C_4 plants, but because they can photosynthesize so efficiently, they are numerous and successful in areas where there are long periods of hot, dry weather. Some of the most important food crops are C_4 plants, including maize (corn), sorghum, and sugar cane. Crabgrass, too, is a C_4 plant, which explains why it can take over a lawn of ordinary grass in no time at all when the weather is hot and dry.

Desert-dwelling plants have evolved yet another interesting way of dealing with the competing needs to conserve water and acquire carbon dioxide. The succulent plants, such as cactus, pineapple, and other fleshy species, live in such hot dry areas that their stomates must remain closed all day to prevent water loss. They open only at night. But during the night there is no light for photosynthesis. Carbon dioxide enters the plant during the night and is stored as part of a four-carbon compound, similar to the C_4 plants. During daylight hours, the stored CO_2 is released behind closed stomates. These types of plants are called **CAM plants** (which stands for crassulacean acid metabolism, named for the plant family Crassulaceae in which it was discovered). Only about 2 percent of plant species are CAM plants, and all are found in desert habitats.

Oxygen, too, is a requirement for plants, just as it is for animals. Oxygen is used in aerobic metabolic pathways in the mitochondria that produce ATP, described in Chapter 9. But two important ways in which plants differ from animals make the need for oxygen less urgent. First, plants have generally lower energy needs than animals because they do not have muscle tissue making high oxygen demands for movement. Second, oxygen is a byproduct of photosynthesis, and green plant tissues usually make more O_2 than their mitochondria need. The same adaptations that allow CO_2 into the leaf are useful for removing excess oxygen or for obtaining atmospheric O_2 in the rare instances when that is necessary. Some tissues, namely bulky tissues such as potato tubers or carrot root, are not exposed to the air, however, and do not make O_2. Tiny air spaces in these tissues contain the oxygen required for aerobic metabolism. Species that live in standing water or in areas with frequent floods often have hollow stems that carry air to their submerged stems and roots, ensuring an adequate supply of oxygen. Rice seedlings growing in wet paddies rely initially on anaerobic pathways to supply their energy needs (Chapter 9), but they rapidly develop hollow stems to carry oxygen to the roots so that they can switch to the more efficient, aerobic metabolism. In hollow-stemmed species, the architecture of the plant reflects a need to bring oxygen to submerged tissues. For most land plants, however, body architecture can be related to the need for water.

12.2.3 Plants Need Water

Water is essential. It is the medium in which life evolved, and that aquatic beginning is evident in all the ways in which terrestrial life still depends on it. Water is the solvent for the many chemical reactions that occur in cells. It provides the electrons used in photosynthesis; it is the medium in which gametes come together in reproduction; it is the solvent in which substances are transported throughout the plant body; it is evaporated from leaf surfaces to cool the plant on a hot summer day. The consequences of too little water are dire indeed. They are often fatal. The successful invasion of land involved ways of more efficiently coping with much less water. Today's land plants have elaborate ways of bringing water into their bodies from their surroundings.

Roots

**Fast Find
12.2c
Root
Growth**

Plants use **roots**, or the descending portions of the plant that anchor it into the ground, to absorb moisture (and minerals) from the soil in which they grow. Root structure is exquisitely suited for this purpose. First, plant roots are often enormous compared with the parts of the plant that appear aboveground. Some plants have root systems with one deep **primary root** that extends downward for one or several meters or, in the case of some desert plants, for tens of meters to reach moisture deep in the dry earth.

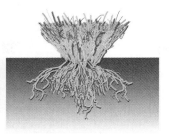

Taproot system (carrot)   Fibrous root system (grass)

Figure 12-12. Different kinds of root systems.

The primary root gives rise to side branches, called lateral roots. This system is called a **taproot** system (Figure 12-12). Other plants have shallow, laterally directed root branches, called **adventitious roots**. These roots absorb moisture from areas around the plant. This kind of lateral system, in which most of the root branches are roughly the same diameter, is called a **fibrous root system**. Both taproots and fibrous roots penetrate the soil in search of water and minerals.

A second feature that maximizes water and mineral absorption is found on the root surface. Tiny hairs, called **root hairs**, extend from the outermost layer of root epidermis cells (Figure 12-13). These delicate extensions of the epidermal cells increase the surface area of the root manyfold—an adaptation that permits more water to enter the root cells and, eventually, the plant body. In 1937, the botanist J. H. Dittmer estimated the total length and surface area of the root system from a single rye plant growing in just a few liters of soil. He found that the combined length of the roots, if lined up end to end, was over 600 kilometers, or nearly 400 miles. The combined surface area was nearly 650 square meters! Root hairs can account for nearly two-thirds of the capacity of the root to absorb water. It is no wonder that gardeners learn to treat roots gently when they transplant plants; yanking a plant from the soil destroys these delicate structures and much of the absorptive surface.

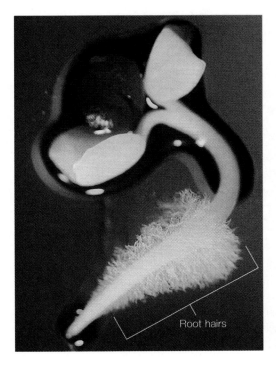

Root hairs

Figure 12-13. Radish seedling showing root hairs. Root hairs greatly increase the surface area for water absorption (magnified 30 times).

Roots of Friendship The roots of many plant species have an additional feature that aids in the uptake of water and inorganic nutrients, called **mycorrhizae**. Mycorrhizae are symbiotic associations between a plant's smallest rootlets and certain kinds of fungi. What kinds of plants have mycorrhizae? What different kinds of interactions can occur between rootlets and fungi? In a true symbiosis, both members benefit. How do fungi benefit from mycorrhizal relationships?

A third adaptive feature is the arrangement of cells that make up the inside of the root. The internal architecture of the root provides a conduit for moving water to all parts of the plant (Figure 12-14). The epidermal layer, complete with root hairs, provides the absorptive surface, and from there water moves from cell to cell through the bulk of the root, called the **cortex**, by diffusion. Air spaces dot the fleshy root cortex, and water moves both through and between the cells on its way to the **vascular tissue**, or the column (or columns) of cells that carry the water and minerals to the upper parts of the plant. A single sheetlike layer of cells, called the **endodermis**, forms a ring around the vascular tissue. The connections between endodermis cells are watertight, forcing water to move into the vascular tissues *through* the cells of the endodermis. In this manner, the endodermis cells regulate water movement into the plant body. (This arrangement also prevents pathogens from entering the vascular system.)

Water and mineral absorption are just two of several functions of roots, however. The fleshy root tissues of some plants serve as storage sites for the carbohydrates and other organic materials made in the leaves. Humans take advantage of this food storage in root crops such as carrots, sugar beets, and turnips. Roots act as anchors, too, holding plants firmly in place. Finally, in some plants, important molecules are made in the roots, including some of the hormones that regulate and integrate the life processes of the entire plant. These important substances are discussed in Section 12.4.

For a plant living on land, light, oxygen, and even carbon dioxide are relatively abundant or easily obtained. These essentials are more readily available than they are in the aquatic habitat. Water is one key substance that limits how and where plants grow on land, and nitrogen is another. Nitrogen is essential not just for land plants, but for all living things. And while nearly 80 percent of the air we breathe is nitrogen gas, it is

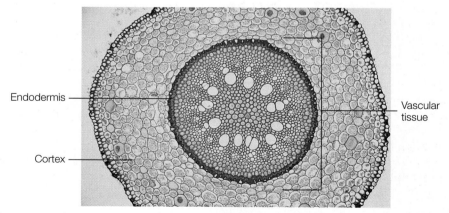

Endodermis

Vascular tissue

Cortex

Figure 12-14. Cross section through a root of a corn plant (magnified 80 times). Vascular tissue in the center of this root is formed from columns of cells that run from the root to the highest leaves. The red ring of cells surrounding the vascular tissue is endodermis. Also visible in this photograph is the root cortex (green cells) and the epidermal layer of cells (brown).

of little use to plants—or any of us, for that matter—in this form. Why do living systems need nitrogen, and how do they get it?

12.2.4 Plants Need Nitrogen and Other Nutrients

We learned in Chapter 4 that the cellular machinery is largely made of protein, and protein is made of amino acids. If you recall, all of the amino acids contain the element nitrogen. Recall, too, that the nucleotides of nucleic acids also contain nitrogen. Without nitrogen there could be neither protein nor nucleic acid; there could be no life.

Nitrogen is enormously abundant on Earth. The nitrogen gas that forms nearly 80 percent of the air we breathe is only a fraction—about 7 percent—of Earth's total nitrogen. It is estimated that another 93 percent is locked up in the rocks that form the planet's crust, and a tiny fraction of a percent is found in organic matter as part of the proteins and nucleic acids of living things. If it is so abundant, then how does nitrogen limit the distribution of plants and, indirectly, of animals? The answer has to do with the different forms in which nitrogen occurs. Only certain forms of the element can be used by living systems.

Nitrogen locked away in the rocks of the Earth's crust is occasionally freed by the natural processes of weathering and erosion, but this accounts for only a tiny fraction of what is needed by living things. Nitrogen in air occurs as N_2, and for reasons we will soon see, this form, too, is unavailable to most living things. For animals, the only usable form of nitrogen is *organic* nitrogen, that is, nitrogen that is part of a carbon-containing compound such as an amino acid. Hence, animal diets must include protein, the macromolecule made of amino acids. Adult humans, for example, must replace about 40 grams of protein every day, lost mainly as feces; sweat; menstrual fluid; and shed skin, hair, and nails. Many of us depend on animal products to meet our protein needs, but the animals we eat depend upon nitrogen from plants, and ultimately, so do we. While animals use organic nitrogen to make most of the amino acids, there are nine amino acids that we cannot make, called the **essential amino acids**.[1] These we must get ready-made from our diets, either directly from the plants that make them or indirectly from the flesh of animals that have eaten plants. So the question becomes Where do plants get nitrogen?

Unlike animals, plants can use nitrogen in inorganic forms, such as ammonia or nitrate (NH_3 or NO_3^-). Soil contains these compounds in limited supply, and plant roots absorb them and turn them into amino acids and nucleotides for use in their own life processes. But how do ammonia and nitrate enter the soil? When plants and animals die, bacteria and fungi attack the dead organic matter and use the energy-rich compounds as a source of chemical energy in a process called **decomposition**. As a result of decomposition, nitrogen is released into the soil in the form of nitrate and ammonia. Plants can absorb these nitrogen compounds and convert them into amino acids and other nitrogen-containing compounds.

'Round and 'round it goes: plants get nitrogen from soil; animals eat plants; plants and animals die; microbes decompose the bodies and return nitrogen to the soil. It is all part of the **nitrogen cycle**, an important natural system that keeps nitrogen available to living forms. The nitrogen cycle is described in some detail in Chapter 15, in the context of global ecology. The cycle, however, is not perfectly efficient. When organic matter decomposes, some of the nitrogen is returned to the atmosphere as N_2 and must

---

[1]The nine essential amino acids are histidine, isoleucine, leucine, lysine, methionine, phenylalanine, threonine, tryptophan, and valine. Most plant proteins are deficient in one or more of the essential amino acids, and a diet based on a single plant food can lead to amino acid deficiencies and their associated diseases.

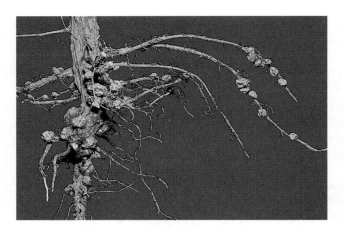

Figure 12-15. Nodules on the roots of a soybean plant house nitrogen-fixing bacteria.

be replenished. If it were not, life would soon come to a halt. Inert nitrogen gas from the air must be brought into the cycle to keep it turning. How is nitrogen from the air introduced into the nitrogen cycle?

Nitrogen gas in the air we breathe occurs as two atoms of nitrogen connected by covalent bonds, written as N_2. Alternatively, we can write it showing the covalent bonds between the two nitrogen nuclei, N N. Note that there are *three* bonds between the two nitrogen atoms. This is a very strong bond that neither plants nor animals have the ability to break. There are a few living things, all of them prokaryotic, that have evolved the ability to break this triple bond. Without them, plants and animals would soon run out of nitrogen. The process in which N_2 is used by certain microbes to make organic compounds is called **nitrogen fixation**. The organisms that can fix nitrogen are called the nitrogen-fixing microbes.

The nitrogen-fixing microbes include both free-living species and species that form close **symbiotic** relationships with other organisms. The free-living forms are widespread in soils and sediments; on leaf and bark surfaces; and, in some cases, in the intestinal tracts of animals. Breaking the bond between nitrogen atoms in air is energetically expensive, and these free-living forms generally grow slowly because most of their energy is spent fixing nitrogen, so less is available for growth. Symbiotic species have a special, intimate relationship with their hosts in which the host provides chemical food energy to the bacterium in return for nitrogen in a usable form. This beautiful relationship is best exemplified by the match between some of the legumes, such as alfalfa, clover, peas, beans, and peanuts, and their symbiotic bacteria. As legume seedlings develop, their roots secrete substances into the soil that attract bacteria called rhizobia.[2] Rhizobia move to the root, digest the cell wall of the host plant root hairs, and enter the root. Inside the root they grow into the cortex (Figure 12-15), causing plant root cells to divide by mitosis. The result is a lump, or **nodule**, in which the nitrogen-fixing bacteria live comfortably. They have access to sugars produced by the plant and are protected from oxygen in the soil and atmosphere. A plentiful supply of sugars is important because it takes nearly the same amount of ATP for a bacterium to convert one molecule of N_2 to two molecules of NH_3 as it takes for the plant to make one molecule of sugar from carbon dioxide; that is, it takes 16 molecules of ATP (see Chapter 9).

Nitrogen fixation by soil-borne microbes is an essential step in the global cycling of nitrogen between living things and the environment. All living organisms depend upon it. Green plants form a crucial link in this cycle by absorbing nitrogen (as nitrates

---

[2]The rhizobia are a group that includes three genera of bacteria: *Rhizobium*, *Bradyrhizobium*, and *Azorhizobium*. These genera are further divided into several species. Some species are quite particular, affiliating with only one host plant; others can infect a small range of host plants.

and ammonia) from the soil and from their associated microbes, and by converting it to organic nitrogen (amino acids) that can be used by animals.

In addition to light, water, gases, and nitrogen, good plant nutrition depends upon other, less obvious nutrients as well. Just as we need minerals in our diets, such as iron to make hemoglobin and calcium for bones, plants require an assortment of minerals for their body structure and metabolic machinery. The list of required plant minerals includes some that animals need, such as phosphorus, calcium, iron, sulfur, potassium, copper, zinc, and magnesium (Table 12-1). Other mineral needs are specific to plants, such as molybdenum and nickel, which are required only in tiny amounts but are needed nonetheless. Here, as we shall see, is yet another way in which plants play an essential role in the biosphere.

Plants meet their mineral needs in the same way they meet their water needs, that is, by using their roots. Roots penetrate a large volume of soil, often to considerable depths, from which they extract water. When water enters the plant through the roots, it brings with it dissolved minerals. Minerals move up the body of the plant from deep in the soil and are transported to the stems and leaves. In this way, plants are like miners. They take minerals from deep in the soil and bring them to the surface. Animals may eat the plants, providing for some of their own mineral needs in the process. Also, plants may drop leaves or die. When plant parts decay, minerals are returned to the soil, but in this case they are on the surface where they are available to soil organisms. Without the constant retrieval from the deep soil by means of roots, essential minerals would soon be washed away by rain and weather. Mineral "mining" is yet another of the many vital activities of plants upon which all life depends.

Table 12-1 The nutrient needs of plants and humans.

	Plants	**Humans**
Essential macronutrients	carbon	carbon
	hydrogen	hydrogen
	oxygen	oxygen
	nitrogen	nitrogen
	potassium	calcium
	calcium	phosphorus
	magnesium	magnesium
	phosphorus	sodium
	sulfur	chlorine
		potassium
		iron
Essential micronutrients	chlorine	zinc
	boron	iodine
	iron	fluorine
	manganese	copper
	zinc	selenium
	copper	chromium
	nickel	manganese
	molybdenum	
	potassium	
Beneficial elements*	sodium	molybdenum
	silicon	
	cobalt	
	selenium	

*Some may be toxic at high concentrations.

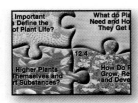

Piecing It Together

Plant form and function are best understood in terms of their needs:

1. Plants need light to provide energy for photosynthesis. Leaves are plant organs that are exquisitely adapted to capture light.

2. Plants need carbon dioxide (CO_2) as a raw material for making sugars by means of photosynthesis. But some of the features that enable plants to acquire CO_2, such as open stomates, also cause them to lose water. Stomates can open and close, depending on the balance between the need for CO_2 and the need to retain water. C_4 and CAM plants have evolved metabolic adaptations to balance these needs, especially in dry habitats.

3. Plants need water. Water is the solvent in which all cells function, and for plants, it is also a raw material for photosynthesis. Roots are plant organs that absorb water from the ground.

4. Plants need nitrogen and other minerals. Some plants have symbiotic relationships with microbes that can fix atmospheric nitrogen and convert it to a form that can be used by other living things. Most living things depend on nitrogen fixation accomplished by symbiotic microbes on plants.

12. 3 HOW DO HIGHER PLANTS SUPPORT THEMSELVES AND TRANSPORT SUBSTANCES?

As you engage in the different activities of your life—sleeping, eating, running to class, solving problems—the demands that your various tissues make on your body's resources vary. Active muscles need energy and oxygen; a full gut requires a copious supply of blood. Plant lives, too, are characterized by a variety of different activities. For example, photosynthesis occurs mainly in the leaves during daylight hours; water and mineral absorption occurs in the roots, especially after a rain; growth, reproduction, and seed development occur in certain tissues at specified times of the year. The needs of plants for different resources vary with their many different activities. The sites at which resources are taken in, processed, or produced do not always coincide with the tissues that need them most. The animal circulatory system moves substances throughout the body from where they are abundant to where they are needed. But what about plants? Plants, too, have a kind of circulation, but as we will see, it differs in many important ways from that of animals. How are substances moved within the plant body?

12.3.1 Translocation Occurs in Conducting Tissues

The process by which materials are moved in bulk throughout the animal body is called circulation. The analogous term for the movement of substances in plants is **translocation**. In the late 17th century, an Italian scientist named Marcello Malpighi did an experiment that provided insights into plant translocation. Malpighi removed a complete ring of bark from around the trunk of a tree, a technique called *girdling*. After a few weeks, he noticed that the bark just above the girdle became swollen with fluid. The bark just below the girdle became thinner. Juices from the upper parts of the tree were accumulating above the girdle. These juices, called sap, accumulated to a greater extent in the summer when the tree's leaves were actively photosynthesizing than in the winter when the leaves were gone. Later the sap was analyzed and found to be a complex mixture rich in sugars. Malpighi correctly concluded that the tissues near the periphery of the tree trunk carry sugars from the leaves, where they are made, downward to the lower parts of the tree. Girdling did not affect the movement of fluids upward from the roots. Upward movements of fluids must be taking place in different tissues than the downward movement of fluids, but Malpighi did not know where.

Malpighi's experiments provided evidence that plants have conducting tissues that carry substances to the various parts of the plant body. In green leaves, in the presence of light and air, the sugars produced by means of photosynthesis enter the sap and travel to other parts of the plant that are not green, such as trunk tissue and roots, where they provide energy for growth and metabolism. At the same time, deep roots collect water and minerals that are transported to even the highest leaf on the tallest tree. We now know a great deal about the structure and mechanism of conducting tissues in plants.

Vascular plants, which include the ferns, gymnosperms, and angiosperms, have two main kinds of **vascular tissues**, that is, tissues adapted for carrying fluids. Vascular tissues are complex and contain several different types of cells. Some are specialized for carrying sap; others provide support for the plant or aid in exchanging substances between conducting cells and the other cells. The vascular tissue that carries sugar-rich fluids is called **phloem**.

**Fast Find
12.3c
Phloem
Transport**

The conducting cells of phloem are connected end to end, forming long narrow tubes of cells called **sieve tubes** (Figure 12-16). The individual cells within a sieve tube are called **sieve-tube members**. These living cells have clusters of large pores on the cell walls of the abutting ends that resemble the holes in a sieve. Sugar-rich sap moves freely from cell to cell through these pores to all parts of the plant. Many such tubes are packaged side by side, making bundles of phloem that go all the way from the leaves to the roots. **Companion cells**, which are adjacent to sieve-tube members, aid in moving substances into and out of the sap, sometimes at the expense of metabolic energy.

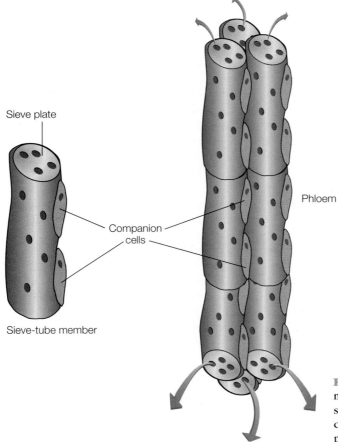

Sieve plate

Companion cells

Phloem

Sieve-tube member

Sap movement

Figure 12-16. Phloem is made of columns of perforated sieve-tube members and their companion cells. The companion cells provide nourishment for the sieve tube cells.

Where does the force for the movement of sap originate? Animal circulation, as we saw in Chapter 10, depends on a pump—the heart—to generate force to move fluids throughout the body, but plants have nothing comparable to a heart. You might guess that gravity moves sap downward, but gravity alone is not sufficient to force sap through the many living cells that form a sieve tube. In fact, it is sometimes necessary to move sugar-rich sap upward, if, for example, new leaves at the top of a plant need sugars from lower leaves to fuel their growth. In these cases, gravity would work *against* the movement of sap. Instead, sap moves mainly by pressure gradients that are created between sites of photosynthesis and sites where sugars are unloaded and used or stored. As sugars are produced in green tissues, or the *source* tissues, companion cells pump them into the sieve tubes of phloem. Water, too, enters the sieve tubes as it follows the sugars by osmosis. Fluid entering the phloem creates pressure against the walls of the sieve tubes that pushes the sugar-rich sap along. Meanwhile, other tissues remove sugars from sap to fuel their metabolic needs. These sugar-consuming tissues are called *sink* tissues. As sugars move out of the phloem at sink tissues, water follows by osmosis and the fluid pressure is lowered. Pressure gradients between source and sink tissues, therefore, move sap throughout the dense, living phloem to all parts of the plant.

But what about the upward movement of fluids? Recall that roots are the organs that provide plants with most of their water and minerals. The direction of water movement, therefore, is mainly upward, and the forces that move water and the vascular tissues that carry it are different from those that move sugar.

12.3.2 Transpiration Moves Water and Minerals

Fast Find 12.3b Transpiration

Have you ever run barefoot across a lush green lawn on a sunny day? The cool grass is always a welcome relief after running on hot pavement. Why, if the grass and the pavement sit under the same hot sun, is the grass so much cooler than a paved surface? The answer is evaporative cooling. The evaporation of water through the stomates of plant leaves is called **transpiration**, and it has everything to do with the way in which plants move water.

The same structural features of leaves that serve to maximize their exposure to the sun for photosynthesis also tend to make them hot. If they become too hot, they will be damaged. Transpiration is one way in which plants avoid overheating. When water evaporates, heat energy is absorbed from the surface and the surroundings and used to convert liquid water into water vapor. Leaves of grass and other broad-leaved plants use evaporation to keep cool, much as sweating (followed by evaporation) keeps us cool on a hot day. This explains why grass is cooler than pavement. A plant growing in a temperate region of the world may lose more than 100 times its own weight in water during its lifetime by transpiration. In some tropical plants, water loss may be even higher.

But how is all this water moved upward against gravity from the roots into the stem and leaves? This was the question that the German botanist Eduard Strasburger asked in the late 19th century. Strasburger wondered whether the movement of water up through a plant required living tissue, or whether it would happen even if the vascular tissue were dead. To answer this, he cut the trunk of a small oak tree close to the ground and put it in a solution of poison acid. The acid moved into the trunk and throughout the tree, killing the living cells as it went. When most of the cells had been killed, Strasburger replaced the acid with a bath of red dye. The red color spread throughout the tree, eventually tingeing even the tallest leaves pink. Strasburger had demonstrated that the upward movement of fluids does not require living cells. Even dead tissue can carry water from the ground to the highest leaves.

The upward movement of water and minerals occurs without benefit of a heart-like pump, indeed, without living tissue. The forces responsible for moving water up through a plant come from *negative pressure* that is created in the leaves when they lose water by transpiration. Imagine drinking water through a straw. As you draw first air, then water, from the straw, you create a negative pressure in the straw. Liquid moves up into the straw from below to replace the air and water that you have removed with your mouth. In a similar way, water and minerals from the roots move up into the xylem to replace the water that is lost to the atmosphere by evaporation at the leaves. The negative pressure generated by transpiration can be quite high, up to −30 atmospheres. (An atmosphere is a unit of pressure equal to that of the air at sea level.) This is the *negative* equivalent of the amount of positive pressure a deep-sea diver would experience at a depth of 985 feet!

One reason transpiration can move water so well has to do with the properties of water itself. Water molecules are **polar**, meaning the electrical charges of the atoms within the molecule are not evenly distributed across the molecule. Although a water molecule does not carry an overall electrical charge, one end is a bit positively charged and the other end has a bit of a negative charge. Because opposite charges attract, water molecules are attracted to each other and tend to stick together, with the positive end of one water molecule adhering to the negative end of another. This property gives water a great deal of cohesion, or **tensile strength**. This and other properties of water are reviewed in the Chemistry Review section of the CD. The tensile strength of water allows it to be pulled upward through a plant in a thin, unbroken column. But it also means that air bubbles or breaks within the column can disrupt the flow of water. Too many air bubbles can be fatal to a plant.

What is the nature of this tissue that can carry fluids when dead? **Xylem** is the main water-conducting tissue of vascular plants. Like phloem, xylem is made of elongated cells arranged end to end. But unlike the cells in phloem's sieve tubes, the cells of xylem are dead at maturity. All that remains of them are hardened cell walls joined at the ends to form a series of continuous, fluid-filled conducting tubes. Xylem extends throughout the plant, from the deepest roots to the tips of the highest branches. No cell is more than a few millimeters away from the xylem.

In flowering plants, xylem consists of several cell types (Figure 12-17). Some of these cells, the **fibers**, are long and thin, with thick cell walls. They provide structural support for the stem. Others, called tracheary elements, are less elongated. These cells have pits and perforations in their cell walls that allow water to move freely among them,

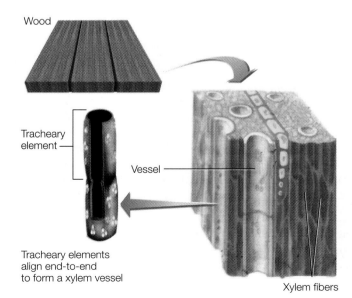

Wood

Tracheary element

Vessel

Tracheary elements align end-to-end to form a xylem vessel

Xylem fibers

Figure 12-17. Xylem is the main component of wood, which contains several cell types. Xylem fibers give wood its strength. Tracheary elements form conducting tubes for carrying water.

both end to end and sideways. In a mature plant, these cells form an extensive, inter-connected network of nonliving tubes that facilitates rapid transport of large volumes of water.

In some kinds of plants, the cell walls of the xylem cells are thickened and hard-ened by a resinous polymer called **lignin**. Lignified xylem, called wood, provides mechanical support for trees and woody shrubs. Support, however, is a requirement for more than just the trees and woody shrubs. Any plant taller than a ground-hugging moss needs some kind of mechanical support to counter gravity and keep the stem standing.

12.3.3 Plants Need Mechanical Support

Reaching for the sun means growing upward, and that means opposing gravity. The taller the plant, the stronger its stem must be to keep it upright. Water-filled cells that function as a kind of watery skeleton can support short, lightweight stems. Taller stems require specialized support tissue.

When we studied cell structure in Chapter 4, we saw that plant cells have a rigid cellulose wall surrounding the more flexible plasma membrane. The plant cell wall plays a critical role in providing mechanical support for soft tissues, but only when the cell inside is swollen with water. Water brought up from the roots moves into the cells of the stem and leaves by osmosis. This creates fluid pressure, or hydro-static pressure, that pushes against the cell walls. This hydrostatic pressure is called **turgor pressure**. Turgor pressure makes the cells rigid. As long as the plant has plenty of water, the cells remain swollen and the stem, leaves, and other soft tissues are supported.

Plants that rely on turgor pressure for support face two limitations. First, they must have access to plenty of water. Have you ever neglected your houseplants for several days? The result is a wilted plant or, in other words, a plant whose cells have lost their turgor and hence their mechanical support. The same is true for plants in the wild. For this reason (and the reasons we learned in Section 12.2.3 above), many nonwoody plants can only grow in moist areas.

Size is the second limitation for plants that rely on turgor pressure for support. Water itself is quite heavy, and plants that rely on hydrostatic support cannot grow to be too large or they begin to bend under their own weight. Soft-stemmed plants, therefore, do not grow to great heights.

But many woody plants do attain great height. Trees, for example, are among some of the largest organisms on Earth. The features of wood that enable trees to grow tall and strong and at the same time remain flexible in the wind are the same features that humans exploit when wood is used as a building material. Wood is mostly xylem tissue that has accumulated over the years and been strengthened with lignin. The lignin-fortified cell walls of the dead xylem combine to make a material that is strong, lightweight, and flexible—perfect for supporting stems of great height in changing weather. Even so, the stems of woody plants must get both taller and wider to fully support the bulk of a large tree. This means that stems must have the ability to grow in two directions. Stems grow both upward in length and outward in diameter. As we will see in the next section, plant growth means getting bigger in all directions.

 Monocots and Dicots The vast world of the flowering plants is divided into two subdivisions, the monocots and the dicots. What features distinguish these two groups? How do the arrangements of vascular tissues differ in the stems of monocots and dicots?

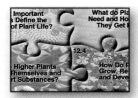

Piecing It Together

All large multicellular organisms, including plants, must have some way of transporting substances throughout their bodies. In some plants, the same tissues that are responsible for moving water also provide support.

1. Translocation is the name given to the movement of fluids within the plant body. Sugar-rich fluids, called sap, which are made in leaves and other photosynthetic tissues, move in phloem, which is the living vascular tissue found near the periphery of the stem. Water and minerals are moved within xylem, which is the vascular tissue usually found nearer to the core, or center, of the stem. Mature xylem is primarily composed of dead cells that form a hollow, interconnected network of tubes.

2. Plants rely on differences in pressure to move fluids. Sap is moved from photosynthetic, or source, tissues where fluid pressures are higher to nonphotosynthetic, or sink, tissues where they are lower. Water is translocated upward from roots to stems and leaves by means of negative pressure generated by transpiration from the leaves.

3. In soft-stemmed plants, mechanical support is provided by turgor pressure, or the force generated when water enters cells by osmosis and pushes against rigid cell walls. In woody plants, mechanical support is provided by xylem reinforced with lignin.

12.4 HOW DO PLANTS GROW, REPRODUCE, AND DEVELOP?

Have you ever known someone to hang a hammock by nailing a hook into the bark of a tree? If that person were to return several years later, the tree would undoubtedly be taller and perhaps even broader, but the hook itself would be no higher than it was the day it was placed. The reason has to do with the way that plants grow.

12.4.1 Plant Growth

Unlike most animals, plants continue to grow as long as they live. This type of growth is called **indeterminate growth**. As the plant gets longer and wider, it can exploit more soil for water and nutrients, more sunlight for photosynthesis, and more air for carbon dioxide. Animals, on the other hand, *move* through their environments gathering resources as they go. Animal growth is **determinate**; that is, they stop growing when they reach a certain size and shape, beyond which movement may become unwieldy. These different ways of growing reflect different strategies for acquiring the necessities of life.

Like animals, plants grow when cells divide and enlarge. In plants, however, only the cells of a specific kind of tissue, called **meristem tissue**, undergo mitosis to give rise to new cells. The locations of meristem tissues explain why initials carved in a tree trunk never get higher. Where exactly are meristems found within a plant? For the answer, we must take a closer look at the anatomy of a plant.

Basic architecture of flowering plants includes the belowground parts, called the **roots**, and the aboveground parts, called the **shoot**. Stems and their branches, leaves, flowers, and fruits are all part of the shoot. The great variety of different kinds of flowering plants, obvious to even the casual observer, hides a key similarity. The shoots and roots of flowering plants have three main groups of tissue systems: **vascular tissues**, whose structure and role in translocation are discussed in Section 12.3; **dermal tissues**,

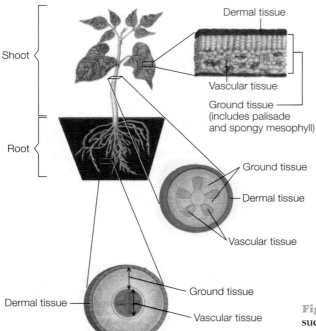

Dermal tissue

Shoot

Vascular tissue

Ground tissue
(includes palisade
and spongy mesophyll)

Root

Ground tissue

Dermal tissue

Vascular tissue

Dermal tissue

Ground tissue

Vascular tissue

Figure 12-18. The three main tissue types in plants: dermal, ground, and vascular.

mentioned in Section 12.1.3, include the epidermis that protects and helps to water-proof a plant's surface; and **ground tissues**, which make up most of the bulk of the plant (Figure 12-18). Meristems are not strictly part of any of these three tissue systems, but they give rise to all three.

Cells of the meristems are unspecialized and can divide by mitosis throughout the life of the plant. Cells that arise in the meristems become specialized when they take up their role as vascular, dermal, or ground tissues. Let's begin with meristems responsible for primary growth.

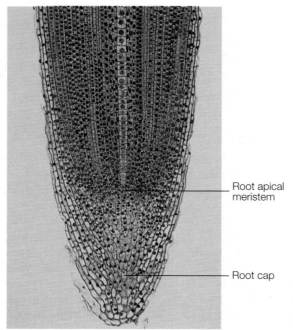

Root apical
meristem

Root cap

Figure 12-19. Section of an onion root tip showing the root apical meristem near the root cap (magnified 78 ti,mes).

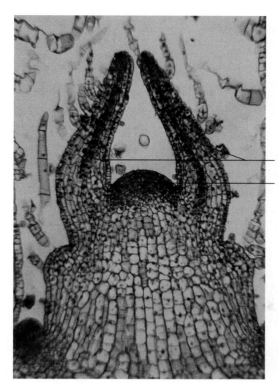

— Developing leaves

— Short apical meristem

Figure 12-20. Section through the growing tip of a Coleus shoot, showing the bud apical meristem (magnified 300 times).

Apical meristems, found at the tips of shoots and roots, are responsible for lengthwise growth, called **primary growth**, of both roots and shoots. Figure 12-19 shows that the root apical meristem is formed at the base of the root cap, or the tissue at the tip of the root. Some of the cells derived from the dividing root apical meristem move toward the tip to replace the root cap cells that have been stripped away as the growing root pushes through the soil. Others add to the growing root, pushing down on the root cap and elongating the root. As new cells continue to be added to the growing root from the meristem (arrow), older cells remain in place, becoming longer (but not appreciably wider) and more specialized at positions above the root cap. The outermost cells differentiate into dermal tissue; the innermost cells give rise to vascular tissue; and the cells in between become ground tissue. The processes of cellular elongation and differentiation (Figure 12-19) are under the influence of plant hormones (discussed in Section 12.4.2) and the genetic program of the cell.

In the shoot, dome-shaped masses of dividing cells (Figure 12-20) in the buds form the apical meristem tissues. The progeny of these cells elongate and differentiate as they remain in place behind the growing tip, much as those of the root apical meristems, but they give rise to shoots instead of roots. During this growth and elongation, some meristem cells are left behind where they act as **axillary bud** meristems at the base of the leaves. Axillary buds are growing shoot tips found in the angle formed by the leaf and stem. Axillary buds, however, usually grow more slowly than **terminal buds**, or those found at the apex of the plant, for reasons that we will discuss in the next section. The rapid growth of terminal buds as compared with axillary buds explains the "Christmas tree" shape of conifers and some other kinds of trees and plants, but it doesn't explain why tree trunks get thicker. For that, we must examine two different kinds of meristematic tissue.

Secondary growth brings about an increase in the *girth*, or diameter, of stems and roots. Secondary growth is most marked in woody plants such as trees, shrubs, and some vines, whose stems last from year to year, and it is usually much slower than primary growth. Figure 12-21 shows the arrangement of the three tissue groups—dermal, vascular, and ground tissues—in a stem that is starting secondary growth. Vascular tissues, the xylem and phloem, are arranged in bundles situated around the periphery of the stem.

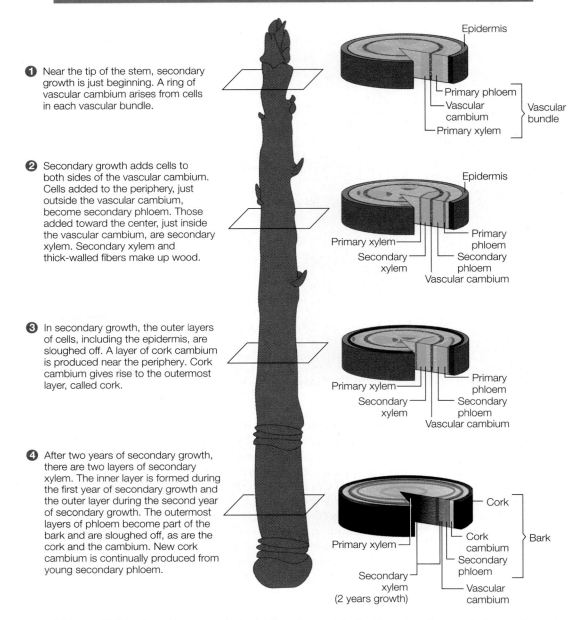

1 Near the tip of the stem, secondary growth is just beginning. A ring of vascular cambium arises from cells in each vascular bundle.

2 Secondary growth adds cells to both sides of the vascular cambium. Cells added to the periphery, just outside the vascular cambium, become secondary phloem. Those added toward the center, just inside the vascular cambium, are secondary xylem. Secondary xylem and thick-walled fibers make up wood.

3 In secondary growth, the outer layers of cells, including the epidermis, are sloughed off. A layer of cork cambium is produced near the periphery. Cork cambium gives rise to the outermost layer, called cork.

4 After two years of secondary growth, there are two layers of secondary xylem. The inner layer is formed during the first year of secondary growth and the outer layer during the second year of secondary growth. The outermost layers of phloem become part of the bark and are sloughed off, as are the cork and the cambium. New cork cambium is continually produced from young secondary phloem.

Figure 12-21. Secondary growth results in an increase in the diameter of a stem; it also results in the formation of rings in the stems of woody plants.

Between the primary xylem and primary phloem is a ring of meristematic tissue called **vascular cambium**. As a result of cell division and specialization of the vascular cambium, new bundles of both xylem and phloem are produced; these are called secondary xylem and secondary phloem, respectively. Secondary growth pushes the first layer of xylem—the primary xylem—toward the inner part of the stem and the primary phloem toward the outer rim of the stem.

In trees and other long-lived woody plants, each growing season results in the production of a new layer of secondary growth. Layer upon layer of secondary xylem, whose cell walls have been thickened with lignin, make up the wood of a tree trunk. The rings that can be seen on the cut ends of logs represent annual growth rings of secondary xylem tissue. The innermost parts of the trunk contain heartwood, made up of the oldest layers of xylem. These dead cells are filled with sticky resins that reinforce the stem and make the inner wood resistant to decay. Closer to the periphery

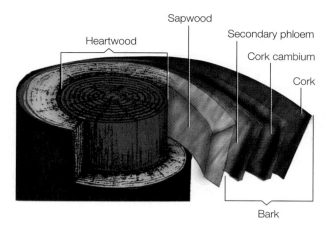

Figure 12-22. Bark is a combination of cork, cork cambium, and secondary phloem. Heartwood is older xylem that has become filled with resin. Sapwood is the younger xylem that conducts water throughout the plant.

of the trunk, the lighter colored sapwood is younger secondary xylem tissue. Sapwood is xylem tissue responsible for moving water and minerals throughout the tree.

The layers on the outer surface of the vascular cambium do not accumulate in the same way that secondary xylem does. When secondary growth begins, the outermost layers of epidermis and ground tissue are sloughed off and a new layer of meristem, called the **cork cambium**, is produced from cells of the ground tissues. Cork cambium produces a new layer of cells called **cork**. When mature, cork cells are dead, and their cell walls fill with a waxy substance that forms a waterproof protective covering. The bark of a tree is a combination of dead cork with its living cork cambium and living secondary phloem (Figure 12-22).

Primary and secondary growth in woody plants, as well as many other plant processes, are regulated by a group of hormonelike chemicals produced within the plant itself. These growth substances have traditionally been called hormones because, like animal hormones, they are produced by the plant and released in response to a stimulus from the environment. As we saw in Chapter 11, animal hormones tend to have only one or a few effects on only one or a few processes or functions. Plant growth substances, by contrast, can influence many aspects of plant physiology in many different ways, as we will see in the next section.

12.4.2 Plants Have Hormones

Charles Darwin is rightly remembered for his revolutionary contributions to our understanding of how organisms evolve by natural selection. Many people do not realize, however, that Darwin was also a prolific experimental biologist. One of his many other findings demonstrated how plants respond to light. It was well known that positioning a light to one side of a growing plant causes it to bend toward the light. This growth response is called **phototropism** (Figure 12-23*a* and *b*). Charles Darwin and his son, Francis Darwin, found that when the growing tips of grass seedlings were covered with tiny black caps, they were no longer able to bend toward a light source. Additionally, when the Darwins buried the seedlings in fine black sand so that only the tips were exposed to the light, they found that the entire plant bent toward the light. Their conclusions, in their own words, were that "some influence is transmitted from the upper to the lower part of the plant, causing the latter to bend." That influence, others would discover much later, was a hormonelike growth substance called **auxin**. In this case, auxin produced at the tip was exerting its effect on parts of the plant below the growing tip.

Auxins

Auxins are a class of molecules, not just one substance. All plants produce auxins. One effect of auxins is to stimulate cell elongation. When light is coming from one side, auxins diffuse down the stem to the shaded side. When those cells elongate more rapidly than those on the lighted side of the stem do, the stem bends toward the light. Similarly, auxins play a role in causing the growing root of a plant to bend downward, a response called **gravitropism** (Figure 12-23c). In gravitropism, cells that are facing upward, away from the direction of gravitational pull, elongate more rapidly than those that are situated downward. The result is that the root bends into the earth.

Tropisms are just a few of the functions of auxins in plants. Auxins from terminal buds also inhibit the growth of the axillary buds on the side branches. Auxins promote apical dominance; in other words, they cause the lead shoot to grow faster than the side shoots, resulting in the pyramid shape that is most familiar, perhaps, in evergreens. For shrubs in which a bushier shape is preferred, horticulturists remove the terminal buds. This practice removes the growth-inhibiting influence of auxins from the terminal bud and hastens growth of the axillary buds, causing the plant to grow outward instead of upward. The plant grows in a more pleasing, bushier form.

Auxins have other widespread effects on plant growth and development, as well. Auxins are involved in fruit development and in preventing fruits from dropping prematurely. You can learn more about auxins and the ways in which they are used in the agricultural industry by exploring the *BioInquiry* web site.

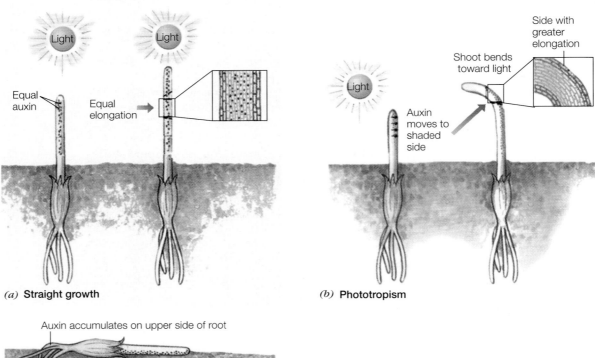

(a) **Straight growth**

(b) **Phototropism**

(c) **Gravitropism**

Figure 12-23. Tropisms. *(a)* With the sun directly above, auxins are distributed equally to all the cells of the growing stem. They all elongate at the same rate, and the shoot grows straight. *(b)* When light hits the stem from one side, auxins from the tip move to the shaded side of the stem. The cells on the shaded side, bathed in auxins, elongate more than the cells on the lighted side. This causes the stem to bend toward the light, demonstrating phototropism. *(c)* A seedling placed on its side accumulates auxin on the upper side of the root. The cells on the upper side, bathed in auxins, elongate more than those on the lower side. This causes the root to bend downward, in the direction of gravity's pull, demonstrating gravitropism.

A Family of Auxins What are some of the kinds of auxins? What other roles do auxins play in plants? How are auxins being used in the agricultural industry to increase the yield of crop plants?

For many years it was believed that auxins were the only hormones found in plants. As more was learned, however, it became obvious that other substances must be present in plants to coordinate and regulate their growth and development. One such class of plant hormones was discovered in early part of the 20th century in a Japanese rice paddy.

Gibberellins

Rice seedlings in Japan can become infected with a fungus that causes them to grow tall and spindly, a form that weakens them so that they topple over before they can mature and produce rice. The Japanese agronomist, Eiichi Kurosawa, studied this fungus, called *Gibberella*. Kurosawa found that he could mimic the effects of *Gibberella* by applying an extract (some of the juices) from the fungus directly to the leaves of uninfected rice seedlings. The active ingredient in the fungal extract, later called **gibberellin**, caused growth not only in rice seedlings, but in other plant species as well. It was later found that gibberellin occurs naturally in the fungus and also in rice and many other species of plants. The gibberellin that came from the fungus destroyed rice not because it was a foreign substance, but because the fungus provided too much of it.

Gibberellins are actually a class of about 100 very similar chemical compounds. They are produced at the tips of both stems and roots of many plants, where they stimulate growth. But gibberellins are most concentrated in seeds, where they facilitate growth of the plant embryo within the seed and emergence of the young shoot and root, a process called **germination**.

In temperate climates, many kinds of plants, called **deciduous** plants, enter a state of suspended growth as the winter approaches. Leaves drop to the ground, buds become enclosed in protective coats, and seeds enter a state of dormancy during which they cannot sprout. These seasonal changes protect the plant and enable it to safely survive until the climate is once again favorable for growth. Dormant seeds often need special environmental conditions, such as light, changes in temperature, or water, to germinate. Treatment with gibberellins can cause otherwise dormant seeds to germinate in the absence of these environmental cues. Naturally occurring gibberellins in seeds probably provide the link between the appropriate environmental cue and the metabolic changes that begin the process of germination. These and other features of gibberellins have made them an important tool for the agriculture industry.

Abscisic Acid (ABA)

Winter cold is not the only stress to which plants may be subjected. A normal day in the hot sun, for example, might raise the plant's temperature to dangerous levels, or it might limit water availability. Plants respond to such stresses by producing a hormone that slows growth, of either the whole plant or just part of it. This substance, discovered in the 1960s, was given the name of **abscisic acid**, abbreviated **ABA**. The term **abscission** refers to the process whereby leaves or fruit are separated from the rest of the plant at the end of a growing season. ABA appears to be involved in abscission (its exact role in the process is still poorly defined), but its main purpose is to inhibit growth. When water is scarce, the levels of ABA increase; likewise when temperatures are very high or very low. And high levels of ABA are correlated with seed dormancy as well. ABA and gibberellins, therefore, act antagonistically: ABA inhibits growth and gibberellins stimulate it. For many seeds and plants, growth is regulated not by either ABA or gibberellins alone, but by the balance between the two.

Ethylene

In earlier centuries, city streets were illuminated at night by street lamps burning coal gas. The pipes that carried gas to the lamps often leaked, emitting coal gases into the city air. As early at 1864, it was reported that gas from leaky pipes was causing leaves to fall from shade trees. But it was years before the culprit gas was identified as ethylene, a byproduct of burning coal. It was many more years before ethylene gas was identified as a natural growth-regulating substance produced by nearly all flowering plants.

Ethylene gas triggers a variety of aging responses in plants, among them fruit ripening. Fruits ripen when small amounts of ethylene are released, causing them to soften and change color. As fruits age, cells break down and more ethylene is released, hastening the process. Thus it is true that one bad apple can spoil the whole barrel, as the gas released by the bad one diffuses throughout the barrel and begins the aging process in all the others. Growers take advantage of ethylene to hasten ripening in tomatoes and other fruits. The fruits are picked while they are still green and sturdy, then ripened in storage bins into which ethylene gas is pumped. Conversely, fruits such as apples can be picked in the fall and preserved until the following summer by storing them in areas in which carbon dioxide is circulated. Carbon dioxide inhibits ethylene, and the gas circulation prevents naturally produced ethylene from accumulating.

One effect of ethylene is to activate enzymes that digest the cell walls of plant cells. The breakdown of cell walls explains why ripe fruit is softer than green fruit. It also helps to explain two other roles of ethylene: abscission, or autumn leaf fall, and the browning of spots where leaves or other tissues are injured or attacked by pathogens. Both autumn leaf fall and browning are examples of planned death—adaptations that kill parts of the plant so that the rest of the plant can survive. In autumn leaf fall, the organs most susceptible to water loss, the leaves, die and are shed. Thus in winter, when the ground is frozen and roots cannot obtain water, water loss is minimized and the tree does not dry out. Areas of dead tissue, or brown spots, that arise where a plant is injured or attacked result from overproduction of ethylene at the point of attack. A ring of dead tissue is formed that surrounds the pathogen, making it difficult for the attacker to penetrate into surrounding living tissues.

Ethylene, therefore, is a chemical product of plants that enables them to respond to their environment by aging or even planned death. Like the other plant hormones that have been discovered so far (and there will likely be more in future years), ethylene can have multiple and complicated effects on growth and development.

 Help Dividing Another important class of plant hormones is the cytokinins. What are cytokinins and what do they do? How are cytokinins used in plant research?

12.4.3 Plants Reproduce Sexually

The various phases that individuals pass through in their lifetime are collectively called a **life cycle**. We are all familiar with the phases of the human life cycle, beginning with birth and childhood, followed by puberty and the attainment of sexual maturity, possibly including parenthood, and ending with old age and finally death. It is helpful when trying to understand the reproductive strategies of organisms that are very different from us to examine and compare life cycles. The life cycles of plants have some features in common with our life cycle, and other features that are very different.

**Fast Find
12.1c
Alternation
of Generations**

In Section 12.1.3, we saw one important difference in the life cycles of plants compared with that of humans and most animals with which we are familiar: plant life cycles are characterized by alternation of generations (see Figure 12-3). Cells in both the haploid phase and the diploid phase of the life cycle divide by mitosis, resulting in a two-part life cycle: a haploid phase called the gametophyte and a diploid phase called the sporophyte. In some kinds of primitive land plants, the gametophyte is the same size as or larger than the sporophyte. In angiosperms, with which we shall be mainly concerned, the gametophyte is reduced to just a few cells that occur within the reproductive structures of the much larger sporophyte. A closer examination of sexual reproduction in flowering plants will illustrate this relationship between gametophyte and sporophyte.

Sexual Reproduction in Flowering Plants

As a group, the angiosperms are so diverse that generalizations about any aspect of their biology are difficult and rife with exceptions. For example, the flowers of most kinds of angiosperms have both male and female organs. Some kinds have separate male and female flowers on the same plant, and still other kinds contain only female flowers or only male flowers. Nonetheless, in spite of the exceptions, it is informative to make a few generalizations, beginning with the structure and function of a "typical" flower. Our typical flower has both male and female sex organs.

Male reproductive structures on flowers are called **stamens** (Figure 12-24). These structures are long filaments, each with a swollen **anther** at the tip. The anthers con-

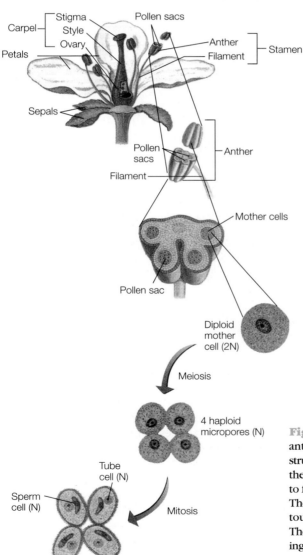

Figure 12-24. Pollen grains form in anthers, part of the male reproductive structures called stamens. Mother cells in the anthers divide by meiosis, giving rise to four haploid cells called microspores. The microspores become enclosed in a tough outer coat and divide by mitosis. The result is pollen grains, each containing a sperm nucleus and a tube cell. The sperm nucleus divides again by mitosis, yielding two sperm nuclei (not shown).

Figure 12-25. Scanning electron micrograph of a pollen grain from ragweed. The sculptured surfaces of pollen grains are highly specific to individual species. Taxonomists use these features to identify the species of different pollens (magnified 35,000 times).

tain specialized cells that divide by meiosis. In Chapter 5 we saw that meiosis is the kind of cell division in which one diploid cell gives rise to four cells, each of them haploid. Within the anther, many such cells divide by meiosis, each one resulting in four haploid cells. Each of these haploid cells is called a **microspore**. The microspore is the first cell of the male gametophyte generation.

Microspores develop resistant outer coats that protect them while they make their often hazardous journey to female flower parts. Inside its protective coat, each microspore divides by mitosis. One of these cells, called the tube cell, eventually forms the passageway into the female organ. The other divides by mitosis and gives rise to two sperm. The tiny two-celled structure, enclosed in its tough coat, is known as a **pollen grain**. The outer walls of pollen grains are often elaborately sculptured (Figure 12-25). Some of the airborne, spiny-coated pollen grains are the cause of hay fever symptoms.

Flowers contain the female reproductive structures as well as the male. The **carpel** is a club-shaped structure found at the center of the flower. Its tip forms a sticky surface for receiving pollen, called the **stigma**. At the base of the carpel is the ovary, or female sex organ, containing one or several **ovules**. Each ovule contains a **megaspore mother cell** that divides by meiosis. The ovule also includes one or more layers of cells surrounding and protecting the megaspore mother cell (Figure 12-26). When the megaspore mother cell divides by meiosis and gives rise to four haploid daughter cells, three of the daughter cells usually degenerate and die. The one that survives is the **megaspore**. The megaspore, like the microspores of the male, is the first cell of the female gametophyte generation.

The megaspore undergoes several mitotic divisions, producing a haploid multicellular structure—a gametophyte—housed in a layer of diploid protective cells from the parent plant. One of the haploid cells of the gametophyte is the egg, the cell that is fertilized by one of the sperm cells from a pollen grain. The zygote formed by this fertilization becomes the embryo and, ultimately, the new plant—or sporophyte—of the next generation. Other cells of the female gametophyte form **endosperm**, a nutritive tissue that feeds the developing embryo within the seed. Endosperm develops when the second sperm cell from the pollen grain contributes its nucleus to two nuclei in the female gametophyte, called polar nuclei, creating an unusual *triploid* nucleus. This kind of event, in which fertilization happens twice for each embryo that is created, occurs only in plants, and mainly in the angiosperms. It is called **double fertilization** (Figure 12-27).

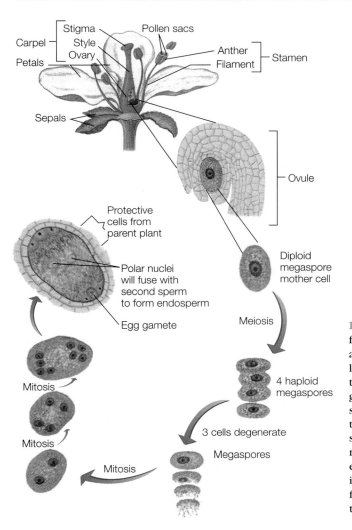

Figure 12-26. The female flower parts house the gametes and female gametophyte. A diploid megaspore mother cell in the ovule divides by meiosis, giving rise to the one megaspore mother cell and three cells that degenerate. The megaspore mother cell divides by mitosis three times, yielding eight nuclei, only one of which is the egg. Cells of the ovule form a protective coat around the female gametophyte.

Pollination

How does pollen get from the male anthers to the female ovary? It might be tempting to assume that pollen simply falls from the anthers to the sticky stigma of the same flower, a short jump away. But self-fertilization between a sperm and egg of the same flower is not as advantageous as cross-fertilization between the sperm of one plant and the egg of a different plant because it does not create as much genetic variety.

Many species have evolved reproductive processes that reduce the probability of self-fertilization and the negative effects of inbreeding. Some species produce flowers with only carpels or only stamens, but not both. Other species release pollen before their carpels are fully mature. The architecture of grasses increases the likelihood that the wind will carry pollen away before it has a chance to fall back on the same grass flower. The stunning array of flower colors, odors, and forms are all evolutionary adaptations that attract animal pollinators, such as insects, birds, and bats, to carry pollen from one flower to another. Bright colors and sweet smells advertise nectar, the sugary sweet fluid flowers produce. Pollinating animals visit flowers in search of nourishing nectar to drink. As they reach into the flower for nectar, the pollinators inadvertently pick up pollen on their bodies and just as inadvertently deliver it to the next flower they visit when the pollen rubs off onto the sticky surface of the stigma. The most successful plants on Earth are those that have evolved ways to attract animals that deliver pollen from flower to flower.

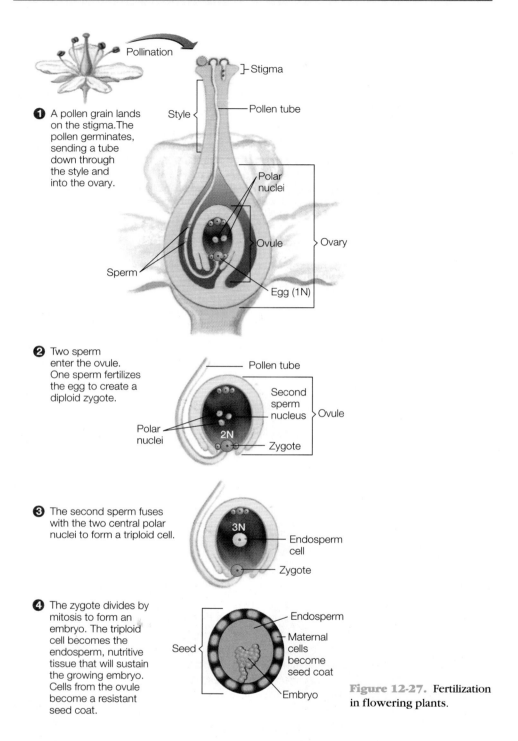

1 A pollen grain lands on the stigma. The pollen germinates, sending a tube down through the style and into the ovary.

Pollination

Stigma

Style

Pollen tube

Polar nuclei

Ovule

Ovary

Sperm

Egg (1N)

2 Two sperm enter the ovule. One sperm fertilizes the egg to create a diploid zygote.

Pollen tube

Second sperm nucleus

Ovule

Polar nuclei

2N

Zygote

3 The second sperm fuses with the two central polar nuclei to form a triploid cell.

3N

Endosperm cell

Zygote

4 The zygote divides by mitosis to form an embryo. The triploid cell becomes the endosperm, nutritive tissue that will sustain the growing embryo. Cells from the ovule become a resistant seed coat.

Endosperm

Maternal cells become seed coat

Seed

Embryo

Figure 12-27. Fertilization in flowering plants.

When pollen from one plant has been successfully delivered to the stigma of another, by wind, water, or animal, it germinates, sending a pollen tube down into the carpel (see Figure 12-27). This tube is a pathway for the two sperm cells to enter the female carpel and travel to the egg. One sperm nucleus fertilizes the egg cell, creating a zygote; the other, along with two female nuclei, gives rise to the triploid endosperm. A new and delicate, genetically distinct plant—a sporophyte—has been produced.

For many plants, the timing of fertilization may not coincide with the best time to germinate. It may be advantageous for the new plant to wait for the next growing season, for example. It also may be advantageous for the embryonic plant to grow in a different place, away from the parent plant on which it was conceived. The seed and fruit are adaptations that enable the newly formed embryo to wait for better conditions before it begins life on its own.

Evolving Together Coevolution is a kind of evolutionary change in which two different interacting species evolve in relation to each other. Flowering plants and their insect pollinators are a good example. We have seen that flowers produce nectar to attract insect pollinators, but in what ways have insects evolved to get nectar? What other adaptations of flowers help to attract animal pollinators?

Seeds and Fruits

Almost immediately after fertilization, the ovule of the parent plant, containing the new embryo and its supporting endosperm, matures into a **seed**. The cells that form a protective coat first around the female gametophyte, and then around the embryo and endosperm, dry into a hard, protective seed coat. Enclosed in its seed coat, the embryonic plant develops until a shoot and root have differentiated, and the three kinds of tissues—dermal-, ground-, and vascular tissues—are present. At this point, growth and development cease and the embryo becomes dormant. It remains dormant under the growth-inhibiting influence of ABA until conditions become favorable for germination. The embryo is protected from desiccation and damage by its seed coat.

Meanwhile, the ovary that surrounds the seed, or some other parental structure, may develop into a **fruit** (Figure 12-28). Many fruits are thickened ovaries that provide

Figure 12-28. Fruits contain the seeds of flowering plants, with fleshy, often sweet-tasting tissue derived from the ovary of the parent plant. Animals may eat the fruits of some plants, digesting the fleshy tissue and expelling the seed in a place distant from the parent plant. In this way, seeds are dispersed to new areas.

an additional layer of protection for the embryo and aid in dispersing the seeds to new places. Fleshy fruits can be sweet tasting and colorful, an appealing treat that attracts animals, including humans. Often the seeds of a fleshy fruit are discarded by the animal, or they are swallowed and pass through the digestive tract protected from digestive enzymes by the tough seed coat. In either case, the animal leaves behind the seed containing the plant embryo, which has been moved to a new place courtesy of the animal consumer. Once again we see how plants and animals have coevolved.

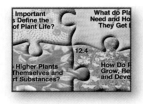

Piecing It Together

Growth and reproduction are among the hallmarks of life. In all organisms, these critical activities are highly coordinated and carefully regulated. Plants are no exception.

1. Plant growth is indeterminate. Cells divide by mitosis within meristem tissues found at the tip of the shoots and roots and in the periphery of the woody trees and shrubs. The meristems responsible for primary growth are apical meristems; those that increase the girth of a woody plant are the vascular cambium and the cork cambium.

2. Plant hormones coordinate and regulate plant growth and other aspects of their physiology. Auxins and gibberellins stimulate growth, ABA inhibits growth, and ethylene promotes maturity and sometimes death. Plant hormones, however, can have multiple effects on many tissues.

3. Plants reproduce sexually; that is, zygotes that give rise to embryos and ultimately new plants are formed by the fusion of male and female gametes. In angiosperms, flowers are the sex organs in which gametes are produced by male and female gametophytes.

Fast Find 12.4a Plant Growth

4. The female sex organ is the carpel, which houses first the ovule, then the female gametophyte, and finally the embryo. The male sex organ is the stamen. Anthers atop stamens contain cells that give rise to the male gametophyte, the pollen, which produces male gametes.

5. Pollen is transported from the anther of one flower to the carpel of another by wind, water, or animal pollinators. Flowers attract animal pollinators with sweet smells, bright colors, and nectar.

6. Pollen carries the sperm that fertilize the female egg. After fertilization, the female ovule develops into a seed, which is a plant embryo and its nutritive tissue enclosed in a tough seed coat. Seeds remain dormant until conditions for growth are appropriate. Seeds are also the forms in which plants are dispersed. The ovary, and sometimes other parts of the parent plant, forms the fruit.

Where Are We Now?

On an island in Samoa, the *taulasea*, or medicine woman, carefully peels bark from a branch of a tropical coral tree. The healer applies the wet bark to the inflamed skin of her patient, and a chemical compound in the bark reduces the inflammation and the pain. Coral tree bark is just one of over a hundred plants that the taulasea uses to treat her village patients. Included in her dispensary are plants used to treat snakebites, headaches, fevers, toothaches, coughs, earaches, stomach ailments, even cancer. Most of these traditional medicines are remarkably effective.

Scientists in the growing field of **ethnobotany**—the study of plants and people— have begun to focus on the medicinal uses of plants by native peoples around the world with the hope of finding new therapeutic agents for treating disease. Plants synthesize an enormous variety of complex organic compounds that influence the physiology of other organisms, including humans. Most of these compounds probably evolved as chemical defenses against predation or infection. Until the 1950s, pharmaceutical companies relied heavily on plants as sources of new medicines. Over 100 commercially available drugs, and about 25 percent of all medicines prescribed in the United States, are derived from vascular plants. Some familiar ones are aspirin, which reduces fever and inflammation; codeine, which eases pain and suppresses coughing; and several drugs for heart function, including reserpine and digitalis.

With the advent of new procedures for synthesizing drugs in the 1960s, however, pharmaceutical companies lost interest in exploring plants for new medicines. Molecular biology and synthetic chemistry replaced more tedious ethnobotanical methods for identifying sources of new drugs. These modern approaches are still widely used, but the discovery of several promising plant products has led many researchers to believe that the best approach to finding new drugs might be a combination of ethnobotany and modern techniques.

Ethnobotanists begin with extended visits to societies that have a tradition of using plants in medicine. Visiting scientists, trained in both anthropology and botany, live in accordance with local customs and learn to speak the language. With permission from the society's leaders, they apprentice themselves to the local healer. An ethnobotanist might spend several years as an apprentice to a healer, taking careful notes on the plants the healer uses and how they are used. With the guidance and approval of the healer, the ethnobotanist collects plants that have healing properties. Samples of these plants are sent to modern research laboratories for analysis.

Several laboratories in the United States, including the National Cancer Institute in Bethesda, Maryland, have the facilities to determine quickly and accurately whether a plant has bioactivity, or the potential to influence animal cells. Researchers may look for a plant's ability to stop the growth of tumor cells, or to interfere with the infectious ability of a virus such as HIV, the virus that causes AIDS. When a plant displays promising activity, chemists try to isolate the chemical compound responsible and determine its molecular structure. Many compounds that have been identified are similar to drugs that are already available; others have undesirable side effects. Most are not pursued as potential medicines.

Occasionally, however, a novel compound is identified that has an important biological effect. One example is the antiviral compound called prostratin. Prostratin was isolated from the wood of the *Homalanthus* tree from the rainforest of Samoa. Healers in Samoa use a tea, made by steeping the wood in hot water, to treat yellow fever. When *Homalanthus* wood was tested at the National Cancer Institute, it was found to have strong activity against HIV in a test tube. The compound responsible was identified as prostratin. This compound is currently being tested as a future commercial drug.

Learning from indigenous healers and collecting plant specimens from their forests are raising some serious ethical questions. How are the interests of the healers and their communities being protected? Ethnobotanists are sensitive to these issues. In the case of prostratin, for example, researchers have guaranteed that some portion of any profit earned from the sale of prostratin will be returned to the people of Samoa. Most ethnobotanists are making similar commitments to their healer teachers. To many of the healers, however, protecting their forests from destruction is more important than earning profits. Many drug companies and foundations are joining the global movement to protect the world's rainforests. It is both environmentally sound and good business to preserve these areas of rich biological diversity.

REVIEW QUESTIONS

1. What is the difference between a heterotroph, a chemoautotroph, and a photoautotroph? Which probably evolved first?

2. How did the early cyanobacteria change the chemical composition of the atmosphere? How did that change influence the evolution of life?

3. What evidence indicates that eukaryotic cells evolved from a symbiotic relationship between two or more prokaryotes?

4. What are the advantages to being multicellular as opposed to unicellular? How did multicellularity in plants contribute to the successful invasion of land?

5. Do humans exhibit alternation of generations? Why or why not? In what way is a human analogous to a plant sporophyte?

6. What features of seeds make seed-bearing plants particularly well adapted to life on land? Do all land plants make seeds?

7. What, in biological terms, is a flower?

8. Describe three ways in which leaves are adapted to absorb light.

9. Explain the conflict between a plant's need to acquire carbon dioxide from the atmosphere and its need to prevent water loss by evaporation. How have plants evolved to meet these two competing needs?

10. Describe three ways in which roots are adapted to absorb water.

11. What is nitrogen fixation? What organisms accomplish nitrogen fixation? Why is it not possible for plants or animals to use the abundant nitrogen in air?

12. How do plants contribute to global cycling of nitrogen and other minerals?

13. What force moves water upward through the body of a vascular plant? What force moves sap throughout the plant body?

14. What are the two main kinds of plant vascular tissues? How are they the same? How are they different?

15. Is a nonwoody plant likely to grow to be very tall? Why or why not?

16. Explain why plant growth is indeterminate and the growth of most animals is determinate.

17. What is a meristem tissue? Where are the meristem tissues found in plants?

18. Distinguish between primary and secondary growth. Do all plants exhibit primary growth? Do all plants exhibit secondary growth?

19. Auxins and ABA have opposite effects on plant growth. Explain.

20. While grocery shopping, you buy two bunches of unripe bananas. You place one bunch under an inverted bowl, and you leave the other bunch out on the countertop. Which bunch will ripen first? Explain.

21. Where is the gametophyte found in flowering plants?

22. What is the difference between a seed and a pollen grain?

23. What is double fertilization? What are the two different products that result from double fertilization?

Chapter 13

Behavior: How Do Animals Interact with Other Animals?

Ants are so much like humans as to be an embarrassment.
They farm fungi, raise aphids as livestock, launch armies into wars,
use chemical sprays to alarm and confuse enemies, capture slaves.
They exchange information ceaselessly. They do everything
but watch television. —Lewis Thomas, 1974

—Overview—

Who fails to be captivated by kittens playing with yarn or amazed at reports that honey-bees have a language? When we see a large dog loose in our neighborhood, we're drawn to its behavior. What is it doing? More importantly, what is it going to do next? In our pre-history, similar questions about the behavior of both predators and prey were matters of survival. Those who could accurately predict bear behavior could avoid injury. Those who could accurately predict antelope behavior feasted. Today, interest in animal behavior is less a matter of survival, but it is still engaging.

In this chapter, we will look at animal behavior—what animals do and how they do it. In some respects, studying behavior is easier than studying other topics in biology. We can easily watch a cat stalk a mouse, but seeing it digest the mouse is another matter. But the study is not as straightforward as it might seem. There are decisions to make. What aspects of cat and mouse interactions should we record? How can our observations be measured? How do we determine the significance of our observations?

As a discipline, animal behavior is firmly rooted in biology's other major fields, especially evolution, genetics, and physiology. We shall see that behavioral patterns evolve; they generally become more complex as animals become more complex, but not always. Behavioral patterns are inherited just as color patterns are. And behaviors invariably involve physiologic processes: food is digested; information about the environment is gathered, analyzed, and responded to; movement usually requires the use of muscles.

Chapter opening photo—Animals behave in a variety of ways.

411

Animal behavior is important and relevant. Ultimately, it will help us understand human behavior. We see that among humans, males tend to be more aggressive than females, who in turn tend to be more nurturing than males. Where did such behaviors come from? Do other animals show similar tendencies? Are there exceptions? And perhaps most importantly, if human society decides to change those patterns, can we, as a species, do so? The study of animal behavior may help answer some of these questions.

The study of animal behavior is also important by itself, without references to humans at all. What animals do and why, how, and when they do it is a fascinating, challenging study in its own right.

13.1   HOW IS ANIMAL BEHAVIOR STUDIED?

Left alone, young boys begin to wrestle. So do litters of wolf puppies. Why do they do it? We imagine the boys are just having fun. Are the puppies having fun? Puppies, as they play, do a lot of mock biting. No one gets hurt, but they bite mainly around the neck, throat, and hind legs. So do adult wolves when they hunt. They may first hobble prey by breaking large tendons in hind legs and eventually go for the throat. At least a part of wolf puppy play teaches predatory skills useful to adults.

We observe similar play in fox puppies, but with a twist. In lean times, the levels of their play intensify. In extremely lean times, stronger pups kill and eat weaker ones. We may be appalled at such behavior, until we realize that in lean times it ensures the survival of at least a few young foxes. Here is an important concept concerning animal behavior frequently missed by nonscientists: human moral values do not apply to animals. Animals do what they do, with no sense of good or evil.

Are there any similarities between the play of boys and wolf puppies? Perhaps. In addition to practicing hunting skills, wolf puppies establish social bonds and dominance hierarchies that define pack structure for years to come. They also learn to handle competition. When young boys wrestle, are they, too, bonding, expressing dominance, and handling competition (Figure 13-1)? Female wolf puppies wrestle as much as males. Why don't young girls wrestle more?

By now we know that scientists start by asking questions whose answers depend on observations and measurements carefully collected and thoughtfully analyzed. What kinds of questions do animal behaviorists ask? According to Niko Tinbergen, a Dutch-born British **ethologist**, there are four general types: (1) What are the mechanisms that cause behavior? (2) How does the behavior develop? (3) What is its survival value? (4) How did it evolve? As in other branches of biology, even the most obvious answers need to be examined critically. We know how the process of science works in such fields as evolution, genetics, diversity, and physiology. How is it applied to the study of animal behavior?

Figure 13-1. Animals play in order to practice and hone their survival skills. Do humans play only because it is fun? How are these forms of play similar?

13.1.1 The History of Animal Behavior

Interest in animal behavior is as old as humanity. Ancient Greek scholars thought that the minds of different types of animals progressed in a natural order like other characteristics. They envisioned a linear progression of mental states: simple minds in simple animals such as sponges; more advanced ones in more advanced animals. The highest expression of mental state was reserved for humans (or, more particularly, man). This idea persisted a long time.

The study of animal behavior was strongly influenced by Darwin's *The Origin of Species*. He provided a mechanism that could explain how mental states and other properties could vary between species. Natural selection, it was seen, could influence and change behavioral patterns just as easily as other animal characteristics. In the 1880s, a protégé of Darwin, George J. Romanes, listed emotions by the order in which he imagined they evolved. According to his scheme, fear and surprise appeared in the segmented worms, and shame and deceit did not exist until dogs and primates appeared (Figure 13-2). Interest in animal behavior grew, and by the early 1900s, the field had split into two major branches.

One branch, **ethology**, developed primarily in Europe. To ethologists, what is striking about behaviors is that they are fixed and seemingly unchangeable.

◆ Kittens and puppies play in characteristic but different ways. Present a kitten with a ball of yarn and invariably it draws back its head and bats the yarn with claws extended. Kittens are generally silent as they play, and their tails twitch. Puppies, by contrast, are most likely to pounce flat-footed on a ball of yarn. They bite and bark, and their tails wag. Two mammals in the same stage of life, with the same **stimulus**, react differently.

◆ The timing of migratory behavior in birds is quite precise. In spring in the Northern Hemisphere, as days begin to lengthen, the behavior of many birds in Central and South America changes. After months of occupying relatively fixed territories, they become restless. Then, they migrate to nesting grounds hundreds of miles from their wintering grounds. These movements are precise. Each year, for example, swallows who winter in South America arrive back at Capistrano, California, within a few days of March 19th.

◆ Some animal behaviors defy logic. If an egg rolls out of a goose's nest, the female retrieves it. Typically, she reaches beyond the egg and rolls it back toward the nest with the underside of her bill. It may take several minutes to complete the act. If, after retrieval has been initiated, the egg is removed from under her bill, the female continues the response as if the egg were still there.

Ethologists came to believe that, ultimately, even the most complex animal behaviors could be broken down into a series of unchangeable stimulus–response reactions. They became convinced that the details of these patterns were as diagnostic of a particular group of animals as were anatomical characteristics. For well over half a century, their search for and description of innate patterns of animal behavior continued.

Meanwhile, mainly in North America, the study of animal behavior took a different tack, developing into **comparative behavior**. What was of interest to comparative behaviorists was where a particular behavior came from, that is, its evolutionary history, how the nervous system controlled it, and the extent to which it could be modified. In 1894, C. Lloyd Morgan, an early comparative behaviorist, insisted that animal behavior be explained as simply as possible without reference to emotions or motivations because these could not be observed or measured. In Morgan's research, animals were put in simple situations, presented with an easily described stimulus, and their resultant behaviors described.

Behaviorism was an important development in comparative behavior. Behaviorists extended the idea that animal behavior should be restricted to only those elements that

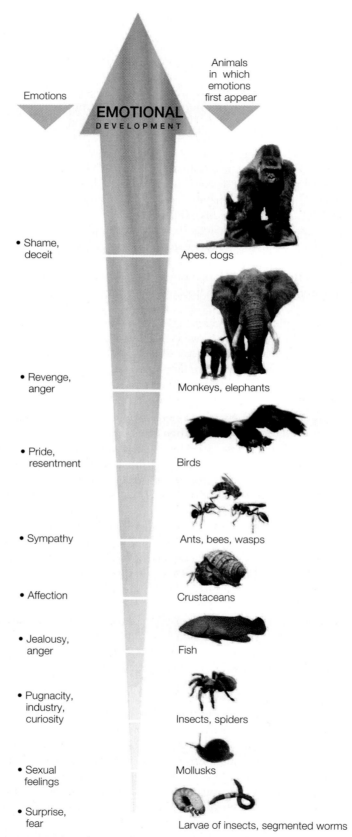

Emotions

EMOTIONAL
DEVELOPMENT

Animals
in which
emotions
first appear

- Shame,
  deceit

Apes. dogs

- Revenge,
  anger

Monkeys, elephants

- Pride,
  resentment

Birds

- Sympathy

Ants, bees, wasps

- Affection

Crustaceans

- Jealousy,
  anger

Fish

- Pugnacity,
  industry,
  curiosity

Insects, spiders

- Sexual
  feelings

Mollusks

- Surprise,
  fear

Larvae of insects, segmented worms

Figure 13-2. The evolution of emotions as imagined by George Romanes, a contemporary and protégé of Darwin. Basically, he thought that, as animals became more and more complex, so did their emotions.

can be directly observed. Here, studies of stimulus–response reactions and the importance of simple rewards to enforce and modify animal behavior were stressed.

Not surprisingly, comparative behaviorists worked most comfortably in the laboratory, where test experiments could be controlled. Their focus was narrowed to a few animals that could be kept at a laboratory; white rats, pigeons, dogs, and monkeys became favorite subjects. Comparative behaviorists stressed the idea that animal behavior could be modified, whereas their ethologist colleagues thought it was innate and unchangeable. Inevitably, the two approaches led to major disagreements.

13.1.2 Controversy between Nature and Nurture Split Early Investigators of Animal Behavior

To early ethologists, the major driving force in behavior was **instinct**, or behaviors that are inherited and unchangeable. Moths move toward light because they inherit the mechanisms to respond this way to light. Although dogs have more options available to them, they bark at strangers for much the same reason.

Not so, countered the comparative behaviorists; learning and rewards are more important factors than instinct in animal behavior. Geese are not born with the ability to recover lost eggs; they learn to do so. If their behavior seems silly to humans (as when they continue to recover an egg no longer there), it's because their ability to learn is limited. There were too many examples of behaviors modified by experience for comparative behaviorists to put their faith in instincts.

The arguments came to a peak in the 1950s and became known as the nature versus nurture controversy. Consider how differently an ethologist and a comparative behaviorist would interpret the begging behavior of a hatchling bird. The first time a hatchling bird is approached by its parent, it begs for food. All baby birds of a particular species beg in exactly the same way. Obviously, said the ethologists, they inherited the ability and the tendency to beg. Baby birds didn't have to learn the behavior; they were born with it—a clear example of innate, unchangeable behavior. Rubbish, countered the comparative behaviorists, parent birds teach their young to beg by stuffing food in their open mouths. Later experiments showed that before hatching, birds make and respond to noises of their nest mates and adults. Isn't it possible that young birds could learn to beg prenatally?

It was hard for ethologists to accept that innate behaviors could be modified by learning. It was equally difficult for comparative behaviorists to accept that genetic factors could dominate learning experiences (Figure 13-3). The controversy raged for over a decade. It stimulated numerous experiments as each side tried to discredit the other, and as a result, our understanding of animal behavior advanced considerably. Another result of the controversy was that distinctions between the two fields narrowed; comparative behavior-

Figure 13-3. Is hunting behavior in wolves learned or is it instinctive? When this one-year-old wolf detected a lemming under the snow, it crouched, stalked stealthily forward, and pounced on its hapless prey. This was surprisingly adept hunting behavior for a young wolf who had been raised by humans, whose parents lived in cages, and who was experiencing its first contact with live prey.

ists realized that both innate factors and learning are important. A bit of both nature and nurture can be found in almost all complex behaviors. Let's look at an example.

One of the most famous series of animal behavior experiments were those conducted by Ivan Pavlov in the 1920s. He observed that dogs begin to salivate whenever they are presented with food. Pavlov was a physiologist, more interested in digestion than dog behavior. He noticed that if he rang a bell whenever he presented food, after a while, his dogs started to salivate at the sound of the bell, even if food was not presented (Figure 13-4). Can you see both innate and learned elements in this experiment? The ability and tendency to salivate at the sight of food is inherited. The same response at the sound of a bell is learned. Both innate and learned factors are important to behavior and in some cases cannot be separated. Today, scientists believe that both natural endowments and environmental factors work together to shape behavior.

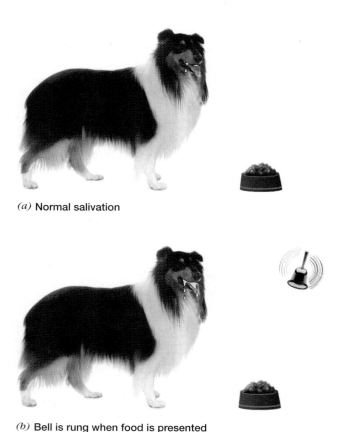

(a) Normal salivation

(b) Bell is rung when food is presented

(c) Salivation at the sound of ringing bell

Figure 13-4: In the 1920s, Ivan Pavlov conducted experiments that became classics on learning in dogs. *(a)* Normally, when dogs see food they salivate. *(b)* In Pavlov's experiments, bells were rung whenever dogs were presented with food. *(c)* Eventually, his conditioned dogs salivated whenever they heard bells ring, irrespective of the presence of food. Notice that some elements of this behavior are innate, whereas others are learned.

At about the time ethologists and comparative behaviorists were acknowledging each other's validity, albeit grudgingly, a new behavioral field emerged: **sociobiology**. It links the study of animal behavior with the study of ecology and evolution. At first, sociobiology was controversial, but many of its concepts have moved into the mainstream of animal behavior thinking.

13.1.3 Animal Learning Illustrates the Importance of Nurture

**Fast Find
13.1a
Animal
Learning**

Learning is defined as changes in behavior that result from experience and that cannot be explained in terms of increasing maturity, sensory adaptation, or fatigue. Note in our definition that not all observed changes in behavior result from learning. Which of the following scenarios do involve learning? How are the other responses explained?

◆ A person walks into a darkened, cluttered room and stumbles over a chair. She stands a few moments until she can dimly see other obstacles. Only then does she move across the room avoiding other obstacles.

◆ Her roommate, who has lived in the cluttered apartment for several years, knows the room well. She walks confidently across the darkened room without pause and avoids all clutter.

◆ One night the roommate comes home, bone tired after an exhausting day, and trips over the clutter.

◆ A 13-year-old boy trips over the clutter even with the lights on, a tendency not seen later when he is 16.

Learning is a complex behavioral response often showing elements of both inherited capabilities and external experiences. Often, learning is grouped into categories of increasing complexity. For each category, note the relative importance of nature and nurture.

Habituation is learning *not* to continue a response due to a lack of reinforcement. For example, tubeworms normally withdraw into their tubes whenever a shadow passes over them. Shadows can indicate the approach of predators. But in the laboratory, the behavior can be modified. Tubeworms *will* cease to withdraw if the shadow is encountered frequently with no dire consequences.

In **classical conditioning**, an animal learns to give a response, normally associated with one stimulus, to a different stimulus. Pavlov's dogs are a classic example. Normally, dogs salivate at the sight of food. After a period of conditioning in which a bell is rung every time food is presented, Pavlov's dogs salivated at the sound of a bell, even when no food was presented. In classical conditioning, the animal's behavior needs reinforcement. If no food is presented when bells are rung, or if food is presented without the accompaniment of bells, dogs lose the response, a phenomenon known as **extinction**.

Operant conditioning occurs when the frequency of a learned behavior increases because of rewards. For example, put a hungry pigeon in an empty box, and it will peck at various things. Eventually, the pigeon pecks at a red button that causes a food pellet to be delivered to the pigeon. The food reward increases the chances that the pigeon will peck at the red button again. Each time the pigeon pecks at the red button, it receives a food reward. Within a half-hour, the pigeon pecks mainly at the red button. Similar observations are seen in rats (Figure 13-5).

Latent learning covers situations in which animals learn new behaviors with no apparent or immediate rewards. Learning through exploration is often sited as an example. If a new log is placed in its territory, a chipmunk explores it for a time, but if no food is found, the chipmunk moves on. When a cat comes into the yard, the chipmunk heads directly for and hides beneath the log.

Insightful learning is illustrated by the sudden appearance of different behaviors that solve previously unsolvable problems. A chimpanzee tries to reach a banana outside her cage, beyond reach. In the cage with her are two sticks with which she tries to

Figure 13-5. Other classic experiments in learning involved rats. An untrained rat is placed in an apparatus in which food is presented whenever a particular button, or similar object, is pushed. Soon, the rat learns to push the correct object whenever it wants food.

reach the banana. She seemingly loses interest in the banana, but continues to play with the sticks. When she discovers that the two sticks can be interlocked, she instantly uses the now longer stick to pull the banana within grasp.

Learning how to learn is called a **learning set**. A pigeon that has learned to get food by pecking a red button is placed in a situation in which food can be obtained only by pecking a blue button. Its learning time is less than (1) the time it took to first learn to peck the red button, and (2) the time it takes a naive pigeon (one that doesn't know to peck buttons for food) to learn to peck a blue button. Presumably, the first pigeon's previous learning experiences make it easier to learn in a different situation.

Social learning is learning from others. The young of many animals, especially carnivores, spend considerable time with their parents, learning where and how to obtain food (Figure 13-6). For another example, place a bird feeder in your yard and watch what happens. Usually, the birds pay no attention to it for several days. Then, one bird approaches the feeder. Almost immediately, other birds in the neighborhood do too.

Figure 13-6. In the early 1920s, certain birds in England learned that they could peck holes in the caps of milk bottles left on doorsteps early in the morning and obtain a free meal. The behavior quickly spread throughout Europe. This is an example of social learning.

Birds, even if they are not of the same species, watch each other and learn through what they observe.

We've already mentioned that **play** is one way young animals learn necessary skills. Sometimes older animals play, too. Why? Play helps animals of all ages to develop *physical attributes* (strength, endurance, coordination), *social bonds* (grooming, reproductive displays, dominance hierarchies), and *cognitive skills* (perceptual abilities). Think of two kittens playing with a ball of yarn. How is each of these categories important in their play?

How We Learn Can you think of other examples of animal behaviors that help animals develop in these categories? Can you think of examples of your own or other human behavior that fit into these categories?

Can Animals Think? Can animals feel? Can animals think abstractly? Can animals reason symbolically?

Piecing It Together

1. Ethology is the study of animal behavior often done under natural conditions.

2. Comparative behavior is the study of animal behavior often done under tightly controlled laboratory conditions.

3. The controversy of nature (inherited tendencies) versus nurture (experiences and learning) as determinants of animal behavior divided the field. The modern synthesis recognizes that both factors are important.

4. Sociobiology is the study of animal behavior that stresses evolutionary and ecological processes.

5. Learning, that is, changes in behavior that result from experience and are not explainable in terms of increasing maturity, sensory adaptation, or fatigue, can be categorized as follows:

   ◆ Habituation is learning not to respond because of a lack of reinforcement.

   ◆ Classical conditioning is when an animal learns to give a response normally associated with one stimulus to a different stimulus.

   ◆ Operant conditioning is learning to increase the frequency of an action because of rewards.

   ◆ Latent learning is when animals learn new behaviors with no apparent or immediate rewards.

   ◆ Insightful learning is the sudden appearance of solutions to previously unsolvable problems.

   ◆ Learning set is learning how to learn.

   ◆ Social learning is learning from others.

   ◆ Play is behavior that appears to be trifling, that is not directed to any immediate needs or goals, but often results in learning. Learning through play increases development (1) physically (strength, endurance, coordination), (2) socially (grooming, social bonds, reproductive displays, dominance hierarchies), and (3) cognitively (perceptual abilities).

13.2 HOW IS ANIMAL BEHAVIOR RELATED TO THE OTHER BRANCHES OF BIOLOGY?

No branch of biology stands alone. In preceding chapters, we have seen numerous examples of the interrelatedness of different fields. For instance, in the last three chapters, we saw how physiological functions and processes are the result of genetics and evolution. The CD shows how that the mammalian four-chambered heart evolved from the three-chambered heart of reptilian ancestors. Animal behavior is also connected to other branches of biology, especially genetics, evolution, and physiology.

13.2.1 The Genetics of Behavior

At first, it may seem unlikely that genes can affect behavior. Genes, after all, are a cell's chemical instructions for how to make proteins. How can the presence or absence of a protein have any effect on how an animal behaves?

Perhaps the most straightforward example can be found in the single-celled protist, *Paramecium*. This protozoan moves by means of cilia, which work like tiny oars. When cilia push one way, the paramecium moves forward. If it encounters a barrier, the cilia reverse, the paramecium backs away, turns approximately 30°, and resumes forward motion. The direction of movement is determined by calcium ions in the cell. Normally, calcium ion concentrations inside the cell are low because calcium ion gates in the cell's membrane are closed (you can review how such gates work in Chapter 4), and the cell moves forward. When a barrier is encountered, calcium ion gates open, and the calcium ions dissolved in surrounding water enter the cell. This causes the cilia to reverse and the paramecium to back up. The calcium gates are proteins whose structure is genetically dictated (Figure 13-7). Two mutant forms of *Paramecium* are known: *pawns*, which lack the ability to back up altogether because their cell membranes have no calcium gate proteins, and *paranoiacs*, which back up excessively because, once opened, their gates only partially close, presumably due to abnormalities in the gate protein structure. Variations in the behavior of *Paramecium* can be explained in terms of variations in protein, and thus gene, presence and structure.

What about higher animals? Normal *Drosophila* (fruit flies) can learn to associate an unusual odor with an unpleasant event, such as a mild electrical shock. When shocked, fruit flies fly, but when exposed to an odor alone they do not. As you might expect, flies exposed to both simultaneously fly. After of period of training, flies will fly when exposed to odor alone. A mutant fly, known as *dunce*, fails to make the association. Normal flies can remember that the odor was accompanied by unpleasantness, whereas dunces cannot. These mutant flies lack a gene that controls the production of an enzyme that destroys a chemical common in many cells, cyclic AMP. Too much cyclic AMP interferes

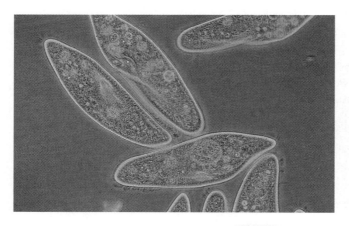

Figure 13-7. Strictly speaking, *Paramecium* are protists, not animals. Although they lack nervous systems and sense organs, they are capable of rather complex behavior. At least some of these behaviors are controlled by surface proteins in their cell membranes (magnified 260 times).

Figure 13-8. A blue-footed booby tends its chicks. These birds spend most of their time at sea, coming to land only to breed.

with communication between certain of the fly's neurons, which affects its memory. Normal flies have the enzyme and thus low concentrations of cyclic AMP. Their neurons communicate normally, and they remember, "Odor! Fly!" Mutant *dunce* flies cannot. Once again, an animal's behavior has been influenced by the actions of genes and proteins.

Blue-footed booby chicks (Figure 13-8) are restricted to their nests for the first few weeks of life. They are fed exclusively by their parents, who provide a diet of regurgitated, partially digested fish. As the chicks get older, they while away times between parental visits by tossing sticks and catching them by the ends. They are, in effect, practicing a behavior that they have never seen or been taught and that will be essential when, as adults, they manipulate fresh-caught fish prior to swallowing. Where does this behavior come from? The behavior must be inherited and, if so, must have a genetic basis. How the genes and their proteins work in this case is not known.

The relationship between genes and complex behavior seen in higher animals is not straightforward. For one thing, just because a gene determining behavior is inherited, it doesn't mean it will be expressed. A blue-footed booby chick raised in a laboratory devoid of sticks could not practice tossing them. This suggests a relationship between genetic tendencies and environment in the development of this behavior.

Shy people often have shy parents and shy children. Does this mean that there is a shy gene? Or do shy children emulate shy parents and thus learn to be shy? We have to conclude that in many instances of complex animal behavior both nature and nurture are important, and separating their influences may not be possible.

How Humans Behave What about human behaviors? Studies show that children of alcoholics are at least twice as likely to become alcoholics as are the children of nonalcoholics. Is this evidence for a genetic basis for alcoholism? Or could alcoholism be learned behavior? Are the two explanations mutually exclusive? How might the controversy be resolved? What effects would resolution of this issue have on the treatment of alcoholism? What is the latest thinking in this area?

Some of these questions could be asked about a proposed genetic basis for homosexual behavior. These are highly controversial and, for some, emotion-packed issues. In confronting the questions and answers, be sensitive to feelings, yours and others.

13.2.2 The Evolution of Behavior

If some behaviors have a genetic basis, and if evolution involves a population-wide shift in allele frequencies, then we can predict that behaviors evolve. In fact, examples have been observed in nature. A species of garter snake in California occurs in two populations. The coastal snake population feeds primarily on banana slugs; inland snakes feed primarily on tadpoles and small fish. Coastal snakes sense banana slugs by flicking their tongues; it's how snakes sense odors. They respond appropriately to cotton swabs soaked in essence of slug. Inland snakes ignore banana slugs, even when no other food is available. The ability to sense slugs is a trait with a genetic basis. Offspring of females from one population crossed with males from the other show intermediate abilities to sense slugs.

How might such behavior evolve? Let's assume that the inland population was the original population, because banana slugs are not native to North America. Sometime after the slug's introduction, garter snakes began to move into coastal regions. A few possessed a rare allele that allowed them to detect the odor of banana slugs. They could take advantage of a new food source, which became abundant in coastal regions. These garter snakes survived and left more offspring than did those that rejected slugs. Thus, the slug-detecting allele dominates in coastal populations, but remains rare in inland populations.

Optimality theory predicts that behaviors will evolve that promote the greatest fitness. Costs and benefits are important. Natural selection favors a particular behavior if the benefits from it outweigh the costs, and if it results in increased numbers of offspring or relatives. Let's look at an example.

Many birds establish and defend territories during their breeding seasons. This territoriality has numerous benefits and costs. On the benefit side, territories may reduce strife, predation, and spread of disease, and they also increase food supplies, mate attractions, and protection of mates and young. The costs of territoriality include energy expended while defending it and the risk of injury and exposure to predators. Territoriality evolves in those species in which benefits exceed costs. There are numerous variations among birds:

◆ Mockingbirds defend nesting and feeding territories from all other songbirds.

◆ Cardinals defend nesting and feeding territories from other cardinals, but ignore other songbirds.

◆ Robins establish nesting territories, but feed communally with other robins.

◆ Some male hummingbirds defend exclusive feeding territories (occasionally consisting of nothing more than a single blossom) from all other males but will permit females to feed if they submit to breeding behavior first (even in the nonbreeding season!).

◆ Osprey defend only the immediate area around their nests from other ospreys and other raptors.

◆ Many herons, gulls, terns, and sea birds nest in dense colonies where only the immediate nest area is defended.

What determines which behavior evolves in which species? Apparently, the particular needs and environmental challenges associated with each species are the determining factors. What is optimal for one species is not optimal for another. Each species lives in a unique environment and has had a unique evolutionary history; therefore, each species is unique, not only in its anatomy and physiology, but in its behavior.

The costs and benefits of a particular behavior often depend on what others of the population are doing. For instance, in order to breed, many male frogs typically attract the attention of females by croaking at night (Figure 13-9). How often the male must croak to be successful depends on what other male frogs are doing. If a male is the only one present, he can get by with only infrequent croaks: enough to attract females, not enough to attract predators. If there are many males present around a pond's edge, each with his own territory, each male had better croak more often to attract a mate, lest he get lost in the crowd.

What behavior evolves in any given situation? The most **evolutionarily stable strategy (ESS)** is thought to evolve, that is, the one that results in the most offspring (or relatives). By definition, an ESS cannot be replaced or bettered once most individuals

**Fast Find
13.2a
Game Theory**

Figure 13-9. At certain times of the year, male frogs and toads croak to attract females. Such behavior is not without risk, because croaking may attract predators in addition to females.

of a population adopt it (unless, of course, environmental conditions change). **Game theory** has been used extensively to predict which ESS will evolve under a given set of conditions. Go to the CD to see how this works.

13.2.3 The Physiology of Behavior

The behavior of an animal is often closely tied to its physiology. Behavior often involves satisfying some physiological need. When a predator seeks prey, its nervous and digestive systems are intimately involved. The amount of food in the digestive system, especially in the stomach, is part of the information the nervous system processes and interprets as hunger, which stimulates food-seeking behavior in many animals.

Reproductive behavior is also often intimately tied with physiological state. For example, many mammals are reproductively active only during certain periods of the year. During colder seasons the testicles of many temperate and northern males are relatively small, recessed into the body cavity, and inactive. During these seasons, males are also generally gregarious, tolerant of other males, and nonterritorial. As the warm seasons approach, physiological state and behavior change. Testicles enlarge, descend out of the body cavity, and begin to actively produce sperm and male hormones. Now males become generally intolerant of other males, more territorial, and solicitous of females. Other kinds of changes in the environment, especially lengthening days, stimulate production of certain hormones in the brain that, in turn, control changes in both the animal's physiology and behavior.

The endocrine system is known to have profound effects on the behavior of many vertebrates. These effects are classified as either organizational or activational. **Organizational effects** are permanent and often appear early in life. For example, early in the embryonic development of many mammals, the presence of certain hormones sets the embryo on a track toward becoming male or female. These hormones not only affect physical characteristics but behavioral tendencies as well (Figure 13-10). **Activational**

Figure 13-10. During most of the year, male dall sheep are tolerant of other males and, indeed, often socialize in large groups. During the breeding season their behavior changes. Males forgo feeding and devote themselves to fighting with other males and attracting females. Such changes in behavior are controlled by hormones.

effects of hormones tend to be of shorter or more periodic duration and appear, generally, later in life. For example, a female dog in estrus (heat) advertises her sexual receptivity both physiologically (producing certain odors and secretions) and behaviorally (soliciting male attention). Sexually competent males respond. Typically, they become overly protective of the female, devote nearly all of their attention to her, forgo feeding and other behaviors, and engage in play activities with the female that are not observed under other conditions. Both males and females are responding to hormones produced only during the reproductive season.

Piecing It Together

1. Animal behaviors are partly under genetic control. The tendencies and ability to perform specific types of behavior are inherited from parents and passed on to offspring. As with other characteristics, they may be expressed in certain individuals and hidden in others, and they are affected by recombinations and mutations. Other genetic factors such as allelic dominance, incomplete dominance, independent assortment, linkage, and sex linkage, also apply to behavioral characteristics.

2. Behavioral characteristics are also subjected to and shaped by natural selection. We discussed in Chapter 2 how behavior is one of the mechanisms that can isolate otherwise sympatric populations and is thus important in evolution.

3. Optimality theory predicts that behaviors will evolve that promote the greatest fitness. Often, the costs and benefits of a particular behavior depend on the behavior of other individuals in the population.

4. The behavioral traits most likely to evolve are thought to be those that result in the most offspring or relatives, that is, in an evolutionarily stable strategy (ESS). By definition, an ESS cannot be replaced or bettered once most individuals in a population adopt it. Game theory is used to predict which behavior qualifies as an ESS.

5. The behavior of an animal is often intimately tied to its physiology. Behavior often involves satisfaction of physiological needs. Though every organ system is at one time or another involved in behavior, the endocrine and nervous systems are most intimately tied to behavior.

6. Hormonal effects on behavior can often be classified as either organizational (long-term, permanent effects often appearing early in life) or activational (short-term, periodic effects that appear later in life).

13.3   HOW DO INDIVIDUAL ANIMALS BEHAVE?

What guides animals in their behavior? This is a difficult question that is tempting to answer from an evolutionary point of view. We could say that foxes benefit rabbit populations by culling out sick, injured, old, and unfit individuals. And rabbits benefit foxes by producing enough individuals to feed these predators. Although there are elements of truth in these statements, it is important to remember that an animal's behavior is directed toward its own survival, not toward the betterment of its species or the benefit of others in its biological community. Each individual does what it can to successfully cope with its particular environment for as long as it can, within its own genetic capabilities and physiological demands. In so doing, animals move from place to place, seek food, avoid predators, and reproduce. Let's take a closer look at how they do so.

13.3.1 Animals Move through Space

When asked what distinguishes animals from plants, many would respond, that animals move from place to place. Certainly there are exceptions, but the ability to move through space is, indeed, a characteristic common to many animals. The intensity of movements varies from rather simple responses to environmental change to highly complex, once-in-a-lifetime migrations.

Perhaps the simplest of animal movements are **kineses**, undirected movements whose intensity is stimulated by unfavorable environmental conditions. Wood lice, also known as roly-poly bugs, are intolerant of dry conditions and congregate in moist places, under rocks or logs, for instance. As their environment begins to dry, they begin to move about. Their movements are not directed toward or away from anything. They simply move, at random, until they happen into a moist area. Then, their movements cease or slow down. These simplest of movements keep wood lice in favorable environments (Figure 13-11).

Like plants growing toward light, some animals experience **taxes**, that is, directed movements toward or away from a stimulus. Moths move toward light. Earthworms move away from light (assuming their environment is neither too wet nor too dry). In many marine invertebrates, phototaxis produces daily movements. During the day, many marine algae show a positive phototaxis toward light and move toward surface waters. At night, they become negatively phototaxic and avoid even artificial lights. Some taxes vary seasonally. The protist *Euglena* moves toward light in the summer but avoids it in the winter.

Other complex animal movements involve seeking food, mates, or shelter or avoiding predators. These movements will be discussed more fully in later sections of this chapter.

Migrations are complex, periodic movements of animals between habitats. Usually when we think of migrations we think of birds annually migrating vast distances between nesting and wintering grounds. But migrations are not always annual. Some are daily movements, whereas others span several years. Also, migrations are not restricted to birds; some mammals, fish, insects, and marine invertebrates migrate regularly. Furthermore, some seasonal movements are not particularly extensive. Migrations for many robins involve little more than a seasonal shift in habitats. In summer, robins nest in shrubs or

Figure 13-11. Wood lice live best in moist conditions. In moist surroundings, they move relatively little. Whenever their environments begin to dry, wood lice increase their rates of movement. They move randomly until moist conditions are once again encountered. In moist surroundings, rates of movement once again become slow. Such undirected, random movements are termed kineses.

low trees and feed in open grasslands. During winter, they move into deep forests and live in large flocks. This migration may only involve a few miles. Many alpine mammals migrate annually from high-altitude areas in summer to valleys in winter. But certainly the most fascinating migrations are those that span continents or oceans (Figure 13-12). All of these periodic movements are migrations.

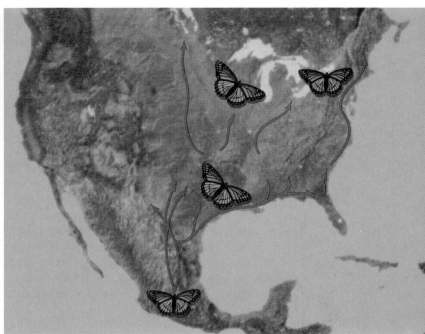

(a) Spring migratory path of the Monarch butterfly

Figure 13-12. Migratory movements of monarch butterflies are the most complex and extensive of any known insect. (a) In the population of eastern North America, millions of monarchs overwinter in a few isolated locations in central Mexico. With the coming of spring they migrate northward. As they migrate, females lay eggs on newly sprouted milkweed plants. These eggs hatch, and the larvae become adults by early- to mid-summer and also drift northward. Eventually, a few of the original migrants and many of their first-generation offspring reach the northern tier of the United States and southern Canada, laying eggs on milkweed plants as they go. Eggs laid late in the season overwinter. They compose the bulk of the next year's population in northern latitudes. (b) The fall migration southward occurs from September to November. The bulk of the millions of south-moving migrants are second-generation offspring plus some surviving first-generation offspring of adults who overwintered in Mexico. (Map©1998 Monarch Watch—www.monarchwatch.org.)

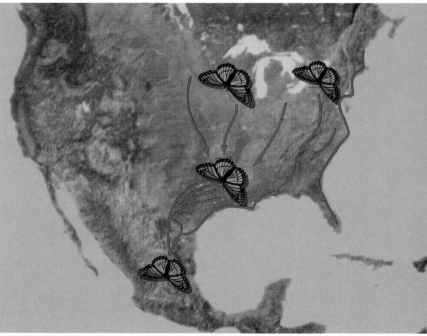

(b) Fall migratory path of the Monarch butterfly

Similar to periodic migrations are once-in-a-lifetime migrations of animals, typified by salmon. These fish hatch in specific areas of freshwater streams or rivers. After a time, they move into marine waters for several months to a few years. Then, they return to their areas of birth to mate, lay eggs, and die. For each species of salmon, the amount of time spent in freshwater, at sea, and on migration is tightly regulated.

Migrations, like other behaviors, have costs and benefits. Chief among costs are the time and energy spent gaining familiarity with multiple habitats. Within their habitats, animals learn the location of food supplies, shelters, and predators. Nonmigrators need to learn these things for one habitat only. Seasonal migrators must learn them for at least two habitats: wintering grounds and summering grounds. There are benefits too. Migrators move into areas where food supplies are seasonally abundant and where competition for breeding sites is low. They avoid the necessity of making seasonal adjustments to changes in food supplies and environmental conditions. For example, migratory insect eaters move to areas where insects are still available and environmental conditions are still summer-like.

Considerable research has been done on how animals orient on long migrations. There are three levels of orientation. **Piloting** involves the sequential recognition of landmarks. Salmon find their way back to their natal streams by smelling chemicals in the water. **Compass orientation** involves moving in a particular direction for a prescribed length of time or distance. Return trips involve movement in the reverse direction for similar spans of time or distance. Most birds use compass orientation when migrating, as do cabbage white butterflies. **True navigation** involves finding one's way by means of an internal map coupled with an internal compass. Homing pigeons have been shown to have such abilities.

Fascinating Journeys For some particularly fascinating examples of animal migrations, go to the WWW and search for monarch butterflies, gray whales, caribou, Neotropical songbirds, or migratory waterfowl. There are others you can find on your own.

13.3.2 Animals Forage for Food

Getting enough to eat is one of the most demanding of animal activities. Basically, the problem is easily stated: obtain sufficient nutrients and energy, but expend minimal energy and time. Each species has its own dietary preferences. Some are strict specialists. Pandas prefer nothing more than fresh bamboo. Generalists eat a wider variety of foods. Sparrows eat many kinds of seeds and insects. Also, each species has its own specific strategies for obtaining food. These can be categorized into three broad types, filter feeding, herbivory, and carnivory.

Filter feeding

Water is frequently loaded with tiny bits of potential food, called **plankton**, or nutrients in solution that many animals tap into with minimal expenditures of energy. Sponges, corals, tubeworms, barnacles, and many other invertebrates park themselves in one place for life and create small currents that bring water to and through apparatuses that trap nutrients for consumption. This is **filter feeding** (Figure 13-13). Most filter feeders restrict periods of activity to times when food is most likely to be present, when predators are least likely to be active, or when their environments are least stressful. Thus, corals feed mainly at night; coastal clams feed mainly when the tide is in.

Some vertebrates also filter feed, but they are considerably more mobile than filter-feeding invertebrates. Baleen whales gulp huge mouthfuls of plankton-rich waters and pass them through special structures that filter foods. Whale sharks and manta rays feed similarly but with different types of filter mechanisms.

(a) *(b)*

(c) *(d)*

Figure 13-13. For organisms living in nutrient-rich aquatic environments, filter feeding is an efficient means of obtaining food. Some filter feeders are sessile, staying in one place through much of their lives, letting food come to them. Other filter feeders are more active, moving to areas where waters are seasonally rich in food. In all cases, filter feeders pass water through a filtering mechanism that removes food, often in the form of plankton, and passes the water out. Coral polyps *(a)* filter with their tentacles. Barnacles *(b)* filter with their legs. Clams *(c)* draw water into their bodies with their siphons and filter with their gills. Baleen whales, such as these humpback whales *(d)*, have highly modified teeth (baleen) that serve as filters.

Herbivory

Many animals eat plants, for which there are several benefits. In most ecosystems, plants are more numerous than animals, and their immobility makes them vulnerable to predation. These factors minimize the time **herbivores** (plant eaters) need to spend searching for food. (But see Chapters 12 and 14 for discussions of how plants fight back.) Thus, it is relatively easy for plant eaters to find, approach, and consume food. On the debit side, plant tissues are generally harder to digest than animal tissues. As a result, plant eaters generally spend more time acquiring and processing food than do carnivores.

Carnivory

Finally, a number of animals are **carnivores**: they eat other animals, for which costs and benefits are generally opposite those of eating plants. Their prey is generally less abundant and harder to catch than plants, but animal tissues are also generally more nutritious than plant tissues. Basically, animal eaters must find prey, approach it without being detected, capture and subdue prey without being injured, and consume it without attracting attention of competitors. Let's see how it is done.

A coyote hunts in a grassy field. Methodically and quickly it works back and forth, stopping briefly to sniff out areas where prey is most likely. Suddenly, a rabbit flushes. After a startled pause, the coyote pursues half-heartedly, it seems. The rabbit may be vulnerable in the open, but it streaks toward a brush pile where it can find protection. Just as it arrives, a second coyote that lies in ambush grabs it. After a brief struggle, the rabbit is subdued. The second coyote feeds first; then it shares the kill with its companion.

What can we learn from this example of the mechanics of predation? Optimality theory, discussed earlier, predicts that strategies evolve that are most efficient for a given predator in a given environmental setting. Generally speaking, predators choose prey items that are most abundant, easiest to obtain, most nutritious, or some combination of these factors. In this case, the coyotes hunted rabbits. Next they had to decide where to hunt. Optimality theory again predicts that predators seek food in areas most likely to be productive. These coyotes hunted in a grassy field. Perhaps they had learned through previous experience that this particular area has rabbits that could be easily obtained. Perhaps they assessed habitat quality through a combination of other factors, instinct or sensing prey presence, for instance. Perhaps they learned through experience that a cooperative hunting strategy enhances success.

An important variation on the carnivory theme is **scavenging**. Animals that scavenge eat dead animal material. Some animals, such as vultures, eat nothing else. Other carnivores become seasonal scavengers. Arctic foxes in summer are strict carnivores, eating only freshly killed food. In winter, when food is much more scarce, Arctic foxes eat any animal they find, dead or alive.

How long do predators exploit a particular prey species or hunting area? Answers vary among predators and situations. Some predators exploit an area for a particular length of time and then move on. Others remain until a specific number of prey items are obtained. Still others continue to exploit an area as long as their rate of success stays high. Incidentally, these factors work equally well for herbivores in choosing what, where, and how long to seek their food.

13.3.3 Animals Avoid Becoming Food

Even as they seek food, secure territories, or advertise for mates, animals must remain vigilant to the approach of predators (Figure 13-14). Even stealthy footfalls silence a whole pond of male frogs. Avoiding predators requires the full range of each animal's adaptations, and the relationship between anatomy and behavior is profound.

Fast Find 13.4a Beehive Structure

Perhaps the simplest way to avoid predators is to avoid detection. The immediate response of many animals when a predator is detected is to freeze. In brightly colored song birds, this phase is usually temporary, passing quickly into some more appropriate

Figure 13-14. Animals must be constantly vigilant to avoid predators. This is especially noticeable in social animals, such as these suricates, where individuals typically take turns watching for danger.

(a)

(b)

Figure 13-15. In temperate regions, where environments change seasonally, some mammals and birds change basic coloration to remain cryptically colored. In summer, the long-tailed weasel *(a)* is basically brown. In winter, the same animal in northern latitudes *(b)* is mostly white.

**Fast Find
13.3a
Cryptic
Coloration**

response. Among animals with **cryptic coloration**, that is, animals that resemble their backgrounds, the freezing behavior may be much more extensive. Thus, a newborn fawn may lie motionless in grass until actually touched by a predator.

Body coverings of many animals blend into their environment. Somber browns and grays, in solids, stripes, or spots, are common colorations for cryptically colored terrestrial mammals. There are exceptions, of course. Some brightly colored insects blend perfectly with the flower petals they hide among. Many fish and aquatic mammals and birds have white bellies and dark backs. Seen from below, white bellies blend into sky light from surface waters. Seen from above, dark backs merge with inky depths. Some animals change coloration to match the season or moment. Thus, many temperate zone mammals and birds are dark in summer and white in winter (Figure 13-15). Some squid change basic coloration and patterns in as little as a few seconds to more closely match surroundings. Finally, some fish and insects are partially transparent, so wherever they go, they blend in.

Many animals avoid predators by resembling some object they are not. When many butterflies close their wings, they look like dead leaves. Other insects look like pieces of bark, bird droppings, plant seeds, or twigs. Some do more than look the part. Stick insects arrange their bodies and legs to look like sticks supporting twigs and gently sway to and fro to resemble twigs wafting in the breeze. They do this even when no breeze is blowing.

To gain maximum surprise, most predators approach prey from the rear whenever possible. As a defense, some butterflies and caterpillars have large spots that look like eyes and extensions that look like antennae at the rear of their wings or bodies. False eyespots and antennae fool predators into approaching from the wrong end and give prey advanced warnings. If the eyespots seem large, predators may avoid contact altogether (Figure 13-16).

Some animals have chemical or physical defenses. Skunks release noxious odors that irritate a predator's eyes and mucous tissue. Monarch butterflies taste bad. Some frogs are downright poisonous. Wasps and bees have stingers. All these animals advertise potential danger by bright colors and conspicuous behavior. Skunks are striking with their bold, black and white stripes allowing them to move through temperate woods unmolested, saving energy and effort. Poisonous frogs are even more brightly colored and sit by day fully exposed in their habitats while their nonpoisonous, drab-colored relatives hide furtively nearby under leaves or pond weeds.

Wasps and bees are also brightly colored, black with either yellow or orange stripes. Predators that learn to avoid one, avoid all. This is **Müllerian mimicry**, when different species, each of which is dangerous, announce their undesirableness by advertising similar characteristics.

Figure 13-16. *(a)* With wings closed, the question-mark butterfly cryptically resembles a dead leaf. If this cryptic coloration fails to fool predators, the butterfly opens its wings *(b)*. A sudden flash of color is intended to startle.

(a)   *(b)*

Robber flies also benefit from black and yellow coloration. They too are avoided, even though they have no stingers or other noxious characteristics. This is **Batesian mimicry**, when otherwise desirable species benefit from false advertising, that is, from resembling species that are dangerous (Figure 13-17).

Even when actually detected and approached by predators, most animals are not totally defenseless. Perhaps their simplest response is to take flight, but this puts them into a direct race with predators. Only if they can outrun predators do they survive. Such contests represent significant expenditures of time and energy for the prey. Thus, it is not unusual for both predator and prey to spend considerable time studying each other. Prey are assessing: How serious is the predator? How capable is it? Is it essential to run?

Many prey send messages to predators that they are fully capable of escape and may, indeed, inflict injuries. A healthy moose when approached by a pack of wolves paws the ground, snorts, shakes its antlers, and perhaps charges. Many cats, when confronted with a dog, arch their back, stiffen and swell their tail, lay back ears, hiss, expose teeth, and flash claws. The message is clear: Don't mess with me! It's not unusual for dogs to back away from cats in such confrontations.

Many animals try to startle predators, as we saw earlier with the question mark butterfly. Ring-necked pheasants remain motionless in tall grass until practically stepped on by predators. Then they explode with cries and loud flapping, startling predators just long enough to escape.

Group behavior protects many animals from predators. A flushed covey of quail or a stampeding herd of gazelles confuses predators by making it difficult for the predator to focus on a single individual. A herd of musk oxen form a circle around their young and present a ring of horns to a pack of wolves. Flocks of crows mob a lone hawk or owl to harass them and to discourage them from feeding—at least on crows.

Figure 13-17. *(a)* Many wasps and bumble bees (like the one pictured here) share a basic coloration of black and yellow that warns predators of potential stings. This is Mullerian mimicry. *(b)* Robber flies have no stingers but are avoided by predators because they look like bumble bees and wasps. This is Batesian mimicry.

(a)   *(b)*

13.3.4 Animals Reproduce

Reproduction is of fundamental importance to all organisms. Through reproduction, successful characteristics are passed on to the next generation. It should not be surprising, then, that reproductive behavior in animals is complex. The territorial aspect of reproductive behavior was discussed above in Section 13.2.2. Courtship displays, as a form of communication, are discussed below in Section 13.4.2. Here we focus on two other aspects of animal reproductive behavior: mating strategies and parental care.

In mating strategies there are almost as many variations as there are species, but three patterns can be discerned: monogamy, polygyny, and polyandry.

Monogamy is when only one male breeds with only one female. In some cases, the bonds between the pair form once during the lives of the partners and remain intact until one or both partners die. In other words, these animals mate for life. More commonly, pair bonds form for only one mating season. This is **serial monogamy**. How long the pair stays bonded varies greatly among species. Male eider ducks stay with females only long enough to breed and then move on, leaving females to tend to eggs and young. Male Canada geese stay with the female, assist in incubation, and feed the young until they disperse. The next year they find a new mate. Pair bonds between osprey also break up when the young disperse. They spend winters separately and some reunite in spring. Monogamy is much more common among birds than mammals, perhaps because, among mammals, gestation and lactation are solely female activities. The few mammals that are monogamous have young that stay with parents for prolonged periods after birth.

Polygyny occurs when one male mates with multiple females; this is perhaps the most common mating system among vertebrates. Reasons why may center on the relative energetic value of sperm and eggs. Typically, eggs have higher value than sperm; that is, they are larger, less numerous, and less mobile than sperm. Each egg, therefore, represents a relatively large portion of a female's reproductive effort, and it is to her advantage to nurture and provide for each egg produced. By contrast, males have much less invested in individual sperm; therefore, males can afford to produce many sperm and disperse sperm more widely than females can disperse eggs.

What do moose, lions, and elephant seals have in common? One of the answers is a form of polygyny in which individual males defend and mate with groups of females; this is particularly common among deerlike mammals and marine mammals. This form of mating favors males (1) who are considerably larger than females; (2) who often sport special characteristics to assist them to attract, keep, and defend females; and (3) who are obviously aggressive, especially during the mating season. Male elephant seals may be two or three times the size of females, and male moose have antlers. The costs to males of this mating system are particularly high. Displays and fights among males are common, and so are injuries. It is not uncommon for polygynous males with harems to forgo eating and other essential activities altogether during the mating season. Mortality rates among such males are typically high both during and after the mating season. Furthermore, relatively few males actually mate. In some cases, the majority of males fail to reproduce.

In some species, males defend valuable resources rather than the females themselves. These territories are vigorously defended from other males, while sexually receptive females are actively solicited. This form of polygyny is relatively common among hummingbirds. Males defend prime flowering plants or individual blossoms.

Still another variation on the polygyny theme is seen among prairie chickens. In the spring, large groups of males gather to defend symbolic territories at hot spots called **leks**. Each male's territory is quite small and possesses no resources of value (Figure 13-18). For several days or weeks, males make noises and display actively among themselves, defend themselves and their leks from other males, and solicit as many females as possible. After the mating season, these males take no further part in the reproductive effort.

Polyandry is when individual females mate with numerous males. This is the rarest of mating strategies. Among birds, polyandry is invariably accompanied by sex role rever-

Figure 13-18. Breeding territories defended by male prairie chickens are quite small and provide no useful resources to either sex other than opportunities to breed.

sal. Among phalaropes—small shorebirds that are typically marine most of the year and who become terrestrial only to nest—females are more brightly colored and more aggressive than males. Typically, the females arrive on the nesting grounds before males, and they establish and defend nest sites. When males arrive, they are actively solicited by females. After mating, males stay with females until clutches of eggs are laid. Soon after, females leave and males incubate the eggs and raise the young. Relieved of their responsibilities, females establish new territories, mate anew, and leave again.

Parental care is another activity associated with reproduction. Again, there are almost as many variations as there are species. Parents inherently are in conflict with their offspring. The time and resources devoted to a particular offspring cannot be spent on activities that benefit parents or in raising additional offspring. Two conflicting tendencies are at work here: to produce as many offspring as possible, but not to produce so many that each one gets too few resources to survive to maturity. Obviously, the amounts of time and energy required depend on each species' particular needs and environmental requirements. The range of possibilities include the following, with decreasing number of young produced as we move down the list:

- Parental care for many marine invertebrates involves nothing more than releasing gametes simultaneously with other individuals of their population and at seasons of the year when resulting larvae are most likely to be successful.

- Female monarch butterflies most commonly lay eggs on the undersides of leaves (to protect them from weather) of milkweed, a plant upon which their caterpillars can feed.

- Females of some snakes retain eggs inside their bodies until they hatch. They then give birth to living young, who must be totally independent upon hatching.

- Female crocodiles mound rotting vegetation around their freshly laid eggs to supply heat for developing embryos. Then they defend the nest site from other crocodiles and predators until the young hatch. After they hatch, females transport the young to protected waterways and continue to defend them until they disperse (in a few weeks, perhaps). They do not provide the young with food.

- Female eider ducks arrive at the nesting sites after the males. Soon after mating, the males leave. Females incubate eggs, lead hatchlings to water, and protect them from cold and predators until they can fly. The young must find their own food.

- Male ostriches mate with several females who lay their eggs in a communal nest. After eggs are laid, most of the females disperse, and the eggs are incubated and cared

Figure 13-19. Among ostriches, males assume the major responsibilities of raising chicks. Typically, males mate with several females, who deposit eggs in a communal nest. These eggs are incubated mainly by the male. After hatching, he, assisted by a few of the females, keeps the brood together, protects them from harm, and leads them to food and water.

for mainly by the male. After hatching, the male keeps the communal brood together, protects the hatchlings from predators and weather, and leads them to food until they can provide for themselves (Figure 13-19).

◆ In many other bird species, males defend territories while females incubate eggs. After eggs hatch, both parents provide food until the young fledge. Thereafter, young disperse and must provide for themselves.

◆ Within a pack of wolves, typically only one pair (the alpha, or lead, male and alpha female) breed. All members of the pack hunt, bringing food to and otherwise caring for the young until they are old enough and strong enough to join the pack.

◆ Human young require the greatest attention and effort of any species. They must be cared for for several years. Care is provided primarily by parents or foster parents, assisted by others in the community.

Once the young are raised, they must become independent. This often means that the parents withhold or cease to provide food. The next step is for the young to disperse from the breeding area. This often involves a gradual breakdown of the parent–offspring bond. Grizzly bear cubs stay with their mothers for a period of two years, during which time the female provides cubs with protection from other bears, predators, and adverse weather. She also provides nearly all of their food. During their second summer, the cubs find some of their own food. Gradually, they begin to wander away from the female and other cubs. Mothers, meanwhile, become less attentive to their cubs, stop providing milk early in the second summer, and may actually drive cubs away late in the season (Figure 13-20).

After dispersal, young animals typically experience a period of traveling while they fine-tune their survival skills. For many, especially birds and mammals, the period immediately following dispersal is the most critical period of their adult lives. Mortality rates are usually highest during this time. Some young leave the territories of their birth, never to return. Others are allowed to share the territories of their parents. Social animals sometimes show both patterns. For example, among African lions, young males disperse while young females stay with their natal group. In other animals, females disperse while males stay put.

Figure 13-20. Young grizzly bears stay with their mothers for two years, during which she protects them and teaches them what they need to know to be successful.

Piecing It Together

1. Animal behavior is directed toward the survival and reproduction of individuals, not the improvement of species or biological communities.

2. Animals move from place to place to position themselves in environments and habitats where they are most likely to be successful.

3. Kineses are undirected movements stimulated by unfavorable environmental conditions.

4. Taxes are directed movements toward or away from a stimulus.

5. Migrations are periodic movements between habitats.

6. While on migration, animals must orient. There are three general types of orientation:

   ◆ **Piloting:** the sequential recognition of landmarks.

   ◆ **Compass orientation:** moving in a particular direction for a prescribed length of time or distance.

   ◆ **True navigation:** finding one's way by means of an internal compass and map.

7. There are three broad strategies for feeding among animals:

   ◆ **Filter feeding:** animals filter dissolved nutrients or plankton out of the water. Invertebrate filter feeders are usually immobile, whereas vertebrate filter feeders are highly mobile.

   ◆ **Herbivory:** when animals eat plants.

   ◆ **Carnivory:** when animals eat other animals; scavenging is a specialized form of carnivory.

8. Although some animals avoid predators by running away when confronted by them, most animals utilize other means, conserving energy and time. These means include:

   ◆ **Cryptic coloration:** resembling the background of their environment.

   ◆ **Misleading coloration:** resembling an object they are not.

◆ Announcing their noxious or dangerous attributes (bad tastes, stingers, and so on) with bright colors and deliberate displays.

◆ **Müllerian mimicry:** when two species resemble each other, each of which is dangerous (several species of wasps and bees are black and yellow).

◆ **Batesian mimicry:** when a nondangerous species gains benefits from resembling dangerous animals.

9. When detected by predators animals can either take flight, fight back, startle, threaten, or engage in group behavior.

10. There are three major mating systems seen among animals:

◆ **Monogamy:** one male mates with one female.

◆ **Polygyny:** one male mates with multiple females.

◆ **Polyandry:** one female mates with multiple males.

11. Time and effort devoted to parental care varies widely. Some animals do nothing more than place numerous eggs in favorable environments. At the other extreme, both parents devote extensive care to a few young for extended periods of time.

12. After the young are raised, they generally become independent and disperse.

13.4 HOW DO GROUPS OF ANIMALS BEHAVE?

With the possible exception of some aquatic invertebrates, all animals must interact with other members of their own species At a very minimum, pairs interact to breed. For some, solitary existences are broken only long enough to mate. At the other extreme are certain species of dolphin, in which individuals are never alone, spending their entire lives surrounded by and interacting with others (Figure 13-21).

Some animals are only seasonally gregarious. In winter, blackbirds in eastern North America feed during the day in flocks of over 100. At night they form huge flocks of several thousand to over a million to roost and spend the night. In summer, these same birds live in pairs, actively defending breeding territories from other blackbirds.

In this section, we will look at how animals live in groups, communicate among themselves, and care for others.

13.4.1 Some Animals Live in Groups While Others Live Alone

As you might expect, there are both costs and benefits associated with social living. Chief among the costs is that resources must be shared. In environments where resources are limited, such sharing may not be possible, and here we find most animals living essentially solitary existences. Where resources are seasonally or regionally abundant, sociality evolves. For example, honeybees, among the most social of insects, can visit enough patches of flowers and make enough honey during warmer months to survive, as a group, through winter months.

There are other costs to sociality. Predators may be attracted to groups of potential prey. Hawks and owls in southeastern United States go to where large flocks of blackbirds roost in winter. Here these raptors gather to feed on individual blackbirds from the group. Parasites and diseases are another cost of social living. They spread more easily among individuals living close together. Another cost is that social interactions take time and effort that could be spent in other pursuits by those living alone.

What are the benefits of social living? In some cases, groups of animals are more efficient in finding food than are lone individuals. A flock of 100 birds has 100 pairs of eyes with which to find elusive food. Groups can sometimes exploit prey unavailable to individuals. For example, an individual dolphin might find individual fish difficult to

(a)

(b)

(c)

Figure 13-21. At least during the breeding season, all animals must interact with others of their own kind. For social animals, such interactions are considerably more extensive. *(a)* Leaf cutter ants, *(b)* Pacific double-saddle butterfly fish, *(c)* snow geese, *(d)* bottle-nosed dolphins.

(d)

catch. But schools of dolphins can herd fish into dense aggregates and then feed at will. Although more predators may be attracted to groups, it is easier for an individual to hide, as one among many, than when alone. In some cases, group living makes finding mates and raising young easier. The whole pack raises a young wolf. Isolated female wolves that raise litters have significantly fewer pups that reach adulthood than do females raising pups as part of a pack.

So, when does sociality evolve? Optimality theory is once again applicable: sociality evolves among those species and in those environments where benefits exceed costs.

13.4.2 Animals Communicate, Especially in Groups

Communication is essential to any social group. It is the glue that binds members together; it molds group behavior. Communication is not easily defined. First and foremost, information must be conveyed, but is that enough to qualify as communication? A brightly colored male bird, singing loudly on a spring day, conveys a great deal of information. To other males it says, "This territory is occupied." To females it says, "Here am I." Unfortunately, to predators it may say, "Here is food." To a person, the singing bird says, "It is spring." Is all of this information communication? No. To qualify as communication, two additional elements must be present. The communication should lead to modifications in the recipient's behavior that benefit the sender, and the sender should intend to send the information. Whereas all four examples above may lead to changes in the recipient's behavior, only the first two benefit the sender *and* include the sender's intent.

Almost any mode of behavior can communicate. Sounds are particularly important for many species. Dogs bark; birds sing; frogs croak; crickets chirp; and many invertebrates click, whistle, or sing. Behaviors that are intended to be seen are important to many animals (Figure 13-22). A dog defending its yard from an intruder stares, stands stiffly, and raises tail and neck hairs. Can you see the three elements of communication in the dog's behavior?

Odors are the basis of much animal communication. Female moths release **pheromones** on the night air to attract males. In canids, which include dogs, wolves, and their relatives, a complex system of glands near the anus convey much information on topics such as sex, sexual receptivity, willingness to fight, age, and perhaps individual identity. It is no wonder, then, that active sniffing is a common greeting among canids. Many animals mark perimeters of territories with odoriferous chemicals secreted in urine or by special glands.

Touch is another means of communication. Many mammals and birds must be in physical contact with others of their group when lounging. An extreme case of touch-based communication is social grooming in primates. Much of their time is spent combing through the fur of another to seek, remove, and eat parasites. This is one of the most important ways that their social bonds are established, maintained, and reinforced.

Unusual forms of communication are found in animals living in specific environments. Animals who live in perpetually dark environments often generate their own light. These include phosphorescent fish living in ocean depths, glow worms in caves, and fire-

Figure 13-22. In addition to singing loud and obvious songs, many male songbirds, such as this cardinal, are brightly colored, especially during the breeding season. Much of their time is spent establishing and defending territories. Bright colors, loud singing, and aggressive behavior are used to discourage competitors and lure predators away from their nests.

flies at night. Certain frogs and insects tap out messages in soil or wood to attract mates. Similarly, water striders tap out vibrations on the surfaces of ponds to attract mates; a different pattern of vibrations warns away competitors. Finally, certain fish living in highly turbid waters generate and communicate by electric fields.

Animals communicate a wide spectrum of information with their behaviors. Some of the easiest for us to recognize are those used in **agonistic encounters**, or actions involved in conflict. This type of behavior occurs both within and between species. Thus, male lions use the same sort of displays to keep both hyenas and female lions away from downed prey until they have eaten. Male moose shake their heads, snort, paw the ground, and charge at both packs of wolves and, during the breeding season, other male moose. Usually, agonistic encounters between members of the same species end in no physical contact. A great deal of bluffing is involved. Two grizzlies meet in an open meadow. They stand on hind legs and look at each other. They may "woof," snap jaws, and make short charges. Finally, they stand sideways to each other. "This is how big I am. Do you still want to fight?" At each step in these encounters, each bear is assessing the other and may back down. Such behavior is not restricted to mammals (Figure 13-23).

Often, agonistic behavior within a social group leads to the establishment of **dominance hierarchies**. For instance, within a group of chickens in a barnyard, one individual is dominant. One hen feeds whenever she wants, with no interference from other chickens. Any chicken who approaches the leader is promptly pecked into submission and retreats. Another chicken in the flock defers to the leader, but pecks all others interfering with her feeding. Another chicken occupies the number three position in the flock, and so on, down to the lowest ranking individual who defers to all others and feeds last. Similar hierarchies, or pecking orders, as they are commonly referred to, are observed in many social animals. Within a wolf pack, for example, two systems are normally established, one for each sex.

Dominance hierarchies create group stability and efficiency. Pecking orders among chickens allow the group to feed with minimal disruption. As long as food is abundant, everyone is fed, eventually. When food is limited, only the dominant feed. Under severe stress, the most highly adapted survive. Within the wolf pack, breeding is typically restricted to the highest ranking individuals. Functionally, the dominance hierarchy is a form of population control for wolves. Removal of a leader from such a group often leads to a temporary period of social instability. Soon, a new pecking order is established with renewed cohesiveness.

Animals that are about to mate also communicate. Often, these encounters start out with behaviors very similar to agonistic encounters. Two animals approach. The male behaves aggressively. If the female is not interested, she may respond aggressively or withdraw. But if she is receptive, she responds with a different kind of behavior: submission. This results in a different kind of response in the male: tolerance. Back and forth the

Figure 13-23. Agonistic behavior is not restricted to mammals and birds. Such behavior is also found in other vertebrates and some invertebrates. This male Siamese fighting fish defends its territories with bright colors and fierce behavior.

Figure 13-24. White-tailed deer use their tails to warn others of danger.

behavior goes, becoming surprisingly elaborate in some cases. Finally, they mate. Especially in birds, these displays do more than establish pair bonds. The stimulus–response give and take of mating rituals is necessary to produce the hormones required for gametes to mature.

Communication is also the basis for individual and group identification. "Birds of a feather flock together" is a well-known example. The statement is true only if "birds of a feather" means similar, but different, species because most bird flocks are mixtures of species. The cohesiveness of flocks is maintained through vocalizations and similar behaviors.

Group identity is communicated chemically among many social insects. An intruder ant is immediately attacked if the aroma it carries doesn't match those of the other members of the group. At the family level, members of a wolf pack use chemical, visual, and behavioral clues to identify an intruder to their territory.

Safety is one of the potential benefits of sociality. This has led to two forms of behavior: **alarm calls** and **distress signals**. Often the first member of a group to spot a danger sounds an alarm. But this behavior potentially has unwanted consequences. It could draw the attention of the predator to the sender. Typically, auditory alarm signals are short, explosive bursts of sound that convey information to group members but are hard for predators to locate. Other alarm signals do not involve sound. White-tailed deer raise their white tails to warn other deer (Figure 13-24). Mule deer release an odor for the same purpose. Vervet monkeys have three separate alarm calls that distinguish between approaching snake, bird, and mammal predators.

Many animals, when trapped or cornered by a predator, vocalize loudly. Among social birds and mammals, this distress signal often results in drawing others toward the threatened individual. A flock may mob or otherwise confuse the predator into releasing the prey.

13.4.3 Some Animals Appear to Behave Altruistically

On an early spring morning, in a clearing deep in the woods, a group of male turkeys gather. They gobble loudly, spread their tails, and strut stiffly around the clearing. It's the breeding season, and their intent is to attract females. Surely, such a large group will be successful, but only a few dominant males have any chance of mating. Also, more than females may be attracted. Such noisy behavior may well attract predators. With little chance of reproducing and increased risk of predation, what are these subordinate birds

Figure 13-25. During the breeding season, all male turkeys in a given social group display and attract females, but only a few—the dominant males—breed with females. Of what benefits can such behavior be to nonbreeders?

doing in the clearing? Why are they displaying? Indeed, doesn't natural selection predict that such behavior would be selected against (Figure 13-25)?

Most young birds when they fledge disperse, but not always. Young Florida scrub jays not only hang around their parents breeding territories, they assist their parents in raising the next clutch of offspring. Seemingly, they sacrifice their own opportunities to mate to ensure that their parents will be successful and their younger brothers and sisters will survive. What benefits do the assistants derive from such an arrangement?

The first prairie dog to spot an approaching golden eagle sounds an alarm, and the whole colony scampers off to the safety of underground burrows. The colony as a whole benefits from such behavior, but what about the individual who sounds the alarm? Isn't it likely that she's the one noticed by the predator? Wouldn't it be more prudent to hunker down and let the eagle take another member of the colony?

**Fast Find
13.4a
Beehive
Structure**

Members of a honeybee colony are mostly female: one queen and a larger number of workers who gather pollen and nectar and tend the colony. Only the queen reproduces. The workers, who rarely lay fertile eggs, perform acts beneficial to the colony but have little chance of reproducing. Earlier we said that behaviors evolve when benefits exceed costs. How can life in a hive possibly benefit worker honeybees?

Arguably, no other type of behavior has been harder to explain than apparent **altruism**, or cooperative behavior by which individuals appear to sacrifice their own safety or opportunities to reproduce for the benefit of others. Is this an exception to natural selection? Or are there hidden benefits to altruistic behavior? Let's take a closer look at the above examples.

The first clue to understanding honeybees is to note that all workers in a hive are sisters. Worker honeybees are genetically nearly identical. For solitary bees, environments are harsh and predation is high. Even if they were capable of reproducing, their chances of success would not be high. By dedicating her labors to the good of the hive, she enhances the survival of individuals with which she is genetically similar.

Kin selection, introduced in Chapter 2 and mentioned again in previous discussions of sociobiology, may at least partially explain altruism. Young Florida scrub jays assist near relatives—parents and younger siblings. Similarly, members of a prairie dog colony are likely to all be members of a large, extended family. Within a given geographic area, most male turkeys are related. In each case, seemingly selfless behavior enhances survival of individuals closely related to the altruist. In other words, survivors and altruist share many genes and characteristics.

Other factors may also be important in the development of altruism. Remember that a good deal of predator behavior is learned. Eagles may learn that a particular colony of prairie dogs is consistently vigilant. In the absence of hunting success, they may move on. By minimizing the eagle's chance of success, the alarm-sounding prairie dog enhances the possibility that it will not be preyed on in the future.

Florida scrub jays live in an environment where breeding territories are particularly limited. It is unlikely that a dispersing young bird will be able to find a suitable, unoccupied territory. Furthermore,the territory on which they were raised is a known commodity; it was at least successful for raising them. They are likely to be familiar with its resources and challenges. Their parents are not going to live forever, so in the event that one or both die, helper birds will be able to step in and assume roles of active reproducers.

Once again, we see that particular behaviors evolve when benefits exceed costs. This is as true of altruistic behaviors as it is for other behaviors. We also see that both benefits and costs are not always obvious.

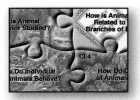

Piecing It Together

1. Nearly all animals must interact with others of their kind, at least for purposes of reproduction. Some are seasonally gregarious, living alone part of the year and in social groups for the rest of the year. For others, living in groups is the norm, and social interactions are a nearly constant aspect of their behavior.

2. There are both costs and benefits associated with sociality. Costs include sharing resources, attracting predators, more easily spreading diseases and parasites, and time spent interacting with others. Potential benefits include more easily locating food, driving off predators, attracting mates, and raising young.

3. Sociality evolves in those species and environments where costs exceed benefits.

4. Animals living in groups communicate. Communication involves conveying information that (1) the sender intends to communicate and (2) leads to modifications in the recipient's behavior that benefits the sender.

5. Almost any mode of behavior can be used for communication. Common modes include sound, visual displays, odors, release of chemicals including pheromones, touch, vibrations in soil or water, and electric fields.

6. Types of communication include:
   - **Agonistic encounters that often lead to dominance hierarchies**
   - **Reproductive displays that communicate reproductive state or receptivity**
   - **Group identification**
   - **Alarm calls and distress signals**

7. Altruistic behavior is cooperative behavior in which the individual apparently sacrifices his or her own well-being for the good of the group.

Where Are We Now?

The last hundred years has seen a virtual explosion in our understanding of animal behavior. Thousands of scientists have studied tens of thousands of species. Generalizations have been drawn and relationships described. Sociobiology represents a new direction in which the field is moving. It places special emphasis on ecological relationships, evolutionary history, and genetic basis of behavioral traits. In 1975, E.O. Wilson, a pio-

Figure 13-26. Behavior in humans is extremely complex. Certain aspects of human behavior resemble behavior seen in other animals. Other aspects may be unique to humans.

neer in the field of sociobiology, predicted that this new field would completely absorb ethology and comparative behavior. His assertion sparked heated discussion among behavioral scientists. Much of the controversy revolved around the extent to which sociobiology could explain human behavior.

One of the first things a male lion does when it takes over a pride of females is kill all of the existing cubs. This stimulates the lionesses to come into estrus, and soon the male is occupied with raising a new set of cubs, directly related to him. Such behavior is abhorrent to humans, and yet among our own kind we observe a higher incidence of child abuse among foster fathers compared to natural fathers. To what extent is the behavior of human males "hard-wired" in their genes? Is the behavior of foster fathers explainable in terms of kin selection, so firmly established in other species? It may be tempting to conclude yes, but many behavioral scientists would argue caution. We should not extrapolate from one species to another without extensive background studies.

Particular caution is needed when interpreting human behavior from observations based on other species (Figure 13-26). We may not be unique biologically, but we certainly are unique in terms of technology and culture. A few other animals use tools, but none to the extent of human technology. Other animals live in groups, but none has the complexity of group interactions as human culture. Technology and culture can and do affect behavior.

Human behavior, say opponents of sociobiology, is much more complex than a simple expression of genetic tendencies, ecological settings, and evolutionary history. Proponents counter that the basis of the behavior is in our genes, our environments, and our ancestors. This argument is certainly reminiscent of the nature versus nurture controversy that endured earlier in the 20th century. In time, additional studies and discussions among scientists have dulled the edges of this controversy. Wilson's predictions have not proven true, but several sociobiological concepts, including inclusive fitness, kin selection, and altruism, have become part of mainstream behavioral biology. These discussions continue, and perhaps as a result of the controversy, the field of animal behavior is more vibrant, relevant, and challenging than ever.

The Future of Behavior What is the current status of sociobiology?

REVIEW QUESTIONS

1. Why do young mammals play?

2. How were the early branches of animal behavior, ethology, and comparative behavior different? How were they similar?

3. What are some examples of animal instincts?

4. How does sociobiology differ from the classical study of animal behavior?

5. Briefly summarize the two positions in the nature versus nurture controversy.

6. Briefly differentiate between the following forms of learning:

   habituation   insightful learning
   classical conditioning   learning set
   operant conditioning   social learning
   latent learning

7. Can behavioral traits evolve? Give an example.

8. Can behavioral traits be passed from generation to generation? Give an example.

9. How do physiological needs influence behavior? Give an example.

10. Lions stalk their prey in packs, while leopards are solitary and ambush their prey. How would the optimality theory explain these differing strategies in the large cats of Africa?

11. What is an evolutionarily stable strategy? Give an example.

12. Differentiate between kineses, taxes, and migrations. Give examples of each.

13. How are piloting, compass orientation, and true navigation different? How are they similar?

14. Basically, how do the strategies of obtaining food differ in filter feeders, herbivores, and carnivores?

15. Briefly discuss the ways in which animals can avoid becoming food for some other animal.

16. How are Müllerian mimicry and Batesian mimicry similar and different? Give examples of each.

17. Distinguish between monogamy, polygyny, and polyandry. Give examples of each.

18. Briefly discuss the costs and benefits of living in a social group.

19. Discuss some categories of animal communication. How are they similar and different?

20. What are the costs and benefits of altruism? Give examples of altruistic behavior in animals. Under what conditions does it evolve?

Chapter 14

Population Ecology: How Do Organisms Interact to Form Populations?

**Ecology is the study of all those complex interactions
referred to by Darwin as the conditions of the struggle for existence.
—Ernst Haeckel, 1870**

—Overview—

Until now, we have viewed biology by peering deep inside individual organisms. We have focused on complex cells, chemicals, and processes that in concert produce, maintain, and reproduce individuals. What we have seen thus far is a series of hierarchical levels leading to individuals. Atoms and elements are organized into molecules and compounds, the largest and most complex of which are biochemicals. These, in turn, are organized into cells, complex entities often made up of smaller parts, or organelles. Cells, we have seen, are the building blocks of life. As with buildings, what is important to life is not only their blocks but how they are arranged. Cells are organized into individuals.

In the next two chapters we will focus on the discovery that life's hierarchy doesn't stop at three levels. Individuals are organized into populations, which are organized into biological communities, which are organized into even higher levels, or ecosystems. Studying these higher levels of organization and the processes and factors that explain them is the subject of ecology, one of biology's youngest and most complex branches.

14.1 WHAT IS ECOLOGY?

For much of the 20th century, ecology was little noted outside of academic circles. Then came the 1970s. Within the course of a few years, environmental awareness blossomed, and ecology was swept into the limelight. Books, television programs, newspaper arti-

Chapter opening photo—Humans seem to live everywhere.

(a)   *(b)*

Figure 14-1. Any situation in which organisms interact with their environment is of potential interest to ecologists. This includes *(a)* vultures feeding at a lion kill and *(b)* rats feeding in an urban dumpster.

Fast Find 14.1a What Is Ecology?

cles, even movies appeared, claiming to be ecological. "Save the Ecology" became a theme at public demonstrations, often with the underlying assumption that the primary goal was to save pristine wilderness.

But environmental concern is not ecology. Certainly, ecological principles are at work in wilderness. But they are equally at work in situations dominated by human activities. Vultures scavenging a lion kill in East Africa are of interest to ecologists. So are rats scavenging dumpsters in urban America (Figure 14-1). So, what exactly is ecology? And what is it not?

14.1.1 Ecology Is a Branch of Biology

Ecology is the study of organisms in relation to their environments. Organisms cannot exist as isolated entities. Organisms interact with their physical environment (soil, climate, light and dark, day-to-day weather). Organisms interact with others of their kind. Organisms interact with other species including those they try to eat, those that try to eat them, and those with which they compete and cooperate. Ecology is also concerned with the abundance and distribution of animals and plants. This description of ecology may come as a surprise to some. In everyday usage, "ecology" appears to mean something else. It is important to remember the following.

Ecology is not a social cause. Although the word "ecology" may be in mainstream societal consciousness, ecology is not a movement.

"Ecology" is not the same as "environment." Ecology is a branch of biology; environment means surroundings. Does this mean that ecology, the science, has nothing to contribute to environmental issues? Of course not. But it's important to remember the limits of the contributions that ecology can make. How would other organisms and the environment as a whole be changed if California condors became extinct in southern California? Ecologists can answer that. Should we spend money to save the condors from extinction? Such political or moral questions cannot be answered by ecologists alone. With many topics of interest to both environmentalists and ecologists, science and politics overlap, or at least butt up against one another, but they are not the same. Our goal here is to keep separate that which is scientific from that which is political; we want to keep the factual separate from the emotional.

Ecology is not natural history. Fascinating stories of nature have captivated people for thousands of years. Usually they stress organisms living in and reacting to particular environments. Isn't this ecology? Not really. There is a difference between natural history and ecology. Let an example illustrate. One of the most conspicuous large mammals in

Figure 14-2. In Denali National Park, Alaska, grizzly bears in summer spend much of their time digging out Arctic ground squirrels from rocky burrows.

Denali National Park in central Alaska is the grizzly bear. Roads allow visitors and scientists to sit comfortably and safely in buses or observation huts and make extensive observations of bears in their natural habitat. Frequently, especially late in summer, grizzlies are seen digging up ground squirrel burrows. They may spend half a day moving room-sized mounds of soil, rocks, and plant material to catch a small meal. All the while, caribou move past the slaving bears, neither seemingly noticing the other (Figure 14-2).

This is natural history—interesting stories of animals in nature. It leads to an interesting question. Why would a grizzly bear spend so much time hunting ground squirrels when much larger prey is seemingly at hand? Can this question be answered scientifically? Yes. Ecologists might approach this question by considering energy requirements. In the Denali area, the weight of the average grizzly bear is about three times that of the average human. Obviously, larger animals need more energy than smaller ones, but it turns out that grizzly bears need considerably less than three times more energy than humans. In 1945, the animal physiologist Samuel Brody reported a relationship between a mammal's size and its minimal daily energy needs, in the form of an equation, $E = 72W^{0.75}$, where E is the energy needed for one day and W is the animal's weight. Plug in the average human's weight (60 kg) and E is nearly 1,500 kcal—not a bad approximation for our average minimal daily energy needs. Plug in the average Denali Park grizzly bear's weight (200 kg) and the energy required is about 4,000 kcal per day.

The average Arctic ground squirrel weighs in at around 2,400 g, nearly three-fourths of which is water. In late summer when they are preparing for a long winter's hibernation, half of the remaining grams are pure fat while the rest are mainly proteins and carbohydrates. Each gram of fat contains 9 kcal of energy. Each gram of protein and carbohydrate has about 5 kcal of energy. If we run these figures through a calculator, we see that the average Arctic ground squirrel is packed with about 4,200 kcal of energy. All the average grizzly bear needs to do is catch one Arctic ground squirrel each day and its basic, daily energy needs are filled.

Still, one caribou weighing 130 kg could provide a grizzly bear's energy needs for several days. Predators often have a choice of prey. Usually the environment offers a number of potential prey species. Ecologists evaluate how predators choose prey in terms of energy efficiencies. What is the ratio, for example, between energy expended catching a caribou compared with the energy payoff once it is captured? Though it is true that a caribou would supply the bear's energy requirements for several days, caribou are not easy to kill. If a digging bear even glances up, a nearby caribou can easily bound away. To chase a healthy caribou, our grizzly would have to expend a lot of energy on an effort that probably would not be successful.

Hunting Arctic ground squirrels, it turns out, is much more efficient. Burrows have only one entrance. If the bear just keeps digging, it is eventually successful. At the end of the burrow is a 4,200 kcal tidbit. Now we have a plausible explanation for what at first appeared to be bizarre behavior in grizzly bears.

Ecology is science. Notice, until we got quantitative and started thinking in terms of energy needs, efficiencies, and balances, we had little more than an interesting story.

Modern ecology, like other modern sciences, is quantitative. Ecological research uses the same methods, steps, and processes as other branches of biology. It usually starts with observations gathered in the field but moves quickly from description to experimentation. Testable hypotheses are proposed. At its best, ecological research provides measurements for testing hypotheses. Measuring specimens is easy enough: length, weight, degree of coloration. But how does one measure an organism's surroundings? Doing so is the heart of ecology. Data are collected, analyzed, and related to what is already known from previous studies. Ecologists strive to get to the point where predictions can be reliably made. This step often pushes existing ecology to its limits. Great progress has been made in recent years, and ecology's track record is constantly improving.

Ecologists need at least a basic understanding of nearly every other branch of science. Biochemistry, taxonomy, physiology, geology, and statistics are especially important to ecologists. In addition to understanding other sciences and branches of biology, ecologists have gathered their own store of knowledge, some of which will engage us for the rest of the chapter.

14.1.2. Ecology Grew from Natural History

Ecology—as a science—has a relatively brief history, encompassing only the 20th century. In a sense, however, interest in organisms and their surroundings is as old as humanity itself. The earliest humans needed to know when and where to find game, edible plants, and other life-supporting materials. Later, at the dawn of history, as humans switched to pastoral lifestyles, knowledge we would call ecological was probably essential: Where will crops grow best? Where are reliable sources of water? How can we discourage pests? What would it take to domesticate chickens, pigs, and cattle? These were the kinds of questions that captivated early farmers. By the Middle Ages, a vast store of ecological folk wisdom existed throughout the world.

Starting in the 1300s, opportunities to study natural history flourished, particularly among Europeans. World travelers brought back to Europe extensive collections of specimens. These collections launched Linnaeus, Darwin, and others on their life's work. Interest went beyond the specimens themselves. Where did they come from? Under what conditions did they live? How did they behave? Travelers and explorers were eager to supply answers. By the time of Darwin, Europeans and North Americans had an extensive literature of natural history. Unfortunately, accuracy varied among these accounts, and readers sometimes had no way of distinguishing reality from fiction in accounts of whales, sea monsters, elephant seals, mermaids, kangaroos, and unicorns.

By the last half of the 19th century, the stage was set to move beyond folk wisdom, stories, and traveler's tales. Darwin had interpreted differences between populations of organisms as arising in response to natural selection. Soon after he wrote, others, including Haeckel, whose quote heads this chapter, began intellectually to build on Darwin's foundation. What aspects of the environment frame and influence natural selection? This question, rooted in a rich tradition of natural history, led to the science of ecology.

The growth of many important ideas in biology can be likened to a tree, with one unifying idea emerging from an extensive historical system of roots. Like as not the trunk is a single individual—Darwin, Mendel—or small group—Schleiden, Schwann, and Virchow. No such individual emerges as the founder of ecology. Rather, ecology can be likened to a bush (Figure 14-3). Near the end of the 19th century, out of a tangled root system of natural history, anecdotes, and folk wisdom, four main stems emerged that led to modern ecology:

1. Early botanists developed *plant biogeography*. Along with the extensive plant collections in Europe's museums and universities, gathered over a 500-year period, were extensive notes on where each specimen had been collected and the environmental conditions in that place. Over the years, accuracy and completeness improved. From these collections, predictions followed. Expect to find cactuses in deserts. Expect to

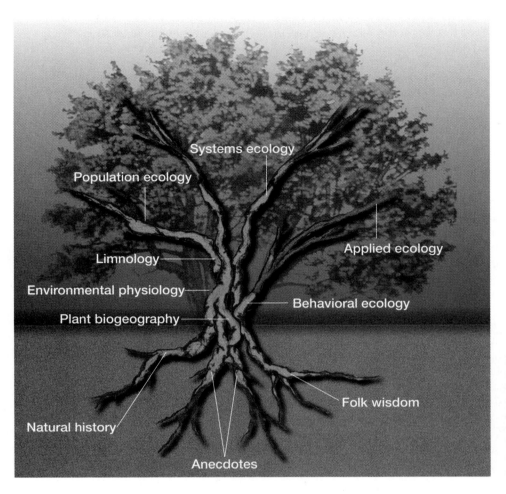

Figure 14-3. The history of ecology is more like a bush than a tree. Its origin cannot be attributed to any single person or idea. Rather, the science of ecology has been the work of numerous scientists working primarily in the 20th century.

find few trees in areas receiving less than 12 inches of rain per year. Expect to find orchids sprouting from branches on tropical trees. There were exceptions, of course, just as there are for many generalizations in biology. But as generalizations, the statements held true.

2. Meanwhile, other early ecologists focused on what today we would call *environmental physiology*. How does living in a particular environment affect an organism's structure and function? Most polar animals are white. Many desert plants store water and have extensive root systems. Marine mammals use tear glands to get rid of excess salt. These are examples of how physiology is attuned to environment.

3. In Europe, during the early 1900s, a group of biologists launched *limnology*, extensive studies of freshwater ponds and lakes. They introduced the idea that whole systems, comprising many different species and many different physical factors, interacted and could be simultaneously studied. These studies became another branch of early ecology.

4. Still another group of biologists were interested in *behavioral ecology*. How does living in a particular area with a particular set of characteristics affect the way animals behave? Within a few decades of the 20th century, such studies focused on a diverse group of insects, birds, and mammals.

By the end of the 19th century, these four somewhat dissimilar kinds of studies along with certain studies of marine biology, economic biology, and latter-day natural history, became lumped into ecology, a new branch of biology. The scientists conducting these diverse inquiries were themselves diverse, coming from different backgrounds, nationalities, and scientific traditions. It may not be surprising, then, that they failed to share a common vision. What is this new science of ecology? There was, at first, little agreement.

Indeed, one of the first serious disagreements was on how the word should be spelled, ecology or oecology. Discussions became so heated that the field nearly split in two.

Behind this rather trivial seeming argument was a more important one. How should ecology best be studied? One group took a holistic approach, feeling that whole systems should be studied as units. By and large, these were the botanists and the limnologists. Their studies of plants, particularly plant biogeography, and of water systems emphasized important associations between organisms and environments. Zoologists, on the other hand, tended to be more reductionist in approach. They believed that the best way to understand complex biological systems was to gain intimate and complete understanding of each component population of the system. These early ecologists studied individual species; their work grew out of the tradition of natural history studies.

Arguments between the two camps raged and continue to do so. We might like to think that disputes in science are worked out in sophisticated, intellectual ways. Alas, such is often not the case. Scientists are nothing if not human, and disputes often get rancorous and personal. Examining these disputes, although amusing, teaches us little about ecology. One important point can be made: ecology is no stranger to controversy.

At the beginning of the 21st century, three of the main branches of ecology are population ecology, systems ecology, and applied ecology.

1. *Population ecology* focuses on dynamic changes occurring in one population or species. Here relatively narrow questions are answered in great detail. These specialists might, for example, follow and predict changes in specific deer populations in temperate deciduous forests of North America. This chapter will focus on factors important to population ecologists.

2. *Systems ecology* is concerned with the dynamics of complex ecological communities and is holistic in approach. Complex systems, with all their many interacting parts, are studied as units by systems ecologists concerned with "big picture" concepts and questions. They might, for example, be interested in looking at the flow of energy through a temperate deciduous forest community from browse plants to deer to wolves and other predators. In Chapter 15 we will examine factors particularly important to systems ecologists.

3. *Applied ecology* is modern ecology's newest branch. This specialty predicts outcomes of human activities and recommends courses of action to mitigate certain of these activities. Within the field there are several subspecialties, the best known, perhaps, being restoration ecology (Figure 14-4). Applied ecologists may be in the limelight, working with economists, politicians, business leaders, and community groups. They might, for example, study what could be done to restore deer habitat in an area that was once temperate deciduous forest. We will discuss specific examples in both Chapters 14 and 15.

Figure 14-4. Restoration ecology is the youngest branch of contemporary ecology. Here a wetlands, damaged by development, is being restored.

Applied Ecology in Action Find examples of projects involving applied ecology that are currently being conducted. What are the overall goals of the project? How are the projects being carried out? Are any such projects being conducted in your area?

The science of ecology continues to change and grow. The Gaia hypothesis views the entire planet as a self-regulating entity that has been likened to a supraorganism. Spiritual ecology attempts to bridge the gap between ecology, environmental awareness, and spirituality. The proper place that these ideas may play in the future science of ecology is being hotly debated among ecologists. Visit our web site for discussion of these controversies.

More on Gaia What is the current status of the Gaia hypothesis? What are the pros and cons of spiritual ecology?

Piecing It Together

Ecology, that branch of biology dealing with the relationship between organisms and their environments, including other organisms, differs from other branches of biology in at least three ways:

1. It is relatively young and, unlike other branches of biology, draws on the work of many founders rather than one or a few.
2. It has captured the imagination of the general public.
3. It is often confused with things it is not. There is a difference between "ecology" and "environment." There is also a difference between "ecology" and "natural history." Both environmentalism and natural history are important; they fulfill our everyday need to know about and respect the organisms with which we share habitats and a planet. But ecology is something different. Ecology is science.

14.2 HOW DO ECOLOGISTS STUDY POPULATIONS?

In a meditation about human life, John Donne wrote, "No man is an island." The same is true for any individual plant or animal. Individuals are always part of something larger, namely, populations. In some respects, populations act like organisms. They require space and nutrients. They move through definite daily and seasonal cycles. They grow, reproduce, and die. But populations have other properties that are unique. They have an age structure, a sex structure, and death and birth rates. They interact with other populations and with environmental pressures.

When ecologists study populations, they ask certain questions: How large is the population and how is it changing? How do populations interact? Can we control changes in a population? Before addressing these questions, let's start with a bit of history.

See How We Grew Consider the growth of the human population. What is the estimated size of the world-wide human population today? Watch how that number changes over the next few days.

14.2.1 The Study of Populations Has Been One of Ecology's Major Tasks

Ecology inherited concerns about populations. In 1798, Thomas Malthus, a British clergyman and economist, published *An Essay on the Principle of Population*. His major thesis was an observation: humanity has an innate and almost unlimited ability to procreate but a limited ability to produce food. Thus, human populations tend to grow and outstrip their ability to feed themselves, which inevitably leads to problems. His rather pessimistic prediction was that if human population growth were left unchecked, the result would be pestilence, war, and famine. If a person's worth can be measured by the amount of controversy his works generate, Thomas Malthus was one of the greatest writers of all times. His essays were instantly controversial because he challenged head-on the prevalent view of his day that humanity was moving inexorably toward social perfection, a utopian state in which all people would be fully provided for. Heaven on Earth was just a matter of time and continued progress. No, said Malthus, humanity is moving inexorably toward misery that can only grow and continue indefinitely. The Establishment did not take his challenge lightly.

Nor do they do so today. Two hundred years after publication, his works continue to be widely quoted, to challenge established policies, and to spark controversy. Certainly, he underestimated humanity's ability to grow food (Figure 14-5). But his overall thesis—that our ability to procreate is nearly unlimited while our ability to produce food sooner or later will meet its limits—concerns some policymakers today.

More important to our story, his concerns went beyond the populations of humans. He observed, "Through the animal and vegetable kingdoms, nature has scattered the seeds of life abroad with the most profuse and liberal hand. She has been comparatively sparing in the room and nourishment necessary to rear them." In this and other passages, he focused attention on populations of animals and plants. Malthus's works were well known to Darwin. Remember that, nearly half a century after the essay's first publication, Darwin observed that "all population's ability to increase is greater than required"—very nearly a restatement of one of Malthus's major points.

After Darwin, population studies, particularly of animals, flourished. In the 20th century, there have been many studies of invertebrates, especially insects (because of their economic importance), farm animals (for the same reason), and game animals (because of an intense need to manage and predict changes in their populations). These studies continue today. Only recently have ecologists begun to study population dynamics in plants.

Figure 14-5. Modern farms are large, expensive to operate, and highly efficient. They have been described as food factories.

14.2.2 The Size of a Population Is Determined by Natality, Mortality, Immigration, and Emigration

Fast Find 14.2a Growth Rate

The size of any population is in large part determined by a balance between several factors, some obvious, some less so. One obvious factor is the number of new individuals being born into the population. This is usually expressed as its **natality**, or **birth rate**.[1] Another factor, working in the opposite direction, is the population's **mortality**, or **death rate**. Theoretically, any population is stable—it neither grows nor shrinks—if these rates are balanced. If not, the population changes. That is, the population's **growth rate** is determined by its birth and death rates. Said more succinctly, $r = (b - d)$, where r is the population's growth rate (also referred to as "little r"), b is its birth rate, and d is its death rate. (What happens to a population if its $b > d$; $b < d$; $b = d$? In each case, what are the values of its "little r"?)

Simple, right? Let's see how things work out in real life. A place in south central Alaska, about equally distant from its two largest cities, Anchorage and Fairbanks, is the home of the Nelchina caribou herd. During its fall migration, the herd comes close to a well-traveled highway, and sport hunters kill many of the animals each year. In the early 1960s, game managers, charged with the responsibility of managing the herd, worried about how this hunting pressure might affect the herd's size. They launched an extensive study of the herd's population dynamics. They determined the birth rate by monitoring the calving grounds in early summer. They already knew a great deal about the natural mortality rate of the population, caused mainly by wolves, bears, wolverines, and severe winter weather. Their results showed the population's "little r" was in fact quite large, indicating that the herd's numbers would probably increase. A large harvest by hunters was thought to be desirable to keep the population from growing too large. For a time hunter success was consistently high. But in the early 1980s, when sport hunters took to the field, few animals were found. What had happened?

Another factor important to population dynamics is movements of individuals between populations. There are two types: **immigrations** bring new individuals into a population, and **emigrations** remove individuals. Some such movements are regular and easy to predict. At the start of the breeding season, nearly all members of certain songbird populations leave winter homes in Central and South America and migrate to nesting grounds in North America. In other cases, predictions are not so easy. Apparently, large segments of the Nelchina caribou herd emigrated to an adjacent herd—the Forty-mile herd (Figure 14-6). Mass movements from the Nelchina herd had not been observed

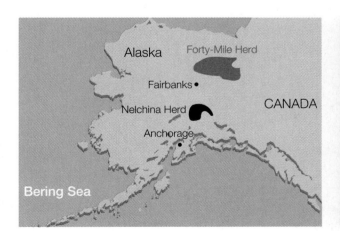

Figure 14-6. The Nelchina caribou herd ranges over a large area of central Alaska. In the early 1980s segments of the herd joined the Forty-mile herd to the northeast in a large emigration that significantly reduced the size of the Nelchina herd. (Map adapted from J.E. Hemming, 1971. The distribution movement patterns of caribou in Alaska. *Wild Tech Bul 1.* ADFG p2.)

[1]Why rates and not absolute numbers? Rates work well in calculations and make comparisons between populations easier. Rates are always expressed over an appropriate time interval, as in "births per year."

before and have not occurred since, but similar migrations have been observed in other herds of caribou. Such movements make predicting future changes in populations difficult at best. At least rewriting our mathematical formula is relatively easy: $r = (b - d) + (i - e)$, where the new terms are i, the population's immigration rate, and e, its emigration rate.

14.2.3 Population Characteristics Are Expressed in a Variety of Ways

A word of caution: Beware when reading about populations. Relevant characteristics such as size and mortality each can be expressed in more than one way.

The most straightforward expression of population size is the **absolute number**. Count all individuals in the population and express the total. This is almost never done, because it is nearly impossible to count each individual. A notable exception is the United States population census, taken every 10 years; it is a massive effort, fraught with errors and assumptions.

More frequently, population size is expressed in terms of **density**, the number of individuals per unit area. Theoretically, determining population density is easy: count all individuals in several representative areas, and determine the average per unit area. (Determine, for example, that the population density of Monaco is over 18,000 people per square kilometer.) Then, to estimate the total population, multiply the density by the total area the population occupies. In practice, especially when dealing with organisms, this is not as easy as it sounds. What is a "representative area"? Where are the population's boundaries?

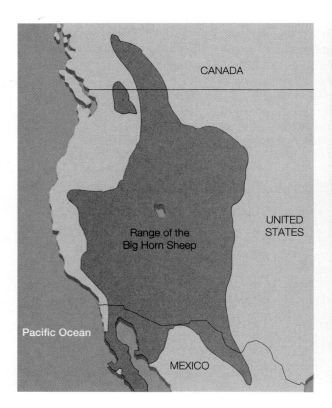

Figure 14-7. Rocky Mountain bighorn sheep roam throughout the mountains of North America from western Mexico to southwestern Canada. (Map adapted from W.H. Burt and R.P. Grossenheider, 1980. *Mammals*, 3rd ed. Peterson Field Guide Series. Houghton-Mifflin.)

This leads to another consideration. According to maps in field guides, such as the one in Figure 14-7, Rocky Mountain bighorn sheep are found throughout all noncoastal mountains in the western United States, extending north into Canada and south into Mexico. This is the animal's **range**. But bighorns are not evenly distributed within their range. Typically, they remain in mountainous ares in summer, descending into adjacent valleys in winter. The mountains and adjacent valleys are the bighorn's **habitat**, or areas in which the species finds all its specific needs. Now, in estimating the species' population size, do you multiply the density by the total area of their range or their habitat? Obviously, the latter would be more useful, but determining the area of the habitat is often as difficult as counting animals.

Mortality is often expressed as the **gross mortality rate**, or the number of deaths in a specified time period divided by the total number of individuals in the population at the start of the period, multiplied by 1,000. (We multiply by 1,000 to get whole numbers rather than decimals, which are difficult to deal with.) An example might be a gross mortality rate of 200 per year per 1,000 individuals. A more accurate picture of what is happening to a population is obtained from **age-specific mortality rates**, in which each age class is considered as a separate population. (An age class is all the individuals of the same age.) From age-specific mortality rates it becomes apparent that, for example, most red foxes die during their first year. Any individuals that survive this critical period have a good chance of living another four or five years. Almost no red foxes live beyond six years. From age-specific mortality rates ecologists construct **life tables**, an example of which can be found on the CD in Section 14.3.

Expression of birth rates is even more variable. Some studies use the number of new individuals born into a population divided by the total population, or the **crude birth rate**. Another way to determine crude birth rates is to divide the number of births by the number of females in the population, because the females bear the new individuals. Another method is to divide the number of births by the number of females of reproductive age, excluding, in human populations, prepubescent and postmenopausal females. Then there are various **age-specific birth rates**.

Confusing, isn't it? The point is this: how population characteristics are expressed is the scientist's or writer's choice and the reader's bane. We will be consistent in this book, but do not expect such consistency elsewhere. Reader beware: Whenever you read a population number, ask, "What does this number mean? What units are being used? How was the number determined?" Things get particularly difficult when you compare one study with another. Are numbers cited in different places comparable? In spite of these difficulties and ambiguities, population studies can provide meaningful insights into the past, present, and future of populations.

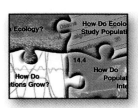

Piecing It Together

1. The study of populations is a major focus of ecology.
2. Factors that are important to the growth of populations include natality, mortality, immigration and emigration.
3. How these factors interact can be expressed in a mathematical equation:

$$r = (b - d) + (i - e)$$

where r is the population's growth rate, b is the birth rate, d is the death rate, i is the immigration rate, and e is the emigration rate.

14.3 HOW DO POPULATIONS GROW?

**Fast Find
14.3a
How Do
Populations
Grow?**

A gardener obtains a few expensive *Salvia* plants from a nursery and plants them in an isolated corner of her yard. For a while, all is well. With careful nurturing, the prize flowers flourish. By fall, they set seeds, and the next year their numbers increase impressively. But the garden is discovered by slugs. At first, their numbers are low and their effects minimal. Then their numbers explode. Seemingly overnight, slugs overwhelm the garden. The flower population falters, then sags. Night after night the gardener is outside with her flashlight picking up slugs and dropping them into paper bags of salt.

Controlling population numbers, that is, promoting those species we enjoy, want, or need while discouraging those we don't, is a consuming interest to many ecologists, amateur and professional (Figure 14-8). Let's consider population-control factors identified in the 1920s by Royal Chapman, an American ecologist. **Biotic potential** is a population's ability to reproduce, a force that promotes population growth. We call such factors **intrinsic factors**, because they are internal to the population. **Environmental resistance**, the second force, consists of factors that limit growth. Most of these factors are external to the population and are called **extrinsic factors**.

14.3.1 Some Populations Grow Exponentially

Especially in their early stages, populations of rapidly reproducing species go through a pattern in which, at first, nothing seems to happen. Then, the growth rate increases, and finally numbers increase rapidly. If the population size is graphed (number of individuals versus time), growth resembles the letter "J" (Figure 14-9). Mathematicians tell us the population's rate of growth is exponential, described mathematically as

$$(dN/dt) = rN$$

in which (dN/dt) are changes in population size with time, $r = (b - d)$ = the growth rate, and N = population size. This is known as the **exponential equation** of population growth.

Let's assume that the slug population in our gardener's yard in early April was 10, that the population's "little r" is 0.4, and that these slugs reproduce once every two weeks. Two weeks into April, the slug population increased by 4 to 14. Now fast-forward to early July when the population is roughly 100. Assume the population's "little

Figure 14-8. Controlling insect pests is expensive and time consuming. *(a)* Large commercial farms need crop-dusting airplanes to efficiently apply pesticides to fields. *(b)* On a smaller scale, suburbanites "dust" their lawns.

(a)   *(b)*

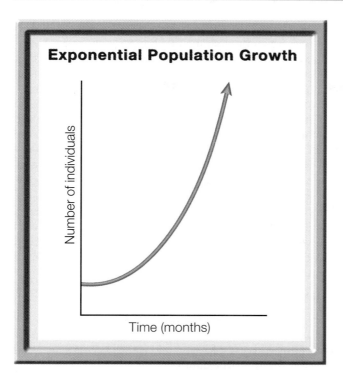

Figure 14-9. Some populations grow exponentially—at least for a time. After a period of relatively slow growth, the population size soars.

r" is still 0.4 (although not necessarily constant, a population's "little r" may remain so), and that the gardener has left them alone (probably not a reasonable assumption). Now in just two weeks the population increases by 40. Fast-forward again to late October, when the population is approximately 1,000 (under the same assumptions). In just two weeks the population increases by 400. In just one year, our slug population could, if left to its own devices, jump from 10 to 45,000! The flowers never had a chance.

Note: With exponential growth, where (dN/dt) = rN, the larger the population, the faster it grows.

14.3.2 Growth of Most Populations Follows an S-shaped Curve

Obviously, in nature populations are seldom "left to their own devices." If the gardener hadn't discovered the slugs and interfered with their exponential growth, something else would have—snakes or shrews, perhaps (Figure 14-10). Predators are one of the factors Chapman had in mind when he spoke of environmental resistance.

Figure 14-10. Shrews are mouse-sized friends to gardeners. These voracious predators can eat an amount of slugs and other garden pests equal to their own weight each day.

What would have happened to the population of *Salvia* in our garden example if there had been no slugs? We can guess that through the first few growing seasons their numbers would have behaved as the slug numbers did. Then, after a period of little change, numbers would increase rapidly, as predicted by the exponential equation. But gardens become full. There is, after all, only so much space for plants. For these plants, environmental resistance is measured in terms of available space. When all spaces are filled, a population can increase no further. Similarly, although factors other than space may be key, a garden filled with *Salvia* can support only so many slugs. Notice that the growth of both these populations when graphed would follow an S-shaped curve (Figure 14-11). Rapid growth would be followed by a flattening off. Numbers would not increase beyond an upper limit. In other words, each population has a **carrying capacity**, defined as the number of individuals the environment can support.

**Fast Find
14.3c
Logistic
Growth**

What does this concept do to our exponential equation? As early as 1838, the French mathematician P. F. Verhulst derived an equation that described S-shaped population growth. He was frustrated, however, by a lack of available field data. Eighty years later, Raymond Pearl and Lowell Reed derived the same equation working with United States census data. In their **logistic equation**,

$$dN/dt = rN[(K-N)/K]$$

dN/dt are changes in population with time, r = the population's growth rate, N is the population's size, K is the population's carrying capacity, $(K-N)$ is a measure of how far removed the population is from its carrying capacity, and $[(K-N)/K]$ is a measure of environmental resistance. This equation is an example of an ecological model that describes how populations grow.

Notice how the equation works. When K is large and N is very small, $[(K-N)/K]$ approximates one. This would be true in the early stages of a population's history. During these stages growth is essentially exponential. But as population numbers approach their carrying capacity, that is, as N approaches K, $[(K-N)/K]$ becomes smaller and smaller until, when $N = K$, it becomes zero. At that point the population ceases to increase. In other words, at that point population size is in equilibrium with the environment. That's what the equation predicts, and that's what we see in nature. Right?

Figure 14-11. Most populations do not grow exponentially for long. After a period of such growth, environmental resistance limits further growth.

Not always. In the same way that a model airplane is not a completely accurate representation of a real airplane, an ecological model is only an approximation of nature. Furthermore, some models are better than others. As a model, the logistic equation is based on the assumption that a population's "little r" is constant. It seldom is, varying from year to year and from age class to age class. In general, growth rates decrease as the population approaches the carrying capacity. Also, the model assumes that the population's carrying capacity is constant. It, too, seldom is. In a good year, the carrying capacity may be considerably higher than in a bad year. Numbers of most populations fluctuate around a median value that can be taken as the carrying capacity. Carrying capacities are seldom accurately known as they are hard to measure. In spite of its limitations, the logistic model gives consistently good first approximations of how population numbers change through time. Often, much good biology comes from explaining why predictions fail to materialize.

14.3.3 A Variety of Factors Limit Population Growth

What exactly is "environmental resistance?" Why are populations limited in their ability to increase in size? For the *Salvia*, as well as for many organisms and populations, available space is a factor that limits population growth when numbers are relatively high. Factors such as space, whose effect depends on whether numbers are low or high, are termed **density-dependent factors** (Figure 14-12).

The effects of predation are often density dependent. As long as the number of rabbits is low, foxes tend to ignore them, seeking more abundant prey—quail, perhaps. With foxes hunting other prey, rabbit numbers increase. Only when rabbits are relatively abundant do they draw the attention of foxes. As rabbit numbers increase further, so does the intensity of predation, which may eventually limit the number of rabbits.

Disease is also often density dependent. Periodic outbreaks of rabies among raccoons in the northeastern United States, for example, are most intense when raccoon

(a)

(b)

Figure 14-12. Food and space are density-dependent limiting factors. *(a)* In this dense stand of trees, all available space is filled with individuals. Although new individuals may replace those that die, further increase in overall numbers is not possible. *(b)* When their numbers become so large that all available food is eaten, the deer may not only fail to increase, the population may decrease.

numbers are relatively high. When population numbers are low, so are contacts between individuals. With little contact, many diseases, rabies included, spread slowly. Only with frequent contact do such diseases spread and reach epidemic proportions.

So far, the examples we have discussed are extrinsic factors. Density-dependent factors may also be intrinsic. Superficially, periodic outbreaks of rabies in Arctic foxes resemble those of raccoons; that is, soon after population densities reach high levels, rabies breaks out, causing declines in the number of foxes. Close study of the foxes, however, revealed an intrinsic factor at work. Microscopic analysis showed Negri bodies (small dark-colored dots) always to be present in the brain tissue of rabid foxes. But examination of large numbers of foxes showed that Negri bodies were also present in the brains of some apparently healthy foxes. As long as the population densities are low, rabies is latent in the Arctic fox. What triggers the rabies is not understood, but when densities surpass a certain level, something apparently changes within individual foxes and the disease is expressed. High population densities are known to be stressful to many mammals, quite often affecting their ability to reproduce.

Some environmental factors that limit population size are apparently **density independent**. Freezing temperatures, for example, kill many flowering plants. If an unusual frost in, say, southern Florida killed all individuals of a species, regardless of population density, we might say that sensitivity to cold is a density-independent factor. Ecologists have argued long, hard, and sometimes bitterly about the importance and, indeed, the existence of density independent factors. For example, frost in Florida might not kill all individuals of the species. Those individuals fortunate enough to live in protected sites might survive. Indeed, freezing temperature may be lethal only when population densities are so high that some individuals live in unprotected sites. If this is true, then the effects of freezing temperatures may, in fact, be density dependent. As we learn more and more about the relationships between species and their environments, we often learn that factors that at first appeared to be independent of density are, in fact, not so.

See if you can think of other density-dependent, extrinsic, intrinsic, and density-independent factors that limit populations; some will be discussed further on the CD.

14.3.4 *K*-Selection and *r*-Selection Underlie Two Patterns of Population Growth

Biologists study population size because they want to predict how it may change. As we have seen, three factors are important: the number of individuals already present, the population's carrying capacity, and its growth rate. The number of individuals changes as the population grows, but the carrying capacity and growth rate are often relatively constant.

Some species have adaptations that permit them to live in a state of equilibrium close to their carrying capacities for long periods. Such species are said to be *K*-selected. Other species have different adaptations that permit them to rapidly increase their numbers when their populations are below their carrying capacities. These species are said to be *r*-selected. *K*-selected and *r*-selected species differ in many ways as summarized in Figure 14-13.

You can probably think of species that fit into each of the two categories. You can also probably think of species that do not fit easily into either category. Where, for example, do humans fit? These two categories can be thought of as the extremes of a continuum, with most species fitting in somewhere between the extremes. The concept of *K*- and *r*-selection is quite useful in categorizing species and predicting how populations will behave. For example, most of our endangered species are *K*-selected. They tend to be sensitive to environmental conditions and intolerant of habitat loss. They tend to have special ecological requirements. Their numbers change slowly. Most of our pests, on the other hand, are *r*-selected. They seem to be able to live anywhere. Their numbers are hard to control and fluctuate widely.

Figure 14-13. Different species have different patterns of population growth. *K*-selected species, such as sperm whales *(a)*, maintain populations close to the carrying capacity. Typically, their population numbers are stable and change slowly. Species that are *r*-selected, such as deer mice *(b)*, maximize "little r," the rate of population increase. Their population numbers are anything but stable and often change rapidly.

(a)   *(b)*

Characteristics of *K*-selected species	Characteristics of *r*-selected species
Live in stable environments	Live in disturbed environments
Are ecological specialists	Are ecological generalists
Have populations stable in size	Have populations that fluctuate rapidly in size
Compete well against other species	Do not compete well against other species
Are restricted in where they can live	Are widely distributed
Take rapid advantage of ecological opportunities	Are slow to respond to ecological opportunities
Are long lived	Are short lived
Have few, relatively large young	Have many, relatively small young
Have long periods of embryonic development	Have short periods of embryonic development
Reach adulthood slowly	Reach adulthood rapidly
Invest intensive parental care in young	Invest little or no parental care in young
Reproduce throughout lifetime	Reproduce once per lifetime

14.3.5 Graphs of Age Structure Allow Predictions of Future Population Growth

Fast Find 14.3d Life Tables

Another technique for predicting how a population's size may change begins with determining information about individuals in each of the population's **age classes**, or groupings based on particular age intervals such as 0 to 4 years. First, the number of individuals in each age class alive at the beginning of some period is determined. Then the number surviving throughout the period is calculated from age-specific birth and death rates. The information on all the age classes is tabulated in a **life table**, thus giving a picture of the makeup of the population. There is an example of a life table on the CD.

The next step in predicting changes in a population is to arrange the life table data into an **age-structure** graph (Figure 14-14). The numbers of individuals in the age classes are arranged in rows, starting with the youngest at the bottom. Notice that within each age class numbers have been separated by sex. Also notice that in general the length of each row—that is, the number of individuals in each age class—decreases with age. Individuals are born into the population only once, into the youngest age class. Mortality, on the other hand, exacts its toll on all age classes.[2]

---

[2]Can you think of a situation in which this might not be true, that is, a situation in which the numbers of individuals in a particular age class could be greater than the age class immediately younger? HINT: there are two possible explanations.

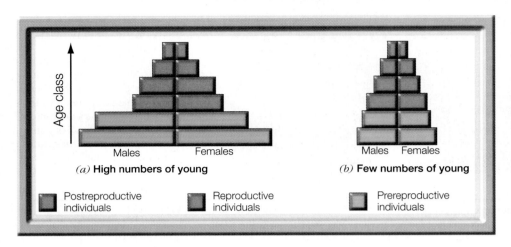

Figure 14-14. Age-structure graphs portray a great deal of information about a population. Each row represents a separate age (or age class). Males are separated from females. Such graphs allow predictions as to future population growth. Populations producing high numbers of young (*a*) are likely to increase in the near future. Populations producing few young (*b*) are less likely to do so.

Notice that in any given age class the number of females may not be exactly equal to the number of males. There are various reasons why this might be so. Some are due to chance. (Flip a coin 50 times. Would you expect to get *exactly* 25 heads?) Other reasons have a more biological basis. In some human populations, for example, infant mortality is higher for females than for males. Or mortality may differ between the sexes during adulthood, as it does in moose populations. During the breeding season, mature male moose gather, and each defends, his harem, which consists of numerous females. Much time is spent intimidating other males. Little time is spent feeding. Such activities increase the mortality rate for reproductive age males.

Finally, notice that the two graphs do not have the same overall shape. Graph (*a*) has a base that is broad, relative to its height, indicating that the population has a large proportion of young individuals. The base of graph (*b*) is narrower relative to its height, indicating that this population has a much smaller proportion of young individuals. From many studies ecologists know that **populations with broad-based, triangular, age-structure graphs tend to increase in size, while those with bullet-shaped graphs tend to be more stable**. In any population, today's young are tomorrow's parents. Numerous young today often means even more young in the next generation. (What would be the shape of the age-structure graph of a population that is expected to decrease in the future?)

For the accuracy of predictions, we need to think carefully about the assumptions on which they are based. Our predictions assume (1) no significant amount of immigration or emigration and (2) constant birth and death rates. How would large scale emigrations affect predictions of growth in an otherwise stable population? What factors of population growth would we need to focus on to stabilize a rapidly expanding population?

14.3.6 Populations Differ in Their Longevity and Survivorship Patterns

There is wide variation in the typical life spans of different species (Figure 14-15). Some insects live only a few days. Annual plants live only a few months. Some mammals live for over a century. Giant tortoises may live for several centuries. Some bristlecone pines have lived for several millennia.

How long would an organism live in an environment where nutrients and water are reliably available and where surroundings are comfortable (whatever that may mean for each species), free of predators, diseases, and other stresses? Under these ideal conditions, organisms would live until their physiological systems fail. That life span is their **physiological longevity**, which differs from species to species. At one extreme are the bristlecone pines whose physiological longevity is so long they may simply live until something kills them. At relatively high altitudes in mountainous deserts in the south-western United States—ideal environments for bristlecone pines—a few individuals have reached ages of over 4,200 years. Albeit old and gnarled, they show no signs of dying. Other organisms die even under conditions ideal for them. Most praying mantises in temperate regions of North America die with the onset of winter. Even those brought into the laboratory or classroom in fall die in midwinter despite efforts to keep them alive. They appear to be programmed, as do most organisms, to die at some age no matter what. Some animals living in zoos come close to living under ideal conditions and to approach their physiological longevity. That animals tend to live longer in zoos than they do in the wild suggests there is another type of longevity at work in natural populations.

Ecological longevity is the age to which an individual might be expected to live in a given environment. In nature, predators, diseases, accidents, harsh physical conditions, or a combination of such factors typically end each individual's life before its physiological time runs out. For example, bristlecone pines that live at lower altitudes from those mentioned earlier do not live for millennia partly because population densities are higher, permitting periodic wildfires to sweep through, killing the closely packed trees.

Humans have come a long way in controlling their ecological longevity. Most of our predators are gone, our disease organisms are under control (temporarily, at least), and our abilities to produce and distribute food expand faster than our population grows (again, temporarily, at least). Ecological longevity, in most human populations, extends accordingly. What is our physiological longevity? No one knows for sure, but the oldest humans alive today may be approaching it. Can our physiological longevity be extended? If it can be, should it be? These are important questions that society needs to answer soon. What do you think?

Species also differ in their patterns of survival. Most plants, invertebrates, fishes, and some reptiles have high mortality in their juvenile stages (seeds and seedlings; eggs and larvae). Relatively few individuals survive these critical early stages. Those that do have a good chance for continued survival. Survivorship of juvenile stages of other reptiles and most birds and mammals is enhanced by parental care. Still, mortality rates are relatively high during early life stages and are highest in the first critical months after young

Figure 14-15. Some organisms, such as this tiger swallowtail *(a)*, live for only a few weeks. Others, such as the bristlecone pine *(b)*, may live for centuries.

(a)   *(b)*

Figure 14-16. In spite of intense parental care, most young mammals and birds die before they reach maturity. Here a great owl is bringing food to its young.

disperse from their parents (Figure 14-16). Later, as reproducing adults, these animals have low mortality; that is, most adults survive. Mortality is high again among the very old, perhaps as they approach their potential physiological longevity.

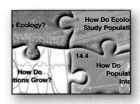

Piecing It Together

1. Two opposing forces affect population size, **biotic potential** (the population's ability to increase) and **environmental resistance** (environmental factors that impede growth).

2. Under ideal conditions, where environmental resistance is low, populations often grow exponentially; the more individuals there are in the population, the faster the population grows. In mathematical terms, this is known as the **exponential equation of population growth**:

$$(dN/dt) = rN$$

where (dN/dt) is the change in population number over time, r is the population's growth rate, and N is the number of individuals present.

3. In nature, populations usually do not grow exponentially, at least not for long. A more accurate description of the growth of most populations is the **logistic equation of population growth**:

$$(dN/dt) = rN\,[(K-N)/K]$$

where K is the population's **carrying capacity**, or the number of individuals the environment can support. The expression $[(K-N)/K]$ can be thought of as a measure of environmental resistance.

4. A number of environmental factors impose limits on population growth. Most are **density-dependent factors**, whose importance increases as the population numbers increase. Density-dependent factors include predators, disease, available space, available food, and other similar factors. A few limiting factors may be **density independent**.

5. Some species are **K-selected**; that is, they possess characteristics that allow them to typically live under stable conditions near their carrying capacities. Other, **r-selected** species possess other characteristics that have them typically living in unstable conditions where they maximize their reproductive potentials. These are extremes of a continuum, with most species living somewhere in between.

6. Populations with broad-based, triangular, age-structure graphs can expect future increases in numbers. Populations with bullet-shaped age-structure graphs are usually stable.

7. In nature, the **ecological longevity** of an organism tends to be considerably shorter than its **physiological longevity**.

8. Several examples of population dynamics can be found on the CD and web site.

14.4 HOW DO POPULATIONS INTERACT?

Fast Find 14.4a How Do Populations Interact?

Everything in biology is a part of something larger, and populations are no exception. Invariably, populations share space with other populations. They interact. They are interdependent. How they interact varies from population to population. Consider a well-known songbird, the American robin (Figure 14-17). In summer, robins nest in our bushes and feed in our yards. To earthworms and insects, they are predators. To cardinals and chickadees, robins are competitors. To tapeworms living in their intestines, robins are habitat. To ecologists, the robin's niche becomes important. A **niche** includes all aspects of the physical and biological environment that are important to a species. Where does the species live? With what other organisms does it share space and other resources? How does it interact with other species? A species' niche includes not only where it lives—its habitat—but its functional role within the community of organisms with which it lives.

The intensity of interactions between populations may be slight or great. We are scarcely aware of the populations of invertebrates that live beneath stones and scraps of wood in our backyards and have little influence on them so long as we leave them undisturbed. We share space and little else. At the other extreme, humans and most of their crop plants are mutually interdependent. Most humans today subsist mainly on crop plants or animals that eat crop plants. Most crops are so changed from their original populations that they could not exist without constant human care and protection. Humans and their crop plants are mutually, totally interdependent.

Let's take a closer look at the interactions between populations.

14.4.1 When Populations Compete, Both Are Harmed

There are at least two kinds of competition with which organisms must contend. Individuals must often compete with others of their kind for food, other resources, space, and mates. We explored competition within species in Chapter 13. In this chapter we

Figure 14-17. American robins are songbirds that feed communally in grassy areas on insects and soft-bodied invertebrates and defend nesting territories in bushes throughout much of North America. This description outlines the niche of the American robin.

will look at the other kind of competition, in which the competition is between species. Competition between species, too, is sometimes for space, resources, and food (but, by our very definition of species, never for mates).

The effects of competition appear to be straightforward. In the 1930s, the Russian ecologist G. F. Gause conducted numerous laboratory experiments on populations of two different species of *Paramecium* grown on culture medium. When grown alone, each species' population showed the expected S-shaped population growth curve. When grown together, both species initially expressed the S-shaped growth pattern, but as time passed, one species survived and the other became extinct. For the next 30 years, similar laboratory experiments demonstrated the same results in other protozoa, yeasts, hydras, water fleas, fruit flies, grain beetles, and duckweeds. So pervasive were the results that in 1960 the ecologist Garrett Hardin put forth the **competitive exclusion principle**, the idea that two species cannot coexist while exploiting the same limiting resource.

But what about natural populations? Does the competitive exclusion principle apply outside the laboratory? Examples are harder to come by. Intense competition between two populations is, after all, a transitory condition. Whenever one population excludes the other, all traces of competition are gone. Species introductions, however, yield evidence that the principle still applies. In the late 1940s, the Hawaii Department of Agriculture released numerous potential insect parasites and predators to control crop-eating pests. Three species of tiny wasps were particularly useful in attacking Hawaiian fruit flies. Over a four-year period, populations of wasps successively replaced each other until by 1951 only one species persisted. Similar studies of introduced wasps that attack scale insects in southern California had similar results.

Two species of barnacle are found on the rocky coasts of Scotland. One species grows in the upper reaches of the intertidal zone, where it is daily exposed to air at low tides. The other species is found farther down on the rocks, where it is less frequently exposed. In the 1950s, Joseph Connell discovered that the shallow-water species could survive in deeper water if the other species was removed. The species' absence there under natural conditions was interpreted as a result of competition.

In the 1950s, Robert MacArthur found what he thought at first was an example of species that could coexist and compete. He was studying warblers in the coniferous forests in the northeastern United States and southeastern Canada. He found five species apparently feeding in the same trees. Detailed observations revealed, however, that competition between the species was minimal. Although there was some overlap, each species had its own preferred region within the trees in which it fed. Intense competition had not resulted in exclusion of species in this case. Instead, resources were partitioned among the species (Figure 14-18).

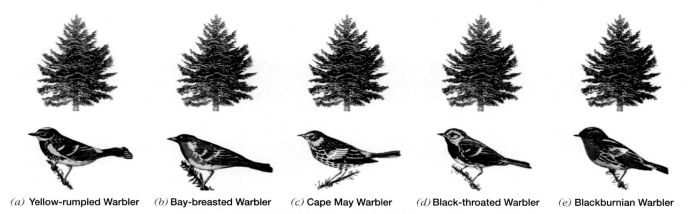

(a) Yellow-rumpled Warbler   (b) Bay-breasted Warbler   (c) Cape May Warbler   (d) Black-throated Warbler   (e) Blackburnian Warbler

Figure 14-18. Several species of warblers forage in evergreen trees in northeastern United States and southeastern Canada. Each species forages in a particular area of the trees, thus minimizing competition between species for foraging sites.

The yellow-rumped warbler is the most cosmopolitan of the five species studied by MacArthur. Its breeding range spans the continent. Notice that, in the trees MacArthur studied, these birds fed in the lower branches. In other regions, where the other warblers are not present, yellow-rumped warblers feed mainly in the tops of trees. This ability to go where there is less competition goes a long way toward explaining why this species is the most numerous and widespread warbler in North America.

Many studies of a variety of organisms verify that competition is often minimized by **resource partitioning**.[3]

So, what are the results of competition between species? First of all, notice that competition is a density-dependent factor. In an ideal environment in which resources are plentiful and population numbers are low, competition is not a factor. Potential competitors can find all they need without interfering with each other. This situation doesn't last for long. As populations grow, interactions between competitors intensify. There are two possible results: one species will exclude the other, or they will partition resources so that competition is minimized and tolerable.

14.4.2 In Predator and Prey Interactions, One Population Benefits at the Expense of the Other

One of the most obvious interactions between populations is a matter of who eats whom. Populations regularly exploit other populations as sources of nutrients and energy. As a result, the fates of predator and prey populations are intimately intertwined. Each is a major determinant of the other's population size. When predators are removed from an area, most prey populations increase, which implies that predator pressure keeps prey populations down. When prey numbers are high, predators may respond with increased reproduction, decreased mortality, or increased immigration. Often, predator and prey populations fluctuate in a predictable fashion. Initially, prey populations increase, followed by increases in predator populations, followed by decreases in prey populations (Figure 14-19).

Remember that predation is a density-dependent factor. For this reason, predation, by itself and under natural conditions, usually does not result in the extinction of a prey species. Here's one way density dependence might prevent extinction: red foxes prey on rabbits only after the number of rabbits becomes relatively high. As rabbit numbers decrease due to predator pressure, the number of squirrels, an alternate prey of the foxes, may increase. Eventually, the squirrels outnumber the rabbits, and the foxes switch to preying on squirrels. (This scenario assumes that rabbits and squirrels are equally available to the foxes.)

Predator–prey relationships can be used to control unwanted populations. In 1839, the prickly pear cactus was introduced into Australia from the Americas. With nothing to control its population, the cactus quickly became a pest. By 1925, it occupied over 240,000 km^2 of once valuable rangeland. In that year, the cactus moth, a predator of prickly pear from South America, was introduced and became established. So successful was the moth that today prickly pear is restricted to few isolated outposts in western Australia. Introductions of predators to control pests must be carefully planned and executed. In Hawaii, introductions of mongooses to control rats have been disastrous, partly because mongooses feed in the day while rats are active at night. Unfortunately, both rats and mongooses thrived and fed on native birds, contributing to the extinction of many species.

---

[3]Potentially competing species often evolve characteristics that minimize competition. On a crowded tidal mud flat, several species of shore bird seemingly compete for food. But species with relatively short bills find their food on the mud's surface. Those with slightly longer bills probe shallow depths, while those with the longest bills probe deeper depths for food.

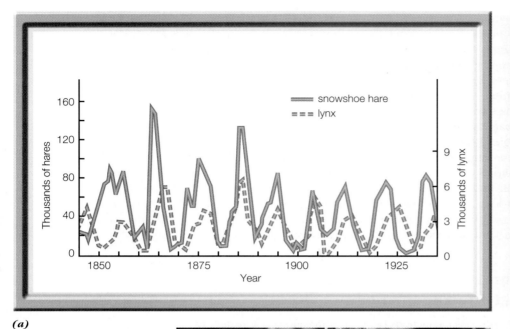

(a)

Figure 14-19. *(a)* In some parts of their ranges, numbers of snowshoe hares *(b)* and lynx *(c)* are interrelated. Large numbers of hares support large numbers of lynx. In the face of intense predator pressure, hare numbers fall. With food lacking, lynx numbers decline. With lynx numbers low, hare populations increase, and a new cycle is initiated. (Graph adapted from MacLulich, 1937.)

(b)

(c)

There are two basic categories of predator–prey relationships: plant–herbivore interactions and herbivore–carnivore interactions. The categories are so different, especially in the ways the predators and prey respond to each other, we will consider each separately.

Plant–Herbivore Interactions

On the face of it, plant–herbivore interactions appear to be relatively simple. Herbivores must simply find plants and eat them. Plants, being virtually immobile, seem pretty much at the mercy of their herbivores. In reality, plant–herbivore interactions are much more complex.

Most herbivores prefer specific types or species of plants and seek them out. In the fall, mule deer, for example, prefer acorns although they can eat a variety of other plants as well. They have an almost uncanny ability to choose nutritious foods. Offer them two samples of soy beans, one grown with fertilizer and one without, and they will choose the most nutritious. Extreme cases of preferred diets are known: koala bears eat only leaves of eucalyptus trees; pandas can survive exclusively on a diet of bamboo.

Herbivores have responded (evolutionarily) to the fact that plant cells are surrounded with thick walls of cellulose that are relatively difficult to digest. The cellulose gets digested because a much larger amount of an herbivore's time, compared to that of a carnivore, is spent eating and processing food. A horse in a pasture spends most of its time feeding, while a pet dog at home spends most of its time sleeping. Herbivore digestive systems are often more complex than those of carnivores. Cows and other hoofed animals, for example, have multiple stomachs.

The eyes of herbivores are typically widely separated, sometimes on the sides of their heads. This gives them a wide field of vision which is useful in detecting their own predators but yields poor depth perception. At least in securing food, herbivores have little need for good depth perception as their food stays put.

The food supplies of many herbivores vary seasonally. Seasons of scarcity follow seasons of abundance. In many regions, dry seasons follow wet ones. In regions beyond the tropics, winters follow summers. Herbivores respond in several ways:

◆ Many put on extensive layers of fat during seasons of abundance that at least partially carry them through seasons of scarcity.

◆ Some migrate. In Africa, herds of wildebeests and zebras are constantly moving to areas where food is available. The same is true of caribou and reindeer in the Arctic and was true of bison in the Great Plains of North America.

◆ Other herbivores hibernate or estivate, lowering their food requirements during seasons of hardship.

◆ Finally, some herbivores respond to seasons of scarcity by making do with foods of relatively low nutritional value. Many nonmigratory songbirds in regions with cold winters switch diets from primarily insects to primarily seeds. Other herbivores, especially mammals, survive for months on nothing more than dead grass or leaves. Their diet is almost exclusively cellulose. To exploit these foods, they need the assistance of other organisms living in their intestinal tract—a type of interaction between organisms we will discuss more fully later.

Plants respond to herbivores. One of the most common responses is simply to produce enough tissue to feed herbivores and still survive. Many grasses, for example, seem to have an unlimited ability to grow new tissue. Graze (or mow) the grass, and it grows right back. There are limits, of course, but grasses sustain heavy grazing and still survive. So, too, do many trees and bushes survive by replacing lost leaves.

Many plants are less passive in their responses to herbivory. When faced with intense herbivore pressure many "fight" back. A typical scenario might be as follows: An herbivore begins to feed on a plant's leaves. In response to either the release of chemicals from damaged plant cells or herbivore saliva, the plant produces hormones that stimulate other leaves to produce chemicals that deter further feeding (Figure 14-20). Some such chemicals are distasteful and avoided by herbivores. Thus certain willows discourage feeding by snowshoe hares. Other deterrent chemicals affect the herbivore's physiology. Some, for example, interact with insect hormones, especially those of larvae, to slow or stop growth. Slowed-down insect larvae are more susceptible to predators or may fail to develop into adults before winter. Some deterrent chemicals are downright toxic to herbivores: chemicals in the leaves of poison ivy and the sap of milkweed are examples.

Figure 14-20. Plants are not always passive in their response to herbivory. Proteins of oak leaves, for example, react with tannins in the leaves and become indigestible to caterpillars.

Of course, evolution being what it is, herbivores respond in turn. Any herbivore that can tolerate a particular plant species' chemical defenses will thrive. Thus a number of birds feed on poison ivy berries. Also, monarch butterfly larvae feed on milkweed with impunity. Not only can they tolerate the highly alkaloid milkweed sap, they accumulate it in their own tissues, thus becoming distasteful (and toxic) to their predators. Remarkably, as adult butterflies they retain the alkaloids, and thereby continue to repel predators.

These relationships are highly complex. In some cases, herbivores able to tolerate defensive plant chemicals are attracted only to the plants producing such chemicals. This may be the primary reason why many plants produce defensive chemicals only after being damaged by herbivores. Not producing the chemicals when the plants are healthy is a deterrent to defense-tolerant predators.

Herbivore–Carnivore Interactions

The relationships between herbivores and the carnivores that feed on them may seem deceptively simple. As we saw in Chapter 13, carnivores must find, approach, and secure prey; herbivores must sense the presence of predators and avoid them (Figure 14-21). As with plant–herbivore interactions, herbivore–carnivore interactions in nature are generally far from simple. Energy is precious to both predator and prey, and we can understand much of the interactions between carnivores and their prey from the perspective of energy expended and energy obtained.

Compared with plant material, animal material is generally easier to digest, and the energy and nutrients are more highly concentrated. This simplifies life for the carnivore. Among the larger ones, except when they are feeding young, most must obtain prey only once every few days. In between feedings, life is relatively easy. They sleep, rest, or socialize. Small carnivores may need to feed more often, but not as often as herbivores. Herbivores feed nearly continuously.

As was pointed out earlier in the chapter, it is to a carnivore's advantage to secure the most food while expending the least energy. Often, this means concentrating on the prey that is most abundant or the most available. To hunt squirrels in an environment where rabbits are more abundant is a waste of a fox's energy, assuming that both are equally "catchable." Where rabbits are more abundant, it is more efficient to hunt rabbits. Even if a fox happens upon a squirrel, it passes it by. Carnivores often seem to formulate a prey image that is what they then seek to the exclusion of other possible prey.

What vastly complicates the task is that some herbivores are highly mobile and potentially dangerous when cornered. A healthy moose is quite capable of successfully defending itself against a pack of hungry wolves. To push their point is not only a waste

Figure 14-21. The puma has seen the rabbit and must now approach it. The rabbit has seen the puma and must now elude it.

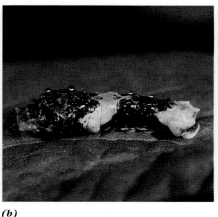

(a)   *(b)*   *(c)*

Figure 14-22. Many insects resemble something they are not. *(a)* These treehoppers mimic thorns; *(b)* this caterpillar resembles a bird dropping; *(c)* this caterpillar looks like a stick.

of the wolves' energy, it is to invite serious injury. Thus a predator spends time studying a particular prey animal before attacking it. Carefully, predators assess chances of success, looking especially for signs of weakness, such as individuals that are sick, injured, old, or very young. Only when faced with starvation do they chance challenging healthy individuals.

Energy expenditures are important to prey species as well. The zebra that runs every time it sees a lion wastes much time and energy. Lions are not always hunting. When first detected, prey often freeze, stare, and spend time studying the carnivores' intent. Usually it is the carnivore that initiates the chase; the prey responds.

Being inconspicuous is obviously important to both predator and prey. Lions are the color of the dead grass they must often skulk through approaching prey. But it is prey species that often master the art of hiding or resembling something they are not. By doing so, they avoid notice. The spotted fawn lies stock still at the wood's edge, looks like a pile of dead leaves, and emits no odor. A herd of running zebras is a blur of black and white stripes, making it difficult to focus on a particular individual. Drab-colored female songbirds resemble pieces of dead bark as they sit on nests. Opossums, in a dangerous situation, play dead. Many of their potential predators eat only fresh-killed prey and thus move on from an animal playing 'possum.

Mimicry, or looking like something in the environment that predators would avoid, reaches its highest expression among insects (Figure 14-22). Some insects not only look like twigs, they gently wave back and forth (even when there is no wind). Some caterpillars have large spots on their tail ends that look like huge eyes. Some insects look like lichens, dead leaves, or bird feces.

Some potential prey possess weapons so dangerous or distasteful that predators avoid them at all cost. Porcupines have spines that can cause predators real pain. Skunks emit a spray that not only smells bad, it tastes worse and irritates eyes. Monarch butterflies, discussed above, are toxic to predators. So are several species of frogs in Central and South America. Notice that many of these potentially dangerous species are brightly colored, which advertises their presence and helps predators avoid them. Among bees and wasps, black and yellow are common colors. Any predator that knows to avoid one stinging bee or wasp avoids all the look-alikes. Because all are harmful, this is an example of **Müllerian mimicry**, discussed in Chapter 13.

This leads us to another form of mimicry. Any insect that happens to be black and yellow is avoided whether it has a stinger or not. This includes several black and yellow flies that look like bumblebees. There are several mimics of monarch butterflies, including viceroys and queens. Most are edible. All are avoided. The edible flies and butterflies are examples of **Batesian mimicry**, also discussed in Chapter 13.

14.4.3 Some Populations Form Intimate Associations

Interactions between 2 pair of species can become so intimate that one or both become dependent on the other. Such relationships are termed **symbioses**. Some are extreme. Certain species of fig are totally dependent on certain species of wasp for pollination. These wasps can pollinate no other flower and are totally dependent on the figs for food. Generally, the relationships are not so extreme. Many flowering plants are dependent on insects for pollination, but the flowers can be pollinated by various pollinators that can, in turn, pollinate several different kinds of flower. Basically, there are three types of symbiosis: parasitism, commensalism, and mutualism.

Parasitism

In **parasitism**, as in predation, one species benefits at the expense of another. Major differences between predation and parasitism are matters of degree and intensity. Generally, parasites are much smaller than their hosts (Figure 14-23). Often parasites are species specific. That is, tapeworms of dogs cannot infect humans, and the tapeworms of humans cannot infect dogs. There are important exceptions, however. *Trichinella spiralis*, a roundworm, causes trichinosis in humans, pigs, bears, and walrus. Some species of protozoan that causes malaria infect both humans and monkeys. When influenza epidemics are not ravaging human populations, the viruses apparently take refuge in birds, where they mutate.

Usually, parasites do not quickly kill their hosts (as do predators). Using an economic analogy, whereas predators live off the principle, parasites live off the interest. Still, for most species, parasitism can be a major contributor to mortality, if not directly, then indirectly. Heavy parasite loads can greatly affect survival. Foxes suffering from mange, which is caused by parasitic mites, lose so much fur they become susceptible to cold, wet, and wind. A free-ranging dog may simultaneously support dozens of populations of tapeworms, flukes, roundworms, and protozoans in its brain, heart, lungs, liver, bladder, stomach, and intestines.

Parasitism, like predation, generally does not result in extinction of its host species. Host numbers may become greatly reduced as a result of parasitism; as they do, it becomes increasingly more difficult for the parasite to spread. A near exception is the blight infestations of American chestnut trees in the northeastern United States.

Commensalism

In **commensalism** one species benefits while the other is seemingly unaffected (Figure 14-24). An often cited example is that of sea anemones growing on the backs of crabs. The crab provides its guests with space on which to grow (space is at a premium

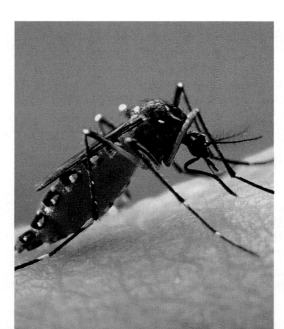

Figure 14-23. Mosquitoes are parasites of humans.

Figure 14-24. Cattle egrets benefit from a commensal relationship by feeding on insects flushed by elephants.

for many substrate-dwelling creatures in marine environments), moves them from place to place, and may even provide food scraps on which they feed. What possible benefit could the crab derive out of the relationship? Another example might be epiphytes or plants, such as orchids, growing on the trunks or branches of trees. The epiphytes live closer to vital sunlight and do not parasitize the trees.

Clear-cut examples of commensalism, right? Not necessarily. What if the sea anemone provides camouflage for the crab, making it look more like a rock than a crab? What if the load of epiphytes, becomes so great, as it frequently does on South and Central American trees, that they break? The point is, if we look closely at what appear to be commensalisms, we often find hidden benefits or costs.

Some clear-cut examples of commensalism might be plants such as cockleburs and sticktights that passively disperse their seeds on the fur of animals and the clothing of people, and certain mites that hop onto the bills of hummingbirds and hitch a ride from flowers on which the birds are feeding to other flowers.

Mutualism

Intimate relationships in which both species benefit, in which partnerships in life are formed, are **mutualisms**, and they are some of the most fascinating and important relationships known in biology. We have already mentioned several examples of mutualisms: flowers and their pollinators, mycorrhizal fungi and the plants on whose roots they attach, nitrogen-fixing bacteria and their legume hosts, flowering plants and the animals that eat their fruits and disperse their seeds, and lichens. You can probably think of others (Figure 14-25). Notice that there are two

Figure 14-25. Oxpeckers feed on external parasites of water buffalo. Each species benefits the other.

broad types: **obligatory mutualism**, in which both partners are mutually and exclusively dependent on each other (figs and their wasps, for example), and **facultative mutualism**, in which both benefit when together but are capable of living independently (for example, corals and the algae that live in their internal spaces).broad types: **obligatory mutualism**, in which both partners are mutually and exclusively dependent on each other (figs and their wasps, for example), and **facultative mutualism**, in which both benefit when together but are capable of living independently (for example, corals and the algae that live in their internal spaces).

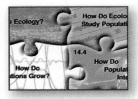

Piecing It Together

Species are not independent entities; frequently they interact. Types of interactions include:

1. **Competition**, in which both species are negatively affected.
2. **Predator–Prey**, in which one species benefits at the expense of the other. There are two kinds of interactions: **plant–herbivore** and **herbivore–carnivore**.
3. **Symbiosis**, in which intimate interrelationships between species include:
   Parasitism, whereby one species benefits at the expense of the other.
   Commensalism, when one species benefits while the other is unaffected.
   Mutualism, when both species benefit.

What about humans? Who are our competitors, predators, prey species, parasites, commensals, and mutualists? In terms of these categories, what is our relationship with domestic dogs and cats?

Where Are We Now?

The study of interactions among populations continues to be one of contemporary ecology's major emphases. Our need to control populations—including our own—grows daily. In the next chapter, after we have explored more about community ecology, we will discuss the impacts of human populations and activities on other species. Suffice it now to say that the control of human populations, by methods that are equitable, humane, and ethical, is one of the major challenges facing humanity. Our need to do so will intensify in the 21st century.

Human activities and interests impact nearly every other species on Earth. Increasingly, we spend more and more of our resources on the study and management of critical species—a trend that will continue. If you are looking for a career for the next 50 years, the study, control, and management of biological populations has great potential.

Often, our goal is to reduce population numbers to manageable levels. Pests claim an estimated 40 to 50 percent of our crops. Periodic outbreaks of insect or fungal infestations wreak havoc on the crops of entire regions of developing countries. We are, literally, at war with our pests. Our strategy to fight this war is a complex set of principles and techniques called **integrated pest management**, which is carefully thought out combinations of pesticides, biological agents, crop rotation, pest-resistant cultivars, and anything else that works. The hope is that using more than one approach will create sufficient environmental resistance to keep pest numbers in check. But it's a war we are barely winning. Pests evolve characteristics by which to overcome elements of control almost as fast as we can develop them. Whenever they detect a chink in the bulwark of our techniques, pests respond with exponential growth.

 Fighting Back Discuss some examples of how integrated pest management (IPM) is being used to control pests. How might IPM techniques be used by homeowners in controlling household and yard pests?

Similar is our need to control **alien species**, those that become established in regions in which they are not native (Figure 14-26). Purple loosestrife is a lovely wild-flower in its native Europe. In North America it has become a scourge to wetlands. Zebra mussels are relatively unimportant in their native Caspian Sea. Soon after their appearance in the Great Lakes in 1986, they became a serious nuisance, clogging water intakes of power plants, among other things. Now they are spreading south, west, and east. North America is currently hosting an estimated 4,500 nonnative species, and new introductions are reported almost daily. Where do they come from? Most are introduced accidentally when they hitch rides on boats, trains, planes, trucks, and cars. The pet and garden industries bring new species to our homes and gardens, from which they may escape or be released. Most do not become established, but some do. Some are useful, welcome additions. Honeybees are not native to North America, nor are most of our ornamental plants, nor are ring-necked pheasants. Many other nonnative species, once established, especially away from our homes and cities, become pests. In the absence of factors that normally control their populations, they thrive. Sometimes their numbers increase exponentially. Sometimes they outcompete and replace native species. Sometimes they overgrow whole communities. Worldwide concern with alien species is growing. A major challenge in the 21st century will be to develop techniques to stop their spread and control their populations.

 The Aliens Have Come What are some of the alien species in your region? How serious a problem are they? What is being done to control them?

Often, our concern is not to reduce population numbers, but to increase them. Management of game species is an important and growing branch of applied ecology. So is the management of endangered species. In both cases, overall management goals center on maintenance of viable populations, often in human-dominated regions. Throughout

(a)

Figure 14-26. Many of the worst pests in the United States are not native to North America. *(a)* Kudzu, a vine that is destroying millions of acres of native plants in the southeastern United States, is native to Japan. *(b)* Norway rats, which are very destructive, come from Europe.

(b)

much of the 20th century, management concentrated on single species or even populations. Game laws focused on harvest limits. The Endangered Species Act focused on populations that were fast approaching extinction. Often, concentrating on single species or populations did not have the desired affects. In the 1960s, for example, it was realized that the bald eagle—symbol of the United States—was becoming endangered in much of North America. Laws were passed forbidding the killing of any eagle for any reason. Still, numbers dwindled. New laws were passed restricting the use of DDT in the United States. This pesticide had been found to accumulate in the tissues of predatory birds, including bald eagles, and to adversely affect their ability to produce viable eggs. Still, eagle numbers failed to recover. Eagles needed more than protection. They needed habitat. New laws and regulations preserved nesting sites, restricted human activities around those sites during breeding seasons, and established areas in which eagles could roost and hunt. Only then, when not only were individuals protected, but so was their habitat, did eagle numbers begin to recover.

Management emphasis has shifted from focusing on isolated species and populations to preservation of habitat. Our attention, too, must shift from populations to larger entities of which they are a part. In the next chapter we address the ecology of whole systems.

REVIEW QUESTIONS

1. Briefly, what is ecology? How is the term "ecology" different from "environment"? How is the science of ecology different from natural history?

2. How are population ecology, systems ecology, and applied ecology similar? How are they different?

3. Who was Thomas Malthus and what was the major thesis of his paper, *An Essay on the Principle of Population*?

4. What four factors determine the size of a population? Which one or ones result in population increases? Which one or ones limit population size?

5. What is a population's growth rate? (Write this answer in words and in a mathematical equation.)

6. How does immigration differ from emigration?

7. What is the difference between a population's absolute number and its density?

8. Describe the range, habitat, and niche of an organism of your choosing. Now do the same for humans.

9. Describe biotic potential and environmental resistance. Give examples of each.

10. How do extrinsic factors differ from intrinsic factors? Give examples of each.

11. What is the meaning of "exponential growth" in a population? Why doesn't this type of growth continue over the long term in nature?

12. What is the meaning of "logistic growth" in a population?

13. What are density-dependent limiting factors? Give some examples.

14. What are density-independent limiting factors? Give some examples.

15. What is meant by the "carrying capacity" of a population?

16. How do *K*-selected species differ from *r*-selected species? Give some examples of each. Are humans *K*-selected or *r*-selected?

17. What is the difference between physiological longevity and ecological longevity?

18. Give some examples of organisms that interact as predator and prey; as competitors; as parasite and host; as commensals; as mutualists.

19. Briefly describe the competitive exclusion principle.

20. What are two probable outcomes of competition between closely related organisms?

21. What are two general types of predator–prey relationships? Give examples of each.

22. Give examples of Müllerian and Batesian mimicry.

Chapter 15

Ecology: How Do Organisms Interact with Their Environments?

**Three hundred trout are needed to support one man for a year.
The trout, in turn, must consume 90,000 frogs, that must consume
27 million grasshoppers that live off of 1,000 tons of grass.**
—G. Tyler Miller, Jr., 1971

—Overview—

Populations, as we saw in the last chapter, are important to ecologists. Indeed, some of ecology's most interesting and challenging questions center on populations. What's happening to the numbers of songbirds in North America? How can we control the number of insect pests in our homes and gardens? Deer are wonderful and exciting animals to have around. How can we keep them from becoming pests in suburban America?

But the story of ecology doesn't stop with populations. Populations are components of something larger, namely ecosystems. Within any ecosystem, populations interact with each other and their surroundings, both biological and physical. All populations are not equal. One or a few may come to dominate a particular ecosystem. Others have few members and remain inconspicuous. We have seen that populations are dynamic. So are ecosystems. It is to the study of these complex systems that we now turn our attention.

Ecosystems are highly complex, made up of all the organisms and all the physical conditions and processes in a particular area. Ecosystems are also organized. Plants trap energy from the sun and tap nutrients from the soil, making both available to other organisms. Nutrients and energy are passed from organism to organism as plant eaters eat the plants and are in turn eaten by meat eaters. Eventually, nutrients are returned to the soil, to be recycled by other organisms.

No less than any other species, humans depend on healthy, natural ecosystems. From them we get our basic necessities: the air we breathe and much of the food we eat. Healthy ecosystems moderate our climates and purify our water. But some human activ-

Chapter opening photo—Organisms interact in a coral reef community.

ities interfere with natural ecosystems. Inadvertently, we damage their capacity to sustain us. One of the biggest challenges facing humans in the next century is to learn to live in peace with, rather than fighting, Earth's ecosystems.

For studying these complex systems we have that branch of biology called systems ecology. Compared with other branches, it is one of the youngest, most complex, and most controversial. It may also be one of the most ambitious. Like the scientists in other branches of biology, systems ecologists start with observations and seek generalizations that would tie together large sets of observations. At its best, systems ecology moves beyond descriptions and mechanisms. Often it requires synthesis, or the bridging of gaps between seemingly unrelated areas of knowledge. In studying how organisms interact and are affected by each other, systems ecologists deal with nearly every other branch of biology. Taxonomy, physiology, genetics, and evolution are particularly important. In studying how organisms affect their physical environments and how the environments affect the resident organisms, systems ecologists bridge biology and the physical sciences. Here aspects of geology, meteorology, climatology, biochemistry, and biophysics are of particular interest. An understanding of statistics and computer science is also essential to systems ecologists.

Ecologists have come to understand that organisms never exist as isolated entities; rather, they are components of systems. It has been said that a California condor is "5 percent feathers and bones and 95 percent habitat." Certainly the numbers can be challenged, but the concept is inescapable. Organisms are inseparable from their surroundings.

15.1 WHAT IS AN ECOSYSTEM?

**Fast Find
15.1a
What Is
an Ecosystem?**

Nothing in biology exists as an isolated entity. Each thing is part of something larger. Molecules make up cells and cells make up complex individuals. Ecology starts with the realization that being a part of something larger doesn't stop with individuals. Individuals exist in groups called populations that are, in turn, the integral parts of biological communities. And so it goes.

Furthermore, biological parts invariably interact. Just as many types of cells depend on other cells for an organism to function, many organisms are social, depending on others of their kind. In the last chapter we learned that populations also interact and are interdependent. Flowers need bees for pollination. Predators need prey. And in the long run, prey benefit from predators who reduce their numbers. Living things are part of the environment, namely its **biotic** component.

Organisms interact with more than each other. They interact with the physical, or **abiotic**, environment. For example, plants depend on just the right combinations of soil nutrients, temperature, light, and available water to grow and blossom. Some songbirds in temperate regions react to changes in day length by migrating. The interacting biotic and abiotic components in a particular place are an **ecosystem**.

15.1.1 An Ecosystem Comprises Interacting Environmental Components

The idea of ecosystems is a profoundly important concept to the field. To think of a pond as an ecosystem is to think of all the organisms in the pond, from the cattails standing in its shallow waters to microscopic green algae floating in its water, from the heron standing among the cattails to the fish it seeks as food to the tiniest microbe wriggling in its bottom mud. But the pond as ecosystem includes more than organisms. The water's temperature, its clarity, and the chemicals dissolved in it are included. So are the particle size and composition of bottom mud, annual and daily fluctuations in temperature and light, and how completely and rapidly water flows through the pond. Inherent to the concept of ecosystem is the realization that all these components interact. Water temperature controls the number of microbes in a pond, which determines water clarity, which determines how well herons see fish. Few systems in all of nature are as complex as ecosystems.

To simplify our task, we will, in later sections, look at the abiotic components separately from the biotic components. But it is important to remember that, within any ecosystem, organisms continuously interact with other organisms and with their physical environment.

Notice that nothing in the definition of ecosystem suggests size. Ecosystems come in almost any size. We can, and on occasion will, think of the entire planet as an ecosystem. Usually we think of smaller areas as ecosystems. A tree could be an ecosystem. A cave could be an ecosystem. So could an island; a grove of trees; an entire river and its tributaries. Can anything be an ecosystem? Not exactly. Let's look at some more characteristics of ecosystems.

15.1.2 Ecosystems Are Definable, Open, and Dynamic

Ecosystems have definable boundaries. Ponds make good examples. It's easy enough to tell where a pond begins and where it ends. Furthermore, ecosystem boundaries have relevance to organisms. North Dakota would not be an ecosystem. Its boundaries are definable politically, but not biologically.

Ecosystems are open. Energy for the pond comes from the sun which is not normally considered part of the pond. Water flows into the pond, particularly following rain, as runoff, from ground water, and from incoming streams. Water evaporates from the pond and may flow out in streams. Flowing water carries nutrients in and out of the pond. If the heron catches a fish and then flies off, it, too, removes nutrients from the pond. If it defecates while fishing, it adds nutrients to the pond. A pond, like any other ecosystem, is an open system.

Ecosystems are dynamic. The conditions within ecosystems change continuously. The heron is part of the pond only while it is physically present. If the heron is successful in catching food, the fish population changes. Water temperature and, indeed, all physical aspects of the pond change seasonally, daily, and perhaps even more frequently (Figure 15-1).

Figure 15-1. A pond is a good example of an ecosystem. It has easily recognized boundaries, yet the pond, like other ecosystems, is not closed. Energy for the pond comes from the sun. Water readily enters by precipitation, runoff, and incoming streams. Water also leaves by evaporation and, in all likelihood, an exit stream. Similarly, nutrients enter, flow through, and exit the pond. Within the pond, populations of organisms interact among themselves, with other populations, and with their physical environment. The pond is a constantly changing, dynamic ecosystem.

Figure 15-2. Materials cycle within ecosystems. When a tree dies, it decays. In time, nutrients locked within its tissues return to the soil and become available to other organisms.

Materials tend to cycle within ecosystems. In the next section we will see another way in which ecosystems are dynamic when we consider how certain physical matter, essential for life, cycles in ecosystems (Figure 15-2).

Landlocked Lake Great Salt Lake, in Utah, is an example of a water body lacking outflowing streams and rivers. How has this affected conditions in this lake? How does this lake lose water?

Piecing It Together

1. An ecosystem is an interacting biological community and its physical environment.
2. Ecosystems are complex.
3. Components of ecosystems, biotic and abiotic, interact continuously.
4. Properties of ecosystems include these:

   ◆ They come in almost any size.

   ◆ They occupy definable spaces.

   ◆ They are open systems, with energy and matter flowing in and out freely.

   ◆ They are dynamic, showing both short-term and long-term changes.

   ◆ Within them, materials tend to cycle.

15.2 HOW DOES THE PHYSICAL ENVIRONMENT AFFECT ORGANISMS?

No organism lives everywhere. Even humans, the most cosmopolitan of species, are basically restricted to terrestrial environments (roughly 20% of Earth's surface) and are mostly concentrated in tropical and temperate regions. All other organisms are even more restricted in where they live. Why can't a specific organism live everywhere? Explaining the distributions of organisms is one of ecology's most important tasks.

In earlier chapters, we discussed some of the factors determining where organisms live. Darwin observed that organisms possess characteristics, or adaptations, that enhance their survival in a given environment. Remember the rabbit's long legs, the fox's complex behavior, and the flower's ability to attract pollinators. Natural selection, Darwin said, favors individuals with the most appropriate adaptations and eliminates those that lack them. Such statements focus on the individual. Ecology shifts the emphasis. What, exactly, is natural selection?

How, exactly, does the environment affect organisms?

15.2.1 The Physical Environment Limits Where Organisms Live

Physical factors can be limiting. Part of the question of why organisms are restricted in where they can live can be answered in terms of **limiting factors**. An organism cannot live where it cannot find the things it needs. Many essential limiting factors are components of the physical environment. Let's use plants as examples.

As we saw in Chapter 12, plants need nutrients, many of which come from the soil, including nitrates, phosphates, and a suite of micronutrients. The amount of nutrients available to plants varies widely. Some soils are rich in phosphorus. Others lack nitrates. Additionally, how much of a given nutrient a plant needs depends partly on the species. Some species need more phosphorus than others. If a particular plant needs more phosphorus than is available in the soil, its growth is affected. A field of wheat growing in such soil may produce a disappointing yield. For want of a few grams of phosphorus, a tree that could be a luxuriant producer of apples can be stunted and unproductive. Given enough time, natural selection may favor those individuals best able to cope with low-phosphorus soil and productivity may improve. But natural selection is at best slow, and there are limits to what it can do. In the meantime, because all plants need at least some phosphorus, lack of phosphorus in the soil limits plant growth.

As early as 1840, the German chemist, Justus von Liebig, observed the importance of phosphorus for field crops. He expanded this observation into his law of the minimum: "the perfect development of a plant," he said, "is dependent on the presence of [certain inorganic nutrients]; for when these substances are totally wanting its growth will be arrested and when they are only deficient it must be impeded." In time it was seen that Liebig's statement was not only true of inorganic nutrients but also many other physical factors. Roses grow best when exposed to full sunlight throughout most of the day. Limited sunlight is as detrimental to their growth as limited phosphorus is to wheat. Swamp plants require soils saturated with water. Their growth is stunted in soils that are only partially or intermittently saturated, and they do not grow at all in dry soils. Tropical plants are similarly sensitive to temperature.

But too much of a good thing is a bad thing. Too much phosphorus in soil is toxic to even the most phosphorus-dependent plant. Tropical plants suffer if air temperatures exceed 45°C (113°F). Thus, Liebig's law of the minimum can be extended into a **law of toleration**: organisms do best when the availability of potential limiting factors lies within a range of values, somewhere between barely enough and too much. The range of a potential limiting factor has three points important to an organism's well-being: the *minimum* below which the organism cannot live; the *maximum* above which life is again adversely affected; and the *optimum*, the level at which it does best (Figure 15-3).

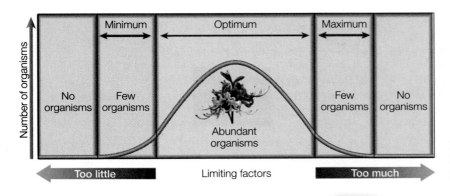

Figure 15-3. Organisms do best within a range of conditions that is optimal for them.

Species differ greatly in terms of what is optimal. An optimal temperature for one species may be too hot for another and too cold for still another. The width of ranges for limiting factors also differs. Some species can tolerate wide ranges in temperature, whereas others are restricted to quite narrow ones. Ecologists use special terms to describe wide and narrow tolerances. The prefix "eury-" denotes wide tolerance (eurythermal means "tolerates wide temperature ranges"; euryhaline means "tolerates wide ranges in water salinity"), and "steno-" denotes narrow ranges. Are humans eurythermal or stenothermal?

15.2.2 The Physical Environment Supports Life through the Cycling of Water and Nutrients and the Flow of Energy

The physical environment is the source of the water, nutrients, and energy upon which all living things depend. The way water and nutrients move through ecosystems differs from the way energy moves. Water and nutrients cycle; energy does not cycle. Think of life in a pond to get a sense of how nutrients cycle. Typically, nutrients dissolved in the pond water feed floating algae and cyanobacteria and are then passed along as the algae and cyanobacteria are eaten by tiny invertebrates, who feed small fish, who feed larger ones, and so on. As organisms die or excrete wastes, the nutrients pass through other organisms that break down the organic material, use some of the nutrients, and release some back into the water. Nutrient cycles involve more than organisms; they also involve the physical environment.

For living things, four of the most important cycles in the physical environment are the water, carbon, nitrogen, and phosphorus cycles. Each cycle has its own set of steps and organisms that are important to it. In general, movement is from the physical environment into organisms, from one organism to another, and back to the physical environment. For example, carbon, an essential nutrient for all organisms, is obtained by plants from the atmosphere as carbon dioxide; through photosynthesis and other processes, plants incorporate the carbon into their tissues. The carbon is passed on whenever animals eat plants and other animals eat those animals. Through respiration, plants, animals, and other organisms use the carbon-containing tissues and compounds, liberating and utilizing energy and returning carbon dioxide to the atmosphere.

Fast Find 15.2a Nutrient Cycles

Some of the cycling material in the physical environment may pass into a reservoir, or sink, and become unavailable to organisms for eons. There are several reservoirs for carbon. Some carbon passes from organisms into deposits of peat, coal, and oil. There it waits, perhaps millions of years, for combustion, either natural or human induced, to release it back into the atmosphere. Some carbon becomes incorporated into carbonate rocks, where it is unavailable to organisms for millions of years until erosion releases it. The oceans are a third carbon sink. Carbon, in solution, sinks. It stays deep in the ocean until chance currents bring it back near the surface where it can reenter the cycle.

Figure 15-4 summaries the water cycle. See the CD for more information and for detailed presentations of the cycles of carbon, nitrogen, and phosphorus. Because such cycles have both biological and geological phases, they are sometimes called **biogeochemical cycles**.

Energy does not cycle in ecosystems. It is, of course, passed from organism to organism. Plants, green algae, and cyanobacteria are the only organisms that can trap energy from the sun, store it in biochemicals, and thus make it available to the rest of the ecosystem. A few bacteria living deep in oceans can similarly convert geothermal energy into biochemical energy that then becomes available to other organisms. As the organisms that trap energy from the sun or Earth are consumed, energy passes from organism to organism. But energy does not cycle. Remember the laws of thermodynamics? (If you don't, review Chapter 9.) Some energy is lost at every transformation. Whenever any fuel is burned, only a portion of the energy can be utilized or stored. The rest is lost as heat—not only from organisms, but from the entire ecosystem and eventually from Earth itself. Thus, energy passes through ecosystems in a sometimes convoluted, but always one-way trip.

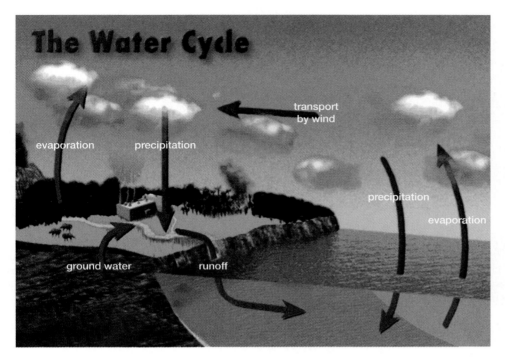

Figure 15-4. The water cycle is an example of materials cycling through ecosystems. A large portion of the cycle does not involve living things. Water evaporates from oceans and other bodies of water and precipitates back to Earth's surface. The cycle becomes more complicated when water is consumed by organisms. Plants soak up water from soil; animals consume it directly. Water can then be passed from organism to organism as they feed. No matter how circuitous its path through organisms, eventually water is passed back to the physical environment to rejoin the basic, physical cycle. How do humans participate in the water cycle?

15.2.3 Light, Temperature, and Wind Affect Organisms

Light, temperature, and wind—like water, nutrients, and energy—are parts of the abiotic environment that affect organisms. They limit where organisms live, provide them what they need, and affect them in many other ways. In this section we look at each of these factors and also their interplay as summed up in climate. We also give more attention to the importance of water to organisms.

Light

The difference between night and day is so obvious we tend to take it for granted. But the differences in light in the course of a 24-hour period have profound effects on many life forms. Morning glories bloom early in the morning and drop their flowers by late afternoon. Some water lilies close their flowers as evening darkens. Owls replace hawks as the nighttime avian predators in many locations. Late at night, mice and cockroaches forage in places where humans wish they wouldn't.

Except in the tropics, the amount of light present in the environment varies with season: days lengthen in summer and shorten in winter. The period of daylight varies with distance from the equator. North of the Arctic Circle (and south of the Antarctic Circle), the sun remains continuously above the horizon for at least 24 hours in summer. Farther north (or south), the period of continuous sunlight is considerably longer. At Point Barrow, the northernmost point in continental North America, the sun stays continuously above the horizon for 82 days. In winter there are similarly long dark periods.

Lengthening periods of light in summer affect plant growth by extending the amount of time plants can carry on photosynthesis. Thus, in all regions north (or south) of the tropics, plants grow more in summer. In south-central Alaska, for example, vegetables such as turnips and cabbage grow to enormous size in summer. There are latitudinal effects of light intensity in addition to the effects of season. At the equator, the sun shines directly at Earth's surface. Here light intensity is highest. Because of the shape of the Earth, sunlight, away from the equator, approaches at an angle. The angular approach spreads the light out, lessening its intensity and increasing the chances for reflection, especially off shiny surfaces such as snow, ice, and water.

Figure 15-5. In most ecosystems, light intensity varies from time to time and place to place.

Light intensity also varies with local topography and vegetation. Trees, bushes, large rocks, hills, and mountains create areas of shade. North of the equator, northfacing slopes receive less sunlight than south-facing ones, as do south-facing ones south of the equator. Organisms living under the forest's canopy are exposed to less light than those living in the open (Figure 15-5). Finally, during the day, light intensity varies depending on cloud cover.

Temperature

It's hard to separate the effects of light from those of temperature. Light, absorbed by matter (soil, water, atmosphere, and so on) is transduced into heat, which is measured as temperature. In general, areas and times receiving the most light have the highest temperatures. Thus, summers are generally warmer than winters; daytime temperatures are generally warmer than nighttime temperatures; sunny days are generally warmer than cloudy days.

A region's temperature, however, does not correspond exactly to the sunlight it receives for two reasons. First, large air masses move vast distances, carrying heat with them. Especially in temperate regions in spring, a sunny warm day can be followed by a sunny cold one as cold air moves in over the region. Second, a region's temperature is strongly affected by its proximity to large bodies of water. One of the important properties of water is its tendency to change temperature slowly. (Review the properties of water in the Chemistry Review, if you need to). Jump in a lake on the first warm day of spring, and you find the water still winter cold. Conversely, ocean waters hold warmth well into winter.

The ecological effects of this property of water are profound. Regions near large bodies of water have less extreme temperatures than those lying away from water. Temperatures are more extreme in continental centers than along coasts.

Available Water

How much water is available is another environmental factor important to organisms that varies widely from place to place. In oceans and permanent bodies of fresh water, water is constantly available. It is nearly so on Mt. Makahanaloa in the Hawaiian islands, where an average of more than 5 cm (2 inches) of rain falls every day. Near the other extreme, deserts east of Jerusalem receive an average of 3 mm of rain a year. In some deserts, even less water falls. Between these extremes, every possibility can be found.

Wind

The importance of wind depends on the kind of organism, time of year, and how much the wind blows. Many terrestrial plants depend on wind for reproduction. It distributes their pollen and later their seeds. Wind speeds up evaporation. A day of strong winds can completely dry out an otherwise damp environment. Several days of strong winds dry out and kill lateral buds of vascular plants. Thus, in mountainous regions where strong winds blow consistently and persistently from one direction, trees may completely lack branches on the upwind side of trunks (Figure 15-6). In northern regions, bushes such as willows and alders may grow no taller than the winter snow pack. Within the snow pack, a plant's delicate buds are protected from wind. Above the snow pack, winds kill buds and thus limit growth.

Climate

The sum total of weather events in a region is its climate. Light, temperature, wind, and available water interact as climate. Water availability is often related to light and temperature. As mentioned above, sunlight is most intense at the equator, where it warms soil, water, and air. Heat evaporates water, and there is plenty of heat near the equator. Warm air is less dense than cold air and so rises. As it does, it cools and drops its moisture. This movement of air masses happens on a global scale and produces a world-girdling belt of warm, moist tropics. The rising air must be replaced, creating currents of air (winds) moving along the surface toward the equator from both north and south (Figure 15-7).

The rising air, which must go someplace, flows away from the equator north and south. Eventually, the air, now devoid of moisture, falls, creating a belt of dryness. Here, in two broad belts, one 30 degrees north of the equator and the other 30 degrees south, are most of Earth's deserts.

At about 60 degrees north and south of the equator a similar, less pronounced pattern prevails. Air again rises, drops its moisture, and flows toward the poles, where the dry, cold air falls.

Figure 15-6. Wind is a powerful environmental factor in some ecosystems. Over the course of these trees' lives, leaf and branch buds were killed by drying from wind blowing persistently from the same direction, producing trees whose branches are all on its downwind side.

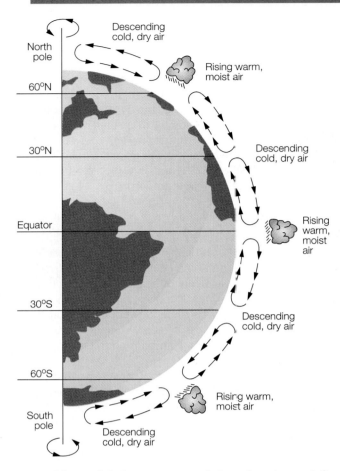

Figure 15-7. Earth's climates are largely determined by patterns of air mass movement. At the equator warm air rises. As it does, the air cools and moisture precipitates. In the Northern Hemisphere, the now dry air flows northward and, at about 30° N latitude, drops to Earth's surface, producing a belt of semi-arid grasslands and deserts. Two additional cells of similar circulation lie north of this latitude. Circulation patterns are similar in the Southern Hemisphere. These gross patterns of air flow are strongly modified by other factors including local topography.

These global movements of air and moisture define Earth's major climate patterns. Moist belts alternate with dry ones. Average temperatures decrease as distance from the equator increases. These generalizations are greatly affected by two factors: the spinning of the Earth and local topography.

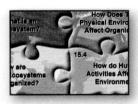

Piecing It Together

1. Organisms are restricted in where they can live by limiting factors—environmental factors whose presence, absence, or abundance makes it possible for particular organisms to survive.

2. Many limiting factors are physical. These include soil nutrients, available energy, available water, temperature, light, and wind.

3. For each potential limiting factor, organisms do best within a range of values. Too much or too little of any of these factors in a given environment can limit a given organism.

4. Tolerance ranges may vary between species, populations, and individuals.

5. Nutrients cycle through ecosystems. They are taken from the physical environment, passed from organism to organism, and returned to the physical environment.

6. Four of the most important biogeochemical cycles are those for water, carbon, nitrogen, and phosphorus.

7. Energy passes, but does not cycle, through ecosystems. At each step as energy is picked up from the environment and passed from organism to organism, some is lost to the system, usually as heat.

8. Factors that make up climate include temperature, light, precipitation, and wind. The distribution of these factors, and thus the climate, varies from region to region.

15.3 HOW ARE ECOSYSTEMS ORGANIZED AND HOW DO THEY DIFFER?

To inquire further about the organization of ecosystems, we must shift our focus from the physical, or abiotic, components to the biotic components. The biotic components of ecosystems are **biological communities**—populations of many species living together in a definable space. At first glance, we may see little evidence of organization in a biological community. Certain animals and plants happen to be living together in a particular area. What more can there be? A great deal, it turns out. At every level and without exception, organisms in a biological community interact and are interrelated.

15.3.1 Biological Communities Are Organized in Trophic Levels

One kind of interrelatedness within a biological community—perhaps the most obvious—is who eats whom. A newly hatched caterpillar feeding on a fresh young leaf of a dogwood tree attracts the attention of a migrating songbird. Before the bird can finish feeding, a sharp-shinned hawk swoops in and takes the bird. Here is an example of a **food chain**. A plant is eaten by a plant eater who is eaten by an animal eater who is eaten by another animal eater. In the same community, more food chains can be found. A mouse eats a flower and later that night is eaten by an owl. A grasshopper, feeding on newly sprouted grass, barely escapes an attacking robin by hopping into dense vegetation next to a pond. Here a frog waits. A twist of its body and a flick of its long tongue secures a meal of grasshopper. All this movement attracts a snake who catches the frog. But before the snake can complete its meal a red-tailed hawk pounces on it and delivers it to three hungry nestlings. Notice that our food chains are getting complex.

Fast Find 15.3b, c, d Food Chains

One view of a biological community is that of a set of food chains, each of which links plants to herbivores to carnivores. Food chains do not go on and on. They may have as few as two links and, in extreme cases, as many as eight. Usually, food chains have four or five links—we'll see why later.

Food chains are not independent entities. They, too, tend to interact. The hawk that ate the songbird could have eaten the robin. The robin could have eaten the caterpillar, and so on. To get a better idea of how food chains interact, go to the CD and fit together the elements of a typical biological community.

What started out to be a simple chain of relationships quickly became a web of interactions. Food chains become **food webs** (Figure 15-8). Their complexity is awesome. Yet, generalizations emerge. Complex, biological communities do have organization. Typically, food chains and webs start with plants (or algae in aquatic environments), the foundation of the community. The next level is organisms that eat plants, or **herbivores**, ranging in size from microbes to tiny insects to grasshoppers to mice to deer. The next set of organisms in the community is animals that eat other animals, or **carnivores**. First we find carnivores that eat herbivores and finally carnivores that eat other carnivores.

Trophic Structure of a Biological Community

We see that food webs and chains in a biological community appear to be organized in levels that indicate feeding relationships, called **trophic** (from the Greek word for food) **levels**. Plants feed herbivores that in turn feed carnivores. To use another set

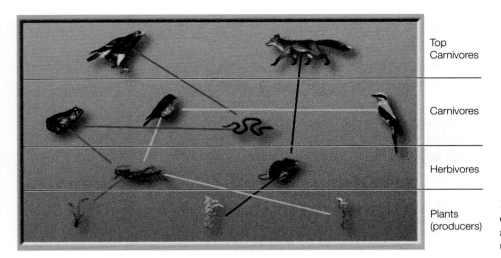

Figure 15-8. Food webs display the complexity of ecosystems. Food chains are interlocked into a web-like structure of interrelationships.

Top Carnivores

Carnivores

Herbivores

Plants (producers)

of terms, **producers** (green plants and algae) feed **primary consumers** (herbivores) who feed **secondary consumers** (carnivores).

At the base of the trophic structure are the producers, organisms that acquire nutrients from the soil and trap energy from the sun. Some they use. But some nutrients and energy are combined and stored in the plant's tissues and thus become available to the primary consumers. In turn, the primary consumers use some nutrients and energy to carry out their activities but also store some nutrients and energy in their tissues. These become available to secondary consumers.

A stable community is like a pyramid, with more producers than primary consumers and more primary consumers than secondary consumers. Does this agree with your experience? Go into a natural area and look around. Are most of the organisms you see producers? Next, look carefully for primary consumers (animals, especially insects). Where are the secondary consumers?

A question soon arises as to how the groups should be measured. The simplest way would be to count the components. Count the individual plants and that number should be greater than the total number of herbivores and so on. Usually you would get a nice symmetrical pyramid. But not always. Consider a tree as an ecosystem. Here there is only one producer and perhaps thousands of insect primary consumers. Our pyramid takes on an inverted appearance. Inverted pyramids appear unstable.

A more satisfactory method of expressing the trophic structure might be to obtain the total weight of the organisms in each group, in other words, to determine each group's **biomass**. Surely, the number of kilograms of tree exceeds the number of kilograms of primary consumers, which in turn will outweigh secondary consumers. It turns out that even biomass pyramids can be inverted. In certain aquatic ecosystems, floating algae make up the producer trophic level, followed by small invertebrates in the primary consumer trophic level. Algae reproduce very rapidly. Sometimes, just a few kilograms of algae can reproduce fast enough to feed a much larger biomass of invertebrates.

The most satisfactory trophic pyramid—one that never inverts in a stable situation—is expressed in terms of energy present per unit time. The amount of energy stored in a day (or whatever time unit is most appropriate) in the producer trophic level is always greater than the amount of energy present in a day in the primary consumer level, which is always greater than the amount of energy present in the secondary consumer level. Notice that we are focusing here on available energy.

Figure 15-9 shows the structures of several stable biological communities. Plants or algae form the base of the community. The next highest tier comprises the organisms that eat plants, followed by those that eat the plant eaters, and so on.

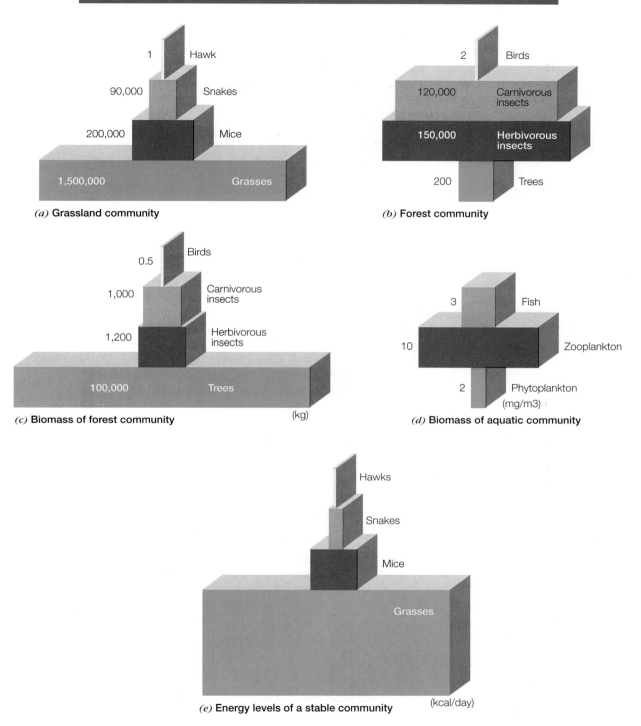

(a) Grassland community

(b) Forest community

(c) Biomass of forest community (kg)

(d) Biomass of aquatic community

(e) Energy levels of a stable community (kcal/day)

Figure 15-9. The trophic structure of stable biological communities can be portrayed in a variety of ways. (*a*) In a grasslands community, the number of individual grasses is far greater than the number of individual plant eaters. (*b*) However, in a forest community, where a small number of large plants support a large number of plant eaters, such graphs have an inverted shape. (*c*) A biomass graph of the same community depicted in (*b*) shows a more typical pyramid shape, in which the biomass of 200 trees far exceeds the biomass of plant-eating insects. (*d*) In some highly productive aquatic environments, biomass graphs may also have an inverted shape, because of the small size and rapid reproductive rates of algae and the large size of the animals that eat them. (*e*) Energy graphs never invert for stable biological communities.

The Energy in Trophic Levels

We can say more about the energy present at various trophic levels. For almost all biological communities, energy enters through green plants or algae. By the process of photosynthesis, they trap sunlight and store its energy in organic chemicals. The total amount of energy trapped by all of the plants and algae in a community is termed the community's **gross primary productivity**. Some of this energy is used by plants as they grow, produce flowers and seeds, and do the things plants do.

Fast Find 15.3e Food Web

That portion of gross primary productivity stored by plants in their tissues is the community's **net primary productivity**. This is the energy available to the next trophic level. The energy consumed by the primary consumers, the herbivores, is the community's **gross secondary productivity**. Again, some is used and some is stored. The stored portion is the community's **net secondary productivity** and is the energy available to secondary consumers, the carnivores.

Fast Find 15.3f Food Web Disturbance

Obviously, any given trophic level cannot consume all of the energy available to it. If the herbivores ate all the plants, the whole system would crash. Again, remember the first law of thermodynamics. As energy is passed from one organism to another or from one trophic level to another, some is invariably lost. As a rule of thumb, only about 10 percent of the gross productivity of one trophic level gets passed on to the next (Figure 15-10). Thus, for every 100 kcal of energy trapped by plants, only 10 kcal passes on to herbivores and only 1 kcal passes on to carnivores.

At the higher levels the community is running out of energy, which limits how many trophic levels a community can support. Usually there is one more level above carnivore or secondary consumer, often called top carnivore or tertiary consumer. In extreme cases there can be as many as eight levels. In Antarctica, orcas (killer whales) eat leopard seals that eat penguins that eat big fish that eat little fish that eat krill that eat little invertebrates that eat phytoplankton. Most systems stop with four levels. Why? Because they run out of energy. Remember: Energy does not cycle. At every level, some is lost.

While thinking about energy, we need to ask, What is a top carnivore? Are humans top carnivores? (Have you ever eaten a carnivore?) What about lions? Lions eat wildebeest that eat grass. Lions may be carnivores, but, because they don't eat other carnivores, they are not, by our definition, top carnivores. What about grizzly bears? In their communities they can eat anything they want to. And do. Much of their diet is plant material, especially ripe berries in early fall. Why is it so difficult to find clearcut examples of top carnivores, usually the fourth trophic level of biological communities? Could it be that at this level energy is so limiting that top carnivores can't be choosy? To survive at the highest trophic level, they must be able to eat anything they find. Notice,

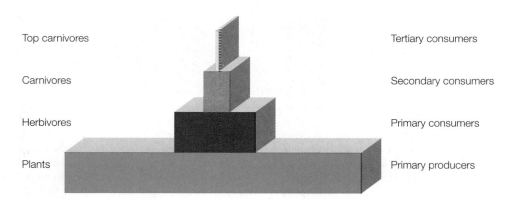

Figure 15-10. Two sets of terms are used by ecologists to describe trophic levels. Plants are the primary producers of the ecosystem. They trap energy from the sun and store it mainly in molecules of glucose. Herbivores, or primary consumers, take energy from the producers to live and ultimately pass it on to the next level, the carnivores, or secondary consumers when they are eaten. Top carnivores (tertiary consumers) get their energy by eating secondary consumers. This graph is not drawn to scale; only about 10 percent of the energy assimilated by one level is passed on to the next higher level.

too, that top carnivores tend to be the largest animals in their community. (Why? Can you think of exceptions to this statement?) Therefore, their absolute energy needs are larger still—another reason not to be picky.

Community Members That Don't Fit the Model

Grizzly bears, humans, and other so-called top carnivores are best called omnivores. They tend to eat anything. They don't fit easily into our model of the trophic structure of biological communities. The problem is, ecologists jokingly say, that some organisms haven't studied ecology and don't know where they belong. Remember that the model is a generalization scientists apply to help them understand ecosystems. Exceptions—the bane of all biology—abound.

Top carnivores are not the only organisms that don't fit our trophic pyramid model of community structure. Where do fungi, bacteria, earthworms, and vultures fit? These are the **detritivores** or scavengers and decomposers, organisms that eat dead stuff. All of them have the same overall function: they make nutrients locked in carcasses and wastes available to the ecosystem again. These are the community's recyclers (Figure 15-11). Here is a completely different set of food chains and webs that may not be layered like the ones we've discussed. These food webs also start with plants. When plants die, decomposers—especially certain insects, fungi, and bacteria—break down and consume their remains. The same is true when individuals from higher trophic levels die. The particular species of insect, fungi, or bacteria may be different, but all dead organisms are attacked by decomposers. Especially at the higher levels, the process may be initiated and hastened by scavengers. Hence, vultures feed on large animal carcasses. Look closely at a dead rodent and you will find certain beetles that specialize in such food. Earthworms eat soil, digesting nutrients from organic material in the soil, and passing the residue on to microbes. Termites eat dead wood. Detrital food webs are important components of ecosystems. Well over half of all plant material is not consumed by herbivores, but rather ends up being broken down by detritivores. It's good that they do this. Imagine for a moment an ecosystem with no decomposers or scavengers.

What role do humans play in detrital food webs? Do we ever feed here? (Hint: The beef we eat is "aged." What does that mean? What are microbes really doing when they turn milk to cheese and grapes to wine?) How are the interactions of populations discussed in Chapter 14 (competition, predator–prey interactions, symbioses) related to trophic levels?

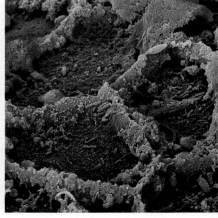

(a)   *(b)*   *(c)*

Figure 15-11. The detritivores, generally not widely admired by humans, play crucial roles in all ecosystems. They derive nutrients and energy by consuming other organisms that have died. In the process they recycle nutrients, a vital process in any ecosys-tem. (*a*) A turkey vulture dines on a dead mammal. (*b*) Mush-rooms find organic matter in the soil nourishing. In doing so they compete with bacteria (*c*), here growing on leaf litter (magnified 3240 times).

15.3.2 Some Ecosystems Are Organized Vertically

A forest ecosystem can be described as a series of layered biological communities (Figure 15-12). Unlike trophic levels, which are conceptual layers, those in a forest are physically real. Organisms in the **canopy**—the sometimes interlocking branches of the tallest trees—are likely to be different from those in the **subcanopy**—smaller trees whose tops form a layer of branches below the canopy. Lower down, a layer of **shrubs** may overlay the forest floor, or **surface layer**, which overlays the **subsurface layer** of soil. Each of these layers has plants or plant parts different from those of other layers. Each layer, too, may house a unique assemblage of animals. In the tropics, extensive communities of plants and animals are restricted to the canopy, whereas other communities are found in other layers. For regions outside the tropics, generalizations are harder to make. Animals may wander widely from layer to layer. In temperate North America, squirrels are typical canopy mammals. But they feed extensively on the forest floor and in open areas, albeit never far from trees. Mice and foxes are generally considered creatures of the ground. But white-footed mice and gray foxes climb extensively in the temperate forests of North America.

There are more differences than just the organisms. Canopies have direct access to sunlight, high and variable temperatures, and wind. In lower layers, sunlight is generally filtered, temperatures are more moderate, and wind is nearly nonexistent. On or beneath the forest floor, nutrients are readily available; in the canopy, nutrients come at a premium. Layering in forests is more than a community phenomenon; it applies to the entire ecosystem.

Study a pond or lake and find similar organization in which biological communities are restricted to particular regions. Here, physical factors are even more important in determining where organisms live. Light is particularly important. Whenever light strikes a watery surface, some is reflected. Hence, the amount of light available to organisms in aquatic environments is always less than that available to surrounding terrestrial environments. Some light penetrates the surface. How deep it goes depends on several factors. Some light is absorbed by water and is transduced into heat, at which point it ceases to be light. Even in water with no impurities, light intensity decreases with depth, and finally there may be a depth below which no light penetrates. Nutrients in solution and particles floating in water also absorb light, limiting its penetration.

Light determines where aquatic plants and other photosynthetic organisms live. Near shore, light penetrates to the bottom. Here, in the **littoral zone**, relatively large plants, anchored to the bottom, thrive. This is the best of all worlds for lake and pond plants. Nutrients are relatively abundant in the bottom muds while oxygen is abundant in the

Figure 15-12. Many ecosystems are layered, as is most easily seen in forests. The topmost layer, the forest canopy, is made up of interlocking branches of the community's tallest trees. Below the canopy are, in turn, the subcanopy, shrub, surface, and sub-surface layers. Each layer provides habitat for a subcommunity.

Figure 15-13. Marine ecosystems are complex. Typically, they are viewed as a series of zones rather than layers as in terrestrial ecosystems.

water. The downside of living in the littoral zone? Competition is usually high and predators are abundant. Away from shore, light penetrates surface waters but does not reach the bottom, creating the **limnetic zone**, in which the producers are algae that soak up sunlight and nutrients and produce oxygen. They feed primary consumers, which are typically microscopic or slightly larger invertebrates, which, in turn, are the food of secondary consumers (usually fish). The tertiary consumers are the larger fish, reptiles, birds, and mammals. In a deep enough lake, below the limnetic zone is the **profundal zone**, an area of no light, often rich in nutrients and low in oxygen. Under these conditions plants cannot grow. Animals are scarce, being mostly fish seeking respite from the dynamics of the rest of the lake. A lake's bottom and subsurface are the **benthic zone**, even richer in nutrients and poorer in oxygen. Here detritus from other regions accumulates, feeding an often rich diversity of scavengers and decomposers.

Similar light- and depth-dependent zones and regions are found in marine ecosystems. Their complexity is even greater than that of terrestrial and freshwater ecosystems (Figure 15-13). This should not be surprising when you remember that marine areas constitute over three-fourths of Earth's surface.

15.3.3 Biological Communities Change from Time to Time and Differ from Place to Place

**Fast Find
15.3h
Succession**

When viewed carefully, biological communities in terrestrial environments give an appearance of great diversity. Next to a marsh may be an upland meadow. Next to that may be several stands of trees. One stand may be mainly oak trees. A stand of maples gives way to a stand of beeches. Patchworks of biological communities cover many terrestrial areas. What causes this pattern to develop?

For one thing, biological communities constantly change. A fish dies, a flower germinates, a stream changes course. The changes that occur in one area may not be the same as changes occurring in a neighboring area. This contributes to the patchwork pattern of communities. Larger, more significant changes also occur. Some take immense amounts of time and are somewhat predictable.

Let's imagine we can watch an area that a glacier moves across, scraping away all plant life and soil. Then the glacier retreats, exposing bare rock completely devoid of life. At first, nothing grows in this hostile environment. The physical conditions are too extreme. Seasonal and daily temperatures vary widely. When sun shines on bare rock, temperatures soar. A cloud blocks the sun and temperatures plummet. Temperatures fall even further at night. Water conditions are also extreme. During and right after a rain, water is plentiful. But there is nothing in bare rock to hold water. It runs off freely, and within a few minutes after the rain stops, bare rock is dry. The wind blows unhindered across the rocks surface.

Few forms of life can tolerate conditions found on bare rock, but there is one type that can. Lichens are curious organisms, indeed. As we learned in Chapters 8 and 14, they are really two forms of life living in an intimate association called **mutualism**. In lichens, algae live in association with fungi. Each brings to the relationship characteristics the other lacks. Algae, being photosynthetic, can trap and store the energy of sunlight. Fungi produce powerful enzymes that etch rock, liberating and utilizing its nutrients. Lichens are unusually tolerant of extreme environmental conditions. If things get too tough, they simply shut down metabolically and wait until conditions improve. Lichens are uniquely suited for life in the bare rock environment. After a while, bare rock supports a community of lichens, their spores brought in by wind or water.

Animal life is sparse in lichen communities. A few spiders may be the only permanent residents. They depend for food on insects blown in from neighboring areas. An occasional bird may fly in, briefly seeking insects or spiders.

With time, the environment changes. Rapid and frequent heating and cooling causes rocks to expand and contract. Sooner or later, they crack. Water speeds the process because when water freezes it expands. Water, caught in cracks, freezes and exerts great pressure. The rocks crack more. Big rocks are broken into smaller ones, and so on, until some rock is pulverized. Rock fragments and powder gather in cracks, mixing with nutrients leeched by lichens, becoming a primitive soil. Even primitive soil holds water, moderates temperatures, and shelters organisms from wind.

The stage is set for a new community. Mosses grow on primitive soil. Their spores can be present for years, but as long as environmental conditions lie outside their range of tolerance, they do not successfully germinate. Perhaps they await a key nutrient or what is for them enough water or the right range of temperatures. Conditions among cracked rocks and lichens change, and when they are proper, moss spores germinate and grow.

The moss grows as a tangled mat of vegetation that soon spreads beyond the cracks. The mat spreads laterally, covering lichen, blocking access to sunlight, and replacing the lichens as the dominant form of life. Moss mats may completely cover rocks and further moderate the physical environment. Moss mats hold even more water than primitive soils. Within the mat, temperature changes are less extreme, and winds are blocked. When winds are stopped, they drop whatever debris they might be carrying, thus enriching the mosses' soil. As mosses grow and die, nutrients brought in with the wind are mixed with those liberated from rocks by lichen and those built by mosses. A more advanced soil develops.

The number and variety of organisms tolerant of the new conditions grow. Insects and other invertebrates feed on moss and on each other. Primitive food chains and webs develop. Again, the community is changing.

Eventually, conditions become tolerable for flowering plants to germinate. A community of grasses and wildflowers appears. These plants find the nutrients they need in the soil but must be tolerant of full sun and, as adults, wind. Their roots penetrate deeply, jam into cracks, further loosening rock fragments, and liberating more rock nutrients. Aboveground, their presence creates shade at soil level and further slows the wind. Their presence sets the stage for still another community, usually one in which shrubs are the dominant plant life. In time the shrubs set the stage for a community of small trees, which in turn may become a community of larger trees, which often changes, sometimes several times, into different communities of even larger trees. Each of these communities has its own assemblage of animals.

In **ecological succession**, one biological community replaces another, triggering changes in the physical and biological environments that set the stage for still another community (Figure 15-14). Where does the process stop? Ecologists are not entirely sure. They used to speak of "climax communities" as the imagined endpoint of the process. More recent work suggests that perhaps the process never ends. Biological communities may continue to change indefinitely, albeit at an ever slowing rate of change.

To see succession, it is not necessary to scrape ground all the way to bare rock. A serious fire may kill all of the trees and undergrowth in a forest. After the fire, the first community to appear is often one dominated by wildflowers. Succession starts there and proceeds through predictable steps to the stage of forest that burned.

At least this is what usually happens. Sometimes the wholesale removal of a complex community so changes environmental conditions that recovery is not possible. This is especially true in semiarid lands. Forests and grasslands pump water from the soil through their tissues and into the atmosphere by a process called **transpiration** (see Chapter 12 for a review of the process). This atmospheric moisture forms precipitation. In effect, the plant community is creating its own rain. Removal of the vegetation eliminates this important source of water. Succession can't replace plants without moisture, and it can't replace moisture without plants. Instead of repeating the old successional pattern, the region slips into a new one based on arid conditions. What was once forest or grassland may become a desert.

Note the tight interplay between organisms and their physical environments. The specific steps of ecological succession depend on the region. Even areas with seemingly similar climates may have different mixes of species in their communities. Thus, the species encountered in temperate regions of Europe differ from those of temperate regions of North America.

And what happens in grasslands? Early successional steps are similar. Lichens pioneer and are replaced with moss and then grasses. But here the process seems to stop.

(a)   *(b)*   *(c)*

(d)   *(e)*   *(f)*

Figure 15-14. Ecological succession is a generally predictable series of biological communities that replace each other in a particular area. Each step in the series is a separate community. Here is a representative series from northeastern North America. *(a)* Lichen have the unusual ability to grow on bare rock. Here lichen are being replaced by moss. *(b)* Eventually a carpet of moss will overgrow and replace the lichen community. Here mosses are being replaced by ferns. *(c)* In most terrestrial ecosystems grasses and other herbaceous plants are the third stage in ecological succession. *(d)* A community of small, woody plants— shrubs or brush—replaces grasses. *(e)* The shrub community is replaced by small trees. *(f)* The sixth stage of ecological succession is a mature community of large evergreen trees, mainly spruces. How might the stages of ecological succession be different in a grassland region?

Figure 15-15. Mountains show another example of ecological succession, based on altitude. Factors in the physical environment account for the differences in biological communities of alpine succession. What are these factors?

In the Great Plains of North America, a combination of dryness and frequent range fires stops succession at the grass community. An even more extreme case is found in the high Arctic of Canada, where bare rock with only a few lichens never gets beyond the first step of succession.

There are several other types of ecological succession. Consider a tall mountain (Figure 15-15). At the summit is mostly bare rock with some lichens. Below is a zone of moss, then a belt of dense shrubs, then one of small trees, followed by larger trees. The valleys are filled with luxuriant forests. Go to a pond and notice the following sequence: submerged vegetation, emergent vegetation, cattails, bushes, small trees, and finally large trees. Watch a beaver pond for many years as it changes from an open pond to a bog to a wet meadow to a swamp to a community of small trees to a community of large trees.

**Fast Find
15.3h
Succession**

So, back to our original question. Why is the land covered with a patchwork of biological communities? There are at least four reasons. First, biological communities change over time and may progress through sucessional stages. Second, every area has an individualized local history. Third, physical conditions may differ between adjacent areas. Finally, some contribution to the patchwork pattern is made by random chance. For example, which tree species starts or succeeds in a particular area is sometimes a matter of chance; that is, whichever species of seed happened to be dropped there grows. Often, one tree becomes a stand of similar trees as the first tree drops seeds at its own base. The stand of trees becomes a biological community, slightly different from neighboring communities where other species happened to grow.

So far, we have looked at patchwork patterns of terrestrial biological communities on a local scale. They can be seen in your own backyard. But such patterns can also be seen on a more global scale.

15.3.4 Biomes Display Global-Scale Variations in Biological Communities

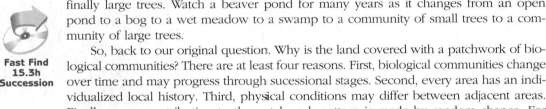

**Fast Find
15.3g
Biomes**

If you see one desert you've seen them all. On the other hand, no two deserts are exactly alike. Only in ecology are two such seemingly opposite statements equally possible. A person familiar with deserts in the southwestern United States would feel at home in the deserts of North Africa or those of central Australia. All deserts are extremely dry. Temperatures vary widely from moment to moment, day to day, and season to season. In all deserts, life is sparse. But look more closely and you see differences. Trees and flowers are not the same species. Wallabies in central Australia are not likely to be confused with jackrabbits in southwestern North America. But they do share some similarities. Both have long hind legs and hop. Both are sand colored. Both tolerate dryness well. Deserts everywhere have plants that share many of the characteristics of cacti, but species differ from desert to desert. As a first approximation, all deserts are the same. It's in the details that they differ (Figure 15-16).

(a) *(b)* *(c)* *(d)* *(e)* *(f)*

Figure 15-16. Biomes of the World. At first glance it may seem that "if you've seen one desert you've seen them all." Certainly, deserts in North Africa resemble those of southwestern United States, which in turn resemble those of Mongolia. But more detailed analyses reveal significant differences. The animals and plants of different deserts, while superficially resembling each other, are usually of different species. The same generalizations are true when comparing specific regions within other biomes—large geographic areas that have similar climates and apparently similar plants and animals. The biomes pictured here are (*a*) tropical rainforest, (*b*) temperate deciduous forest (*c*) taiga, or coniferous forest, (*d*) tundra, (*e*) grasslands, (*f*) desert.

Much the same can be said for temperate forests in North America, Europe, and Asia. Temperate forests everywhere are grossly similar. So are the tundra regions of North America, northern Asia, and the Palmer Peninsula of Antarctica. So are a plethora of grasslands everywhere.

Vast geographic areas that have similar climates and similar biological communities, that is, similar ecosystems, are called **biomes**. Section 15.3 of the CD shows some of the world's great biomes. Note that biomes tend to occur in vast belts girding the Earth at similar distances from the equator. Why? Each biome has its own characteristic climate. Different species of plants and animals in different regions of a given biome may resemble each other and have similar tolerance ranges for environmental conditions.

Piecing It Together

1. Biological communities and ecosystems are complex. But in these seemingly chaotic assemblages of organisms, factors, and processes ecologists recognize several levels of organization.

2. Organisms within a biological community are organized functionally into trophic levels, namely producers, such as plants or algae, followed by primary consumers, secondary consumers, and so on.

3. Generally, only about 10 percent of the energy produced by one trophic level is available for the next higher level. Thus, there is insufficient energy to support more than about four levels in terrestrial communities, sometimes more in highly productive freshwater or marine communities.

4. Biological communities are often organized vertically. In terrestrial communities, the amount of moisture, wind and light are defining factors. Freshwater and marine communities are organized into zones rather than layers. Here, water depth and light conditions are of prime importance.

5. Biological communities constantly change as a result of ecological succession, variations in local history, differences in physical conditions, and random chance events.

6. Within biomes, physical conditions are similar on a global scale. Organisms living in a given biome have similar tolerance ranges and adaptations. Often they share characteristics. Thus, superficially at least, ecosystems within a given biome resemble each other.

15.4 HOW DO HUMAN ACTIVITIES AFFECT THE ENVIRONMENT?

In the relatively short span of 300,000 years, modern humans have become one of the dominant species on Earth. One key to our success is that we are extremely clever at re-creating, and carrying with us wherever we go, key elements of the environment in which we evolved. We are a tropical grassland species. It is in the tropical grassland biome that we are most comfortable. In more scientific terms, in the tropical grassland biome, environmental values are optimal for many of our tolerance ranges. So when we move into a new area, what do we do first? Clear the trees and grow grass. In deserts, we bring in water and grow grass. In grassland biomes, we take away native species and replace them with domesticated grasses. New housing developments everywhere are invariably

Figure 15-17. In our cars, offices, homes, and even within our clothing, humans re-create the environmental conditions of a tropical grassland biome. We strive for optimal temperatures that are stable and warm but not too hot, humidities that are low, and long periods of bright light.

grasslands, devoid of trees. Only after several years, when the community becomes established and feels safe and comfortable, are trees planted.

Wherever we are, we spend considerable resources, time, and effort re-creating optimal, tropical grassland temperatures. Inside our homes, workspaces, cars, and clothing, we work hard to get and maintain environmental temperatures close to 22°C (Figure 15-17). Get in the shower and notice how narrow is our range of tolerable temperatures. The difference between too hot and too cold is just a few degrees and an infinitesimal turn of the spigot.

As a result of our ability to create environments with optimal conditions, humans live almost everywhere in the terrestrial environment. As we saw in Chapter 14, our numbers have grown nearly exponentially, and if the end of our growth is in sight, it is barely so.

15.4.1 Humans Are Dominant in Today's Ecosystems

Our importance to ecosystems goes beyond mere numbers. We are generalists and opportunists. We not only live nearly everywhere but use nearly everything we find. We are catholic in our choice of foods, building supplies, clothing fabrics, and raw materials for tools. We are resource hungry. We interact directly with most organisms. With some we have established mutualistic relations (crops and pets, for example). We attack competitors with a vengeance, be they insects eating our crops or alien plants claiming our yards (to watch a typical suburbanite go after dandelions is to see brutality). Our major predators are currently either extinct or reduced in numbers and under control.

Our impact on Earth's ecosystems is extreme and growing.

At the same time, we are the only species conscious of what we are doing. We are unique in our understanding of consequences. We, more than any other species, can control our actions, plan, and make decisions, collectively and individually. Show a cat a mouse and it will kill it. It will eat the mouse, if hungry, play with it if not, ignore the carcass perhaps. Cats choose between options, but make no conscious choices. But aren't we different?

Our populations are huge and most are growing. So is our need for resources. So, too, is our understanding that uncontrolled growth and resource use can create problems. We also understand that problems require solutions, carefully thought out and fairly implemented. We have special traits and characteristics that give us special responsibilities. We didn't ask for them. We don't always want them. They are not always comfortable or easy to deal with. But they are ours nonetheless, demanding recognition, consideration, and effort.

15.4.2 Science Plays Only a Limited Role in Solving Environmental Problems

Unfortunately, many solutions to environmental problems are not scientific. Scientists, including ecologists, do best with problems that can be precisely defined, about which observations can be gathered and analyzed, and that end with solutions that explain and predict how nature works. Often, environmental problems show none of these characteristics. Solving many environmental problems requires changing human behavior. As we discussed earlier in the chapter, solutions to environmental problems are often moral, cultural, or political; they are seldom scientific. Science can only define the options and implications. This text deals with biology. It can be argued that its scope, like that of science generally, should stop with options and implications. In subsequent

(a) *(b)*

(c) *(d)*

Figure 15-18. Certain human activities cause significant environmental problems. Pictured here are (*a*) land being clear-cut; (*b*) smokestacks polluting the air; (*c*) raw sewage polluting a waterway; and (*d*) a modern landfill.

pages we will, of course, discuss the scientific aspects of various environmental issues. But to stop there is to create an aura of pessimism. Focusing exclusively on the definition of problems leaves the impression that there are no solutions. This can be depressing. On a global scale, there is a widespread feeling among concerned citizens that the environmental situation is hopeless. This is patently not true.

Recently, environmental pessimists have been derisively called "doom and gloomers." Some are "doom and gloomers" out of ignorance. It is easier to focus on problems than to grapple with solutions. Others are scientists unwilling to venture out of their areas of expertise. This is certainly safe and honest. But it contributes to a populationwide malaise.

The fact is that all environmental problems have solutions (Figure 15-18). The solutions may not be easy to understand or implement, but they are there. It is not so much a question of "can we solve these problems" but rather "will we" and "when." With some trepidation, then, we want to go beyond understanding environmental problems to suggesting possible solutions. Please keep in mind that solutions are often controversial and outside the domain of traditional science. It is not our intent to say, "This is the way to solve it." Rather, we are saying "It could be solved this way." Feel free to disagree. We challenge you to disagree and to offer alternatives. We invite you to engage in discussions. We invite you to become knowledgeable, opinionated, and passionately involved.

15.4.3 Can We Come to Grips with Growth of the Human Population?

In some respects, human populations are no different from those of other species. Births, deaths, immigrations, and emigrations are key factors in shaping their growth and development. As with other species, population growth is tempered by environmental resistance. What is unique to humans is the ability to control their environment. We, more than any other species, shape and change our immediate physical surroundings. To an even greater extent, we control our biological environment. We punish our competitors, predators, and parasites into submission and, at least in the economically advantaged societies, have practically eliminated them from everyday life. Worldwide, we reward our positive **symbionts**. Even our commensals thrive, often at the expense of other species. As a result, the growth of human populations is somewhat unique among species. Whereas other species' populations plateau at or near their **carrying capacity**, ours continues to grow exponentially. The human population worldwide at the end of Paleolithic times, roughly 12,000 years ago, is thought to have been around 1,000,000. This is an estimate, of course, based on signs left behind (old campgrounds, hunting sites, kitchen middens, tools, weapons, and the like) and our understanding of how many hunters and gatherers the environment could support. For most humans, a life style that had sustained our species for well over a million years ended with the Paleolithic Age. When things went well, life for the hunter-gatherer was good. When game and food plants were plentiful, they ate well and thrived. Typically, hunters and gatherers ate a surprisingly diverse and balanced diet. Unfortunately, things did not always go well. What if game did not show up? What if food plants did not grow? Suffering and starvation were often the result. Humans like to feel in control. Perhaps hunting and gathering was too much at the mercy of conditions beyond human control or understanding. A new kind of lifestyle was needed.

Apparently first in the Middle East, humans discovered that edible seeds put aside from winter food stores, sown in spring, and nurtured through summer could be harvested in fall. A more reliable food source was obtainable when haphazard gathering was replaced by more deliberate sowing and harvesting. Prices were paid, of course. People became more or less tied to the vicinity of their fields. The number of plant species that could be brought under human control was somewhat limited. Hunting, which continued at least for a time, typically exhausted nearby sources of game. Initially, the advent of agriculture was accompanied by an immediate decrease in the diversity of foods consumed. The bones of many early agriculturists show signs of malnutrition probably related to the people's eating fewer foods. Still, the advantages of agriculture so far outweighed the disadvantages that it quickly spread worldwide. Only the kinds of grass cultivated as cereal crops differed: wheat in the Middle East, Africa, and Europe, rice in Asia, and maize in the Americas. As time passed, the number and diversity of domesticated crop species increased. Animals were added to the list: dogs, goats, sheep, cattle, chickens, horses, pigs, and llamas. Agriculture resulted in an overall, worldwide increase in human numbers. As agriculture spread, the human population slowly increased over the next 2,000 years to a new, stable level of roughly 5,000,000.

Agriculturists were still at the mercy of the environment. With too much rain or too little, crops failed and suffering returned. If conditions got too bad, a community could simply pick up and go somewhere else—into the next valley, perhaps. That was fine, as long as the next valley was unoccupied. As the human population grew, more and more valleys filled. Conflicts between neighbors increased. Productivity of cultivated fields decreased. The stage was set for another change in human society.

The Industrial Revolution brought great changes in the 1700s, especially in Europe. Large numbers of people began living in cities where their chief occupation was to manufacture goods, most of which were not used locally, but exported to other regions. Worldwide trading was nothing new; there was extensive and widespread trading well before Paleolithic times. But by the 1700s, transportation methods had improved, and the extent of trading intensified and distances over which goods were moved increased. Those who lived in cities became totally dependent for food on others in neighboring regions who were still agriculturists. But there was a difference. Agriculturists were not just producing their own food. Excess food was traded for goods manufactured in the cities. Extensive symbiotic relationships developed between peoples—a pattern that continues and intensifies today.

And the number of people soared (Figure 15-19).

◆ In the early 1700s there were around 500,000,000 people worldwide.
◆ Somewhere around 1830 we reached our first billion.
◆ By 1910 we had added another billion.
◆ In the year 2000 there were 6 billion people, worldwide.

More and more people beget more and more people, and the process continues. In the time it takes to read this sentence, the world's human population will increase by 17.

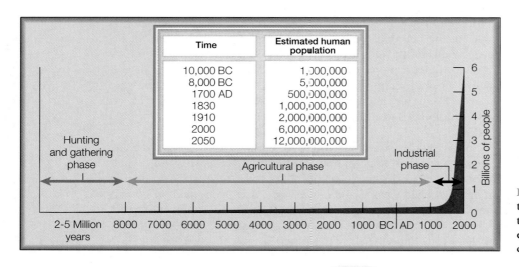

Time	Estimated human population
10,000 BC	1,000,000
8,000 BC	5,000,000
1700 AD	500,000,000
1830	1,000,000,000
1910	2,000,000,000
2000	6,000,000,000
2050	12,000,000,000

Figure 15-19. In recent centuries, the worldwide growth of the human population has been exponential. What is our carrying capacity?

That's not just births. That's births minus deaths equals an increase of 17. Read the sentence again as another 17 people join our ranks.

◆ Every three hours we add the equivalent of a new small village to our numbers.
◆ Every year we add 90 billion new people to the world population.

People are not evenly distributed worldwide. Most humans live in the tropics, in countries economically described as "underdeveloped" or "developing." The human population is increasing rapidly in these countries. Women living in Europe bear an average of 1.5 children (the replacement value is 2.1). In Kenya that average is 8.1. Compare the age structure graphs in Figure 15.20 and guess where the populations are going. An age structure graph of the human population worldwide more closely resembles that of Kenya than that of Austria. It is estimated that the world's human population will double again by the year 2050.

What is our carrying capacity? When will the growth curve of our population begin to resemble that of other species? What factors of environmental resistance will come into play to change our growth from an exponential to an S-shaped curve? These are among the most important unanswered questions of our time.

How might our population growth be controlled? What options can we consider?

Increased Immigration

Traditionally, immigration has been an option of solving our population growth problems. When one valley became too crowded we moved into the next one. When one continent became too crowded, we moved into another one. The move was not always smooth if the next valley or continent was already occupied. In today's world, most valleys and all continents are already occupied. Mass migrations of peoples from the tropics to the temperate regions—South Americans into North America, for example, or Africans into Europe—could equalize distributions, but such movements are not politically popular nor likely to occur.

Technological Increases in Carrying Capacity

In the past, with each new colonization, humans increased their carrying capacity as they expanded their range from a single continent (presumably Africa) to the rest. Con-

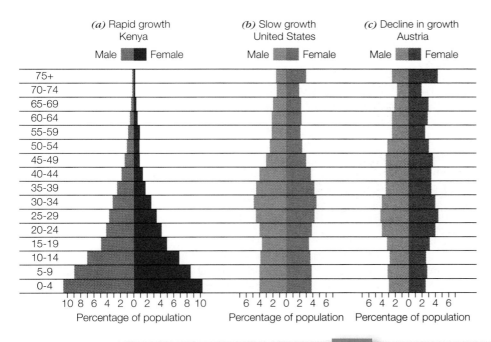

Figure 15-20. Analysis of age structure graphs allows predictions as to future growth patterns of populations. Populations with high proportions of young (*a*) can be expected to grow, as young mature, become reproductively active, and have offspring. Populations with smaller proportions of young may be expected not to change (*b*) or to decrease in size (*c*). What other factors might upset these predictions?

sistently, technology played an important accompanying role in these expansions. Those that possessed the most advanced farming methods and weapons were the ones who most easily colonized new regions. In other words, the immigrations discussed in the previous paragraph at least partially depended on advances in technology.

Why not continue to lean on technology? Even in today's world, humans occupy only about three-fourths of Earth's surface, and beyond Earth is a whole universe of potential planets. Can't technology save us again? Possibly not. In prehistoric times, technological breakthrough involved improved spear points or seed stocks. Today, the required technological breakthroughs are infinitely more expensive. More importantly, those technologies do not now exist, while the problem is immediate. Remember, in the time it takes to read this sentence the world's human population will increase by another 17.

What about technological breakthroughs that would improve food production? If we can produce more food, we can support more people; this would expand our carrying capacity. Indefinitely? Why not? Malthus's predictions of dire consequences have not proven accurate because of his underestimation of advances in agriculture. Even in today's world, starvation is caused not by our inability to produce enough food, but in our inability to distribute it equitably. Optimists tell us that our ability to produce food and other essentials will always grow as our needs grow. Pessimists tell us that we are already approaching and have perhaps exceeded our technological limits. Moderates tell us that current and future advances in technology will buy us much-needed time to seek out and implement more permanent solutions.

Increased Mortality

For most species, the increases in environmental resistance that result in population stability most often involve increases in mortality. Certainly, increased human mortality, on a worldwide basis, could control our population. We could, for example, encourage war, disease epidemics, and widespread starvation. But doing so would go against human nature. Humans have gone a long way to remove or control predators, parasites, and diseases, although increases by disease-causing organisms in immunity to drugs and in virulence suggest that diseases may fight back. Malthus's predictions that pestilence, starvation, and war would control human populations may prove to be accurate, but they are not our solutions of choice.

Decreased "Little r"

We are the only species that has not only the capacity to understand the forces that shape our destiny, but also the capacity to control and change that destiny. Unlike any other species, we can control our ability to reproduce. Decreasing our "little r" is an option available to us. Indeed, the birth rate has already decreased in some countries, notably in Europe, and it may be decreasing in North America, China, and Mexico. Could birth rates decrease in the future in other countries? For this to happen, experts tell us, the following changes are required (listed in descending order of effectiveness):

◆ Increased levels of education.
◆ Increased economic opportunities.
◆ Increased female empowerment.
◆ Accelerated cultural change.
◆ Increased availability of birth control methods.

Note that these are not technological or scientific changes. They are largely economic, political, moral, and cultural changes. Without exception, they are difficult to implement. Still, these possible changes are at hand. Awareness of possible changes, hope, and resolve are the first steps toward a solution. Meanwhile, in the time it takes to read this sentence, the world's human population will again increase by 17.

The Human Population Revisited In Chapter 14 you were asked to visit web sites and determine the current size of the human population. Revisit those sites. How has the human population changed in the few weeks since you last visited these sites?

15.4.4 Are Human Activities Poisoning the Earth?

The heading of this section is an example of the provocative rhetoric often encountered in discussions of environmental issues. Phrases such as "poisoning the Earth" inflame passions and arouse emotions. This is not always bad. This is the way many of us get the adrenaline that spurs us to action. But inflamed passions and aroused emotions are not always useful when seeking solutions and effecting change. Effective action requires an approach that is calm and thoughtful and has the right mix of passion and reason.

Certainly, water and air pollution are serious problems—they are civilization's downside. So many people consuming so many resources and inadvertently producing so much waste is a monumental problem. In many cities, air quality has deteriorated to the point where eyes burn and those with susceptible health problems, such as asthma, are at risk. Water quality, too, is often a problem. Rivers and lakes once teeming with fish have become barren. Waterways once supportive of recreation have become unhealthy and unusable. Are we changing the very chemical nature of our global ecosystem? We seem to have that capability. We also have the capability to reverse ourselves as the example in Figure 15-21 demonstrates.

Figure 15-21. In 1969, the Cuyahoga River, highly polluted with oil, caught fire and burned in Cleveland, Ohio. During the eight days it burned, two bridges were destroyed. Today, after extensive recovery efforts the river is once again clean.

(a)   *(b)*

Dirty Water How serious is the water pollution problem in North America? Is it a problem in the area in which you live? What can be, or is being, done to clean up our waterways?

Let's review another example in which such reversal may be underway. Up until the 1950s, air conditioning in summer was a longed-for dream throughout much of the civilized world. The theory of air conditioning was relatively simple and had been understood for nearly a century: expanding fluids cool and contracting fluids warm. If we could bring fluids into a warm room and allow them to expand by evaporation, they could cool the room. If the fluids were then transported outside and condensed, the heat they carried could be dissipated to the environment. The fluids could then be cycled back into the room and reused. In effect, heat would be diverted from the room to the environment. To minimize costs, a fluid was needed that evaporates near room temperature. One was known—ether—but it is extremely explosive.

Then in the 1950s, chemists discovered chlorofluorocarbons, or CFCs. These substances have many of the evaporative properties of ether and were at the time thought to be virtually nonreactive. So, an industry was born. Within about a decade, air conditioning became a reality. CFCs, substances once unknown in the natural environment, became commonplace. Over the years, more and more were produced and used. Inevitably, some leaked or were allowed to escape into the environment. No problem. Remember, they were thought to be nonreactive. Having relatively low density, they rose. Out of sight, out of mind. Eventually they gathered near the top of the atmosphere and, unfortunately, participated in an unexpected reaction. CFCs react strongly with ozone, a molecule consisting of three atoms of oxygen.

A layer of ozone high above Earth's surface is essential to life on Earth as we know it. Ozone absorbs much of the ultraviolet (UV) radiation streaming in from the sun. It is good that it does so. In humans, UV rays cause suntan, sunburn, and, in some, skin cancer. Its effects on other organisms are less well understood, but they can be serious, even life threatening.

Unknown to anyone, CFCs began to reduce the ozone layer, particularly around the South and North Poles, increasing the amount of UV radiation reaching Earth's surface (Figure 15-22). Within a few decades effects began to impinge upon the quality of life for many humans. Skin cancers increased. For many humans in temperate regions, one of

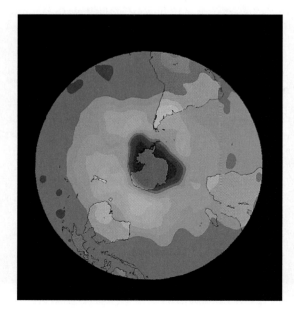

Figure 15-22. In the 1970s, scientists discovered that chlorofluorocarbons (CFCs), widely used as refrigerants, were degrading the ozone layer. Once released into the atmosphere at ground level, CFCs float upward, eventually reaching the stratosphere. Here, they react with and destroy ozone, particularly in polar regions. The ozone layer is vitally important, as it absorbs ultraviolet rays streaming in from the sun. In humans, excess exposure to ultraviolet rays is a leading cause of skin cancer. Since 1989, governments of the world have significantly reduced the use of CFC, and the process of ozone degradation may be reversing.

the joys of summer had been to "soak up some rays" at the beach or in the back yard. No more. We are warned to restrict exposure to direct sunlight and protect our skin with blocking agents.

Without a doubt, here was a problem of monumental proportions. But then a remarkable thing happened. Governments of the world called for a world congress to discuss the problem and its solutions. In 1987 the Montreal Protocol was signed by 27 countries, including the United States, restricting the production, use, and handling of CFCs. Alternatives for the air-conditioning industry were sought, found, and produced. CFCs released before the Protocol was signed will continue for years to float up into the ozone layer, but remarkably, we may be seeing an abatement of the problem. Some evidence suggests that the rate of deterioration of the ozone layer is slowing down. Indications are that fairly early in the 21st century the problem may completely abate. If so, sunbathing may once again be safe. We may well look back on the "hole in the ozone layer" as an early environmental success story.

Maybe we *can* solve environmental problems, if we try.

Holely Ozone How did the governments of the world come together and agree on limiting use of CFC? How did the refrigeration industry respond? Currently, is the hole in the ozone layer getting bigger? Staying the same? Getting smaller?

Dirty Air Is air pollution a serious problem in North America? Is it a problem in the area where you live? What can be, and is being, done to clean up our air?

15.4.5 Will Human-Caused Global Warming Change Earth's Climate?

In an earlier section, we discussed the ecological importance of light and heat to organisms and biological communities. These factors are also significant on a global scale. Sunlight streams into Earth continuously. At each place on Earth, light alternates with dark, but Earth as a whole is always receiving sunlight. Light equals heat. When the sun shines on water, rocks, soil, plants, and gases in the atmosphere, light is transformed into heat. At night, heat is transformed into infrared radiation (IR) and lost to space. We measure heat as temperature.

As Figure 15-23 shows, some of the incoming light never becomes heat. About 30 percent is immediately reflected, mainly off clouds, and has no effect on the temperature of Earth. An additional 20 percent is absorbed by gases in the upper atmosphere. These are briefly warmed, but they quickly lose heat to space. Only about half of the incoming light reaches near enough to Earth's surface to affect its temperature. Any warm body loses heat to a cooler environment in the form of IR. Particularly at night, the warmed Earth loses heat to cold space. So, energy from the sun approaches Earth as light, is absorbed by matter, is transformed into heat, raises Earth's temperature, and is then lost to space. There is another important factor. IR radiation on its way out may be absorbed by atmospheric gases. If so, it is again transformed into heat and adds warmth to Earth.

Here we encounter a delicate balance. If more energy is absorbed, or if less is lost, Earth's temperature rises. Conversely, if less is absorbed than lost, Earth cools. Key factors in this equation are those atmospheric gases that reabsorb IR radiation and add warmth. They include water vapor, carbon dioxide, methane, and a host of others and are called greenhouse gases for reasons about to be explained.

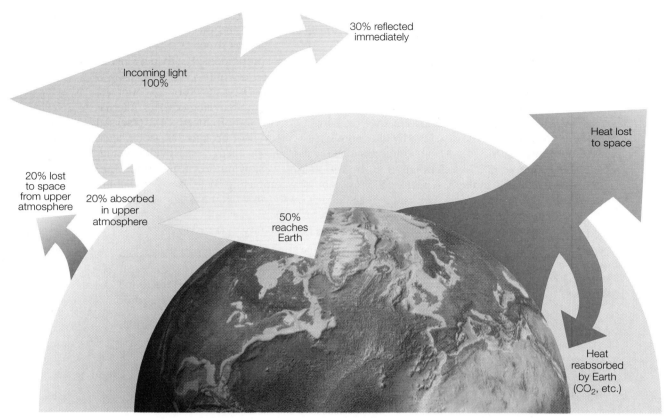

30% reflected immediately

Incoming light 100%

Heat lost to space

20% lost to space from upper atmosphere

20% absorbed in upper atmosphere

50% reaches Earth

Heat reabsorbed by Earth (CO_2, etc.)

Figure 15-23. This diagram summarizes how Earth receives light and thus heat from sunlight. Light constantly streams to Earth from the sun (yellow arrows), is transformed into heat, and eventually is lost to space as infrared radiation (red arrows). Of the incoming light, 30 percent is immediately reflected back to space mainly off of clouds. Another 20 percent of the incoming light (red arrow at the left of the diagrams) is absorbed by and briefly warms the atmosphere before being lost to space. Only 50 percent of incoming light is absorbed by Earth's surface (water, rocks, soil, plants, and so on),is eventually transformed into heat, then transformed into infrared radiation, and lost to space mainly at night. The small red arrow at the right of the diagram represents the infrared radiation that is reabsorbed by greenhouse gases. This energy increases the temperature of the atmosphere, but it is also eventually lost to space. A significant increase in the concentration of greenhouse gases could increase Earth's overall temperature. Many scientists believe that recent increases in these gases, mainly resulting from human activities, could lead to global warming in the 21st century. What factors, natural and human associated, contribute to increased levels of these gases? Can the process be reversed?

Water vapor in the atmosphere is by and large beyond human control. Review the water cycle and note that the amount of water vapor in the atmosphere is self-regulating. The amount of water vapor in the atmosphere should be more or less constant. If more evaporates, more precipitates. Of course, how much water evaporates depends on temperature. This, too, may be self-regulating. If Earth's temperature rises, more water evaporates. This increases cloud cover, which increases the reflection of incoming sunlight, which decreases temperature.

The amount of other greenhouse gases may be more under human control. Carbon dioxide, for example, is added to the atmosphere every time something burns. Almost all human activities burn something. So, as our numbers and our dependence on fossil fuels (coal, oil, gas) increase, so does the amount of carbon dioxide in the atmosphere. In similar ways resulting from human activities, the other greenhouse gases are being added to the atmosphere.

Notice that Earth's atmosphere works something like a greenhouse. In both a greenhouse and the atmosphere, energy enters the system as light and is transformed into heat that is trapped within the system, raising the temperature. Because of this similarity, the resultant increase in Earth's temperature caused by the buildup of atmospheric gases has been called the **greenhouse effect**.

Both our increasing numbers and dependence on fossil fuels have become acutely important factors producing the greenhouse effect since the beginning of the Industrial Revolution. Indeed, calculations suggest that Earth's average temperature could increase several degrees by the year 2050.

Is that a significant rise? Yes. Earth's temperature balance is delicate. A rise of just a few degrees could trigger melting of the polar ice caps, a significant rise in sea levels, widespread coastal flooding, shifts in climate, and an increase in unsettled and extreme weather.

Will this really happen and has it already started? These are extremely important questions. Affirmative answers could have an impact on human affairs everywhere. Unfortunately, they are not easy questions to answer and are quite controversial, even among scientists.

There are many other examples of human activities that contribute to or cause environmental problems. Many have less promising possibilities for solutions than the ozone problem. All do have potential solutions. The WebModules will supply details on some of these. As you examine each of these issues, ask yourself, how serious is this problem? What makes it a problem? What mechanisms and processes are involved? Where is the solution? How can it be implemented? And, most importantly, what can I do?

Hot Earth What is the latest scientific thinking on global warming? Has it started yet? If not, when is it going to?

Piecing It Together

1. Humans dominate ecosystems throughout the Earth. Our increasing numbers and dependence on fossil fuels affect ecosystems worldwide, producing numerous environmental problems, including overpopulation, habitat modification and destruction, destruction of the ozone layer, and water and air pollution.

2. Environmental problems can be solved, but doing so will require worldwide effort, commitment, money, and time.

3. Science can play a limited role in solving environmental problems. It can be used to help identify the problems, evaluate their seriousness, and define options. Solving environmental problems often involves taking into consideration political, economic, moral, and cultural values.

4. Worldwide, the human population has increased exponentially, especially over the last 300 years. It reached 6 billion in late 1999 and is expected to double by the year 2050.

5. In the time it takes to read this sentence the world's human population will increase by 17.

6. Curbing the human population will depend upon the following:

   ◆ Increased levels of education.

   ◆ Increased economic opportunities.

   ◆ Increased female empowerment.

   ◆ Accelerated cultural change.

   ◆ Increased availability of birth control methods.

7. Destruction of the ozone layer was brought about by the large-scale release of chloro-fluorocarbons (CFCs), starting in the 1950s. The solution for this problem took the form of a worldwide ban on the use of CFC. The speed with which the world's governments came together to implement a solution to this problem may serve as a model for approaching and solving other environmental problems.

8. A significant buildup of greenhouse gases in the atmosphere, including carbon dioxide and methane, may be contributing to global warming.

Where Are We Now?

Is there intelligent life on Earth?

In the early 1950s, a clandestine gathering of scientists was called in upstate New York. The purpose was to discuss the possibility that intelligent life might exist elsewhere in the universe (Figure 15-24). The topic was so far outside the mainstream of science that many attendees agreed to come only on condition that their affiliates and colleagues would never know they attended. They met and discussions were spirited.

First, they agreed that life anywhere in the universe should be recognizable. The diversity of life on Earth is so great that it is hard to imagine life anywhere that wouldn't resemble something we already live with. (Try it. Imagine a form of life that in no way resembles some form of life on Earth.) Furthermore, life everywhere should be subject

Figure 15-24. Are we alone?

to and show similar reactions to natural laws and processes. Gravity should have the same importance to life anywhere in the universe as it does here. Life anywhere would require an energy source, utilize and dispose of resources obtained from the environment, accumulate wastes, and form interdependencies with other organisms.

Next, the scientists turned to questions of suitable sites. There are thought to be roughly 10^5 galaxies, each with 10^5 stars. How many of those 10^{10} stars might have planets? And how many of those might be of just the right size and density to hold a gaseous atmosphere but not be so large that gravity would impede life? And how many of those might be just the right distance from a star so as to not be too cold or too hot; have just the right mix of elements; have sufficient water or similar liquid? And so on. Here the scientists were speculating, dealing with educated guesses and generalized statements. Each question whittled down the number of possible sites, but the number was still impressively large. They proceeded until they came to the last question, which they could not answer: Is there intelligent life on Earth?

The scientists noted that one of the characteristics of intelligence should be the ability to continue to exist indefinitely. In order to do so, intelligent life would have to face and solve many of the same problems we are facing. Growth of populations would have to be controlled. So would use, equitable distribution, and recycling of resources and the accumulation and processing of wastes. If life were truly intelligent, these problems would be faced and solved, and numerous examples of life should be expected throughout the universe.

Or is the fate of all life modeled by *Paramecium* in a test tube of nutrient broth? Once a population of *Paramecium* gets started, its numbers increase, for a time unabated. Nutrients are consumed with abandon and are eventually exhausted. Wastes accumulate and eventually become toxic. Numbers reach incredible levels. Eventually, the population crashes. Is this the fate of all life? Are the differences between us and *Paramecium* only those of degree? Our time line may be longer and the path we follow more circuitous, but is the endpoint the same? If so, life may have evolved numerous times and in numerous places, evolved to just beyond where we are now, and then crashed.

Is life capable of an indefinitely sustainable intelligence? Here the discussions ended. Given our current state of knowledge, there is no way to intelligently choose between answers. But we can hope the answer is yes. The title of the book summarizing the meeting, *We Are Not Alone*, suggests the scientists felt a sense of optimism. Surely, intelligence carries with it abilities to curb numbers, control and manage resources, and minimize and recycle wastes. Now, nearly half a century later, we have even more reasons to be hopeful. Now is the time for us to face our ultimate challenges.

It's not a matter of "can we solve these problems" but one of "will we" and "when."

REVIEW QUESTIONS

1. What is an ecosystem?

2. List four characteristics of ecosystems. Explain each characteristic.

3. How is an ecosystem different from a biological community?

4. What does it mean that "ecosystems are dynamic"?

5. What are limiting factors in the physical environment? Give some examples.

6. Describe the water, carbon, nitrogen, and phosphorus cycles. How is each of them important to life?

7. Describe how you participate in the water, carbon, nitrogen, and phosphorus cycles.

8. What factors of the physical environment make up climate? How does climate limit where organisms can live?

9. Describe the law of tolerances. How would you de-

scribe your tolerances for temperature, light, available water, and available energy?

10. What does it mean that "nutrients cycle through ecosystems, but energy does not"?

11. Describe the trophic structure of ecosystems.

12. In the ecosystem in which you live, list some producers, primary consumers, secondary consumers, and tertiary consumers.

13. Why are there usually no more than four trophic levels in terrestrial ecosystems? Why are there sometimes more than four trophic levels in highly productive marine ecosystems?

14. What are food chains? What are food webs? How are they related?

15. What is the difference between net productivity and gross productivity? Between primary productivity and secondary productivity? What is gross primary productivity? What is net secondary productivity?

16. Describe the vertical organization of a typical forest. Describe the typical organization of a pond or lake.

17. What is ecological succession? List the steps that ecological succession goes through in a typical forest biome.

18. What biomes are present in North America? In general terms, where is each located? In which biome do you live?

19. Briefly describe the growth of the human population over the last 12,000 years. Why haven't our numbers plateaued at a carrying capacity like other populations? If the growth in human population is to be controlled, what changes will have to be implemented?

20. How have human activities contributed to the destruction of the ozone layer? How might this problem be solved?

21. What is the greenhouse effect? Explain.

22. How have human activities contributed to the greenhouse effect?

23. What can you do to help solve the environmental problems?

Key terms from the textbook and the CD.

abiotic Abiotic factors are the nonliving components of an ecosystem.

abscisic acid (ABA) A plant hormone that promotes dormancy, leaf fall, and age-related changes in plants, while reducing the effects of growth-promoting hormones.

absolute number The total number of individuals in a population, which is nearly impossible to assess.

absorption To take in and stop radiant energy such as light energy. A characteristic of certain molecules such as chlorophyll.

acellular, or plasmoidal, slime mold A group of simple soil microorganisms (myxomycota) considered to be either fungi or protists, which exists as a plasmodium, or a mass of protoplasm that can spread over several square meters (yards) and does not contain individual cells.

achondroplasia A cause of congenital dwarfism involving malformation of the cartilage at the ends of the long bones.

acid Any substance that releases hydrogen ions (protons) into a solution. This results in a decrease in the pH of the solution.

acquired immunity Resistance to a pathogen achieved through previous exposure or through immunization.

actin A structural protein found in such places as the cytoskeleton and muscle microfilaments.

action potential The transient reversal in electrical charge across the membrane of a neuron or muscle in response to a stimulus.

activational effect One of two types of hormonal influences on animal behavior, of short duration and appearing late in life, usually. See *organizational effect.*

active site The region of an enzyme to which the substrate binds. It must be of the proper shape for the substrate to bind.

active transport The movement of substances across a membrane from an area of lower concentration to an area of higher concentration. Active transport requires energy and transport molecules (transport proteins).

adaption The change in organisms in response to pressures of the environment. Adaptions result in increased fitness. In evolutionary terms, adaption is the outcome of natural selection and results in genetic variability.

adaptive radiation The evolution of divergent species from a single ancestor. Examples include Darwin's finches.

adenine A nitrogenous purine base found in DNA and RNA.

adenosine diphosphate (ADP) ADP results from the "energy-releasing" hydrolysis of ATP, and reforms back to ATP during the ATP cycle. It also stores usable energy in its phosphate bond.

adenosine triphosphate (ATP) ATP is the main energy molecule in living organisms. During cellular metabolism, it is broken down to release usable energy and reformed to store energy. Chemically, it is a nucleotide.

adipose tissue Animal connective tissue made up of cells which contain droplets of fat.

aerobic Describes a process that requires oxygen.

aerobic respiration The oxidation of the products of glycosis in a series of steps that consume oxygen and produce energy for cells.

age-specific birth rate Birth rate for a specific age group in a population.

age-specific mortality rate Death rate for a specific age group in a population.

age structure In a population, the number of people in each age class, arranged in rows with the youngest at the bottom. Age structure is an indication of potential population growth.

agonistic encounter Animal communication that often leads to dominance hierarchies.

AIDS Acquired immune deficiency syndrome (AIDS) is a viral disease of the immune system. It is a result of infection by the HIV virus and is characterized by a decrease in the body's ability to fight disease.

alarm call Animal communication that alerts the group to the presence of danger.

alcoholic fermentation The anaerobic conversion of sugars into ethyl alcohol by the actions of yeasts and other microorganisms.

aldosterone A hormone that is produced when the body is low on sodium. It binds with proteins to accelerate the cell's sodium pump, causing the cell to reclaim sodium.

algae A diverse group of organisms generally classified as aquatic, unicelled, and photosynthetic.

alien species Species that become established in regions in which they are not native.

allele Alternate form of a gene. There are usually two present for each gene.

allelic frequency The percentage of alleles for each gene within a population.

allopatric speciation Evolution of a species in separate and mutually exclusive geographical areas.

all-or-nothing rule Characteristic of nerve cells in which the response, or action potential, is the same no matter how strong the stimulus.

allosteric site A binding site on an enzyme other than the active site. Inhibitor molecules can bind here, which results in noncompetitive inhibition of the enzyme's activity.

alpha chain One of two types of amino acid sequences that make up a hemoglobin molecule. See *beta chain.*

alternation of generations The life cycle of plants is so named because two distinct forms (generations) are present. They are a diploid sporophyte and a haploid gametophyte.

altruism The behavior of an individual looking after the interests of others instead of his or her own.

alveoli Tiny air sacs of vertebrate lungs where gases are exchanged with the circulatory system.

amino acid The building blocks for proteins. There are twenty commonly occurring amino acids found in proteins.

amino group A group of atoms that includes one nitrogen and two hydrogen atoms. It is one of the functional groups found in all amino acids.

amoeboid Resembling an amoeba in shape or movement.

amylase An enzyme that breaks down carbohydrates. Amylase is produced in the salivary glands and pancreas.

anaerobic Describes a process that occurs only in the absence of oxygen, as in fermentation.

analogous structures Physiological features possessed by different species that are similar in function, but that are not related in development or evolutionary origin.

anaphase The phase of cell division when chromosomes move to opposite ends of the cell. During mitosis, the chromatids are separated during anaphase. During meiosis I, homologous chromosomes separate from each other, and the chromatids are pulled apart during meiosis II.

angiosperms The flowering seed plants.

animal A member of the animal kingdom is characteristically multicelled, heterotrophic, and eukaryotic.

animal pole One of two poles of an egg or embryo that, in some organisms, experiences more rapid cleavage following fertilization than the opposite, or vegetal, pole.

anther The terminal portion of the male reproductive organ (stamen) in flowering plants. Anthers contain the pollen sacs.

antibody A protein molecule that is formed in response to an antigen (a foreign substance). Antibodies are part of the vertebrate immune system.

anticodon Three-base sequence of transfer RNA that is complementary to a particular codon of messenger RNA.

anticodon region The three-base region of a tRNA molecule that binds to a triplet codon of mRNA, thus translating a message into an amino acid sequence.

antidiuretic hormone, ADH Reduces the quantity of urine by enhancing the reabsorption of water by kidney tubules.

antigen Any substance foreign to an organism that triggers the proliferation of a specific antibody.

anus The opening through which waste passes out of the digestive system.

apical meristem The growth tissue found at the tips of roots and stems of plants.

archaebacteria Prokaryotic organisms that are thought to be the most primitive form of life and the ancestor of all life. They are capable of inhabiting extreme environments.

arteriole Very small artery.

artery A blood vessel that transports blood away from the heart.

ascus A saclike structure in which spores are produced in the Ascomycota fungi.

asexual reproduction Reproduction without the production and union of gametes (sperm and eggs).

atherosclerosis The accumulation of cholesterol plaque on the walls of blood vessels, resulting in narrowing of their internal diameter and restricted blood flow.

atom The smallest particle of an element. Atoms can exist in combination with other atoms.

ATP cycle The breakdown of ATP into phosphate and ADP with the release of energy, and the synthesis of ATP from ADP and phosphate with the storage of energy. The ATP cycle ensures that there are sufficient quantities of ATP and energy available to a cell.

atrium The chamber of the heart that receives blood returning to the heart from parts of the body.

autoimmune disease Disorder involving the body mounting an immune response against its own tissue antigens.

autonomic nervous system (ANS) The involuntary nervous system, including both sympathetic and parasympathetic systems, which acts on smooth and cardiac muscles and glands.

autosomes Those chromosomes other than the sex chromosomes. Autosomes usually occur as pairs, one of maternal origin and the other of paternal origin.

autotroph An organism that can manufacture its own food. Autotrophs can convert light energy into chemical energy.

auxin Any of various related plant growth hormones that govern cell elongation and differentiation. Auxins are involved in root growth, the development of vascular tissue, phototropism, the development of fruits, and the normal suppression of the growth of lateral buds. Auxins cause plant shoots to curve toward the light by elon-

gating the cells on the shaded side of the shoot.

avirulent "Avirulent" means the same thing as nonvirulent, that is, not causing disease.

axillary bud A plant offshoot growing in the axil, which is the angle between a leaf or branch and the main stem from which it grows. See *terminal bud*.

axon A thick process originating from the cell body of a neuron. It is highly branched at its terminal end, where it may synapse with the next neuron or effector.

bacteria Members of the kingdom Eubacteria are prokaryotic and therefore possess no defined nucleus. They also lack other membranous organelles. Bacteria are economically important as well as the cause of certain diseases.

bacterial chromosome Single, closed loop of DNA that contains all the genes for a bacterial cell.

bacteriophage A virus that has a bacterium as its host. Bacteriophages have DNA as their genetic material, which is enclosed in a protein covering.

base A substance that can remove hydrogen ions (protons) from a solution. Bases are one of the components of nucleotides.

basidium A spore-producing structure found in the Basidiomycota fungi.

basophil A type of white blood cell characterized by a pale nucleus and large granules.

behaviorism A theory that explains animal behavior largely in terms of stimulus and response. Under this broad set of ideas, behavior is held to be motivated and limited by factors in the environment and by physiological factors within the animal. Psychological causes for behavior that cannot be directly observed are regarded with skepticism.

bell-shaped curve A curve that illustrates the continual distribution of some trait or characteristic. It connects all the frequencies for this trait into one continual distribution that resembles the shape of a bell.

benthic zone The bottom surface of a body of water.

beta chain One of two types of amino acid sequences that make up a hemoglobin molecule. See *alpha chain*.

bicuspid valve A valve on the left side of the heart between the atrium and ventricle.

bilateral symmetry A body plan in which there are two corresponding halves with most organs occurring in pairs.

bilayer Phospholipids are arranged in a

bilayer in membranes. They are oriented with their polar regions facing the inner and outer surfaces and their nonpolar regions facing inward.

bile A liver secretion that emulsifies lipids prior to their digestion.

bile duct Passage from the gallbladder to the duodenum.

binary fission A method of asexual reproduction in which a single organism splits into two identical individuals.

binomial nomenclature The standard system of double Latinized names used to designate an organism, consisting of a generic name (genus), followed by a specific name (species), for example, *Ursus arctos*, for brown bear.

biocatalyst Enzyme that is necessary for most of the reactions of metabolism. Biocatalysts alter the rate of a reaction by lowering the activation energy. In the process, they are not chemically altered and may be used again.

bioenergetics The flow of energy in an ecosystem or the transformation of energy in living organisms. Also, the study of these topics.

biogeochemical cycle The movement of elements between living organisms and their nonliving environment. Cycle that has both biological and geological phases, such as the water cycle.

biological clock A biological timekeeping process that allows certain physiological functions to occur in a regular, often predictable, cycle.

biological communities Populations of many species living together in a definable space.

biomass The total mass of all the living organisms in a food chain or a food web.

biome A major regional community of plants and animals with a distinctive climate and fauna.

biosphere The biosphere consists of all the populations, communities, ecosystems, and biomes on the planet Earth. It includes all biotic and abiotic components.

biotic Biotic factors are the living components of an ecosystem.

biotic potential The ability of a population to increase, resulting from maximal rate of reproduction and minimal mortality.

bird A warm-blooded, feathered vertebrate, belonging to the class Aves. Birds' wings are modified forelimbs.

birth rate, or natality The number of births within a population over a set period.

blastula The first recognizable form of ani-

mal development. It is a hollow ball of cells with an opening called a blastopore.

blending theory An obsolete theory theory explaining how traits are inherited. According to the blending theory, factors determining traits are not discrete units but permanently blend together. Once blended, factors can never separate.

blood The liquid portion of the circulatory system. Blood mostly consists of water, but also contains a variety of cells and other molecules. Blood is considered a connective tissue.

blood pressure A measure of the heart's effectiveness. Blood pressure indicates the force generated by pumping action of the atria and ventricles.

blood vessels Tubes that carry blood toward and away from the heart.

blueprint A plan for the making of something. For example, DNA serves as a blueprint (template) for the synthesis of new proteins.

bond Attraction between or among atoms. Bonds may be ionic or covalent.

bone A calcified connective tissue that forms the skeleton of vertebrates. Bone is living tissue and can grow and change with age.

bottleneck A drastic decrease in the size of a population with a resulting decrease in genetic variability within a population.

brain The main control center of an animal's nervous system that receives and processes information.

bristle Fine hairlike projections on the surface of cells. Bristles may be used for locomotion or protection.

bronchus A branch of the trachea, or windpipe, leading to one of the lungs in vertebrates.

brown algae The common name for Phaeophyta, an alga occurring mostly in saltwater and that contains the brown pigment fucoxanthin.

brush border Dense covering of minute, fingerlike projections on the surface of epithelial cells in the intestine and the renal epithelium.

bryophytes Any of a group of simple, spore-bearing, nonvascular plants that contains mosses, liverworts, and hornworts.

budding A type of asexual reproduction in fungi and algae with haploid life cycles.

bundle sheath cell The cells that surround the vascular tissues of C_4 plants and that store carbon dioxide in a four-carbon compound.

bushes Individual shrubs, or the layer of

vegetation between the subcanopy and surface layer of a forest.

C4 plant Any of a group of plants, such as corn and sugarcane, that fix carbon dioxide by the C_4 pathway.

Calvin-Benson cycle The light-independent reactions of photosynthesis by which carbohydrates are synthesized as a product of photosynthesis. Carbon dioxide is necessary for these reactions.

CAM plant Any of a group of plants, such as cacti, that fix carbon dioxide by the C_4 pathway in hot, arid climates.

canopy Cover formed by the branches and leaves of the tallest trees in a forest.

capillary Tiny blood vessels that are involved in the exchange of materials between tissues and the circulatory system. They are one cell in thickness to allow for optimal transport.

capsid The protein coat of a virus, which surrounds the nucleic acid.

capsule A stiff outer covering. Capsules include such things as spore-producing structures, the outer covering of many bacteria, and the connective tissue covering of such organs as the kidney.

carbohydrate Macromolecules that contain carbon, hydrogen, and oxygen. Their common name is sugar or starch.

carbon The main element of living things. Carbon is present in every organic molecule. It is widely found because of its ability to form four covalent bonds.

carbon cycle The set of biological processes that circulates carbon between the biosphere and the nonliving environment. Carbon from atmospheric carbon dioxide (CO_2) enters plants through photosynthesis, is fixed into organic compounds, circulates within the biosphere, and eventually returns to the atmosphere chiefly through the respiration of animals, but also through the burning of plants and fossil fuels.

carboxyl group A functional grouping of atoms containing one carbon, two oxygens, and one hydrogen. The carboxyl group is found in all amino acids.

cardiac muscle A specialized muscle tissue found in the hearts of vertebrates.

cardiovascular system The heart and blood vessels.

carnivore An animal that eats other animals.

carpel The female reproductive structure in flowers containing one or more ovules. The carpels together make up the pistil.

carrier protein Integral membrane proteins that regulate the movement of materials across a membrane. Carrier proteins attach to the substance to be transported and by changing their shape move it to the other side of the membrane.

carrying capacity The maximum size that a population can reach without being adversely affected.

cartilage A firm, elastic skeletal tissue composed of cells embedded in a fibrous matrix. Cartilage is a connective tissue.

catalyst A catalyst alters the rate of a chemical reaction but is not chemically changed during the reaction. Enzymes are biocatalysts.

cell The basic unit of life. All living things are composed of cells.

cell body The cell body of a neuron contains the nucleus and many organelles and is where most cell activities take place.

cell cycle The life cycle of a cell. The sequence of events that takes place within the cell from the time it forms until it divides.

cell division The process by which cells make new cells. Cell division may be either mitosis or meiosis.

cell membrane (plasma membrane) The membrane that surrounds all cells. The cell membrane regulates the exchange of materials between the cell and its environment.

cell metabolism All the cellular reactions that occur within living cells. Cell metabolism includes such things as cell respiration and photosynthesis.

cell plate The material laid down across the middle of a dividing plant cell from which new partition walls form.

cell theory The idea that all organisms are made up of cells, and that the cell is the fundamental unit of life. First proposed by Schleiden for plants and Schwann for animals in the late 1830s.

cellular respiration The oxidation of organic molecules to release energy. One product is new ATP, which is then used as the main energy molecule for metabolism.

cellular slime mold A group of simple soil microorganisms (Acrasiomycota) considered to be either fungi or protists. Usually existing as free-living amoeba, they aggregate to form a multicellular fruiting body and produce spores.

cellulose The main component of the cell wall of plants and bacteria. Cellulose is a polysaccharide chain of glucose molecules.

cell wall A rigid structure outside the cell membrane. Cell walls are found in plants, fungi, and some bacteria and protistans. It provides additional protection of and rigidity to cells.

central nervous system The system that includes the brain and spinal cord.

central proposition of evolution The idea that new types of living organisms develop from preexisting types by the accumulation of genetic differences over long periods of time.

centriole Found primarily in animal cells, functions as a site for spindle fiber attachment during cell division. Centrioles usually exist in pairs and are made of microtubules with a structure similar to cilia and flagella.

centromere The constricted region of a chromosome at the junction of the two chromosomes' arms. Spindle fibers attach to the centromere of each chromatid during cell division.

cephalization A type of body plan in which most sense organs and nerve cells are concentrated in the head area.

cervix The entrance into the uterus.

channel protein Integral membrane proteins that are used to transport substances across membranes. They are sometimes referred to as pores.

Chargaff's ratios In any DNA molecule, the amounts of adenine and thymine are the same, and the amounts of cytosine and guanine are the same. This piece of information helped Watson and Crick determine the structure of DNA.

chemical bond An attraction between or among atoms. Chemical bonds may be weak such as a hydrogen bond or an ionic bond, or they may be strong in the case of covalent bonds.

chemical, or bond, energy The energy that is required to link atoms together to form a bond, or that is released when a bond is broken.

chemiosmosis The process that links the movement of hydrogen ions across a mitochondrial membrane with the phosphorylation of ADP to make ATP.

chemisynthesis The use of chemical energy to convert smaller molecules into larger molecules.

chemoautotroph An organism that uses inorganic sources of carbon, nitrogen, and so on, as starting material for biosynthesis and as a chemical energy source.

chiasma The X-shaped structure that is formed at the site of crossover and exchange of DNA between homologous chromatids in meiotic prophase.

chitin A nitrogen-containing polysaccharide that is found in the cell walls of fungi. Chitin is also found in animal structures such as the exoskeleton of arthropods.

chlorophyll Any of a group of green pigment molecules that respond to light energy. The result of trapping light energy is to raise electrons in chlorophyll to a higher energy level from which they may be used to initiate the light-dependent reactions.

chloroplast The cell organelle in which photosynthesis takes place. Chloroplasts have a continuous outer membrane and a highly folded inner membrane. It is in the chloroplast that light energy is converted to chemical energy.

cholesterol A lipid found primarily in animals. Cholesterol is a component of membranes and is a precursor for steroid hormones.

chromatid One half of a replicated chromosome.

chromatin Nucleoprotein material which stains with basic dyes, found in the nucleus of eukaryotic cells and becoming organized into visible chromosomes at cell division.

chromosome The physical carrier of the genetic material (genes). Chromosomes are composed of DNA, RNA, and protein. During sexual reproduction, they are transmitted by the egg and sperm to the next generation.

chromosome replication Chromosome replication occurs during interphase of the cell cycle. It results in each chromosome containing two chromatids joined by a common centromere. One chromatid is the original chromosome; the other is an exact replicate of this chromosome.

chromosome theory of inheritance This theory establishes the relationship of genes to chromosomes. It states that the chromosomes are the physical carriers of genes.

chronic Chronic refers to an illness or a disease that is ongoing. A chronic illness may last for several years.

chylomicron Small lipoprotein particle in plasma and other body fluids that transports cholesterol and various lipids from intestine to adipose tissue.

chyme Food in liquid form that is ready to leave the stomach.

cilia Exterior organelles used for locomotion. Cilia are made up of nine pairs of microtubules that surround a central pair of microtubules.

ciliate A phylum of protozoans having cilia, which are motile, hairlike outgrowths from the surface of a cell.

circulatory system The transport system of fluids within an animal, usually consisting of a pump (heart) and a series of vessels.

class A taxonomic group within a division (plants) or phylum (animals), but above the group's order, genus, and species.

classical conditioning Also known as Pavlovian conditioning. A technique that extends a natural response (like the salivating of a dog) from the original stimulus (the smell of food) to an unrelated stimulus (the ringing of a bell) through repeated association of the two stimuli (ringing the bell and presenting the food).

cleavage The early mitotic divisions that follow fertilization in an embryo.

cleavage furrow Constriction that forms in the midportion of a cell when that cell is dividing.

clonal selection An immune response to the presence of an antigen; the proliferation of clones of matching lymphocytes.

codon A three-base sequence of messenger RNA that is specific for a particular amino acid. Some codons also are important for the initiation and termination of protein synthesis.

coelom The main body cavity in animals. The cavity is formed from layers of mesoderm.

coenzymes Compounds such as NAD, NADP, FAD, and coenzyme A that remain unchanged at the end of an enzyme-catalyzed reaction. These are nonprotein substances that are required by a protein for biological activity. See *cofactor*.

coevolution Evolution of two species in relation to each other, such as predator and prey.

cofactor Any nonprotein substance required by proteins for biological activity, such as coenzymes. See *coenzymes*.

cohort A group of individuals in a population that are similar in age.

colony A group of cells that act together to perform various functions. At times, colonies may appear to be a multicelled organism. Cells within a colony are often specialized to perform specific functions.

color phase A regularly occurring color variety in an organism.

columnar Describes cells that are longer than they are broad, as in cells of the epithelium.

commensalism A type of symbiosis that benefits one species and has no apparent effect on the other.

community A group of populations that interact with each other.

companion cell A plant cell that helps conduct sugars and amino acids through a plant's phloem.

comparative behavior Study of animal behavior, often done under controlled conditions, with an emphasis on its evolutionary history.

compass orientation Method by which animals orient, moving in a particular direction for a prescribed distance or length of time.

competitive exclusion principle The principle that two different species cannot occupy the same ecological niche, because one will eventually dominate and the other will be eliminated.

competitive inhibition When a molecule resembling the substrate of an enzyme binds to the active site of the enzyme. This prevents the binding of the substrate and thus inhibits the activity of the enzyme. The result is the chemical reaction is blocked until the inhibitor is removed.

complementary bases The complementary bases or nucleotides of DNA pair to form the double helix. The three types of RNA are complementary to the DNA from which they are synthesized. The complementary bases are adenine and thymine (uracil in RNA), and guanine and cytosine.

complementary nucleotides The complementary nucleotides of DNA pair to form the double helix. The complementary nucleotides are adenine and thymine (uracil in RNA), and guanine and cytosine.

complementary strands Strands of nucleotides that contain sequences of bases complementary to each other. The complementary bases are adenine and thymine or guanine and cytosine. In RNA, uracil replaces thymine.

complete digestive system Digestive systems with two openings, a mouth and an anus. Characteristic of most animal phyla.

compound A substance composed of several elements. It may refer to the structure of such things as a compound leaf or a chemical compound.

concentration gradient A difference in concentration of a substance across a membrane. Gradients determine the direction of movement of substances across a membrane unless energy and/or carrier molecules are involved.

condensation reaction Chemical reactions in which two smaller molecules are joined together with the loss of water. Condensation reactions are sometimes referred to as dehydration synthesis reactions. The formation of a disaccharide from two monosaccharides is an example.

conducting system Structures including mouth, nose, trachea, and bronchi, that carry air to the lungs.

cone The reproductive organ of gymnosperms. Cones may be male or female. Cone-bearing plants are also called conifers.

conformation The shape of a molecule. Conformation may be influenced by, among other things, temperature and pH.

conidia Asexual spores produced by ascomycote and deuteromycote fungi.

conjugation A form of sexual reproduction in which nuclei are exchanged between two individuals.

connective tissue Tissue characterized by the presence of cells in a noncellular matrix.

conservation genetics The branch of biology concerned with preserving genetic variation in endangered species.

consumers The heterotrophic organisms in a food chain or food web. They consume what the producers have produced.

continuous variation Traits that vary on a continuum within a population. They are determined by more than one gene (polygenes) and may be influenced by the environment.

control center The part of an animal that controls and regulates nervous activities. It may be a complex brain, a ganglion, or a single nerve cell.

convergent evolution The appearance of similar structures in unrelated organisms. This may result from organisms adapting to similar environmental pressures.

copulation Sexual union; the act of mating.

cork External layer of plant tissue composed of dead cells that forms a barrier impenetrable to water. Present on woody stems and derived from the *cork cambium*.

cork cambium A secondary meristematic layer in woody stems that gives rise to the cork layer of the bark.

corpus luteum The ovarian follicle left behind after ovulation.

cortex The outer layer of a structure or an organ. In vascular plants, the tissue in stem and root surrounding and not part of the vascular bundles.

counter current exchange Exchange of liquids or gases by their diffusion between tubes in which their content is flowing in opposite directions.

covalent bond An attraction resulting from the sharing of electrons. The sharing may be equal (nonpolar covalent) or unequal (polar covalent). Covalent bonds are one of the most commonly occurring bonds in living things.

cristae The folds of the inner mitochondrial membrane.

critical limiting factor The environmental factor that is most limiting to a species, influencing where it lives and how well it thrives.

cross-fertilization The fusion of male and female gametes of different individuals, especially of different genotypes.

crude birth rate The number of individuals born into a population divided by the total population.

cryptic coloration Coloration that allows an organism to blend into its background.

cuboidal Describes a type of epithelial tissue in which cells are as tall as they are wide.

cyanobacteria Prokaryotic organisms characterized by the presence of blue-green pigments. They were previously known as blue-green algae.

cyclic photophosphorylation Light-dependent reactions of photosystem I in which electrons in chlorophyll that are raised to a higher level of energy by light energy return through a series of carrier molecules back to photosystem I. In the process, ADP is phosphorylated to make ATP.

cytogenetics The branch of biology that combines the study of cells with genetics.

cytokine A protein factor produced by one cell that affects the growth and division of other cells.

cytokinesis The formation of two new daughter cells at the end of cell division.

cytoplasm The contents of a cell excluding the nucleus. Cytoplasm contains mostly water and is where all other cell structures are found.

cytosine A nitrogenous pyrimidine nucleotide found in DNA and RNA.

cytoskeleton A supportive network of fibers found in eukaryotic cells. It maintains the shape of cells and also is involved in cellular movement.

cytosol The cytoplasm other than the various membrane-bound organelles.

daughter cells Cells that are produced during cell division. Two diploid daughter cells result at the end of mitosis, and four haploid daughter cells at the end of meiosis.

death rate, or mortality The number of deaths in a population over a set period.

deciduous Deciduous plants drop their leaves during colder months.

decomposer An organism that breaks down dead organic material and returns nutrients to the soil. Bacteria and fungi are examples.

decomposition The breakdown of organic compounds into more basic units by the activity of microbes when those compounds are no longer part of a living system.

deduction Reasoning from a known principle to an unknown one. Deduction works from the general to the specific.

defecation The act of voiding the bowels of feces.

deletion Mutation involving loss of part of a chromosome, or a base or bases in a DNA sequence.

dendrite Thin, highly branched process in a neuron. Most stimuli are received by the dendrites of a neuron.

density An estimate of population size based on the number of individuals per unit area.

density-dependent factor Factor that limits a population's growth that is related to the population density.

density-independent factor Factor that limits a population's growth that is not related to the population density.

deoxyribonucleic acid (DNA) The genetic material in all living cells. It is a nucleic acid constructed from building blocks called nucleotides.

deoxyribose A pentose sugar that is one of the components of a DNA nucleotide.

dephosphorylation The removal of a terminal phosphate group from ATP to make ADP and a single free phosphate molecule.

dermal tissue In higher plants, the epidermis and associated tissues form the dermal tissue system.

desalinization The removal of salts from sea water. The result is water that can be consumed by humans and other organisms.

detector A feature of homeostatic control; a system that senses imbalances in the internal environment.

determinate growth Typical of animals, growth of an organism or part of an organism that stops when a certain size is achieved.

detritivores Scavengers or decomposers; organisms that eat "dead stuff."

deuterostome A lineage of animals whose first embryonic opening (the blastopore) becomes the anus, while a second opening becomes the mouth.

diaphragm In some mammals, the dome-shaped muscle between the thoracic and abdominal cavities, used in breathing.

dicot A group of flowering plants characterized by the presence of two cotyledons (seed leaves) in the embryo. Dicots also have netted, veined leaves; flowers in groups of fours or fives; and vascular tissue in the form of a cylinder.

diffusion The movement of substances from an area of higher concentration to an area of lower concentration. Diffusion is a passive process that does not require energy and occurs across the surface of all membranes. Diffusion is also known as simple diffusion.

digestion The process of breaking down food for use by an organism, including the formation of waste. Digestion may begin by mechanical process and requires digestive enzymes.

digestive tract The total system of organs through which food passes while in the body.

diploid When a cell contains two sets of chromosomes it is said to be diploid. One set is maternal in origin and the other is paternal.

dihybrid cross A fertilization in which the parents differ in two distinct characteristics.

directional selection Selection that acts on one extreme of the range of variation in a particular characteristic, and therefore tends to shift the entire population toward one extreme or the other.

disaccharide A sugar molecule containing two monosaccharides or simple sugars.

discontinuous variation When quantifiable physical traits (variation) in a population fall into distinct categories.

disruptive or diversifying selection Selection that favors the extremes and disfavors the middle range of particular traits in a population, tending to cause a split in the distribution of a trait that may eventually lead to two distinct populations, one at either end of the range.

distress signal Animal communication that alerts the group to an animal's distress.

divergent evolution The divergence of genetic characteristics in a population of a species. Some common characteristics remain, but new ones arise in response to specific environmental pressures.

diversity The vast variation observed among living things. It has a genetic basis but is also under the influence of the environment.

division A major taxonomic grouping in plants, corresponding to phylum in animals.

DNA fingerprinting Method by which DNA from various species or individuals can be compared and contrasted. The DNA is fragmented into smaller pieces by enzymes and then separated in an electrical field (electrophoresis).

DNA ligase An enzyme that acts exclusively on double-stranded DNA, joining strands and repairing breaks in the sugar-phosphate backbone. It is used in genetic engineering to splice DNA strands.

DNase An enzyme that digests DNA.

domain The most general (inclusive) category in classifying living things. Most biologists agree on three domains: Archaea, Bacteria, and Eukarya.

dominance hierarchy A social system based on aggression and submission organized into an established ranking of members.

dominant trait A trait that is expressed whenever the allele for it is present.

dorsal The surface that is the upper surface, or back, of an animal.

double fertilization In flowering plants, when fertilization happens twice for each embryo. One sperm cell from a pollen grain fuses with the nucleus of the egg to produce the zygote, and a second sperm cell fuses with two nuclei in the female gametophyte, creating a triploid nucleus.

double helix The shape of the DNA molecule in living cells. This molecular structure contains two complementary strands of nucleotide bases twisted around a central axis.

duct Tube that transports materials from one location to another.

ecological longevity The life span of an individual under a set of environmental conditions.

ecological succession When one biological community replaces another, triggering changes in the physical and biological environments that set the stage for still another community.

ecology The study of ecosystems. Ecologists study the relationship between organisms and their environment.

ecosystem Communities of organisms and their interaction with biotic and abiotic factors that affect them.

ectothermic Describes animals whose body heat is derived externally, such as amphibians and fishes, informally referred to as "cold-blooded."

effector The cell or tissue, such as muscle, that produces a response to stimuli perceived by the receptors and processed by some control center.

effector system A physiological system that carries out a homeostatic response.

egg (ovum) The female gamete. Eggs (ova) are produced by meiosis and contain a haploid set of chromosomes.

ejaculation The expulsion of semen from the penis.

electrocardiogram A medical recording of electrical signals produced by the heart that indicate cardiac activity.

electromagnetic radiation Energy, including light, that moves in waves or particles and has the properties of an oscillating electric charge.

electron Negatively charged subatomic particle. Electrons are contained in energy levels surrounding the nucleus of atoms and convey the chemical properties of an atom.

electron carrier molecule Molecule in the mitochondrial membrane that acts in a chain to transport electrons during cellular metabolism and generate ATP in the process.

electronegativity The ability of any element to attract electrons from other elements in a covalent bond.

electron micrograph A photograph of an image viewed at very high magnification through an electron microscope.

electron microscope Microscope that utilizes electrons instead of light rays in the magnification of objects. The image formed can be viewed on a fluorescent screen or photographed. Magnifications of 300,000 times or more are possible.

electron transport chain A series of molecules that pass electrons, transferring them from higher energy levels to lower energy levels and using the energy to form ATP. These are oxidation-reduction reactions.

electron transport molecule Molecule in the mitochondrial membrane that acts in a chain to transport electrons during cellular metabolism and generate ATP in the process.

element A substance that cannot be separated into similar substances by ordinary chemical means.

embryo The early stage of development of an organism.

emigration The movement of individuals out of a population.

endocrine gland Tissue that produces hormones and releases them directly into the circulatory system. Endocrine glands are ductless glands.

endocrine system The endocrine glands as a whole, controlled by the pituitary gland and neurological activity.

endocytosis Process by which eukaryotic cells absorb outside material by inverting the plasma membrane to form vesicles that encase the material.

endodermis Innermost layer of corticle cells in stems and roots of vascular plants.

endomembranal system The intracellular membrane system of a eukaryotic cell.

endometrium The mucous membrane that lines the uterus.

endoplasmic reticulum (ER) A network of membranes throughout the cytoplasm of the cell. The ER is known as rough endoplasmic reticulum when ribosomes are attached and smooth endoplasmic reticulum when they are not.

endoskeleton Any internal skeleton or supporting structure.

endosperm The nutritive tissue surrounding the embryo in most seeds.

endosymbiotic theory A theory that explains the evolution of eukaryotic cells. It holds that some organelles were once free-living prokaryotic cells but became incorporated into larger such cells, forming a mutually beneficial union.

endothermic Describes animals that metabolically produce heat to maintain a constant temperature, informally referred to as "warm-blooded."

energize The addition of energy to a molecule energizes it. As a result, the molecule is raised to a higher level of energy. Some molecules must be energized to undergo chemical reactions.

energy The ability to do work. Energy is needed by living things to carry out their daily activities. The ultimate source of energy is the sun.

energy carrier molecule A molecule that transfers energy from one reaction to another. These are also called energy transport molecules. Molecules such as NAD, NADP, and FAD are examples.

energy gradient The change in the quantity of energy over distance. An energy gradient goes from higher to lower levels of energy. Electrons may be transported down such a gradient.

energy molecule A molecule that supplies the cell with energy when needed. ATP and carbohydrates are energy molecules.

entomologist One who studies insects. Entomology includes the study of their natural history as well as their biology.

environment The combination of all the physical factors in which life exists. The environment and living things compose an ecosystem.

environmental resistance The environmental factors that limit the growth of a population.

enzyme A biocatalyst that lowers the activation energy of a metabolic reaction. Enzymes are mostly proteins. Their activity depends on their maintaining a specific shape (conformation).

epidermal tissue Tissue that covers an organism. It serves to protect and guard against the loss of water.

epiglottis A fleshy structure at the back of the throat that directs food into the esophagus or air into the trachea.

epithelial tissue Sheets of tightly packed cells that cover the body of animals and line their digestive tracts, glands, and ducts.

equilibrium A state of balance or stasis in which forces acting upon a system cancel one another.

escherichia coli (E. coli) A bacillus (bacterium) that is used experimentally as an important model prokaryote to understand how genes function. It is also found everywhere, including the human digestive system.

esophagus A tube that connects the mouth to the stomach. No digestion occurs in the esophagus.

essential amino acid One of the nine amino acids that animals cannot synthesize and must obtain from their diet.

estuary The mouth of a stream or river affected by ocean tides and containing brackish, or salty, water.

ethnobotany The study of plants and people.

ethology The study of animal behavior, often done under natural conditions.

ethylene A plant hormone that stimulates fruit ripening.

eubacteria A prokaryotic kingdom that includes the bacteria and cyanobacteria.

eukaryote An organism whose cells possess a nucleus. Eukaryotes also have other membranous organelles.

evolution The change in organisms over time. This change involves genetic mutation and the selective forces of natural selection.

evolutionarily stable strategy (ESS) A behavioral strategy that results in the most offspring, which individuals in a population must adopt because most others have adopted it. The group's strategy selection inherently reduces the effectiveness of any alternative strategy, unless conditions change.

excitable cell A body cell that can respond to nerve impulses.

excited When electrons have been raised to a higher energy level. They may then be transferred to some other substance, lost, or returned to the energy level from which they came.

excretion The process of removing metabolic waste. In humans, this occurs in the kidney.

excretory system The system responsible for ridding an animal of nitrogenous wastes and maintaining proper water and salt balance.

excurrent siphon Tube in clams by which water is expelled.

exocytosis A form of transport used by cells to move molecules, particles, and other substances contained in vesicles across the plasma membrane and into the outside environment.

exoskeleton An external covering that is used for protection, attachment of appendages, and attachment of wings. It is a characteristic of the arthropods.

explanation An effort to create a hypothesis, or coherent sequence of causality, by which a phenomenon can be understood.

exponential equation Mathematical description of a population that is experiencing growth limited only by its reproductive rate.

exponential growth A type of population growth that only depends on the number of individuals and their potential to reproduce. Resources are not a limiting factor.

extinction The end of something, such as the loss of a species due to mortality of remaining individuals, or the loss of a conditioned response.

extracellular That which is outside the cell. The extracellular environment is external to the cell.

extracellular fluid Fluid outside body cells, such as intestinal fluid and plasma.

extracellular matrix A network of fibers outside the cell. It is used for such things as cell adhesion, cell-to-cell communication, and other regulatory functions.

extracellular space Space outside the cell.

extract A portion of a substance that has been removed from the main substance. It allows for the study of individual components of a substance.

extrinsic factors External factors that affect some process. They are not a part of the process.

F_1, or first filial, generation The first generation of offspring in a genetic cross.

F_2, or second filial, generation The second generation of offspring in a genetic cross.

facilitated diffusion Diffusion that requires the assistance of transport proteins.

facultative mutualism When two species benefit from a symbiotic relationship but can live independently.

fallopian tube Duct that transports ova from the ovary to the uterus.

family The taxonomic level that groups together members of related genera.

fat Triglyceride that is composed of glycerol and three fatty acids. Fats are formed by condensation reactions.

fatty acid Building blocks for such lipids as fats, triglycerides, oils, and others. Fatty acids may be saturated and thus have fewer double bonds than unsaturated fatty acids.

fauna The living animal species in a given location (habitat).

feedback control The regulation of a process by which a component from a later stage in the process controls its own production by influencing an earlier stage in the process.

female The sex that produces ova (eggs). It also refers to other characteristics of the members of this sex.

female parent The parent that, through her eggs, contributes one haploid set of chromosomes to her offspring.

fertilization The union of the female gamete with the male gamete. The result is the formation of a zygote.

ferns Members of Pterophyta. Seedless vascular plants that reproduce by spores.

fetus The later portion of development that occurs within the mother's body. In humans, includes the last six months of development.

fibrous connective tissue Connective tissue that forms tendons and ligaments.

fibrous root system A root system in plants without a taproot, having many same-sized roots instead.

filter feeding Removing dissolved nutrients or plankton from water by means of straining, as bivalves do.

flagellum Locomotive organelle found in protists and motile gametes (both sperm and eggs). They are long extensions of cells and microtubules that are arranged as nine pairs surrounding a central pair.

flatworms Members of the phylum Platyhelminthes. Flattened, bilaterally symmetric worms, including tapeworms and other parasitic and nonparasitic species.

flavin adenine dinucleotide (FAD) An important molecule in cell metabolism. In oxidation-reduction reactions, the oxidized form, FAD+ is reduced to FADH2. It is an energy (electron) carrier molecule because it passes these electrons on in the electron-transport chain.

flower The reproductive organ of flowering plants. They may be of one sex or may contain both sexes.

flowering plants Plants that reproduce by means of external reproductive organs (stamens and ovaries) encased in flowers (an outer ring of sepals and petals), which form seeds and fruit after pollination and fertilization.

fluid-mosaic model The currently accepted model of the structure of biological membranes. According to this model, the phospholipids are arranged in a fluid bilayer with their polar regions facing the inner and outer membrane surfaces. The membrane proteins are scattered throughout the bilayer in a mosaic pattern.

follicle A small sac or cavity, for example, ovarian follicle.

food chain A linear group of organisms that feed one upon another and thus pass energy along the chain.

food vesicle Vacuole formed by endocytosis.

food web An interconnected group of food chains.

forelimb The appendages nearest the head of an animal, such as front legs or wings. In vertebrates, forelimbs are homologous structures derived from a common ancestor.

fossil The preserved remains of organisms from a former geologic age.

fossil fuels Combustible substances such as oil, coal, and natural gas that have formed from fossil deposits.

founder effect The eventual genetic difference between an isolated offshoot population and the original population from which it came, owing to the non-representative set of alleles in the first generation.

frame-shift mutation A change in the reading frame resulting from the insertion or deletion of nucleotides in the DNA sequence for a protein.

frequency diagram A diagram or graph that shows the distribution of the number of items in a particular category.

fructose A six-carbon monosaccharide, or simple sugar. Fructose is found in many fruit juices, cane sugar, and honey. It bonds to glucose to form the disaccharide sucrose (table sugar).

fruit The developed ovary of a flower containing seeds. The flesh of the fruit protects seeds and aids in their dispersal.

functional group Accessory chemical entities, such as OH, NH^2, CH^3, that help determine the identity and chemical properties of a compound.

fungi Members of the kingdom fungi are eukaryotic and heterotrophic. They serve as decomposers.

galactose Six-carbon aldose sugar. A constituent, with glucose, of lactose. Also found in the pectins of plant cell walls.

gallbladder A small pouch behind the liver that stores bile.

gamete Reproductive cell associated with sexual reproduction. Gametes are ova (eggs) produced by females and sperm produced by males. They contain a haploid set of chromosomes.

game theory Mathematical inquiry into the ways in which strategies of conflict and cooperation between members of a population can result in optimal outcomes.

gametogenesis Gamete formation.

gametophyte The multicelled, haploid form of an organism with an alternation-of-generations life cycle. It gets its name because it produces gametes by mitosis.

ganglion Any group of nerve cell bodies lying outside the central nervous system (except in the case of the basal ganglia, which are inside CNS).

gastrovascular cavity A central body cavity that functions both in circulation and digestion.

gated channel Channels whose openings are controlled by a "gate" or covering. The gate must be opened before transport can occur.

gated ion channels Openings (channels) in the cell membrane that open and close in response to certain stimuli and so regulate the flow of ion through the channels.

gel electrophoresis A technique for separating molecules of protein or DNA according to size by driving them through a gel substance with an electrical field; the smaller molecules migrate farther.

gene The blueprint for the synthesis of a protein. Genes are composed of DNA and are found in all living cells; they may be composed of DNA or RNA in viruses. Genes are mostly contained in the nucleus, but may also be present in such organelles as mitochondria and chloroplasts.

genetic code Information contained in a gene (DNA) for the synthesis of a protein. The code refers to the sequence of bases that ultimately determine the amino acid sequence of a protein.

gene flow The shift in allelic frequencies within a population and between populations resulting from outbreeding or migration of individuals from one population to another.

gene pool The sum of genes in a population, including all genes in all members.

gene therapy Treatment of a genetic disease by alteration of the affected person's genotype, or the genotype of affected cells.

genetic drift Random change in allelic frequencies as a result of chance alone.

genetic equilibrium When a population maintains the same genotypic and allelic frequencies generation after generation.

genetic information Information conveyed by the genes of an organism. Genes encode the blueprints of proteins, as well as the information necessary for making the whole organism.

genetic makeup For a population, the frequencies of alleles and genotypes within that population.

genetic material The genes of an organism. It is DNA in all living things and may be DNA or RNA in nonliving viruses.

genetic recombination When segments of a chromosome are exchanged by crossing over during prophase I of meiosis. Genes usually are exchanged between homologous chromosomes.

genetics The study of inheritance. Geneticists study how traits are passed from one generation to another.

genetic variation Variation in a species that has a genetic basis. It is due to mutations in the genes.

genome The information stored in all of the DNA of a single set of chromosomes.

genotype The genetic makeup of an individual. The genotype is determined by the alleles that are present for each trait.

genotypic frequency The percentage of each genotype in population.

genus One of the taxonomic categories, the group just above species. It is the first name in the scientific name of an organism.

geology Study of the history and structure of the Earth. Geologists are interested in such things as the structure of rocks, dynamic changes in the Earth's structure, and changes in the Earth as revealed by fossils.

germination The beginning of growth of

a seed or spore, usually after a period of dormancy.

gibberellin One of about 50 compounds that promote growth by stimulating cell elongation and division.

gill The respiratory surface of certain aquatic organisms such as fish and tadpoles. Gills exchange gases directly in and out of the water surrounding the animal.

glomerulus In the kidney, a ball of capillaries enclosed in a capsule at one end of the nephron tubule.

glucose A monosaccharide containing six carbons, twelve hydrogens, and six oxygens. It is the building block for many larger carbohydrates.

glucose metabolism A series of reactions that break down glucose into water and carbon dioxide. This results in the release of the energy stored in glucose.

glucose phosphate A molecule of glucose to which is bonded a molecule of phosphate. The added phosphate is usually from ATP and raises glucose to a higher level of energy.

glyceride Any ester of glycerol, including some fats and oils, in which three molecules of a fatty acid combine with one molecule of glycerol.

glycerol A three-carbon alcohol. Glycerol is combined with three fatty acids to form triglyceride.

glycogen A polysaccharide composed of glucose molecules. The principal storage carbohydrate in animals, glycogen is often referred to as animal starch.

glycolipid Lipid that is bonded to a carbohydrate. It is principally found on the surface of membranes.

glycolysis The breakdown of glucose into pyruvate. Glycolysis is first in the series of reactions involved in glucose metabolism. In eukaryotic cells, glycolysis occurs in the cytoplasm.

glycoprotein Protein that is bonded to a carbohydrate. Examples of glycoproteins include antigens, enzymes, and hormones. Glycoproteins also are found on the surface of membranes.

Golgi apparatus (Golgi body, or Golgi complex) A cytoplasmic membrane system that serves as a packaging center for cell products. It collects materials and then packages them into vesicles for transport out of the cell or for use elsewhere in the cell.

gradualism The evolution of a species by an accumulation of small steps over a long period of time. A theory of evolution that focuses on this type of change.

grana In chloroplasts, groups of stacked, disk-shaped structures called thylakoids, whose membranes contain the photosynthetic pigments.

gravitropism Changes in plant growth owing to the effects of gravity.

green algae Common name of the Chlorophyta, photosynthetic protists that are thought to be the precursors of plants.

greenhouse effect The retention of heat in the lower atmosphere by the presence of carbon dioxide, water vapor, and other compounds in the upper atmosphere. Human activities are suspected of contributing to the changes in the global climate brought about by the greenhouse effect.

gross mortality rate The number of deaths in a specified time period divided by the total number of individuals in a population at the start of the period, times one thousand.

gross primary productivity The total assimilation of energy in a plant community per unit of time.

gross reproductive rate The number of individuals born into a population divided by the total population, times one thousand.

gross secondary productivity All the energy consumed by the primary consumers of a community.

ground tissue Plant tissue excluding the dermal and vascular tissues.

growth factor Peptide hormone that regulates the growth of cells and tissues.

growth rate, *r* Increase in the size of a population per unit of time.

guanine A nitrogenous purine base found in DNA and RNA.

guard cells Cells that form the opening of a stoma. The stomata open when the guard cells swell and close when they shrink.

gut The intestinal (digestive) tract of an animal.

gymnosperm Seed plant that produces cones for reproduction.

habitat Where an organism lives, its address.

habituation A phenomenon in which an animal learns not to respond to a repeated stimulus.

haploid Said of a cell that contains one set of chromosomes. The chromosomes may be either maternal or paternal in origin.

Hardy-Weinberg equation The equation for calculating allelic and genotypic frequencies within a population.

heart The pump of an animal's circulatory system.

heartbeat The rhythm made by the contraction of the heart muscles. A single contraction of the heart muscles.

hemoglobin A protein with four subunits that is the main carrier of oxygen in many animals. It has a heme group joined to four amino acids chains.

herbivore An animal that eats plants.

heterotrophic An organism that cannot synthesize its own food. Heterotrophs rely on, that is, eat, other organisms, autotrophs, to do this.

heterozygote Organism that has inherited two different alleles for a same trait from each parent.

heterozygote advantage A case in which a heterozygote for a given pair of alleles is of superior fitness to either homozygote.

heterozygous Having two different alleles for the same trait.

hexose A monosaccharide that contains six carbon atoms. An example is fructose.

hierarchy A ranking that progresses from either lower to higher or higher to lower order. For example, our current classification system is a hierarchy from more to less inclusive.

higher order structure The level of molecular organization that determines the function of a protein.

histology The study of tissues and their organization into organs. Examination of microscopic anatomy is made possible by a variety of techniques that allow thin sections of tissues to be stained and studied.

homeostasis The maintenance of an equilibrium. Feedback control allows organisms to maintain the optimal conditions for physiological processes.

homeostatic control system System that keeps the body's internal environment stable.

homologous chromosomes Chromosomes that determine the same traits. Homologues usually occur in pairs in which one is maternal and the other is paternal in origin.

homologous structures Structures or organs that are similar and have a similar evolutionary origin. Though they have a common origin, homologous structures may have different functions.

homozygous Having the same alleles for a specific trait. The alleles may be both dominant or both recessive.

hormone A chemical messenger produced by an endocrine gland. Hormones are transported in the bloodstream to their target tissue, where they are recognized by specific receptors.

human immunodeficiency virus (HIV) The causative agent of AIDS. HIV is an RNA virus and primarily attacks the immune system.

hybrid An individual whose parents possess different genetic traits or are members of different species.

hybridization The process of producing individuals from the union of gametes that differ in one or more genes.

hydrogen The simplest of the elements, containing one proton and one electron. Two hydrogens join with oxygen to form water. Hydrogen is also important in energy storage and energy transfer.

hydrogen bond A chemical bond that is a weak bond. These bonds join hydrogen of one side group or molecule to oxygen or nitrogen of another side group or molecule. Examples are the bonds between two or more water molecules and the bonds joining complementary strands of DNA.

hydrogen ion A hydrogen atom that has lost its electron (H+). Hydrogen ions are also called protons. The concentration of hydrogen ions is what is measured by pH.

hydrolysis reaction A chemical reaction in which a larger molecule is broken down into two smaller molecules. A molecule of water is used for each bond broken. The breakdown of a disaccharide into two monosaccharides is an example.

hydrolyzed Describes a simple molecule that has undergone splitting by the addition of hydrogen and hydroxyl ions of water.

hydrophilic Said of substances that have an affinity for water. Hydrophilic molecules are usually polar substances and form hydrogen bonds with water.

hydrophobic Said of substances that tend to exclude water molecules. These include such nonpolar substances as lipids. Hydrophobic interactions are responsible for the tertiary level of protein structure and for the basic structure of the phospholipid bilayer of membranes.

hydrostatic skeleton A type of skeleton that helps animals retain their shapes against gravity and water pressure by means of internal pressure, produced by bands of muscle tightening around closed, fluid-filled chambers.

hypertonic When comparing two solutions, the hypertonic solution is the one with the higher solute concentration.

hypha A fine filament in fungi and oomycota that forms a network called a mycelium. Hyphae may be modified to perform various functions such as spore formation.

hypothalamus Part of the brain that controls behaviors, such as eating, drinking, and sexual activity, through the release of pituitary hormones.

hypothesis An educated guess. Scientists make hypotheses in an attempt to explain observations of natural phenomena.

hypotonic When comparing two solutions, the hypotonic solution is the one with the lower solute concentration.

icon A symbol that is used to explain some complex system. Icons are used to simplify complex chemical structures.

immigration The permanent relocation of new individuals into a population.

immune response The specific defense reaction initiated by an antigen invading an organism. The response may be chemical or cellular.

immune system The cells and tissues that allow vertebrates to mount a specific, defensive response to invasion by microorganisms, parasites, and other pathogens.

immunity Resistance to harmful microorganisms, poisons, and other pathogens.

impulse The nerve signal that travels along a neuron. An impulse is a series of changes in the electrical potential across the neural membrane caused by the movement of certain ions.

incomplete dominance A pattern of inheritance in which neither allele is phenotypically dominant to the other.

incurrent siphon The tube of a clam that takes in water.

independent assortment Mendel's second principle, which says that factors (genes) for each trait are independent of those for other traits. Assortment occurs during the random alignment of chromosomes during metaphase I of meiosis.

indeterminate growth Typical of plants, growth that is not restricted by a terminal point of growth, such as from an apical meristem, which forms an unrestricted number of lateral stems, branches, or shoots.

induced-fit model A model that explains how an enzyme and its substrate interact. According to this model, the enzyme assumes its final functional shape when it comes in contact with the substrate(s).

infection The invasion of an individual of one species by that of another. It usually refers to the invasion of a pathogen into individuals of an appropriate host species.

inflammation Localized protective response of the body to tissue damage involving swelling, increased heat, redness, and so forth.

inheritance The passage of traits, determined by genes, from one generation to another.

inhibit To prevent something from occurring. For example, the inhibition of the activity of an enzyme.

innate Something that is inherited, that is, determined by genes.

insect A member of the phylum Arthropoda and the class Insecta. Insects are the largest animal class in terms of the number of species. They inhabit every possible environment.

insightful learning The abrupt solution of a problem previously unsolvable.

instar An insect or other arthropod at a particular stage between molts.

instinct Any behavioral trait that is inherited and not learned.

insulin A hormone that decreases the level of blood glucose by stimulating its uptake by cell.

integrated pest management Control of pests through a combination of chemical, biological, and cultivational methods.

integrating center A homeostatic control system that processes signals from detectors and coordinates a response.

interphase The phase of the cell cycle during which a cell is not dividing. It is during this phase that the cell performs its normal activities.

integral protein Membrane proteins that penetrate the phospholipid bilayer.

interkinesis The stage between the two phases of meiotic cell division.

intermediate filament Protein filaments that form a tough and durable network within eukaryotic cells.

intestine The largely tubular portion of the digestive tract between the stomach and the anus, comprising several organs.

intrinsic factors External factors that affect some process. These factors are not a part of the process.

inversion Chromosomal rearrangement in which the sequence of DNA segment is in reverse of its normal order on the chromosome due to the repositioning of a chunk of chromosomal material.

in vitro Said of an experiment that is carried out outside of a living organism or cell. In vitro experiments are often referred to as "test tube" experiments.

in vivo Said of an experiment that is carried out inside a living organism or cell.

ion A charged atom. Ions may be positive (atoms that have lost electrons) or negative (atoms that have gained electrons).

ion channel Openings (channels) in membranes that allow for the passage of certain ions.

ionic bond A bond (attraction) between two or more ions. Ionic bonds involve at least one positive and one negative ion. They are generally weak bonds.

isotonic Said of two solutions that have the same solute concentration.

isotope A variation of an element that has a different number of neutrons. Many isotopes are radioactive and are used as radioactive labels in biological research.

joints In the skeleton, the points of attachment for cartilage and bone.

jointed-legged animals Arthropods or animals with exoskeletons, bodies, and legs with joints.

K selection Selection of species that are adapted to live in a state of equilibrium near their carrying capacity.

kidney An organ that is responsible for the elimination of metabolic wastes and also water and salt balance. It is made of functional units called nephrons.

kingdom The most inclusive taxonomic group. Most scientists accept that there are six kingdoms: Archaebacteria, Eubacteria, Protista, Fungi, Plantae, and Animalia.

kin selection Selection of genes due to individuals favoring the survival of relatives, other than offspring, who possess the same genes.

kinesis Movement of animals that is random, prompted by environmental deterioration.

kinetic energy The energy of motion, that is, the energy that is doing something or causing an effect on matter.

Krebs cycle A series of reactions that are involved in the metabolism of glucose. These reactions occur in the mitochondria, begin with citrate, and end with oxaloacetate.

labeling (radioactive) A method of making molecules detectable. This is usually done by using radioactive isotopes to replace non-radioactive atoms in the substance to be tested.

lac operon Genetic unit in bacteria that regulates the synthesis of three proteins involved in the uptake and metabolism of lactose.

lactic acid fermentation The formation of lactic acid on the cellular level under anaerobic, or oxygen-deficient conditions, or in the muscles during strenuous exercise.

large intestine The compartment of the digestive system in which water is absorbed into the blood. No chemical digestion occurs in the large intestine.

larva An immature stage in a metamorphic life cycle. Larvae usually are voracious feeders and undergo several molts.

latent learning Problem-solving skills that lie dormant until an opportunity for applying them arrives.

laws of thermodynamics The laws that govern the utilization of energy in the universe. They apply to both living and non-living things. The first law states that energy cannot be created or destroyed. The second law describes the conversion of useful energy into useless energy, which contributes to an increase in entropy, or disorder.

leaf The photosynthetic organ of plants. Leaf cells contain chloroplasts which are the organelles that convert light energy into chemical energy. Leaves also perform gas exchange.

learning The process by which an animal benefits from experience so that its behavior is better suited to environmental conditions. Learned behaviors are those changes that cannot be attributed to fatigue, sensory adaptation, or maturation.

learning set Animal behavior that results from an extension of previous experience, or learning to learn.

lek Area used by animals for communal activity such as mating display, especially among birds.

lethality The capacity to cause death.

leukocyte White blood cell. These cells develop in the bone marrow and thymus and protect the body from infection.

lichen A symbiotic association between certain fungi and green algae, forming a crustlike covering on rocks and trees.

life cycle The life history of an organism. For example, an insect life cycle includes the adult, egg, larval, and pupal stages.

life table A portrayal of the demographic makeup of a population. Life tables may include a variety of data such as age distribution, survivorship, reproductive potential, and so forth.

ligaments In the skeleton, the strong, flexible bands the connect cartilage and bones together.

light dependent Referring to reactions that only occur in the presence of light. An example is the light-dependent reactions of photosynthesis.

light energy A form of energy that consists of electromagnetic waves, which can act on a visual receptor or molecule.

light independent Referring to reactions that do not require light. An example is the light-independent reactions of photosynthesis.

light microscope A microscope that creates an image using light rays that are focused on a specimen by a condenser lens. The light rays that have passed through the specimen are magnified by objective and ocular lenses to form an image of the specimen.

lignin A hard material found in the cell walls of woody plant cells.

limnetic zone Open water of a lake, through which sunlight passes, allowing photosynthesis.

linear Referring to something in a straight line. For example, a linear sequence of amino acids characterizes a protein.

linkage The association of particular traits on a chromosome. Linked traits are evidence for the chromosomal theory of inheritance.

lipase An enzyme that breaks down lipids.

lipid A compound characterized by its insolubility in water. Lipids are soluble in organic solvents such as alcohol. They include the fats (triglycerides), waxes, oils, and steroids.

lipid-anchored protein Protein that is joined by carbohydrate to the surface of a membrane. These may be found on the inner or outer membrane surface and are thought to be involved in the transformation of normal cells into malignant cells, among other things.

littoral zone The near-shore, shallow waters of a lake or ocean where light penetrates and bottom-rooted plants grow.

liver Among vertebrates, a glandular organ that produces bile and has an important role in metabolizing and storing nutrients, minerals, and vitamins. Also detoxifies substances, removes worn-out blood cells, and synthesizes the active form of vitamin D.

locus The location of a gene on a chromosome.

logistic growth Population growth that eventually reaches an equilibrium. Logistic growth characterizes a population that achieves its carrying capacity.

logistics equation An equation used to describe the growth of a population that reaches an equilibrium.

loose connective tissue Sheets of connective tissue that surround and protect organs.

lumen The space within a tissue or tube.

lung The respiratory organ of certain animals. Lungs are internal and are characteristic of animals in terrestrial environments.

lymph The fluid of the lymphatic vessels, which contains lymphocytes.

lymph nodes In the immune system, concentrations of lymphoid tissues found scattered throughout the body that trap invaders and prevent their spread.

lymphocyte A type of white blood cell found in lymph nodes and associated with the immune system.

lymphoid tissue Tissue that produces lymphocytes.

lysosome Membrane sac (vesicle) that contains digestive enzymes found mostly in animal cells. The contents of lysosomes destroy cell debris and rid organisms of dead cells.

macroevolution Evolutionary changes over long time periods that lead to the appearance of new taxonomic groups.

macromolecule A large organic molecule of biological significance. There are four classes of commonly occurring macromolecules found in living things: carbohydrates, proteins, lipids, and nucleic acids.

macrophage Fixed or free-moving white blood cells that can absorb and destroy invading microorganisms and pieces of cell debris.

magnification The enlargement of an object when seen through a series of lenses. Microscopes use lenses to magnify images of objects.

male The sex that produces sperm. It also refers to other characteristics of the members of this sex.

male parent The parent that, through sperm, contributes one haploid set of chromosomes to each of his offspring.

malignant cell Cell that is characterized by uncontrolled cell division. Malignancy also refers to the ability to produce secondary growths, which are called metastases.

malpighian tubule Fine, thin-walled excretory tubes found in large number in the gut of insects.

mammals Members of the phylum Chordata and the class Vertebrata. They are characterized by the presence of mammary glands that are used to feed their young.

mammary gland The milk-producing gland in female mammals.

mantle A modification of the integument of the dorsal surface of a mollusk that is responsible for the secretion of a shell.

marine Said of an organism that lives in saltwater.

marsupial Pouched mammal. Marsupials' young are born immature and migrate to a pouch, or marsupium, in which they continue to develop.

marsupium The external pouch in which marsupial animals suckle still immature young.

mast cell Cells that secrete substances involved in inflammation reactions.

maternal Pertaining to the female parent. A maternal trait or chromosome is one that is inherited from the mother.

medusa The motile, sexually reproductive stage of the life cycle of a cnidarian. They are often called jellyfish.

megaspore The surviving haploid cell in the ovule that is the first cell of the female gametophyte generation.

megaspore mother cell In flowering plants, the spore found in each ovule that divides into four haploid daughter cells, one of which survives to give rise to the female gametophyte generation.

meiosis A type of cell division that results in the formation of two haploid daughter cells. Meiosis is referred to as reductional division.

membrane A thin sheet or layer of soft, pliable tissue that covers a part, lines a tube or cavity, or connects organs or structures of an animal or plant.

membrane potential The electrical potential across a membrane. The potential is the difference in the electrical charge between the outside and inside of a cell.

membranous organelles Organelles that are bounded by a membrane. They include the mitochondrion, endoplasmic reticulum, Golgi apparatus, and chloroplast. Membranous organelles are absent in prokaryotic organisms.

memory cells Specialized lymphocytes that act as the memory of the immune system after exposure to an antigen.

Mendel's law of independent assortment Second of Gregor Mendel's laws of inheritance, which states that the alleles of one gene are passed to offspring independently of the alleles of other genes.

Mendel's law of segregation First of Mendel's laws of inheritance, which states that a heterozygous parent contributes only one of its two alleles for a trait to each offspring and that the allele donated is random.

menstrual cycle Monthly cycle of ovulation among humans and some other primates. If fertilization does not occur, the thickened lining of the uterus is shed along with the unfertilized ovum in a short period of bleeding (menstruation).

meristem Clusters of cells in plants that retain their ability to divide, thereby creating new cells.

messenger RNA (mRNA) A strand of RNA that is complementary to the DNA sequence for a gene. Messenger RNA molecules provide the information required for the synthesis of proteins.

metabolic pathway In cells, enzyme-catalyzed chain-reactions that build up or break down complex molecules.

metabolism All of the chemical reactions that occur in living organisms. Metabolism may be divided into anabolism (synthesis) and catabolism (breakdown) reactions.

metamorphosis A marked change or transition between developmental stages; a type of life history in which an animal goes from one form (stage of development) to a radically different form. For example, an insect that has egg, larval, pupal, and adult stages.

metaphase The phase of cell division during which chromosomes randomly line up in the center of the cell. During mitotic metaphase, individual chromosomes line up. During metaphase I of meiosis, homologous pairs of chromosomes line up.

metastasis The spread of cancerous cells in the body to a new site or tissue.

microbiologist A person who studies microbes, or bacteria.

microevolution Genetic change within a species that does not necessarily result in the emergence of a new species, for example, pesticide resistance in insects.

microfilament A polymer of actin protein. Microfilaments are a component of the cytoskeleton and are involved in maintaining cell shape and carrying out cell movements.

microorganism Organisms that can only be viewed through a microscope. They are usually unicellular.

microscope A device or apparatus that produces enlarged images, thus enabling small objects to be seen. There are several types of microscopes, including light and electron microscopes.

microspore In plants, the haploid cell that gives rise to the male gametophyte generation.

microtubule Hollow, thin, protein cylinders that are found in such organelles as cilia, flagella, and centrioles. They generally are arranged in pairs in a circular pattern around a central core.

microvillus Small, fingerlike projection from the surface of an epithelial cell.

migration Movement of a population into or out of an area, or periodically between habitats.

mimicry, Batesian When one species benefits from resembling the coloration of a dangerous species as a strategy to confound or discourage predators.

mimicry, Müllerian When different species, each of which is dangerous, announce their undesirableness to predators with similar coloring characteristics.

mitochondria Cell organelles in which aerobic respiration takes place. Large quantities of ATP are made in mitochondria.

mitosis A type of cell division that results in the formation of four haploid daughter cells.

molecular biology The study of the components of cells and the role of molecules in living systems. For example, the study of DNA and its role as the genetic material is an active area of molecular biology.

molecule The smallest particle of a substance that can exist separately and retain the characteristic properties of the substance. A molecule is a combination of atoms that forms a specific compound.

mollusks The phylum Mullosca, comprising soft-bodied, unsegmented animals.

molting The shedding of an exoskeleton; the process of growth in organisms with an exoskeleton.

monocot A flowering plant characterized by the presence of one cotyledon in the embryo. Monocots also have parallel leaves, flowers in groups of threes or sixes, and scattered vascular bundles.

monogamy Having one mate for life.

monogenic Describes a trait determined by a single gene.

monohybrid cross A cross in which the parental genotypes are different only for one trait.

monosaccharide A simple sugar that usually contains five or six carbon atoms along with hydrogen and oxygen atoms. Monosaccharides are the building blocks for more complex sugars.

morphology The study of form and shape of organisms.

mosses Simple nonvascular, multicellular plants that are members of the division Bryophyta.

motility Capacity for spontaneous movement. In the digestive system, the movement of food and digestive secretions through the digestive tract.

motor neuron A neuron that sends messages from the brain and spinal cord to the tissues and organs that effect a response.

multicelled An organism comprising more than one cell. Cells in these organisms become specialized to perform different functions.

multinucleated Said of a cell that has more than one nucleus.

muscle An organ responsible for movement. A muscle contains muscle tissue as well as other tissues, such as nervous tissue and blood.

mutant Organism or cell that carries altered genetic material that makes it physically different from its parents.

mutation A permanent change in the genetic material of a cell or organism. Mutations can be inherited from generation to generation.

mutualism A symbiotic relationship in which both organisms receive some benefit.

mycelium (mycelia) A filamentous fiber (hypha) that forms the main body of a fungus. Mycelia sometimes are underground.

mycorrhizal association A relationship between soil fungi and the roots of vascular plants that helps the plant extract water and minerals from the soil.

myelin Multilayered lipid and protein covering formed around axons of many peripheral and central nervous system neurons.

myosin A structural protein that interacts with actin to cause muscle contraction.

naturalist A person who studies natural history. Natural history is the informal study of plants and animals. Naturalists consider classification, ecology, geographic distribution, life cycles, and economic importance of the organisms in which they are interested.

natural selection Theory of evolution proposed by Darwin and Wallace. It states that variations in a population that are favorable survive, while those less favorable do not.

nature and nurture The combination of influences determining a trait. Nature refers to the aspects of a trait that are influenced by genes, and nurture refers to those influenced by the environment.

negative feedback A control mechanism in which a product from a later step in a process inhibits an earlier step to slow down the entire process.

nematocyst A stinging cell of a cnidarian. Nematocysts are released to capture prey and for protection.

nephridia Tubes surrounded by capillaries found in an organism's excretory organs that remove nitrogenous waste and regulate the water and chemical balance of bodily fluids.

nephridial pore Pore in the nephridial tube which allows elimination of wastes.

nephron The functional component of a kidney.

nerve A bundle of several nerve fibers that are enclosed within connective tissue.

nerve impulse A signal that is transmitted by a neuron. An impulse is propagated by changes in the ionic transport of the neural membrane.

nervous system The system of nervous tissue that regulates and controls the activities of an animal.

net primary productivity The portion of energy trapped by all the plants and algae in a community that is stored in their tissues.

net secondary productivity The portion of the gross secondary productivity that is stored by primary consumers and available to secondary consumers.

neurohormone Hormone produced by neurosecretory cells, usually in the brain.

neuron A cell of the nervous system that carries impulses.

neurotransmitter A chemical signal that transmits nerve impulses from one neuron to another.

neutral mutation A mutation that creates no advantage or disadvantage in terms of natural selection.

neutron A subatomic particle located in the nucleus of an atom. Neutrons carry no charge and along with protons constitute the atomic weight (mass) of the atom.

neutrophil A type of white blood cell involved in destruction of invading microorganisms.

niche The position an organism occupies in its habitat and environment.

nicotinamide adenine dinucleotide (NAD) An important molecule in cell metabolism. NAD is involved in oxidation-reduction reactions whereby the oxidized form, NAD+, is reduced to NADH. It is considered to be an energy (electron) carrier molecule.

nicotinamide adenine dinucleotide-phosphate (NADP) An important molecule in cell metabolism. NADP is involved in oxidation-reduction reactions whereby the oxidized form, NADP+, is reduced to NADPH. It is considered to be an energy (electron) carrier molecule.

nitrogen cycle The set of processes by which nitrogen circulates between the atmosphere and the biosphere.

nitrogen fixation The conversion of atmospheric nitrogen gas (N^2) into ammonia (NH^3) by activity of bacteria and cyanobacteria.

nodule A local swelling or thickening as part of an organism's structure. A knob-like outgrowth.

non-competitive inhibition A type of enzyme inhibition in which an inhibitor molecule binds to a site other than the active site of an enzyme. The result is a change in the shape of the enzyme and its active site. The substrate can no longer bind to the active site, and the reaction does not occur.

noncyclic photophosphorylation The formation of ATP by the transfer of electrons from water to NADP+ in the light-dependent reactions of photosynthesis.

nondisjunction An instance when paired chromosomes fail to separate during mitosis or meiosis.

non-polar Substances that have no electrical charge.

non-virulent Referring to being non-infectious or not disease causing.

nuclear membrane (nuclear envelope) A double membrane that surrounds the nucleus. It is selectively permeable and is characteristic of all eukaryotic organisms.

nucleic acid Molecules built from building blocks called nucleotides. Nucleic acids include deoxyribonucleic acid (DNA) and ribonucleic acid (RNA).

nucleolus A cell organelle located within the nucleus of eukaryotic organisms. It is the site of ribosome synthesis.

nucleotide The building block of nucleic acids. Nucleotides consist of phosphate, a pentose sugar, and a nitrogen-containing base.

nucleus The control center of the eukaryotic cell. It contains most of the genetic material of a cell.

nutrient broth A medium used to culture organisms. It provides the necessary ingredients for the life of those organisms.

nutrient cycling The cycling of components through ecosystems.

obligatory mutualism When two species benefit from a symbiotic relationship and are mutually dependent on each other.

observation An instance, or the act, of noticing or recording. This is the first step of the scientific method.

oil A fluid lipid. Oils are insoluble in water but soluble in organic solvents.

omnivore An animal that eats both plants and animals.

operant conditioning Use of rewards to increase the frequency of a learned behavior.

operon A regulatory unit in prokaryotic cells that controls the expression of structural genes.

optimal Said of the ideal condition for some process to occur. It is usually represented by a range of conditions.

optimality theory Theory stating the behaviors that evolve are those that produce the greatest fitness.

order A level in the taxonomic hierarchy in which families are grouped together.

organ A group of tissues that work together to perform similar functions. For example, the vertebrate heart contains muscle, nerve, connective, blood, and other tissues.

organ system A group of functionally related organs.

organizational effect One of two types of hormonal influences on behavior, expressed early, usually, and lasting indefinitely. See *activational effect.*

organelle Cellular structure that performs a specific function. Organelles are found in all eukaryotic cell types.

organic Referring to a molecule that contains carbon.

orgasm Sensory and motor events involving the ejaculation of semen in males and the involuntary contractions of uterine muscles in females at the climax of sexual intercourse.

osmosis The movement of water across a selectively permeable membrane. Osmosis is influenced by the concentration of solute on both sides of the membrane. It is a type of diffusion and does not require energy.

ovary The organ in which ova mature. It is the female gamete-producing organ

ovulation The shedding of an egg or eggs from the ovary.

ovule The part of the ovary in flowering plants that develops into a seed.

oxidation The loss of electrons during metabolism.

oxidation-reduction reactions Reactions involving the transfer of electrons from one molecule to another. One molecule loses electrons (oxidation), while the other gains electrons (reduction).

oxidative phosphorylation A series of oxidation-reduction reactions that passes electrons from higher energy levels to lower energy levels. As a result, ADP is phosphorylated to form ATP. These reactions are often referred to as electron transport.

oxytocin A hormone that is responsible for uterine contractions during the birth process.

pacemaker cells Cells that determine the rhythm or motion of other cells or tissues, as in the heart.

palisade layer Layer of cells just beneath the epidermis of leaves that functions in photosynthesis.

pancreas An organ of digestion that also has endocrine function. It produces a variety of digestive enzymes that are secreted into the small intestine. The pancreas also produces such hormones as insulin and glucagon, which are involved in glucose metabolism.

parasite An organism that must live some of its life inside another organism (host). This type of symbiotic relationship only benefits the parasite and harms the host.

parasympathetic Those parts of the autonomic nervous system controlling involuntary movement of blood vessels, glands, genitals, and so on.

passive transport The simple diffusion of small uncharged molecules, or mediated transport of ions and other molecules, across a biological membrane, requiring no energy.

paternal Pertaining to the male parent. A paternal trait or chromosome is one that is inherited from the father.

pathogen Refers to a disease-causing organism. For example, a bacterium that causes a certain disease.

pedigree A diagram that charts the inheritance of a trait among members of successive family generations.

pentose A monosaccharide that contains five carbon atoms. An example is ribose.

pepsin An enzyme that breaks down proteins.

peptide A small chain of amino acids linked by peptide bonds.

peptide bond A covalent bond joining two amino acids. These bonds are formed by condensation reactions.

peripheral nervous system That part of a vertebrate's nervous system that exists independent of the brain and spinal cord, involving nerves in limbs, organs, and the trunk.

peripheral protein A membrane protein that does not penetrate the phospholipid bilayer.

peristalsis The muscular contractions of the digestive system. Peristalsis moves food from one compartment to the other.

peristaltic wave A succession of muscular contractions along the walls of a hollow muscular structure, such as the digestive tract.

peroxisome Microbody (membrane vesicle) that contains a variety of enzymes. Peroxisomes are involved in the conversion of harmful peroxides into harmless products.

petal One of the parts of the corolla in a flowering plant. Petals are often colored and prominent to attract pollinators or scare prey.

pH A measure of the hydrogen ion (proton) concentration of a solution. It is measured on a scale from 1 to 14 where 7 is neutral, above 7 is basic, and below 7 is acidic.

phagocyte A cell that can absorb and destroy microbes, cell debris, and other foreign matter.

phagocytosis The process by which phagocytes ingest particles of matter.

pharynx A chamber in the back of the mouth. It serves as an entrance to the digestive and respiratory systems.

phenotype The physical appearance of a trait. It is the expression of the genotype. The same phenotype may be produced by different genotypes.

pheromone Chemical signal sent by one organism to another, usually of the same species, by release at a distance of a small quantity. Usually an attractant, although it can also be an alarm.

phloem Plant vascular tissue that transports sugars from the leaves to other parts of the plant.

phosphate A functional group that consists of phosphorus and oxygen.

phosphate bond A bond that joins phosphate to such molecules as ATP. It is easier to break than many covalent bonds and therefore readily releases energy into the cell.

phospholipid A molecule that is similar to triglyceride (fats), but instead of three fatty acid chains, it has two. The third fatty acid is replaced by a phosphate bonded to a polar group. Phosopholipids are the main components of biological membranes, where they occur as a bilayer.

phospholipid bilayer The arrangement of phospholipids in a bilayer in membranes. The phospholipids are oriented with their polar regions facing the inner and outer surfaces, and their nonpolar regions facing inward.

phosphorylation The addition of phosphate to some molecule. For example, ADP is phosphorylated to form ATP.

photoautotroph Organism using light as an energy source and carbon dioxide as the main source of carbon, such as green plants and bacteria.

phototropism The growth response of a plant to light.

photosynthesis A series of energy reactions that converts light energy into chemical energy (light-dependent reactions). Photosynthesis reactions also produce energy-rich carbohydrate molecules (light-independent reactions). These reactions occur in the chloroplasts of eukaryotic plant cells.

photosystem An aggregate of chlorophyll and other accessory pigments in a chloroplast. It is the chlorophyll in a photosystem that responds to light energy.

phylogenetic tree A representation of the key evolutionary divergences of organisms. Such things as DNA sequences, morphological characteristics, and embryological events are used as criteria.

phylum A major taxonomic group that includes classes, orders, families, genera, and species.

physiological longevity The life-span of an individual under ideal conditions.

physiological processes How part of an organism functions.

phytoplankton Microscopic, photosynthetic organisms that live near the surface of seas and bodies of freshwater.

pigment A molecule that appears colored under the proper light conditions. Pigments absorb light of various specific wavelengths.

piloting A method by which animals orient, using sequential recognition of landmarks.

pilus Slender, hairlike structure involved in the transfer of DNA in conjugating bacteria.

pituitary gland A small endocrine gland at the base of the brain, often called the master gland of the body.

placenta The structure in mammals through which the mother and embryo or fetus exchange nutrients and waste.

plankton Microscopic photosynthetic organisms that as a result of photosynthesis supply a large quantity of oxygen to the atmosphere.

plant A eukaryotic, autotrophic, multi-celled organism with cells that possess a cell wall.

plasma In the circulatory system, the fluid portion of the blood.

plasma membrane The membrane that surrounds all cells. It regulates exchange between the cell and its environment.

plasmid A small circle of DNA in a bacteria that is not its own.

plasmodia Membrane-covered portions of cytoplasm that flow along a surface. Plasmodia are multinucleated.

play Behavior that is not directed to any immediate needs or goals but that often results in learning.

pleiotropy Where a single gene produces two or more phenotypic effects.

point mutation A change in a single base pair in DNA.

polar Referring to molecules that have localized areas of charge.

polar covalent bond A bond that results from the unequal sharing of electrons between two or more atoms. Polar covalent bonds are generally strong bonds and are important in biological molecules.

pollen grain The microspore of a seed plant that contains the male gametophyte.

pollen tube The tube that develops when the pollen grain germinates. It contains nuclei that participate in fertilization.

pollination The transfer of pollen from an anther to a stigma. This may be done by wind, insects, birds, or other animals. Pollination may also be done artificially.

polyandry When females mate with numerous males.

polygenic, or quantitative Describes a phenotype influenced by two or more genes at different loci.

polygyny When males mate with multiple females.

polymerase chain reaction (PCR) A technique in which a tiny amount of DNA is duplicated many times to yield quantities large enough to analyze.

polyp The non-motile, vegetative stage of the life cycle of a cnidarian. It may reproduce asexually by budding.

polyploid An organism or cell containing three or more complete sets of chromosomes. Rare in animals but common in plants.

polysaccharide Molecules that contain three or more monosaccharides. Biologically important ones include cellulose, starch, and glycogen.

population A group of individuals of the same species that occupy the same geographical area. Members of a population tend to interbreed.

population genetics The study of how genetic principles apply to interbreeding groups (populations).

pore An opening.

porin Protein in the outer membrane of mitochondria that allows small molecules to pass in or out.

positive feedback A control mechanism in which a product from a later step in a process signals an earlier step to enhance the entire process.

potential energy The energy that is stored or inactive and that could be released to do work.

predator The pursuing or consuming species in a predator-prey interaction.

prediction An explanation of expected outcomes from an experiment made before the experiment is carried out.

prey The organisms eaten in a predator-prey interaction.

primary consumer The first trophic level of a food chain. These are herbivores.

primary growth Growth from the apical meristems of plants, resulting in an increase in the length of shoots and roots.

primary production A measure of the production of food by the autotrophs of an ecosystem. It is a measure of photosynthesis.

primary structure In proteins, the amino acid sequence. In nucleotides, the nucleotide sequence.

principle of segregation The principle proposed by Mendel to explain how traits are inherited. It says that traits are determined by two factors (alleles) that segregate (separate) during gamete production, resulting in one factor present for each trait per gamete.

producers Those organisms that convert light energy into chemical energy. They are the autotrophs.

product The end result of a chemical reaction. Reactants undergo chemical change to form the products.

profundal zone Deep, open water of a lake or ocean where it is too deep for photosynthesis to occur.

prokaryote Type of cell that lacks a nucleus. Prokaryotics also do not possess membranous organelles.

promoter sequence A short segment of DNA to which RNA polymerase attaches at the start of transcription.

propagation The transfer (movement) of an impulse (action potential) down a neuron.

prophase The phase of cell division during which the chromosomes become visible. Also, the time when the nuclear membrane breaks apart and the spindle fibers attach to the centromeres of chromosomes.

protease An enzyme that specifically destroys the properties of proteins.

protein A macromolecule composed of amino acids joined together by peptide bonds (covalent bonds). Proteins are synthesized from information stored in the genetic information of cells (DNA).

protista The kingdom comprising unicellular eukaryotes. Protists may be plant-like, fungus-like, or animal-like. Some protistans may form colonies that appear multicelled.

protocell Name given to a hypothetical stage in the evolution of cellular life forms.

proton Positively charged subatomic particle found in the nucleus of atoms. The number of protons in the nucleus is the atomic number of an atom. The number of protons also equals the number of electrons in a neutral atom.

proton pump An active transport system in a membrane that moves ions from one side of the membrane to the other.

protostome A lineage of animals whose first embryonic opening (the blastopore) forms the mouth of the animal.

protozoan One of the unicellular, eukaryotic protists, which exhibit great variety in size, motility, nutrient source, and life cycle.

pseudocoelom A body cavity in an animal that is not formed from mesoderm and therefore is not a true coelom.

pseudopod A cytoplasmic extension that allows for locomotion of cells and such organisms as amoebas.

pulmonary system The portion of the circulatory system that transports blood from the heart to the lungs and back to the heart.

punctuated equilibrium A theory that proposes phases of rapid evolution of new species, followed by long periods of little or no change.

Punnett square A method for determining the genotypes of offspring produced from the gametes of two parents. It combines all the male alleles with all the female alleles.

pupa A stage in the life cycle of insects. It is the stage during which adult tissues develop from embryonic cells.

pupate When an insect passes through the developmental phase that transforms it from larva to adult.

pure-breeding Said of a trait that breeds the same from generation to generation. It is a homozygous trait. Pure-breeding traits are sometimes referred to as true-breeding traits.

purine One of the two major categories of bases in DNA and RNA. The purines in both DNA and RNA are adenine and guanine.

pus Liquid at an inflammation site containing leukocytes and cell debris.

pyloric valve Muscular valve between the stomach and duodenum.

pyrimidine One of the two major categories of bases in DNA and RNA. The DNA pyrimidines are thymine and cytosine. In RNA, thymine is replaced by uracil.

questioning Part of the scientific process resulting from observation and leading to experimentation.

r selection Selection of species that are adapted to maximizing their reproductive rate.

R group Functional group of atoms that is attached to molecules. R groups convey special properties to those molecules. Some examples of such groups are the amino and carboxyl groups found in amino acids.

radial symmetry A form of symmetry in which a number of similar parts radiate from a central axis. A cut pie has radial symmetry.

radioactive labeling A method of making molecules detectable. The labeling of compounds is usually done by substituting a radioactive isotope for the corresponding nonradioactive atoms in the substance to be tested.

radioactivity A substance that emits radiant energy such as alpha, beta, and gamma particles from atomic nuclei.

range The geographic area in which a population or species lives and, especially, seeks food.

reactant One or more molecules that begin a chemical reaction. They are chemically changed as a result of the reaction.

reaction The process by which reactants undergo chemical change to form some product(s). Energy may be released or stored during a chemical reaction. Energy is also transferred during a reaction.

receptor A molecule or cell that responds to a stimulus from outside or inside of an organism.

receptor protein Protein on the surface of postsynaptic cells that binds the neurotransmitter released by the presynaptic cell during transmission of an electrical impulse.

recessive trait A trait that is not expressed when the allele for it is present with a dominant allele.

recognition site Specific DNA sequences that are recognized and cut by enzymes.

recombinant DNA Laboratory-created DNA molecules that contain DNA sequences from different biological sources.

recombination The exchange of chromosomal segments by crossing over during prophase I of meiosis. As a result, genes are exchanged between homologous chromosomes.

rectum The place where food waste is stored until it is expelled from an animal's body.

red algae Common name for Rhodophyta. Photosynthetic, multicellular organisms living in marine environments, with both plant and protist characteristics.

red blood cells The cells that transport oxygen.

reduction The gain of electrons by molecules during cellular metabolism. Reduction is one half of oxidation-reduction reactions.

reductional division The cell division called meiosis, in which four haploid cells are produced. The cells produced are either sperm, eggs, or spores.

reflex An immediate and unconscious response to a stimulus. A reflex is very quick and in humans passes directly through the spinal cord, initially bypassing the brain.

refractory period The period when a neuron cannot be stimulated. It is a result of hyperpolarization of the neuron.

regeneration The ability to replace (regenerate) lost body parts. Less specialized animals can do this better than more specialized ones.

regulator protein A membrane protein that regulates cellular activities.

regulatory sequence DNA sequence that directs the production of protein by regulating the transcription of genes.

releasing factor A factor that is needed for some other substance to be released, usually into the blood.

release-inhibiting hormone Type of hormone produced by hypothalamus that blocks secretion of other hormones by the anterior pituitary.

releasing hormone Small peptides that are produced by the hypothalamus and induce the pituitary gland to release growth hormones.

renal corpuscle The glomerulus enclosed in a capsule; where blood is filtered.

renal process One of three main functions of kidneys, including filtration, reabsorption, and secretion.

replication Making an exact copy of something. An example is when DNA replicates prior to cell division.

reproduction The formation of new individuals either sexually (through the fusion of gametes) or asexually (by fission or budding from an existing organism).

resolution The ability to see two neighboring objects as distinct objects. Better microscopes have better resolving powers.

resource partitioning The use of resources that are scarce in a given environment by different species at different times, different places, or in different ways.

respiratory system The system that is responsible for the exchange of carbon dioxide and oxygen with the atmosphere.

resting potential The potential across the membrane of a neuron that is not being stimulated.

restriction endonuclease An enzyme of microorganisms that recognizes short DNA sequences and cuts the double helix at a particular location. Widely used in genetic engineering.

retrovirus Virus that has RNA as its genetic material. Retroviruses get their name because they must reverse the transcription process to reproduce.

rhizobia Nitrogen-fixing bacteria residing in root nodules in plants such as legumes.

rhizome A primitive root that is used for support and water and nutrient acquisition.

ribonucleic acid (RNA) The molecule that is necessary for genes (DNA) to be translated. RNA is synthesized in the nucleus of eukaryotic cells. It is also the genetic material of some viruses.

ribose A pentose sugar that is one of the components of an RNA nucleotide.

ribosomal RNA RNA that is found in ribosomes. Its function is not well understood, but it has been implicated in the translation of messenger RNA. The sequence of nucleotides in ribosomal RNAs may be used as a tool for classification.

ribosome The cellular site for the synthesis of proteins. In eukaryotic cells, ribosomes are synthesized in the nucleolus but function in the cytoplasm.

RNA polymerase Enzyme responsible for transcribing RNA.

RNase An enzyme that destroys the properties of RNA.

root Plant organ that supports the plant and takes up water and nutrients from the soil.

root hair Unicellular root outgrowths that are involved in the uptake of water and solutes.

root nodule Structure on some roots that contains nitrogen-fixing bacteria.

root pressure The force behind the movement of water and nutrients up a plant in the xylem. It is caused by the movement of water by osmosis into the tissues of the plant.

rough endoplasmic reticulum (RER) The endoplasmic reticulum that has ribosomes attached to it. This is the site of protein synthesis.

roundworms Members of the phylum Nematoda. A group of slender, unsegmented worms, both parasitic and free-living.

salivary gland Gland that secretes saliva, a watery fluid containing digestive enzymes, into the mouth.

saprophyte An organism that obtains its nutrients by absorbing them from dead organisms. In the process, saprophytes act as decomposers.

saturated Describes fatty acids with a fully hydrogenated carbon backbone.

scavenging Feeding on the remains of animals that died or were killed by other predators.

scrotum A skin-covered pouch that contains the testes and their accessory structures.

secondary consumer The species that makes up the second trophic level of a food chain. This may be a carnivore or an omnivore.

secondary growth In plants, growth from the cambium, resulting in increased diameter of stems and roots.

secretion Production and release of a useful fluid from a cell or gland. In the digestive system, the release of digestive enzymes, acids, and other substances that aid digestion.

seed In flowering plants, the mature ovule, comprising an embryo and nourishment tissue called endosperm.

segmentation The division of an animal's body into distinct units (segments). Sometimes these units are repeated. In digestion, the rhythmic pattern of contraction and relaxation of small regions or segments of the small intestine.

segmented worms Members of the phylum Annelida. A type of invertebrate with distinct and repeated ring-like body parts.

selective permeability A process that restricts the movement of materials across membranes. It is also known as semi-permeability.

self-fertilize A cross in which the same organism is both the male and female parent; that is, both the sperm and the eggs come from the same parent. Many plants are capable of self-fertilizing.

semen Male reproductive fluid containing sperm.

semiconservative replication The mechanism by which DNA replicates. For each molecule formed, one old strand is conserved and a new complementary strand is synthesized.

semilunar valves One-way valves in the ventricles of the heart that prevent blood that has been pumped into the pulmonary arteries or the aorta from flowing back into the ventricles.

seminiferous tubule Small, coiled tube in the testis of mammals where semen is produced.

sense strand The strand of DNA that is transcribed into messenger RNA.

sensory nerve A nerve that conducts sensory impulses into the central nervous system.

sensory system The system that is responsible for receiving stimuli from the environment. It includes sense organs and sensory neurons.

sepal The outermost layer of a flower's casing.

serial monogamy The tendency or practice of taking one mate at a time, but having a succession of mates over an individual's life span.

sex chromosomes The chromosomes that are different in males and females and that determine the sex of an organism.

sex hormones Hormones that initiate and maintain secondary sexual characteristics.

sex linkage The association of a trait with one of the sex chromosomes, the X or Y chromosome.

sexual reproduction Reproduction that requires gametes (sperm and eggs). The sperm unites with the egg to form a zygote.

sexual selection The natural selection of adaptations that improve the chances for mating.

shoot In angiosperms, the system consisting of stems, leaves, flowers, and fruits.

sickle-cell anemia A disease of the blood caused by a point mutation in the beta chain of human hemoglobin. The result is a change in shape of the red blood cells from a normal ovoid shape to a sickle shape.

side group, or residue See *R group*.

sieve tube A tube in phloem made of connected cells that transports nutrients from leaves to other parts, such as roots.

sieve-tube member A food-conducting cell found in the phloem tissue of plants.

signal pathway One of two systems, neural and hormonal, that transmits messages between the homeostatic control systems.

sinoatrial node A compact mass of cardiac muscle fibers that establishes the heart's rhythm. Sometimes called the "pacemaker."

sister chromatids The two copies of a replicated chromosome.

skeletal muscular system The skeleton and associated muscles that are responsible for movement of an animal.

skeleton A hard structure that supports an animal. It also serves as the attachment site for such structures as muscles, wing, and legs. A skeleton may be external (exoskeleton) or internal (endoskeleton).

skin An organ that serves as a cover for animals.

sliding filament theory of muscle contraction Explains the sliding of thick and thin muscle fibers along each other during striated muscle contraction.

small intestine The narrow upper section of the gut that is the longest compartment of the digestive system. It is here that most chemical digestion and nutrient absorption into the blood occur.

smooth endoplasmic reticulum (SER) The network of membranes found in the cytoplasm that, for example, synthesizes lipids and detoxifies harmful substances. The smooth endoplasmic reticulum gets its name because it does not have ribosomes attached to it.

smooth muscle tissue Non-striated sheets of muscle, not under voluntary control, found in alimentary canal, arteries, and in many other organs.

social learning Learning that occurs as a consequence of teaching by other members of a population.

sociobiology The study of the genetic and biological causes of social behavior that stresses evolutionary and ecological processes.

solar tracking When leaves adjust their position to follow the sun throughout the day.

solubility The extent to which a solute dissolves in a solvent. The solubility of a substance can influence its transport across membranes.

solute The solid portion of a solution. Solutes may be dissolved or suspended in the solvent (the liquid portion of a solution).

solution A mixture consisting of a solvent (liquid) and solid material (solute).

solvent The liquid portion of a solution. In living things, it is water.

somatic nervous system Portion of the peripheral nervous system that carries sensory signals to the central nervous system and muscle commands away to the muscles.

speciation The evolution of new species from an existing species.

species The most specific taxonomic group and the second name in the binomial system. Members of a species are closely related and can interbreed.

specificity The precise, unique action of a hormone on its target cell.

sperm Male gametes produced by meiosis and containing a haploid set of chromosomes.

spicule A structure containing silicon or calcium that forms the skeleton of sponges, corals, and certain protistans.

spinal cord An elongated cord of nervous tissue. It extends from the brain down the central axis of an animal. The spinal cord and the brain compose the central nervous system.

spindle apparatus (spindle fibers) An arrangement of microtubules that is involved in the movement of chromosomes during cell division. The spindle apparatus is found in all animals but is absent in many plants.

spiny-skinned animals Members of the phylum Echinodermata. Marine animals with radial symmetry.

spiracle An opening other than the mouth that facilitates respiration in arthropods.

spleen A lymphoid tissue that removes aging or damaged blood cells and destroys them.

sponges Members of the phylum Porifera. Simple, multicellular animals, without muscles or nervous systems, that remove nutrients from water through passive filtration.

spongy tissue Loosely packed layer of leaf cells beneath the palisade layer.

spore A single haploid or diploid cell that is produced by either meiosis or mitosis. Spores function in both sexual and asexual reproduction. They are produced in specialized structures called sporangia.

sporophyte The multicelled diploid form of an alternation-of-generations life cycle. It gets its name because it produces spores by meiosis.

squamous Type of epithelial tissue in which cells are flat and thin.

stabilizing selection Selection that operates against the extremes in the distribution of a particular trait in a population and, therefore, tends to stabilize the population around the mean.

stamen Male reproductive organ of a flower.

starch A polysaccharide built from many glucose molecules. Starch is a storage form of carbohydrates found in plants and other organisms.

stem The above-ground support organ of a plant. The stem also serves as an attachment site for leaves and reproductive organs such as flowers and cones.

sterile Said of anything that is without any pathogenic microorganisms.

sterol (steroid) A type of lipid with long side chains and at least one hydroxyl group. Sterols are a component of cell membranes, the most common one being cholesterol.

stigma The terminal portion of the style in flowering plants. It receives pollen.

stimulus An external or internal signal that triggers a series of events resulting in some response.

stomach A main compartment of an animal's digestive system. Digestion of proteins begins here, and partially digested food is stored here until it passes into the small intestine. Alcohol is also absorbed here.

stomate Opening on the undersurface of a leaf. Stomates (stomata) open to allow for gas exchange.

striated muscle Striated, or skeletal, muscle is named for striations (bands) that can be seen traversing the muscle fibers. The striations are the contractile proteins that allow muscles to contract.

stroma The space between the thylakoid membrane and the outer membrane of chloroplasts.

stromal thylakoid Unstacked thylakoid that contains grana.

style The tall, slender portion of the pistil that bears the stigma.

subcanopy The topmost branches of trees that form a layer below the canopy.

substrate One or more of the reactants of a chemical reaction involving an enzyme. It is what the enzyme interacts with.

subsurface layer The soil of a forest floor.

succession Periodic changes in the composition of a community. The gradual progression goes from the initial inhabitants to some constant, climax community.

sucrose A disaccharide formed from two monosaccharides, glucose and fructose. It is commonly called table sugar.

sugar The common name for a carbohydrate.

surface floor The vegetation covering the ground of a forest.

surface-to-volume ratio The relationship of the surface area of a cell to the volume inside it. It indicates the effectiveness of such things as membrane transport. Cells cannot become too large because too little surface area cannot sustain a large volume.

surfactant Substance produced in the lungs that reduces surface tension.

symbiont One of the individuals who benefits from an instance of symbiosis.

symbiosis An arrangement in which two or more organisms live closely together.

symmetry A repetition of form around an axis or point in the body plan of an organism.

sympathetic Part of the autonomic nervous system that controls such actions as digestion, heartbeat, and excretion.

sympatric speciation Evolution of species that occurs in populations with overlapping distributions.

synapse The point where an impulse is transferred from one neuron to another.

synapsis The pairing of homologous chromosomes during prophase I of meiosis.

synaptic cleft The space between neurons. It is into this space that neurotransmitters are released.

synergism An instance of the combined effect of several actions being greater than the sum of the individual actions.

synthesize To combine more than one thing into one. An example is the combining, or synthesis, of a larger molecule from several smaller ones.

systematics The branch of biology concerned with the evolutionary relationships between taxonomic groups.

systemic system The portion of the circulatory system that transports blood from the heart to the body (except the lungs) and back to the heart.

taproot The long main root of some plants, which grows straight down and from which many lateral roots grow.

target tissue The site of the activity of a hormone. Hormones are produced in endocrine tissue and exert their effects on target tissues.

taxis Movement of organisms or cells toward or away from a stimulus.

taxonomy The study of the classification of living things. A taxonomist places organisms into categories called taxa.

telophase The phase of cell division in which cytokinesis occurs and daughter cells are formed. The chromosomes return to a less coiled condition and the nuclear membrane reforms.

template A model from which something can be synthesized. For example, during DNA replication, one strand serves as a template for a new strand to be synthesized.

tendons Tough bands of white connective tissue that attach skeletal muscles to bones.

tensile strength Resistance of a substance to being pulled apart. Property of water that allows it to be drawn upwards in plants.

tentacle Long, flexible process located near the mouth of an organism. Tentacles may be used for prey capture and directing food into the mouth.

terminal bud The offshoot at the end of a twig or branch. See *axillary bud*.

test cross A method to determine the genotypes of organisms that have the same phenotype. A test cross is performed by crossing each unknown with an individual that is homozygous for the traits being tested.

testing Experimentation that determines the accuracy of scientific predictions or hypotheses.

testis Male gonad that produces sperm and male hormones.

tetrad A unit of four chromatids found in prophase I of meiosis.

theory A hypothesis that has withstood numerous experimental tests.

thorax The chest.

threshold Level of stimulus necessary to trigger an action potential.

thylakoid A disklike sac formed by the chloroplast inner membrane. Thylakoids are the site for the light-dependent reactions of photosynthesis. They are stacked to form a structure called a granum (plural, grana).

thymine A nitrogenous pyrimidine base

found in DNA. Uracil takes the place of thymine in RNA.

thymus gland A gland located between the lungs that plays an important role in immunity.

tissue A group of cells that perform similar functions.

toxin A poisonous substance produced by one organism that has its effect on another organism. Toxins are often produced for protection from harm by others.

trachea The windpipe, which connects the larynx to the bronchi in vertebrates. Also, a tube that connects a spiracle to cells in arthropods.

trait A distinguishing characteristic or feature. Genetic traits are inherited.

transcription The synthesis of RNA from a DNA template. The new RNA synthesized is a complementary copy of the DNA strand, and is used to synthesize a specific protein.

transcription factor Any protein directly involved in regulating the initiation of transcription.

transfer RNA RNA that transfers amino acids to a growing peptide chain during protein synthesis. Transfer RNAs contain the anticodons that recognize the codons of messenger RNA.

transformation The change in the genetic properties of a cell by DNA of another cell.

transforming principle The substance first identified by Griffith when he observed that the genetic properties of certain bacterial cells could be permanently altered by the influence of other cells. The transforming principle was later identified as DNA.

transgenic Possessing genes derived from a different species, as in genetically engineered organisms.

translation The process by which amino acids are assembled into a protein molecule. It involves the ribosomes, messenger RNA, transfer RNA, and other molecules. The messenger RNA determines the sequence of the amino acids in the protein.

translocation Movement in general, especially within phloem of a plant, or among chromosomes.

transpiration The movement of water and nutrients up a plant by a pulling process made possible by evaporation of water from the leaf.

transport protein An integral membrane protein that regulates the movement of materials across a membrane. Transport proteins form channels or pores in the membrane. Sometimes these channels

are gated; that is, they must be opened to allow passage of materials.

transposable genetic elements Variety of DNA sequences that can insert themselves by transposition in various nonhomologous regions on chromosomes and other DNAs.

transposition The movement of a DNA sequence to another position on the same or a different DNA molecule by means of replication and insertion of the copy.

transposon Transposable genetic elements, especially those found in bacteria.

tricuspid valve Valve on the right side of the heart between the atrium and ventricle.

triglyceride A lipid consisting of a glycerol group and three fatty acids. Triglycerides are important as a food reserve, and the amount and proportion of different types of triglycerides may play a significant role in human heart disease.

trophic level A level of energy transfer within an ecosystem. Trophic levels begin with the producers and proceed through various consumers.

true navigation Method by which animals orient, using an internal compass and map.

tube foot An extension of the water vascular system of an echinoderm. Tube feet function in locomotion and grasping of food.

tubule A small tube that carries substances from one location to another.

turgor pressure The internal pressure in plant cells caused by the diffusion of water into the cell, which allows the plant to stand upright.

type specimen The individual chosen for the designation and description of a new species.

ultrafiltration Process by which kidneys filter small molecules.

ultrastructure The structure of cells beyond that seen with a light microscope. It is cell structure as seen by an electron microscope or by other physical or chemical means.

unicelled (unicellular) Said of an organism composed of a single cell. A unicellular organism can perform all the life functions necessary to keep itself alive.

unsaturated Fatty acids with one or more double bond in their hydrocarbon chain.

uracil A nitrogenous pyrimidine base in RNA. It takes the place of thymine, pairing with adenine.

urea A nitrogenous waste product that is the main excretory product in amphibians and mammals.

ureter A tube that runs from the kidney to the urinary bladder.

urethra The duct from the bladder to the outside of the body, which carries urine and also semen, in the case of males.

urinary bladder Sac where urine is stored prior to urination.

urinary system The structures that form and export urine, including kidneys, ureters, urinary bladder, and urethra.

urine A liquid waste product processed in the kidney. It contains water and urea or uric acid.

uterus The portion of the female reproductive tract in which embryonic and fetal development occurs.

vacuole A membrane sac that is used to store materials such as food, pigments, and water. In plant cells, vacuoles may occupy most of the cell's volume.

vagina A muscular, tubular organ in female mammals that is the copulatory organ and birth canal.

variation The difference observed among organisms. Variation has a genetic basis in gene mutations. It provides the raw material for evolution.

vascular Referring to the transport tissues of an organism.

vascular cambium In perennials, secondary meristem that produces new vascular tissue.

vascular plant Plants that have vascular tissues, xylem and phloem.

vascular tissue The tissue used for transport in plants. It includes the xylem and phloem.

vas deferens The duct that carries sperm from the epididymal duct to the ejaculatory duct in vertebrates.

vector A host that carries pathogens from one organism to another. Also, a host DNA used by genetic engineers to introduce a DNA sequence into bacteria for amplification.

vegetal pole One of two poles of an egg or embryo that cleaves more slowly due to the presence of more yoke than at the other, or animal, pole.

vegetative Said of a cell that is growing or not involved in sexual reproduction. Such cells are prominent in the life cycle of protistans and fungi.

vein A vessel that carries blood toward the heart.

vena cava The large vein that carries blood to the heart from the peripheral tissues of the body.

ventilation The oxygenation of blood.

ventral The under surface, or belly of an animal. It is opposite the dorsal surface.

ventricle A chamber of the heart that pumps blood from the heart to other parts of the body.

venules The smallest veins.

vesicle A membrane sac formed usually from the Golgi apparatus. It may contain enzymes that are involved in metabolic reactions or waste products that are stored or eliminated to the outside. A lysosome is an example of a vesicle.

villus A small fold in the membrane of cells lining the small intestine. Villi are projections that increase the surface area of cells that transport substances in and out of the digestive system.

virulent Causing disease. Such things as bacteria and viruses may be virulent.

virus A nonliving infectious agent that needs a host cell to reproduce.

wave A disturbance propagated in some medium such as the air. Examples are sound waves and light waves. A wavelength is the distance between corresponding waves.

waxes Insoluble lipids that form a protective coating on plant surfaces, insect feathers, and animal fur.

white blood cells Cells that are associated with the immune systems. Some white blood cells directly attack foreign substances, whereas others produce chemicals for defense.

wild type The non-domesticated or reference version of a phenotype.

X chromosome One of the sex chromosomes. There is usually a pair of X chromosomes in human females.

X-linkage When a trait is determined by a gene on the X-chromosome.

xylem A plant's vascular tissue which transports water and minerals from the roots to the rest of the plant.

Y chromosome One of the sex chromosomes. In humans, it carries genes that determine maleness.

Y-linkage When a trait is determined by a gene on the Y chromosome.

zygote A fertilized egg. Zygotes are diploid and will divide by mitosis to form multicelled, diploid organisms.

Photo Credits

Chapter 1 *Opener:* Rawlings/Salaber/Gamma Liaison. *Figure 1-1a:* Dennis O'Clair/Tony Stone Images/New York, Inc. *Figure 1-1b:* Claudia Kunin/Tony Stone Images/New York, Inc. *Figure 1-1c:* ©The New York Times. Reproduced with permission. *Figure 1-2:* Hank Morgan/Photo Researchers. *Figure 1-3:* E. R. Degginger/Bruce Coleman, Inc. *Figure 1.4:* Jeff Greenberg/Visuals Unlimited. *Figure 1-5:* Arthur R. Hill/Visuals Unlimited. *Figure 1-7a:* Steve Solum/Bruce Coleman, Inc. *Figures 1-7b, 1-8 and 1-9:* Courtesy Pieter Johnson. *Figure 1-10:* Cousteau Society/The Image Bank.

Chapter 2 *Opener:* John Reader/Science Photo Library/Photo Researchers. *Figure 2.1:* AP/Wide World Photos. *Figure 2.2a:* ©Mary Evans Picture Library/Photo Researchers. *Figure 2.3a:* Jan Lindblad/Photo Researchers. *Figure 2.3b:* Tom McHugh/Photo Researchers. *Figure 2.4a:* ©Science Photo Library/Photo Researchers. *Figure 2.4b:* Volker Steger/Photo Researchers. *Figure 2.5:* Gregory Ochocki/Photo Researchers. *Figure 2.6* (left): Joe McDonald/Visuals Unlimited. *Figure 2.6* (right): F.S. Westmorland/Photo Researchers. *Figure 2.7:* John D. Cunningham/Visuals Unlimited. *Figure 2.8a:* John Gerlach/Visuals Unlimited. *Figure 2.8b:* Arthur Morris/Visuals Unlimited. *Figure 2.13a:* ©Vu/Bruce Berg/Visuals Unlimited. *Figure 2.13b:* Gerald and Buff Corsi/Visuals Unlimited. *Figure 2.15:* Courtesy Warren Abrahamson, Bucknell University. *Figure 2.16:* Johnny Johnson/Animals Animals NYC. *Figure 2.18a:* Gerard Lacz/Animals Animals NYC. *Figure 2.18b:* John Gerlach/Visuals Unlimited. *Figure 2.18c:* ©Vu/Walt Anderson/Visuals Unlimited. *Figure 2-22:* ©Vu/Ken Lucas/Visuals Unlimited. *Figure 2-23:* Derrick Ditchburn/Visuals Unlimited. *Figure 2.24:* Tom McHugh/Photo Researchers. *Figure 2.25:* Gary W. Carter/Visuals Unlimited. *Figure 2.26:* Joe McDonald/Animals Animals NYC.

Chapter 3 *Opener:* National Library of Medicine/Photo Researchers. *Figure 3-4a:* Allan Tannenbaum/Sygma. *Figure 3-4b:* Walter McBride/Retna. *Figure 3-4c:* Armando Gallo/Retna. *Figure 3.5:* ©Meckes/Ottawa/Photo Researchers. *Figure 3.8:* ©Vu/Robert Calentine/Visuals Unlimited. *Figure 3.12:* Simon Fraser/RVI, Newcastle-Upon-Tyne/Photo Researchers. *Figure 3.13:* ©Vu/Robert Calentine/Visuals Unlimited. *Figure 3.15:* Carolyn A. McKeone/Photo Researchers. *Figure 3.17:* Mitchell Funk/The Image Bank.

Chapter 4 *Opener:* Jeremy Burgess/Photo Researchers. *Figure 4-1a:* ©SIU/Visuals Unlimited. *Figure 4-1b:* Ed Young/Photo Researchers. *Figure 4-1c:* Jean Claude Revy/Phototake. *Figure 4-2a:* The Granger Collection. *Figure 4-2b:* ©Visuals Unlimited.

Figure 4-3: Dr. Jeremy Burgess/Photo Researchers. *Figure 4-19a,b:* Dr. Dennis Kunkel/Phototake. *Figure 4-19c:* Carolina Biological Supply/Phototake. *Figure 4-19d* and *4-20a:* David M. Phillips/Visuals Unlimited. *Figure 4-20b:* R. Kessel/Visuals Unlimited. *Figure 4-21:* Dr. Linda Stannard/Photo Researchers. *Figure 4-22a:* A.M. Siegelman/Visuals Unlimited. *Figure 4-22b:* Karl Aufderheide/Visuals Unlimited. *Figure 4-22c:* Newcomb & Wergin/Tony Stone Images/New York, Inc. *Figure 4-25:* Meckes/Gelderblom/Photo Researchers.

Chapter 5 *Opener:* Biophoto Associates/Photo Researchers. *Figure 5.1:* From E. DuPraw, *DNA & Chromosomes*, Holt, Rinehart & Winston, Inc., 1970. Photo courtesy of E. DuPraw. *Figure 5.3a:* ©Vu/David M. Phillips/Visuals Unlimited. *Figure 5.3b:* ©Vu/K.G. Murti/Visuals Unlimited. *Figure 5.4:* Biophoto Associates/Photo Researchers. *Figure 5.6:* Dennis Kunkel/Phototake. *Figure 5.11b:* ©Vu/Cabisco/Visuals Unlimited. *Figure 5.14:* Oliver Meckes/E.O.S/MPI-Tubingen/Photo Researchers.

Chapter 6 *Opener:* A. Barrington Brown/Photo Researchers. *Figure 6-2:* Science Source/Photo Researchers. *Figure 6-14a:* CNRI/Photo Researchers. *Figure 6-14b:* ©Vu/M. Coleman/Visuals Unlimited. *Figure 6-16:* D. Vo Trung/Phototake.

Chapter 7 *Opener:* David Boyle/Animals Animals NYC. *Figure 7-1:* Courtesy Ernie Marx, Oregon State University. *Figure 7-7:* From Matthew L. Warman, Nature Genetics, vol. 17, Sept. 1997, p. 18. Reproduced with permission. *Figure 7.8:* M.W.F. Tweedie/Photo Researchers. *Figure 7-15:* Photo by Clement D. Erhardt, provided courtesy of Victor A. McKusick, Institute of Genetic Medicine, Johns Hopkins Hospital. *Figure 7-16:* G.C. Kelley/Photo Researchers.

Chapter 8 *Opener:* ©Vu/WHOI-D. Foster/Visuals Unlimited. *Figure 8-1:* Rex A. Butcher/Tony Stone Images/New York, Inc. *Figure 8-2a:* Paul Harris/Tony Stone Images/New York, Inc. *Figure 8-2b:* James W. Richardson/Visuals Unlimited. *Figure 8-2c:* Tim Davis/Tony Stone Images/New York, Inc. *Figure 8-4:* Courtesy Department of Library Services, American Museum of Natural History. *Figure 8-7a:* Stanley Flegler/Visuals Unlimited. *Figure 8-7b:* Christoph Burki/Tony Stone Images/New York, Inc. *Figure 8-7c:* ©Vu/A.M. Siegelman/Visuals Unlimited. *Figure 8-7d:* ©Tony Stone Images/New York, Inc. *Figure 8-9a:* Gay Bumgarner/Tony Stone Images/New York, Inc. *Figure 8-9b:* Raymond A. Mendez/Animals Animals. *Figure 8-13a:* Pauer & Cook/Auscape International Pty., Ltd. *Figure 8-13b:* ©Planet Earth Pictures. *Figure 8-15a:* S. Lowry/Tony Stone Images/New York, Inc. *Figure 8-15b: NIBSC/Science Photo Library/Photo Researchers *Figure 8-15c:* David M. Phillips/Photo Researchers. *Figure 8-16:* John D. Cunningham/Visuals Unlimited. *Figure 8-17:* Paul Johnson/BPS/ Tony Stone Images/New York, Inc. *Figure 8-18:* Gary T. Cole/ BPS/Tony Stone Images/New York, Inc. *Figure 8-19a:* Kaz Chiba/ Tony Stone Images/New York, Inc. *Figure 8-19b:* Gay Bumgarner/Tony Stone Images/New York, Inc. *Figure 8-20:* Tim Davis/ Tony Stone Images/New York, Inc. *Figure 8-21:* Hal Beral/Visuals Unlimited. *Figure 8-22a:* Charles Krebs/Tony Stone Images/ New York, Inc. *Figure 8-22b:* Marc Chamberlain/Tony Stone Images/New York, Inc. *Figure 8-23a:* L. West/Photo Researchers. *Figure 8-23b:* George Whiteley/Photo Researchers. *Figure 8-23c:* Jeff Lepore/Photo Researchers. *Figure 8-24:* Rex Ziak/Tony Stone Images/New York, Inc.

Chapter 9 *Opener:* Brian Bailey/Tony Stone Images/New York, Inc. *Figure 9-3:* ©Tony Stone Images/New York, Inc. *Figure 9-14:* T. Kanariki-D. Fawcett/Visuals Unlimited. *Figure 9-16a:* David Hiser/Tony Stone Images/New York, Inc. *Figure 9-16c:* R. Bhatnagar/Visuals Unlimited. *Figure 9-18:* David Newman/Visuals Unlimited. *Figure 9-19a:* Jeremy Burgess/Photo Researchers. *Figure 9-22:* Courtesy James Bassham and Melvin Calvin. From Gerald Karp, *Cell and Molecular Biology*, 2e., John Wiley & Sons, figure 6.17, p. 240.

Chapter 10 *Opener:* Phil A. Dotson/Photo Researchers. *Figure 10-1:* John Walsh/Photo Researchers. *Figure 10-3:* Zig Leszczynski/Animals Animals NYC. *Figure 10-5:* Paul Chesley/Tony Stone Images/New York, Inc. *Figure 10-11:* Tom McHugh/Photo Researchers. *Figure 10-13a:* Wayne R. Bilenduke/Tony Stone Images/New York, Inc. *Figure 10-13b:* Hugh Rose/Visuals Unlimited. *Figure 10-14a:* David M. Phillips/Visuals Unlimited. *Figure 10-14b:* K.G. Murti/Visuals Unlimited. *Figure 10-14c:* A.M. Siegelman/Visuals Unlimited. *Figure 10-15:* ©Psu Entomology/Photo Researchers. *Figure 10-16:* ©Cabisco/Visuals Unlimited. *Figure 10-19:* Courtesy Charles C. Brinton, Jr., and Judith Carnahan. Reproduced with permission of Sarah Wood. *Figure 10-20:* Larry Roberts/Visuals Unlimited. *Figure 10-21:* Jon Bertsch/Visuals Unlimited. *Figure 10-22a:* Rudolf Arndt/Visuals Unlimited. *Figure 10-22b:* Bill Ivy/Tony Stone Images/New York, Inc. *Figure 10-23:* Joe McDonald/Animals Animals NYC. *Figure 10-24:* Dr. Lloyd M. Beidler/Photo Researchers. *Figure 10-25a:* Robert Lubeck/Animals Animals NYC. Figure 10-25b: Tom J. Ulrich/Visuals Unlimited. *Figure 10-26:* Mark Harmel/Tony Stone Images/New York, Inc.

Chapter 11 *Opener:* ©Kevin Somerville. *Figure 11-17a:* Fred Hossler/Visuals Unlimited. *Figure 11-17b:* Biophoto Associates/ Photo Researchers.

Chapter 12 *Opener:* Kevin Schafer/Tony Stone Images/New York, Inc. *Figure 12-5:* Kevin Collins/Visuals Unlimited. *Figure 12-7:* ©Vu/John Gerlach/Visuals Unlimited. *Figure 12-8:* Courtesy Nancy Pruitt. *Figure 12-10:* C. Gerald van Dyke/Visuals Unlimited. *Figure 12-13:* ©Cabisco/Visuals Unlimited. *Figure 12-14:* James Solliday/BPS/Tony Stone Images/New York, Inc. *Figure 12-15:* Ken Wagner/Visuals Unlimited. *Figure 12-19:* John D. Cunningham/Visuals Unlimited. *Figure 12-20:* Jack M. Bostrack/Visuals Unlimited. *Figure 12-25:* ©Vu/George Musil/Visuals Unlimited. *Figure 12-28:* David Parker/Photo Researchers.

Chapter 13 *Opener:* Tom Walker/Tony Stone Images/New York, Inc. *Figure 13-1a:* Jeff Greenberg/Visuals Unlimited. *Figure 13-1b:* Ralph A. Reinhold/Animals Animals NYC. *Figure 13.3:* Courtesy Larry Underwood. *Figure 13-5:* Nina Leen/LIFE Magazine, Time, Inc. *Figure 13-6:* ©Animals Animals NYC. *Figure 13-7:* Philip Sze/Visuals Unlimited. *Figure 13-8:* John Gerlach/Visuals Unlimited. *Figure 13-9:* Mark Boulton/Photo Researchers. *Figure 13-10:* Mark Newman/Visuals Unlimited. *Figure 13-11:* Holt Studios/Nigel Cattlin/Photo Researchers. *Figure 13-12a:* George Lepp/ Tony Stone Images/New York, Inc. *Figure 13-13a:* Hal Beral/ Visuals Unlimited. *Figure 13-13b:* K.B. Sandved/Visuals Unlimited. *Figure 13-13c:* Glenn Oliver/Visuals Unlimited. *Figure 13-13d:* Bill Kamin/Visuals Unlimited. *Figure 13-14:* J & B Photographers/ Animals Animals NYC. *Figure 13-15a:* Robert Maier/Animals Animals NYC. *Figure 13-15b:* Robert Maier/Animals Animals NYC. *Figure 13-16a:* Leroy Simon/Visuals Unlimited. *Figure 13-16b:* A. Kerstitch/Visuals Unlimited. *Figure 13-17a:* Richard Walters/Visuals Unlimited. *Figure 13-17b:* A.H. Rider/Photo Researchers. *Figure 13-18:* Richard Thom/Visuals Unlimited. *Figure 13-19:* Art Wolfe/Tony Stone Images/New York, Inc. *Figure 13-20:* Johnny Johnson/Animals Animals NYC. *Figure 13-21a:* Bryan Mullennix/ Tony Stone Images/New York, Inc. *Figure 13-21b:* Art Wolfe/Tony Stone Images/New York, Inc. *Figure 13-21c:* Tim Davis/Photo Researchers. *Figure 13-21d:* Daniel J. Cox/Tony Stone Images/ New York, Inc. *Figure 13-22:* Tom Tietz/Tony Stone Images/New York, Inc. *Figure 13-23:* ©Varin/Jacana/Photo Researchers. *Figure 13-24:* Erwin and Peggy Bauer/Bruce Coleman, Inc. *Figure 13-25:* Joe McDonald/Visuals Unlimited. *Figure 13-26:* Arthur Tilley/Tony Stone Images/New York, Inc.

Chapter 14 *Opener:* Don Smetzer/Tony Stone Images/New York, Inc. *Figure 14-1a:* Glenn Oliver/Visuals Unlimited. *Figure 14-1b:* Stephen Dalton/Photo Researchers. *Figure 14-2:* ©Vu/ Patrick J. Endres/Visuals Unlimited. *Figure 14-4:* David Frazier/ Tony Stone Images/New York, Inc. *Figure 14-5:* François Gohier/ Photo Researchers. *Figure 14-6:* ©Vu/Steve McCutcheon/Visuals Unlimited. *Figure 14-7:* Bud Titlow/Visuals Unlimited. *Figure 14-8a:* Chuck Keeler/Tony Stone Images/New York, Inc. *Figure 14-8b:* Tom Edwards/Visuals Unlimited. *Figure 14-10:* Tom McHugh/ Photo Researchers. *Figure 14-12a:* Michael Busselle/Tony Stone Images/New York, Inc. *Figure 14-12b:* C.C. Lockwood/Animals Animals NYC. *Figure 14-13a:* François Gohier/Photo Researchers. *Figure 14-13b:* Tom McHugh/Photo Researchers. *Figure 14-15a:* Bill Banaszewski/Visuals Unlimited. *Figure 14-15b:* Gerald & Buff Corsi/Visuals Unlimited. *Figure 14-16:* Art Wolfe/Tony Stone Images/New York, Inc. *Figure 14-17:* S. Maslowski/Visuals Unlimited. *Figure 14-19b:* Leonard Lee Rue III/Animals Animals NYC. *Figure 14-19c:* Daniel J. Cox/Tony Stone Images/New York, Inc. *Figure 14-20:* W.M. Banaszewski/Visuals Unlimited. *Figure 14-21:* Jeanne Drake/Tony Stone Images/New York, Inc. *Figure 14-22a:* Joe McDonald/Visuals Unlimited. *Figure 14-22b:* Thomas Gula/Visuals Unlimited. *Figure 14-22c:* J. Alcock/Visuals Unlimited. *Figure 14-23:* LSHTM/Tony Stone Images/New York, Inc. *Figure 14-24:* Hal Beral/Visuals Unlimited. *Figure 14-25:* Norbert Rosing/Animals Animals NYC. *Figure 14-26a:* Inga Spece/Visuals Unlimited. *Figure 14-26b:* ©Science VU/Visuals Unlimited.

Chapter 15 Opener: David B. Feetham/Visuals Unlimited. *Figure 15-1:* ©Vu/Doug Sokell/Visuals Unlimited. *Figure 15-2:* Roger Cole/Visuals Unlimited. *Figure 15-5:* Michael Busselle/Tony Stone

CD Index

Index